MW00612055

INDUSTRIAL ELECTRICAL WIRING

INDUSTRIAL ELECTRICAL WIRING

Design, Installation, and Maintenance

John E. Traister

McGraw-Hill
New York San Francisco Washington, D.C. Auckland Bogotá
Caracas Lisbon London Madrid Mexico City Milan
Montreal New Delhi Paris San Juan
Sydney Tokyo Toronto

McGraw-Hill

A Division of The ***McGraw-Hill*** *Companies*

1 2 3 4 5 6 7 8 9 0 DOC/DOC 9 0 1 0 9 8 7 6

ISBN-0-07-065329-1

Printed and bound by R.R. Donnelley & Sons.

This book is printed on recycled, acid-free paper containing a minimum of 50% recycled, de-inked fiber.

Contents

Preface

Texts covering industrial electrical wiring have normally included page after page of electrical theory, mathematical calculations, and equations with very little material geared directly toward practical applications. *Industrial Electrical Wiring* is designed to change this; that is, with the scores of texts available covering theory, the author chose to concentrate mainly on the practical applications of industrial wiring — utilizing his more than 35 years' experience to cover practically every conceivable application from small textile mills to huge petroleum refineries.

This book begins with an overview of industrial electrical systems, touches briefly on essentials such as codes, standards, and print reading, then quickly jumps into practical categories such as the design and implementation of actual installations.

The plant engineer and electrical maintenance personnel, obviously, cannot afford to be without this book, but others in the electrical industry will also benefit:

- Consulting engineers
- Electricians who work on industrial electrical systems — either new construction or maintenance
- Apprentice electricians
- Manufacturers' representatives
- Electrical inspectors

In fact, anyone involved in the electrical industry — in any capacity — will find helpful information in this book that will be of use on a daily basis, especially those technicians involved with manufacturing processes.

John E. Traister

INDUSTRIAL ELECTRICAL WIRING

Chapter 1

Introduction

Industrial wiring installations include systems for all types of industrial plants, factories, refineries, and similar facilities. The basic principles of wiring for industrial installations are very similar to other types of electrical installations except that in most cases the currents used in industrial electrical systems will be larger — requiring larger wire and conduit sizes; higher voltage will normally be used in industrial wiring systems; three-phase in addition to single-phase systems will be in use, and different types of materials and equipment may be involved.

Factors to Consider

Factors affecting the planning of an electrical installation for an industrial plant include the following:

1. New structure or modernization of an existing one.

2. Type of general building construction; for example, masonry, reinforced concrete, structural steel frame, etc.

3. Type of floor, ceiling, partitions, roof, and so on.

4. Type and voltage of service entrance, transformer connections, and whether service is underground or overhead.

Figure 1-1: The type of general building construction is one factor to consider when planning an industrial electrical installation.

5. Type and voltage of distribution system for power and lighting.
6. Type of required service equipment, such as unit substation or transformer bank.
7. Type of distribution system, including step-down transformers.
8. Who is responsible for furnishing the service-entrance and distribution equipment, the power company or plant?
9. Wiring methods, types of raceways, special raceways, busways, and the like.
10. Types of power-control equipment and extent of the worker's responsibility for connection to it.
11. Furnishing of motor starters, controls, and disconnects.

12. Extent of wiring to be installed on machine tools.

13. Extent of wiring connections to electric cranes and similar apparatus.

14. Type and construction of lighting fixtures, hangers and supports, and the like.

15. Extent of floodlighting. Type and dimensions of floodlighting supporting poles and mounting brackets.

16. Extent of signal and communication systems.

17. Ground conditions affecting the installation of underground wiring.

18. The size, type, and condition of existing wiring systems and services for modernization projects.

19. Whether the plant will be in use during the electrical installation.

20. Allowable working hours when an occupied building is being rewired.

Figure 1-2: Furnishing power to electric cranes is an item to consider in industrial wiring.

Items 1 through 3 can be determined by studying the architectural drawings of the facility or, in the case of a modernization project, by a job site investigation.

Item 4 is usually worked out between the engineers and the local utility company, while items 5 through 10 are determined from the electrical drawings and specifications.

The electrical drawings as well as the mechanical drawings should be consulted to determine the condition of items 10 and 11. Also consult the special equipment section of the written specifications. In most cases, it is the responsibility of the trade furnishing motor-driven equipment to also furnish the starters and controls. The electrical contractor then provides an adequate circuit to each.

Item 12 can be determined by consulting the special equipment section in the written specifications and also

Figure 1-3: The extent of signal and communication systems must be considered.

by referring to the shop drawings supplied by the trade furnishing the equipment. It may be necessary to contact the manufacturer of the equipment in some cases.

On most projects, crane installation is a specialized category and installed by specialists in this field. All control work for the crane is normally done by the contractor who furnishes the crane. However, electrical workers will be required to furnish a feeder circuit from the main distribution panel to supply power for motors and controls to operate the crane. The extent of this work should be carefully coordinated between the trades involved.

Items 14 through 16 can be determined by consulting the electrical drawings and specifications for the project.

While items 17 through 19 might be covered to a certain extent in the project specifications, in most cases this information is obtained by a job site investigation.

Item 20 should be called out in the general specifications, but it may be necessary to hold a conference with the owners to determine the exact conditions. Item 21 is determined by referring to the local labor agreement, by contacting the local labor organization, or from past experience.

Planning and Coordination

Even with carefully engineered drawings, the person in charge of the electrical installation in an industrial occupancy must still do much planning and coordination to carry out the work in the allotted amount of time. One problem that has existed in the past has been a variation of interpretation of the code requirements by two or more inspection authorities having coinciding jurisdiction over the same job. Therefore, at the planning stage, the superintendent or foreman should meet with all inspection authorities having jurisdiction to settle any problems at the outset.

Many industrial plants will have one or more wiring installations in hazardous locations. Therefore, those in charge should frequently consult the *National Electrical Code®* (*NEC®*) to ascertain that all wiring is

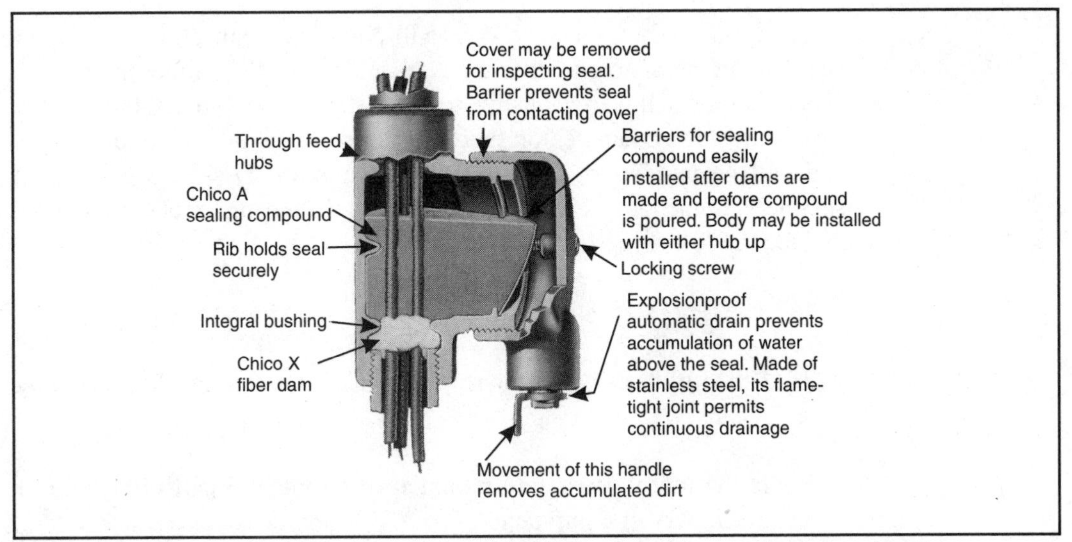

Figure 1-4: Cross-sectional view of a seal, used in wiring systems for some hazardous locations.

installed in a safe manner. Provisions must also be made to isolate the hazardous areas from those not considered hazardous.

Drawings from consulting engineers will vary in quality, and in most cases the wiring layout for a hazardous area is little different than the layout for a nonhazardous area. Usually the only distinction is a note on the drawing or in the specifications stating that the wiring in a given area or room shall conform to the *NEC* requirements for hazardous locations. Rarely do the working drawings contain much detail of the system, leaving much of the design to the workers on the job. Therefore, the electrical foreman must study these areas very carefully, and consult the *NEC* and other references, to determine exactly what is required.

Sometimes electrical contractors will have drafters prepare special drawings for use by their personnel in installing systems in such areas. If time permits, this is probably the best approach as it will save money plus much time in the field once the project has begun. In either case, whether preparing drawings or determining the requirements at the job site, considerable damage can be done to life and property if the system installed is faulty. Explosionproof boxes, fittings, and equipment are very expensive and vary in cost for different types, sizes, and hub entrances, and require considerably more labor than nonhazardous installations.

For other than very simple systems, it is advisable to make detailed wiring layouts of all wiring systems in hazardous locations, even if it is only in the form of sketches and rough notes.

The typical plant wiring system will entail the connection of a power supply to many motors of different sizes and types. Sometimes the electrical contractor will be responsible for furnishing these motors, but in most cases they will be handled on special order from the motor manufacturers and will be purchased directly by the owners or by special equipment suppliers. Some factors involved in planning the wiring for electric motors include the following:

1. The type, size, and voltage of the motor and related equipment.
2. Who furnishes the motor, starter, control stations, and disconnecting means?
3. Is the motor separately mounted or an integral part of a piece of machinery or equipment?
4. Type and size of junction box or connection chamber on the motor.
5. The extent of control wiring required.
6. The type of wiring method of the wiring system to which the motor is to be connected, that is, conduit and wire, bus duct, trolley duct. Is the motor located in a hazardous area? If so, what provisions have been made to ensure that it will be wired according to the *NEC*?
7. Who mounts the motor?
8. The physical shape and weight of the motor.

Obtaining all of the above information will facilitate the installation of all electric motors on the job by ensuring that proper materials will be available for the wiring and that no conflicts will arise between trades.

The installation of transformers and transformer vaults is another type of work that is frequently encountered in industrial wiring. A transformer vault, for example, is representative of the type of installation situation when, within a small area of the building and comprising a specialized section of the wiring system, a relatively small quantity of a number of different items of equipment and material is required. In many instances, even when working drawings and specifications are provided by consulting engineering firms, the vault will not be completely laid out to the extent that workers can perform the installation without further planning or

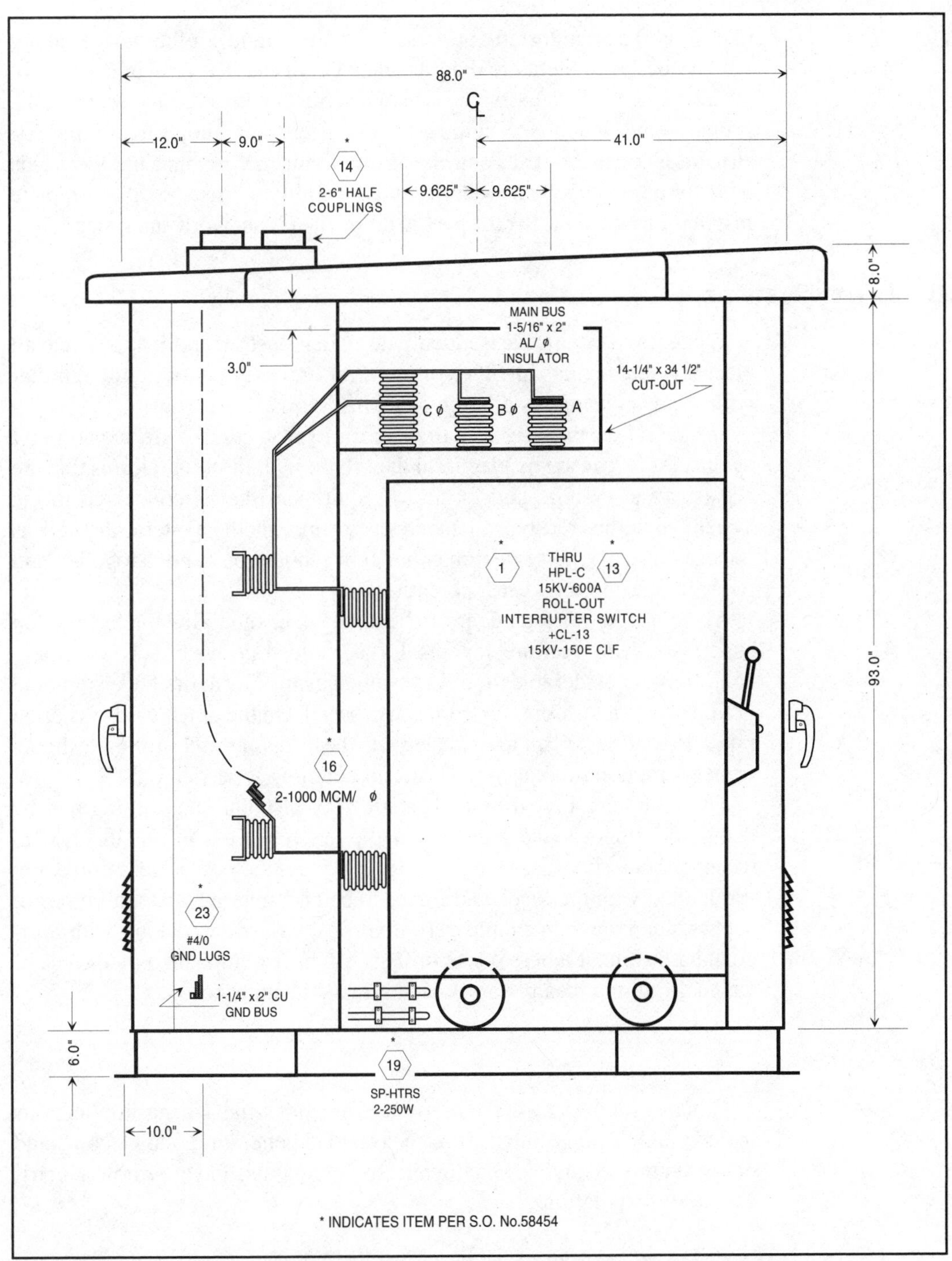

Figure 1-5: Shop drawings greatly facilitate the installation of special electrical equipment.

questions. The major transformers, disconnects, and similar devices along with a one-line schematic diagram may be all that the drawings show. In such instances, the supervisor or worker must make a rough layout of the primary and secondary, indicating the necessary supports, supporting structures, connections, controls and metering wiring, and the like. This calls for experienced knowledge, and the ability to visualize the complete installation on the part of the person doing the layout work and supervision.

Cable Tray Systems

Cable tray systems are frequently used in industrial applications, and all electrical technicians involved in such work should be thoroughly familiar with the design and installation of such systems.

In general, a cable tray system must afford protection to life and property against faults caused by electrical disturbances, lightning, failures that are a part of the system, and failure of equipment that is connected to the system. For this reason, all metal enclosures of the system, as well as noncurrent-carrying or neutral conductors, should be bonded together and reduced to a common earth potential.

There is a frequent tendency to become lax in supplying the installation supervisor with definite layouts for a cable tray system. Under these conditions, considerable time is consumed in arriving at final decisions and definite routings before the work can proceed. On the other hand, it is often possible for one person to predetermine these layouts and save many hours of field erection time, provided careful planning is carried out.

For economical erection and satisfactory installation, working out the details of supports and hangers for the system is the job of the system designer and should not be left to the judgment of a field force not acquainted with the loads and forces to be encountered. Also, all types of supports and hangers should permit vertical adjustment, along with horizontal adjustment where possible. This can be accomplished by the use of channel framing, beam clamps, and threaded hanger rods.

Other Systems

Other systems that are mainly used in industrial wiring applications include high-voltage substations, heavy-load generating plants, crane and hoist systems, ac and dc standby electrical systems, and enormous electrical motor installations.

Chapter 2

Codes and Standards

PURPOSE AND HISTORY OF THE NEC

Owing to the potential fire and explosion hazards caused by the improper handling and installation of electrical wiring, certain rules in the selection of materials, quality of workmanship, and precautions for safety must be followed. To standardize and simplify these rules and provide a reliable guide for electrical construction, the *National Electrical Code (NEC)* was developed. The *NEC*, originally prepared in 1897, is frequently revised to meet changing conditions, improved equipment and materials, and new fire hazards. It is a result of the best efforts of electrical engineers, manufacturers of electrical equipment, insurance underwriters, fire fighters, and other concerned experts throughout the country.

The *NEC* is now published by the National Fire Protection Association (NFPA), Batterymarch Park, Quincy, Massachusetts 02269. It contains specific rules and regulations intended to help in the practical safeguarding of persons and property from hazards arising from the use of electricity.

Although the *NEC* itself states, "This Code is not intended as a design specification nor an instruction manual for untrained persons," it does provide a sound basis for the study of electrical installation procedures — under the proper guidance. The probable reason for the *NEC*'s self-analysis is that the Code also states, "This Code contains provisions considered necessary for safety. Compliance therewith and proper maintenance will result in an installation essentially free from hazard, but not necessarily efficient, convenient, or adequate for good service or future expansion of electrical use."

The *NEC*, however, has become the bible of the electrical construction industry, and anyone involved in electrical work, in any capacity, should obtain an up-to-date copy, keep it handy at all times, and refer to it frequently.

NEC TERMINOLOGY

There are two basic types of rules in the *NEC*: mandatory rules and advisory rules. Here is how to recognize the two types of rules and how they relate to all types of electrical systems.

- Mandatory rules — All mandatory rules have the word *shall* in them. The word "shall" means *must*. If a rule is mandatory, you must comply with it.
- Advisory rules — All advisory rules have the word *should* in them. The word "should" in this case means *recommended but not necessarily required*. If a rule is advisory, compliance is discretionary.

Be alert to local amendments to the *NEC*. Local ordinances may amend the language of the *NEC*, changing it from *should* to *shall*. This means that you must do in that county or city what may only be recommended in some other area. The office that issues building permits will either sell you a copy of the code that's enforced in that area or tell you where the code is sold. In rare instances, the electrical inspector having jurisdiction may issue these regulations verbally.

There are a few other "landmarks" that you will encounter while looking through the *NEC*. These are summarized in Figure 2-1, and a brief explanation of each follows.

Explanatory material: Explanatory material in the form of Fine Print Notes is designated (FPN). Where these appear, the FPNs normally apply to the *NEC* Section or paragraph immediately preceding the FPN.

Change bar: A change bar in the margins indicates that a change in the *NEC* has been made since the last edition. When becoming familiar with each new edition of the *NEC*, always review these changes. There are also several illustrated publications on the market that point out changes in the *NEC* with detailed explanations of each. Such publications make excellent reference material.

Bullets: A filled-in circle called a "bullet" indicates that something has been deleted from the last edition of the *NEC*. Although not absolutely necessary, many electricians like to compare the previous *NEC* edition to the most recent one when these bullets are encountered, just to see what

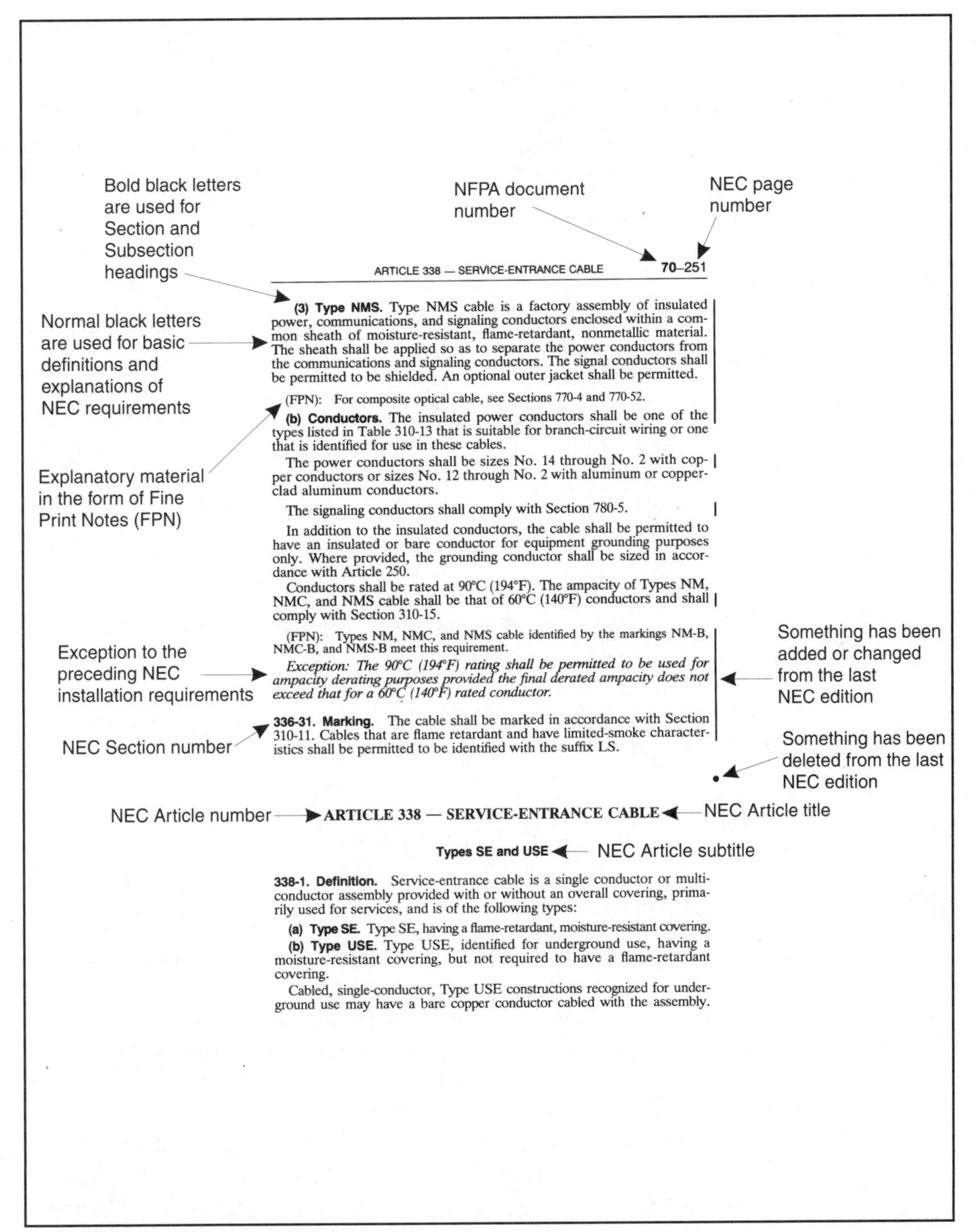

ARTICLE 338 — SERVICE-ENTRANCE CABLE **70–251**

(3) Type NMS. Type NMS cable is a factory assembly of insulated power, communications, and signaling conductors enclosed within a common sheath of moisture-resistant, flame-retardant, nonmetallic material. The sheath shall be applied so as to separate the power conductors from the communications and signaling conductors. The signal conductors shall be permitted to be shielded. An optional outer jacket shall be permitted.

(FPN): For composite optical cable, see Sections 770-4 and 770-52.

(b) Conductors. The insulated power conductors shall be one of the types listed in Table 310-13 that is suitable for branch-circuit wiring or one that is identified for use in these cables.

The power conductors shall be sizes No. 14 through No. 2 with copper conductors or sizes No. 12 through No. 2 with aluminum or copper-clad aluminum conductors.

The signaling conductors shall comply with Section 780-5.

In addition to the insulated conductors, the cable shall be permitted to have an insulated or bare conductor for equipment grounding purposes only. Where provided, the grounding conductor shall be sized in accordance with Article 250.

Conductors shall be rated at 90°C (194°F). The ampacity of Types NM, NMC, and NMS cable shall be that of 60°C (140°F) conductors and shall comply with Section 310-15.

(FPN): Types NM, NMC, and NMS cable identified by the markings NM-B, NMC-B, and NMS-B meet this requirement.

Exception: The 90°C (194°F) rating shall be permitted to be used for ampacity derating purposes provided the final derated ampacity does not exceed that for a 60°C (140°F) rated conductor.

336-31. Marking. The cable shall be marked in accordance with Section 310-11. Cables that are flame retardant and have limited-smoke characteristics shall be permitted to be identified with the suffix LS.

ARTICLE 338 — SERVICE-ENTRANCE CABLE

Types SE and USE

338-1. Definition. Service-entrance cable is a single conductor or multiconductor assembly provided with or without an overall covering, primarily used for services, and is of the following types:

(a) Type SE. Type SE, having a flame-retardant, moisture-resistant covering.

(b) Type USE. Type USE, identified for underground use, having a moisture-resistant covering, but not required to have a flame-retardant covering.

Cabled, single-conductor, Type USE constructions recognized for underground use may have a bare copper conductor cabled with the assembly.

Figure 2-1: Layout of typical NEC page.

has been omitted from the latest edition. The most probable reasons for the deletions are errors in the previous edition, or obsolete items.

Extracted text: Material identified by the superscript letter *x* includes text extracted from other NFPA documents as identified in Appendix A of the *NEC*. For example, ". . . 516-6.^{X} This section shall apply to processes in which combustible dry powders . . . ."

NEC Text Formats

As you open the *NEC* book, you will notice several different styles of text used. Here is an explanation of each.

1. *Normal black letters:* Basic definitions and explanations of the *NEC* requirements.

2. *Bold black letters:* Used for Article, Section, and Subsection headings.

3. *Exceptions:* These explain the situations when a specific rule does not apply. Exceptions are written in italics under the Section or paragraph to which they apply.

4. *Tables:* Tables are often included when there is more than one possible application of a requirement.

5. *Diagrams:* A few diagrams are scattered throughout the *NEC* to illustrate certain *NEC* applications.

LEARNING THE NEC LAYOUT

The *NEC* is divided into the Introduction (Article 90) and nine chapters. Chapters 1, 2, 3, and 4 apply generally; Chapters 5, 6, and 7 apply to special occupancies, special equipment, or other special conditions. These latter chapters supplement or modify the general rules. Chapters 1 through 4 apply except as amended by Chapters 5, 6, and 7 for the particular conditions.

While looking through these *NEC* chapters, if you should encounter a word or term that is unfamiliar, look in Chapter 1, Article 100 — Definitions. Chances are, the term will be found here. If not, look in the Index for the word and the *NEC* page number. Many terms are included in Article 100, but others are scattered throughout the book.

For definitions of terms not found in the *NEC*, check the glossary in the back of this book or obtain a copy of *Illustrated Dictionary for Electrical Workers*, available from Delmar Publishers, Inc., Albany, New York.

Chapter 8 of the *NEC* covers communications systems and is independent of the other chapters except where they are specifically referenced therein.

Chapter 9 consists of tables and examples.

There is also the *NEC* Contents at the beginning of the book and a comprehensive index at the back. You will find frequent use for both of these helpful "tools" when searching for various installation requirements.

Each chapter is divided into one or more Articles. For example, Chapter 1 contains Articles 100 and 110. These Articles are subdivided into Sections. For example, Article 110 of Chapter 1 begins with Section 110-2 – Approval. Some sections may contain only one sentence or a paragraph, while others may be further subdivided into lettered or numbered paragraphs such as (a), (1), (2), and so on.

Begin your study of the *NEC* with Articles 90, 100, and 110. These three articles have the basic information that will make the rest of the *NEC* easier to understand. Article 100 defines terms you will need to understand the code. Article 110 gives the general requirements for electrical installations. Read these three articles over several times until you are thoroughly familiar with all the information they contain. It's time well spent. For example, Article 90 contains the following sections:

- Purpose (90-1)
- Scope (90-2)
- Code Arrangement (90-3)
- Enforcement (90-4)
- Mandatory Rules and Explanatory Material (90-5)
- Formal Interpretations (90-6)
- Examination of Equipment for Safety (90-7)
- Wiring Planning (90-8)
- Metric Units of Measurement (90-9)

Once you are familiar with Articles 90, 100, and 110 you can move on to the rest of the *NEC*. There are several key sections you will use often while installing and servicing electrical systems. Let's discuss each of these important sections.

WIRING DESIGN AND PROTECTION

Chapter 2 of the *NEC* discusses wiring design and protection, the information electrical technicians need most often. It covers the use and identification of grounded conductors, branch circuits, feeders, calculations, services, overcurrent protection and grounding. This is essential information for any electrical system, regardless of the type.

Chapter 2 is also a "how-to" chapter. It explains how to provide proper spacing for conductor supports, how to provide temporary wiring, and how to size the proper grounding conductor or electrode. If a problem is encountered that is related to the design/installation of a conventional electrical system, the solution for it can normally be found in this chapter.

WIRING METHODS AND MATERIALS

Chapter 3 has the rules on wiring methods and materials. The materials and procedures to use on a particular system depend on the type of building construction, the type of occupancy, the location of the wiring in the building, the type of atmosphere in the building or in the area surrounding the building, mechanical factors, and the relative costs of different wiring methods.

The provisions of this article apply to all wiring installations unless specified otherwise in *NEC* Articles and Sections.

Wiring Methods

There are four basic wiring methods used in most modern electrical systems. Nearly all wiring methods are a variation of one or more of these four basic methods:

- Sheathed cables of two or more conductors, such as nonmetallic-sheathed cable and armored cable (Articles 330 through 339)
- Raceway wiring systems, such as rigid steel conduit and electrical metallic tubing (Articles 342 to 358)
- Busways (Article 364)
- Cable tray (Article 318)

Electrical Conductors

Article 310 in Chapter 3 gives a complete description of all types of electrical conductors. Electrical conductors come in a wide range of sizes and forms. Be sure to check the working drawings and specifications to see what sizes and types of conductors are required for a specific job. If conductor type and size are not specified, choose the most appropriate type and size meeting standard *NEC* requirements.

When workers have the choice of selecting the wiring method to use, most will select the least expensive method allowed by the *NEC*. However, in some cases, what appears to be the least expensive method may not hold true in the final results. For example, when rewiring existing buildings where much "fishing" of cable is necessary, workers have found that Type AC armored cable (BX) is usually easier to fish in concealed partitions than Type NM (Romex) cable. Although BX cable is more expensive, the savings in labor usually offset the cost.

Boxes, Cabinets, and Enclosures

Articles 318 through 384 give rules for raceways, boxes, cabinets, and raceway fittings. Outlet boxes vary in size and shape, depending on their use, the size of the raceway, the number of conductors entering the box, the type of building construction, and atmospheric conditions of the areas. Chapter 3 should answer most questions on the selection and use of these items.

The *NEC* does not describe in detail all types and sizes of outlet boxes. But manufacturers of outlet boxes have excellent catalogs showing all of their products. Collect these catalogs. They are essential to your work.

Wiring Devices and Switchgear

Article 380 covers the switches, pushbuttons, pilot lamps, receptacles, and power outlets. Again, get the manufacturers' catalogs on these items. They will provide you with detailed descriptions of each.

Article 384 covers switchboards and panelboards, including their location, installation methods, clearances, grounding, and overcurrent protection.

EQUIPMENT FOR GENERAL USE

Chapter 4 of the *NEC* begins with the use and installation of flexible cords and cables, including the trade name, type letter, wire size, number of conductors, conductor insulation, outer covering, and use of each. The chapter also includes fixture wires, again giving the trade name, type letter, and other important details.

Article 410 on lighting fixtures is especially important. It gives installation procedures for fixtures in specific locations. For example, it covers fixtures near combustible material and fixtures in closets. The *NEC* does not describe how many fixtures will be needed in a given area to provide a certain amount of illumination.

Article 430 covers electric motors, including mounting the motor and making electrical connections to it. Motor controls and overload protection are also covered.

Articles 440 through 460 cover air-conditioning and heating equipment, transformers, and capacitors.

Article 480 gives most requirements related to battery-operated electrical systems. Storage batteries are seldom thought of as part of a conventional electrical system, but they often provide standby emergency lighting service. They may also supply power to security systems that are separate from the main ac electrical system.

SPECIAL OCCUPANCIES

Chapter 5 of the *NEC* covers special occupancy areas. These are areas where the sparks generated by electrical equipment may cause an explosion or fire. The hazard may be due to the atmosphere of the area or just the presence of a volatile material in the area. Commercial garages, aircraft hangers, and service stations are typical special occupancy locations.

Articles 500 – 501 cover the different types of special occupancy atmospheres that are considered to be hazardous areas. The atmospheric groups were established to make it easy to test and approve equipment for various types of uses.

Articles 501-4, 502-4, and 503-3 cover the installation of wiring in hazardous (Classified) locations. Wiring in these areas must be designed to prevent the ignition of a surrounding explosive atmosphere when arcing occurs within the electrical system.

There are three main classes of special occupancy locations:

- Class I (Article 501): Areas containing flammable gases or vapors in the air. Class I areas include paint spray booths, dyeing plants where hazardous liquids are used, and gas generator rooms.
- Class II (Article 502): Areas where combustible dust is present, such as grain handling and storage plants, dust and stock collector areas, and sugar pulverizing plants. These are areas where, under normal operating conditions, there may be enough combustible dust in the air to produce explosive or ignitable mixtures.

- Class III (Article 503): Areas that are hazardous because of the presence of easily ignitable fibers or flyings in the air, although not in large enough quantity to produce ignitable mixtures. Class III locations include cotton mills, rayon mills, and clothing manufacturing plants.

Articles 511 and 514 regulate garages and similar locations where volatile or flammable liquids are used. While these areas are not always considered critically hazardous locations, there may be enough danger to require special precautions in the electrical installation. In these areas, the *NEC* requires that volatile gases be confined to an area not more than 4 ft above the floor. So in most cases, conventional raceway systems are permitted above this level. If the area is judged critically hazardous, explosionproof wiring (including seal-offs) may be required.

Article 520 regulates theaters and similar occupancies where fire and panic can cause hazards to life and property. Drive-in theaters do not present the same hazards as enclosed auditoriums. But the projection rooms and adjacent areas must be properly ventilated and wired for the protection of operating personnel and others using the area.

Chapter 5 also covers residential storage garages, aircraft hangars, service stations, bulk storage plants, health care facilities, mobile homes and parks, and recreation vehicles and parks.

When security technicians are installing systems in hazardous locations, extreme caution must be used. You may be working with only 12 or 24 V, but a spark caused by, say, an improper connection can set off a violent explosion. You may have already witnessed a low-voltage explosion in the common automotive battery. Although only 12 V dc are present, if a spark occurs near the battery and battery gases are leaking through the battery housing, chances are the battery will explode with a report similar to a shotgun firing.

When installing security systems in Class I, Division 1 locations, explosionproof fittings are required and most electrical wiring must be enclosed in rigid steel conduit (pipe).

SPECIAL EQUIPMENT

Article 600 covers electric signs and outline lighting. Article 610 applies to cranes and hoists. Article 620 covers the majority of the electrical work involved in the installation and operation of elevators, dumbwaiters, escalators, and moving walks. The manufacturer is responsible for most of this work. The electrician usually just furnishes a feeder terminating in a disconnect means in the bottom of the elevator shaft. The electrician may also be responsible for a lighting circuit to a junction box midway in the elevator shaft for connecting the elevator cage lighting cable and exhaust fans. Articles in Chapter 6 of the *NEC* give most of the requirements for these installations.

Article 630 regulates electric welding equipment. It is normally treated as a piece of industrial power equipment requiring a special power outlet. But there are special conditions that apply to the circuits supplying welding equipment. These are outlined in detail in Chapter 6 of the *NEC*.

Article 640 covers wiring for sound-recording and similar equipment. This type of equipment normally requires low-voltage wiring. Special outlet boxes or cabinets are usually provided with the equipment. But some items may be mounted in or on standard outlet boxes. Some sound-recording electrical systems require direct current, supplied from rectifying equipment, batteries, or motor generators. Low-voltage alternating current comes from relatively small transformers connected on the primary side to a 120-V circuit within the building.

Other items covered in Chapter 6 of the *NEC* include: X-ray equipment (Article 660), induction and dielectric heat-generating equipment (Article 665), and machine tools (Article 670).

If you ever have work that involves Chapter 6, study the chapter before work begins. That can save a lot of installation time. Here is another way to cut down on labor hours and prevent installation errors. Get a set of rough-in drawings of the equipment being installed. It is easy to install the wrong outlet box or to install the right box in the wrong place. Having a set of rough-in drawings can prevent those simple but costly errors.

SPECIAL CONDITIONS

In most commercial buildings, the *NEC* and local ordinances require a means of lighting public rooms, halls, stairways and entrances. There must be enough light to allow the occupants to exit from the building if the general building lighting is interrupted. Exit doors must be clearly indicated by illuminated exit signs.

Chapter 7 of the *NEC* covers the installation of emergency lighting systems. These circuits should be arranged so that they can automatically transfer to an alternate source of current, usually storage batteries or gasoline-driven generators. As an alternative in some types of occupancies, you can connect them to the supply side of the main service so disconnecting the main service switch would not disconnect the emergency circuits. (*See* Article 700.) *NEC* Chapter 7 also covers a variety of other equipment, systems, and conditions that are not easily categorized elsewhere in the *NEC*.

Chapter 8 is a special category for wiring associated with electronic communications systems including telephone and telegraph, radio and TV, fire and burglar alarms, and community antenna systems.

Once you become familiar with the *NEC* through repeated usage, you will generally know where to look for a particular topic. While this chapter provides you with an initial familiarization of the *NEC* layout, much additional usage experience will be needed for you to feel comfortable with the *NEC*'s content.

The *NEC* is not an easy book to read and understand at first. In fact, seasoned electrical workers and technicians sometimes find it confusing. Basically, it is a reference book written in a legal, contract-type language and its content does assume prior knowledge of most subjects listed. Consequently, you will sometimes find the *NEC* frustrating to use because terms aren't always defined, or because of some unknown prerequisite knowledge.

DEFINITIONS

Many definitions of terms dealing with the *NEC* may be found in *NEC* Article 100. However, other definitions are scattered throughout the *NEC* under their appropriate category. For example the term *lighting track* is not listed in Article 100. The term is listed under *NEC* Section 410-100 and reads as follows:

Lighting track is a manufactured assembly designed to support and energize lighting fixtures that are capable of being readily repositioned on the track. Its length may be altered by the addition or subtraction of sections of track.

Regardless of where the definition may be located — in Article 100 or under the appropriate *NEC* Section elsewhere in the book — the best way to learn and remember these definitions is to form a mental picture of each item or device as you read the definition. For example, turn to page 70-25 of the 1996 *NEC* and under Article 100 — Definitions, scan down the page until you come to the term "Attachment Plug (Plug Cap) (Cap)." After reading the definition, you will probably have already formed a mental picture of attachment plugs.

Once again, scan through the definitions until the term "Appliance" is found. Read the definition and then try to form a mental picture of what appliances look like. They should be familiar to everyone.

Each and every term listed in the *NEC* should be understood. Know what the item looks like and how it is used on the job. If a term is unfamiliar, try other reference books such as manufacturers' catalogs for an illustration of the item. Then research the item further to determine its purpose in electrical systems. Once you are familiar with all the common terms and definitions found in the *NEC*, navigating through the *NEC* (and understanding what you read) will be much easier.

There are many definitions included in Article 100. You should become familiar with the definitions. Since a copy of the latest *NEC* is compulsory for any type of electrical wiring, there is no need to duplicate them here. However, here are two definitions that you should become especially familiar with:

- Labeled — Equipment or materials to which has been attached a label, symbol, or other identifying mark of an organization acceptable to the authority having jurisdiction and concerned with product evaluation, that maintains periodic inspection of production of labeled equipment or materials, and by whose labeling the manufacturer indicates compliance with appropriate standards or performance in a specified manner.
- Listed — Equipment or materials included in a list published by an organization acceptable to the authority having jurisdiction and concerned with product evaluation, that maintains periodic inspection of production of listed equipment or materials, and whose listing states either that the equipment or material meets appropriate designated standards or has been tested and found suitable for use in a specified manner.

TESTING LABORATORIES

Besides installation rules, you will also have to be concerned with the type and quality of materials that are used in electrical wiring systems. Nationally recognized testing laboratories (Underwriters' Laboratories, Inc., for example) are product safety-certification laboratories. They establish and operate product safety certification programs to make sure that items produced under the service are safeguarded against reasonable foreseeable risks. Some of these organizations maintain a worldwide

Figure 2-2: U.L. "listing seal."

network of field representatives who make unannounced visits to manufacturing facilities to countercheck products bearing their "listing seal." *See* Figure 2-2.

However, proper selection, overall functional performance, and reliability of a product are factors that are not within the basic scope of U.L activities.

To fully understand the *NEC*, it is important to understand the organizations that govern it. The following organizations will frequently be encountered and associated with materials and equipment used on almost every electrical installation.

Nationally Recognized Testing Laboratory (NRTL)

Nationally Recognized Testing Laboratories are product safety certification laboratories. They establish and operate product safety certification programs to make sure that items produced under the service are safeguarded against reasonable foreseeable risks. An approved item, however, does not mean that the item is approved for all uses; it is safe only for the purpose for which it is intended. NRTL maintains a worldwide network of field representatives who make unannounced visits to factories to countercheck products bearing the safety mark.

National Electrical Manufacturers Association (NEMA)

The National Electrical Manufacturers Association was founded in 1926. It is made up of companies that manufacture equipment used for generation, transmission, distribution, control, and utilization of electric power. The objectives of NEMA are to maintain and improve the quality and reliability of products; to ensure safety standards in the manufacture and use of products; to develop product standards covering such matters as naming, ratings, performance, testing, and dimensions. NEMA participates in developing the *NEC* and the National Electrical Safety Code and advocates their acceptance by state and local authorities.

National Fire Protection Association (NFPA)

The NFPA was founded in 1896. Its membership is drawn from the fire service, business and industry, health care, educational and other institutions, and individuals in the fields of insurance, government, architecture, and engineering. The duties of the NFPA include:

- Developing, publishing, and distributing standards prepared by approximately 175 technical committees. These standards are intended to minimize the possibility and effects of fire and explosion.
- Conducting fire safety education programs for the general public.
- Providing information on fire protection, prevention, and suppression.
- Compiling annual statistics on causes and occupancies of fires, large-loss fires (over $1 million), fire deaths, and fire fighter casualties.
- Providing field service by specialists on electricity, flammable liquids and gases, and marine fire problems.
- Conducting research projects that apply statistical methods and operations research to develop computer modes and data management systems.

The Role of Testing Laboratories

Testing laboratories are an integral part of the development of the code. The NFPA, NEMA, and NRTL all provide testing laboratories to conduct research into electrical equipment and its safety. These laboratories perform extensive testing of new products to make sure they are built to code standards for electrical and fire safety. These organizations receive statistics and reports from agencies all over the United States concerning electrical shocks and fires and their causes. Upon seeing trends developing concerning association of certain equipment and dangerous situations or circumstances, this equipment will be specifically targeted for research.

SUMMARY

The *NEC* specifies the minimum provisions necessary for protecting people and property from hazards arising from the use of electricity and electrical equipment. Anyone involved in any phase of the electrical industry must be aware of how to use and apply the code on the job. Using the *NEC* will help you to safely install and maintain the electrical security equipment and systems that you come into contact with.

The *NEC* is composed of the following components:

Appendix: Appendix A includes material extracted from other NFPA documents. Appendix B is not part of the requirements of the *NEC* and contains additional material for informational purposes only. Appendix A and Appendix B are located at the end of the code book.

Article: Beginning with Article 90 — Introduction, and ending with Article 820 — Community Antenna Television and Radio Distribution Systems, the *NEC* Articles are the main topics in the code book.

Chapter: The *NEC* includes nine chapters. Chapter 1 — General, Chapter 2 — Wiring and Protection, Chapter 3 — Wiring Methods and Materials, Chapter 4 — Equipment for General Use, Chapter 5 — Special Occupancies, Chapter 6 — Special Equipment, Chapter 7 — Special Conditions, Chapter 8 — Communications Systems, and Chapter 9 — Tables and Examples. The Chapters form the broad structure of the *NEC*.

Contents: Located among the first pages of the code book, the contents section provides a complete outline of the Chapters, Articles, Parts, Tables, and Examples. The contents section, used with the index, provides excellent direction for locating answers to electrical problems and questions.

Diagrams and Figures: Diagrams and Figures appear in the *NEC* to illustrate some requirements of the *NEC*.

Examples: Service and feeder calculations for various types of buildings.

Exceptions: Exceptions follow code sections and allow alternative methods, to be used under specific conditions, to the rule stated in the section.

FPN (Fine Print Note): A Fine Print Note is defined in *NEC* Section 110-1; that is, explanatory material is in the form of Fine Print Notes (FPN).

Notes: Notes typically follow tables and are used to provide additional information to the tables or clarification of tables.

Part: Certain Articles in the *NEC* are divided into Parts. Article 220 — Branch Circuit, Feeder, and Service Calculations is divided into Part A, B, C, and D.

Section: Parts and Articles are divided into Sections. A reference to a section will look like the following:

300-19. Supporting Conductors in Vertical Raceways.

NEC Sections provide more detailed information within *NEC* Articles.

Tables: Tables are located within Chapters to provide more detailed information explaining code content.

See Figure 2-3 for a summary of *NEC* installation requirements for various occupancies.

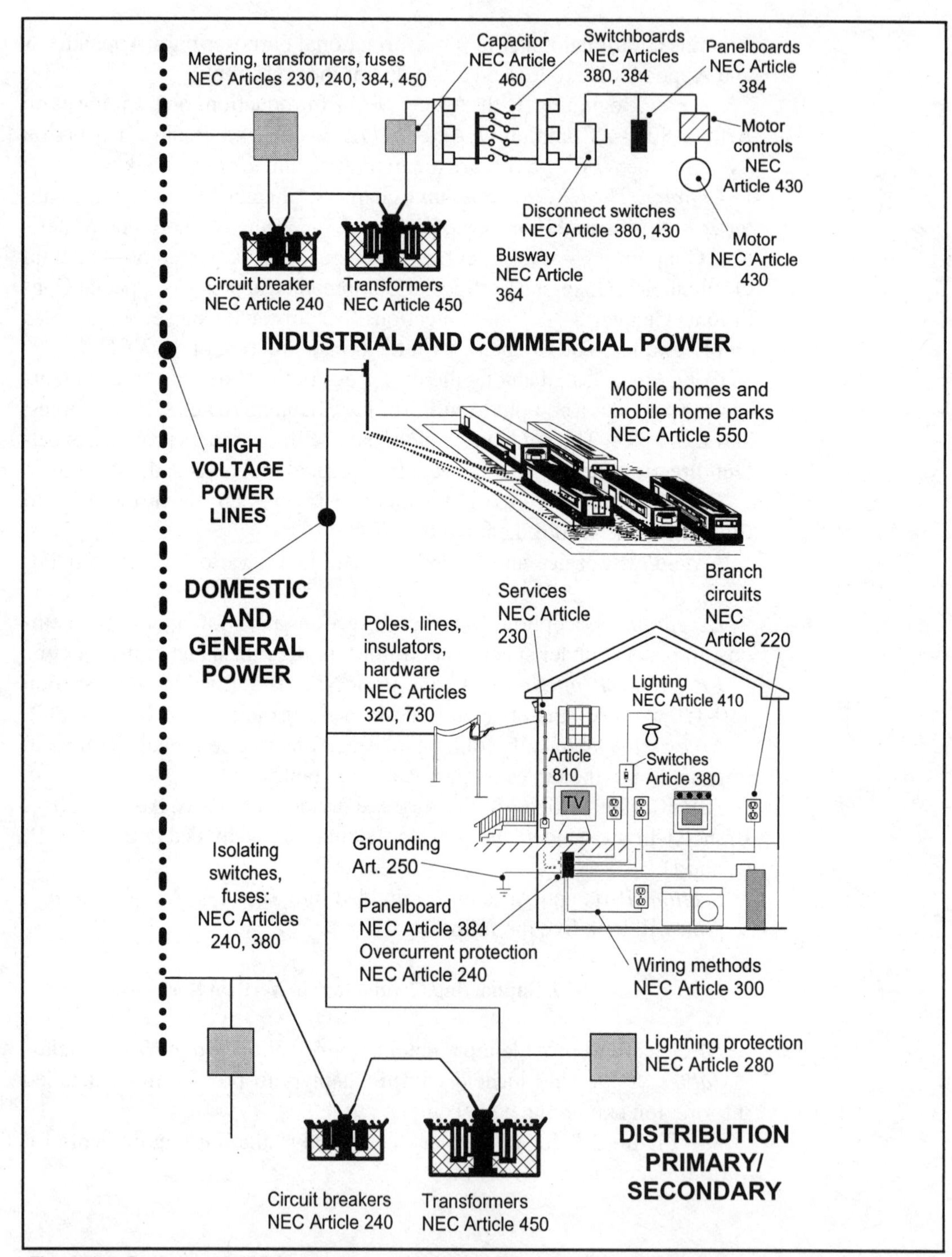

Figure 2-3: Summary of NEC installation requirements for various occupancies.

Chapter 3

Industrial Construction Documents

In every branch of electrical work, there is often occasion to interpret an electrical drawing. Electricians, for example, who are responsible for installing the electrical system in a new installation, usually consult a set of electrical drawings and specifications to locate the incoming electric service, switchgear, main distribution panels, subpanels, motor-control centers, routing of raceways and circuits, and similar details. Electrical estimators must refer to electrical drawings to determine the quantity of material needed in preparing a bid. Electricians in industrial plants consult schematic diagrams when wiring electrical controls for motor applications. Plant maintenance personnel use electrical drawings in troubleshooting problems that occur. Circuits may be tested and checked against the original drawings to help locate any faulty points in the installation.

TYPES OF ELECTRICAL DRAWINGS

An electrical drawing consists of lines, symbols, dimensions, and notations to accurately convey an engineer's design to workers who install the electrical system on the job. Workers should be able to take a complete set of electrical drawings and related written specifications, shop drawings, and supplemental drawings, and without further instruction, install or produce the electrical system as the engineer or designer intended it to be accomplished. An electrical drawing, therefore, is an abbreviated language for conveying a large amount of exact, detailed information, which would otherwise take many pages of manuscript or hours of verbal instruction to convey.

Pictorial Drawings

In this type of drawing, the objects are drawn in one view only; that is, three-dimensional effects are simulated on the flat plane of drawing paper by drawing several faces of an object in a single view. This type of drawing is very useful to describe objects and convey information to those who are not well trained in print reading or to supplement conventional diagrams in certain special cases.

One example of a pictorial drawing would be an exploded view of a manual motor-starter switch used to show the physical relationship of each part so that the starter can be disassembled and reassembled during maintenance. *See* Figure 3-1.

The types of pictorial drawings most often found on electrical construction drawings include:

- Isometric
- Oblique
- Perspective

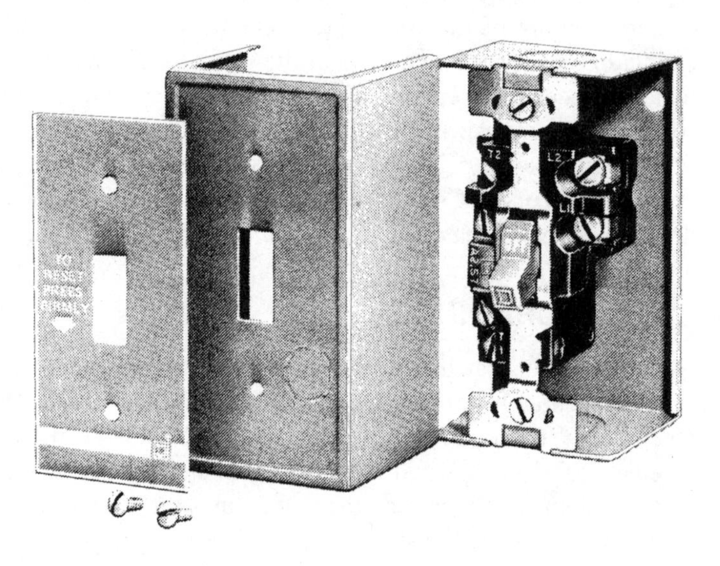

Figure 3-1: Pictorial view of a manual motor-starter switch, showing how the housing cover and plate are to be installed.

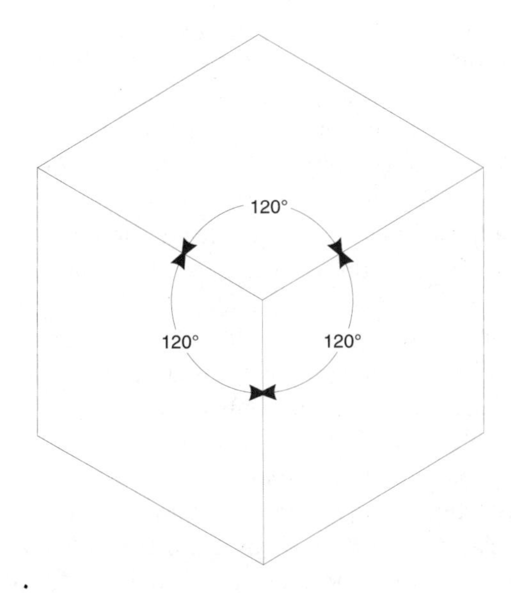

Figure 3-2: Isometric drawing of a cube.

All of these drawings are relatively difficult to draw and their use is normally limited to the manufacturers of electrical components for showing their products in catalogs, brochures, and similar applications. However, in recent times, they are gradually being replaced by photographs where possible.

By definition, an isometric drawing is a view projected onto a vertical plane in which all of the edges are foreshortened equally. Figure 3-2 shows an isometric drawing of a cube. In this view, the edges are 120° apart and are called the isometric axes, while the three surfaces shown are called the isometric planes. The lines parallel to the isometric axes are called the isometric lines.

Isometric drawings are usually preferred over the other two types of pictorial drawings for use in engineering departments to show certain details on installation drawings, because it is possible to draw isometric lines to scale using 30° and 60° lines. The power-riser diagram shown in Figure 3-3 is one practical example of isometric drawings in electrical applications.

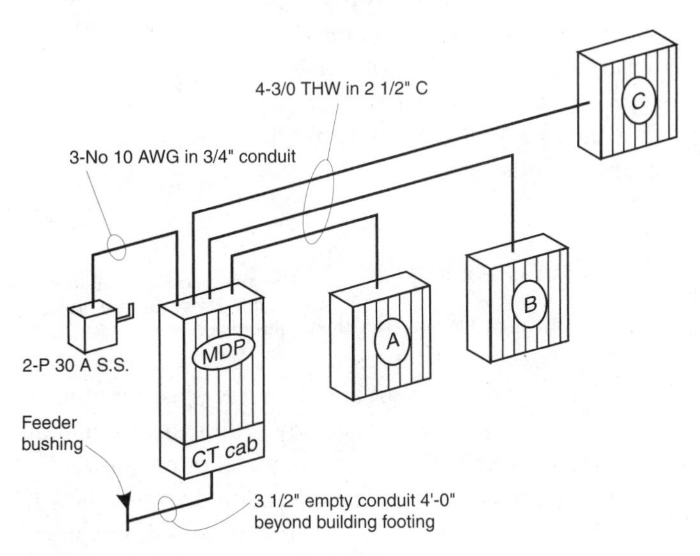

Figure 3-3: Isometric drawing of a power-riser diagram.

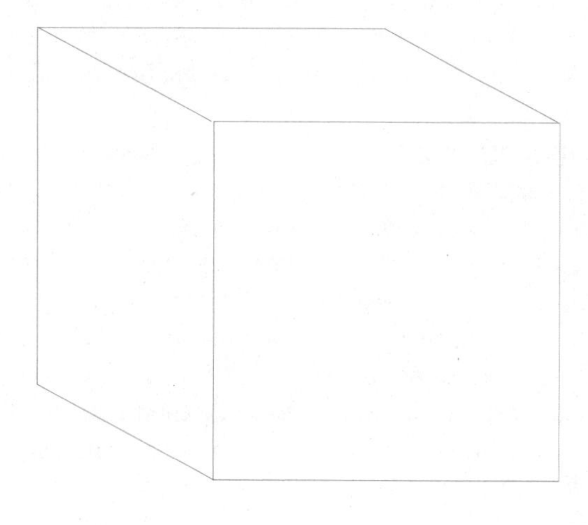

Figure 3-4: Oblique drawing of a cube.

The oblique drawing is similar to the isometric drawing in that one face of the object is drawn in its true shape and the other visible faces are shown by parallel lines drawn at the same angle (usually 45° – 30°) with the horizontal. However, unlike an isometric drawing, the lines drawn at a 30° angle are shortened to preserve the appearance of the object and are therefore not drawn to scale. The drawing in Figure 3-4 is an oblique drawing of a cube.

The two methods of pictorial drawing described so far produce only approximate representations of objects as they appear to the eye, as each type produces some degree of distortion of any object so drawn. However, because of certain advantages, the previous two types are the ones most often found in engineering drawings.

Sometimes — for a certain catalog illustration, or a more detailed instruction manual — an exact pictorial representation of an object becomes necessary. A drawing of this type is called a perspective drawing and such a drawing appears in Figure 3-5.

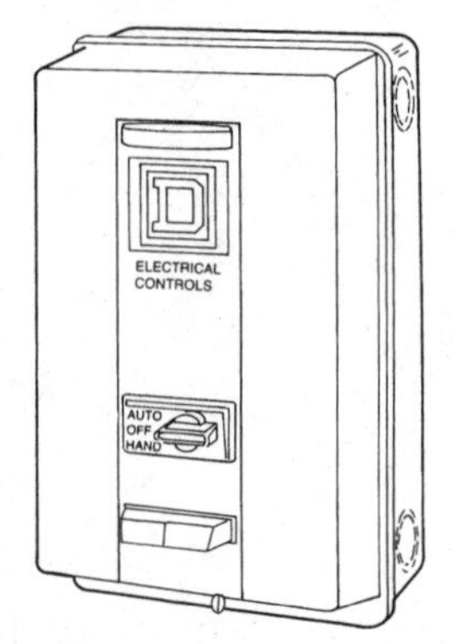

Figure 3-5: Pictorial drawing of a motor starter. Such drawings are typically used for catalog illustrations.

Orthographic-Projection Drawing

An orthographic-projection drawing is one that represents the physical arrangement and views of specific objects. These drawings give all plan views, elevation views, dimensions, and other details necessary to construct the project or object. For example, Figure 3-6 shows the basic form of a transformer vault, but it does not show the actual shape of the surface, nor does it show the dimensions of the object so that it may be constructed.

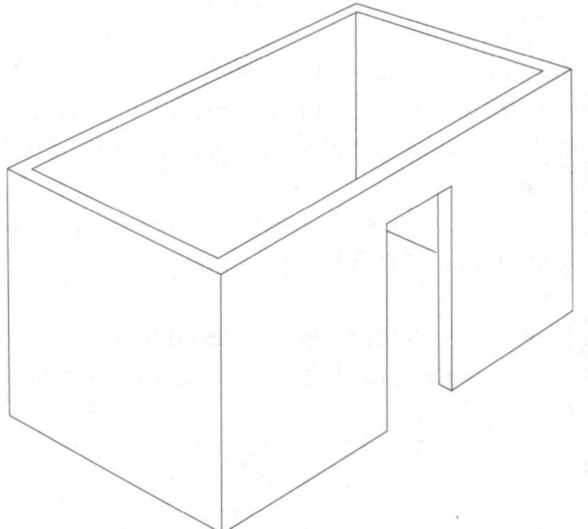

Figure 3-6: Pictorial drawing of a transformer vault.

An orthographic projection of the vault in Figure 3-6 is shown in Figure 3-7. One of the drawings in this figure shows the vault as though the observer were looking straight at the left side; one, as though the observer were looking straight at the right side; one as though the observer were looking at the front, and one as though the observer were looking at the rear of the vault. The remaining view is as if the observer were looking straight down on top of the vault. These views, when combined with dimensions, will allow the object to be constructed properly from materials called for in the written specifications.

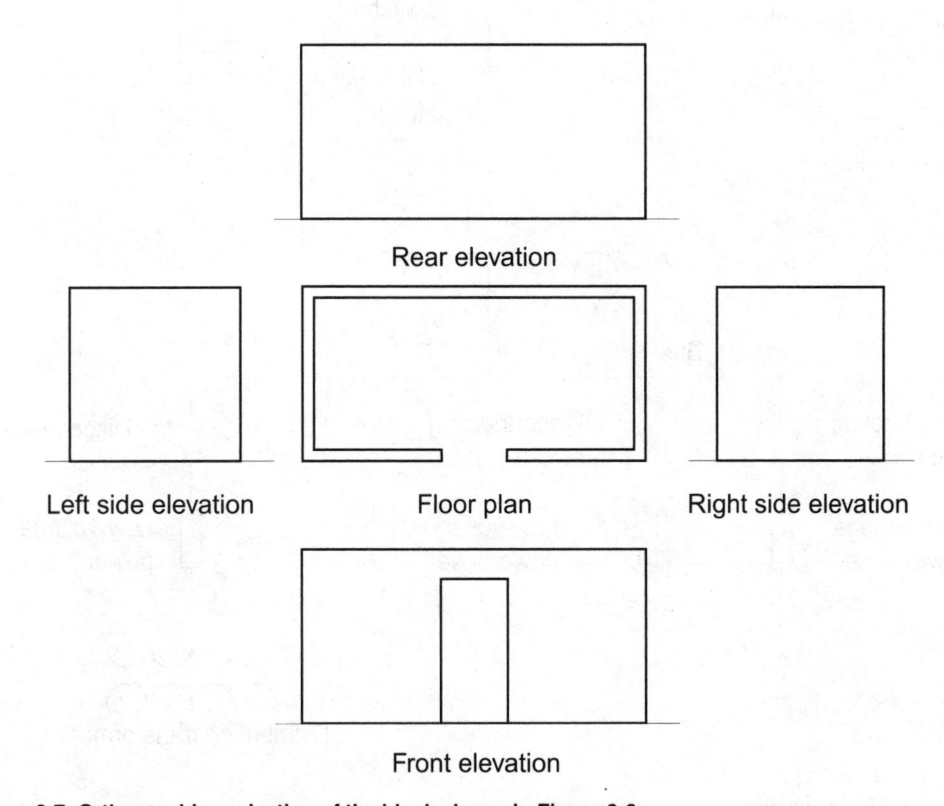

Figure 3-7: Orthographic projection of the block shown in Figure 3-6.

Electrical Diagrams

Electrical diagrams are drawings intended to show, in diagrammatic form, electrical components and their related connections. Such drawings are seldom drawn to scale, and show only the electrical association of the different components. In diagram drawings, symbols are used extensively to represent the various pieces of electrical equipment or components, and lines are used to connect these symbols — indicating the size, type, number of wires, and the like.

In general, the types of diagrams that will be encountered by industrial electrical workers will include, single-line diagrams (Figure 3-8) and schematic wiring diagrams (Figure 3-9).

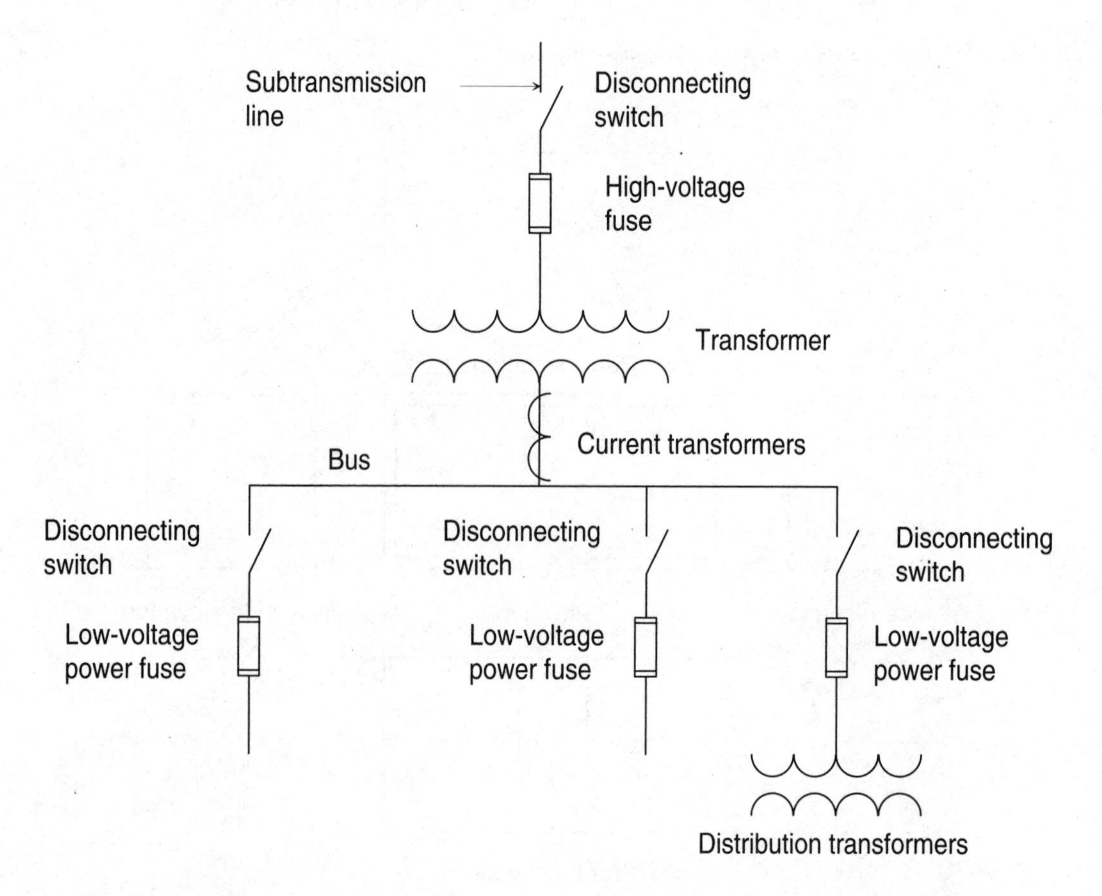

Figure 3-8: Single-line diagram of a substation.

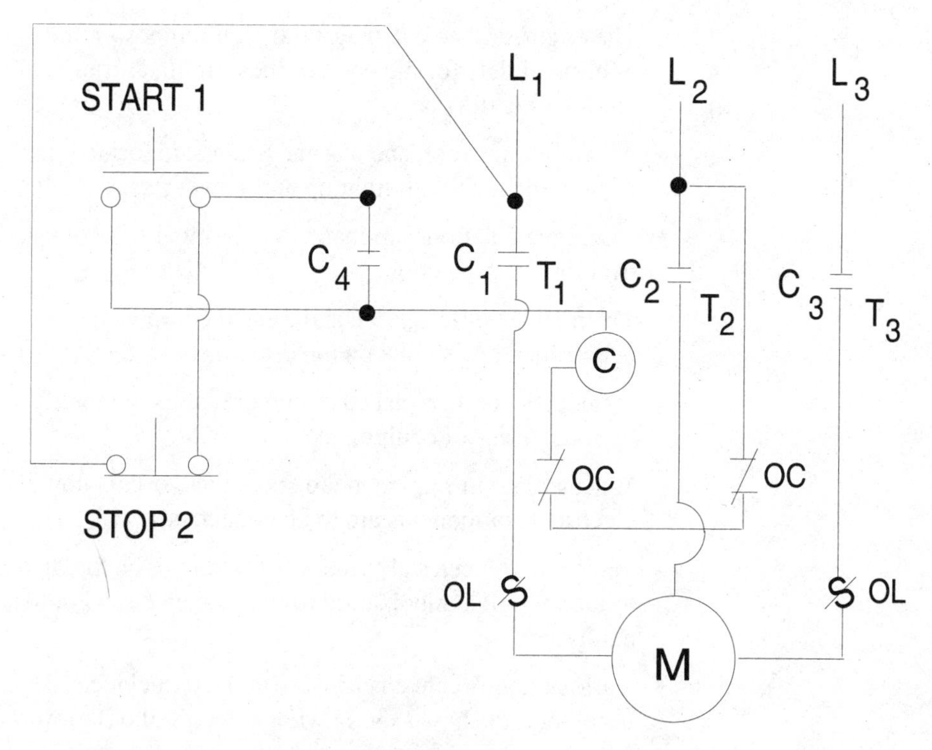

Figure 3-9: Wiring diagram of an ac magnetic motor starter.

ANALYZING ELECTRICAL DRAWINGS

The most practical way to learn how to read electrical construction documents is to analyze an existing set of drawings prepared by consulting or industrial engineers.

Engineers or electrical designers are responsible for the complete layout of electrical systems for most projects, and electrical drafters transform the engineer's designs into working drawings — either with manual drafting instruments or using computer-aided design (CAD) systems. In the preparation of the electrical design and working drawings, the following is a brief outline of what usually takes place:

- The engineer meets with the architect and owner to discuss the electrical needs of the building or project and to discuss various recommendations made by all parties.
- After that, an outline of the architect's floor plan is laid out.

- The engineer then calculates the required power and lighting outlets for the project; these are later transferred to the working drawings.
- All communication and alarm systems are located on the floor plan along with lighting and power panelboards.
- Circuit calculations are made to determine wire size and overcurrent protection.
- The main electric service and related components are determined and shown on the drawings.
- Schedules are then placed on the drawings to identify various pieces of equipment.
- Wiring diagrams are made to show the workers how various electrical components are to be connected.
- A legend or electrical symbol list is shown on the drawings to identify all symbols used to indicate electrical outlets or equipment.
- Various large-scale electrical details are included, if necessary, to show exactly what is required of the workers.
- Written specifications are then made to give a description of the materials and installation methods.

Development of Site Plans

In general practice, it is usually the owner's responsibility to furnish the architect/engineer with property and topographic surveys, which are made by a certified land surveyor or civil engineer. These surveys will show:

- All property lines.
- Existing public utilities and their location on or near the property; that is, electrical lines, sanitary sewer, gas line, water-supply line, storm sewer, manholes, telephone lines, etc.

A land surveyor does the property survey from information obtained from a deed description of the property. A property survey shows only the property lines and their lengths as if the property were perfectly flat.

The topographic survey will show the property lines but, in addition, will show the physical character of the land by using contour lines, notes, and symbols. The physical characteristics may include:

- The direction of the land slope.
- Whether the land is flat, hilly, wooded, swampy, high, or low, and other features of its physical nature.

All of this information is necessary so that the architect can properly design a building to fit the property. The electrical engineer also needs this information to locate existing electrical utilities and to route the new service to the building, provide outdoor lighting and circuits, etc.

Electrical site work is sometimes shown on the architect's plot plan. However, when site work involves many trades and several utilities — gas, telephone, electric, TV, water and sewage — things can become confusing if all details are shown on one drawing sheet. In cases like these, it is best to have a separate sheet devoted entirely to the electrical work as shown in Figure 3-10 on the next page. The project shown in this illustration is an office/warehouse building for Virginia Electric Inc. The electrical drawings consist of four 24- × 36-in drawing sheets, along with a set of written specifications.

The site or plot plan shown in Figure 3-10 has the conventional architect's and engineer's title blocks in the lower right-hand corner of the drawings. These blocks identify the project and the project's owners, the architect and engineer, and they also show how this drawing sheet relates to the entire set of drawings. Note the engineer's professional stamp of approval to the left of the engineer's title block. Similar blocks appear on all four of the electrical drawing sheets.

When examining a set of electrical drawings for the first time, always look at the area around the title block. This is where most revision blocks or revision notes are placed. If revisions have been made to the drawings, make certain that you have a clear understanding of what has taken place before proceeding with the work.

Refer again to the drawing in Figure 3-10 and note the "North Arrow" in the upper left corner. A North Arrow shows the direction of true north to help you orient the drawing to the site.

Glance directly down from the North Arrow to the bottom of the page and notice the drawing title, "Plot Utilities." Directly beneath the drawing title is the drawing scale; a scale of 1´´ = 30´ is shown. This means that each inch on the drawing represents 30 ft on the actual job site. This scale holds true for all drawings on the page unless otherwise noted.

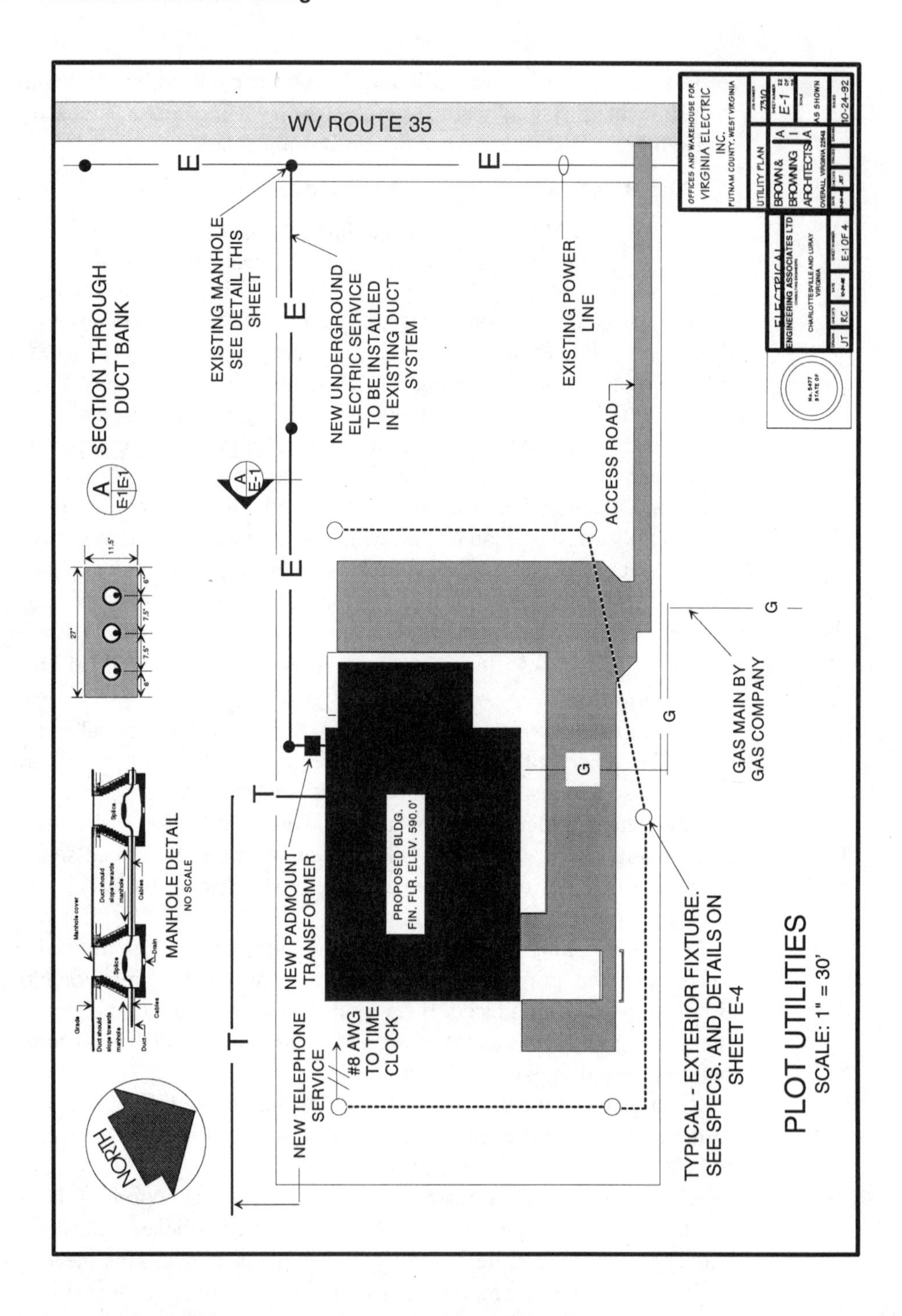

Figure 3-10: Plot utilities and details for an office/warehouse building.

An outline of the proposed building is indicated on the drawing by cross-hatched rectangles along with a callout stating, "Proposed Bldg. Fin. Flr. Elev. 590.0′." This translates to "the finished floor level of the building is to be 590 ft above sea level," which in this part of the country will be about 2 ft above finished grade around the building. This information helps the electrician get conduit sleeves and stub-ups to the correct height before the finished concrete floor is poured.

The shaded area represents asphalt paving for the access road, drives, and parking lot. Note that the access road leads into a highway which is designated "Route 35." This information further helps workers to orient the drawings to the building site.

Existing manholes are indicated by a solid circle, while an open circle is used to show the position of the five new pole-mounted lighting fixtures which are to be installed around the new building. Existing power lines are shown with a light solid line with the letter "E" placed at intervals along the line. The new underground electric service is shown the same way except that the lines are somewhat wider and show darker on the drawing. Note that this new high-voltage cable terminates into a padmount transformer near the proposed building. New telephone lines are similar except the letter " T " is used to identify the telephone lines.

The direct-burial underground cable supplying the exterior lighting fixtures is indicated with dashed lines on the drawing — shown connecting the open circles. A homerun for this circuit is also shown to a time clock.

The manhole detail shown to the right of the North Arrow may seem to serve very little purpose on this drawing since the manholes have already been installed. However, dimensions and details of their construction will help the electrical contractor or his or her supervisor to better plan the pulling of the high-voltage cable. The same is true of the cross-section shown of the duct bank. The electrical contractor knows that three empty ducts are available if it is discovered that one of them is damaged when the work begins.

Although the electrical work will not involve working with gas, the main gas line is shown on the electrical drawing to let the electrical workers know its approximate location while they are installing the direct-burial conductors for the exterior lighting fixtures.

POWER PLANS

Drawing sheet E-1 (Figure 3-11 on the next page) shows the complete floor plan of the office/warehouse building with all interior partitions drawn to scale. Sometimes the physical location of all wiring and outlets

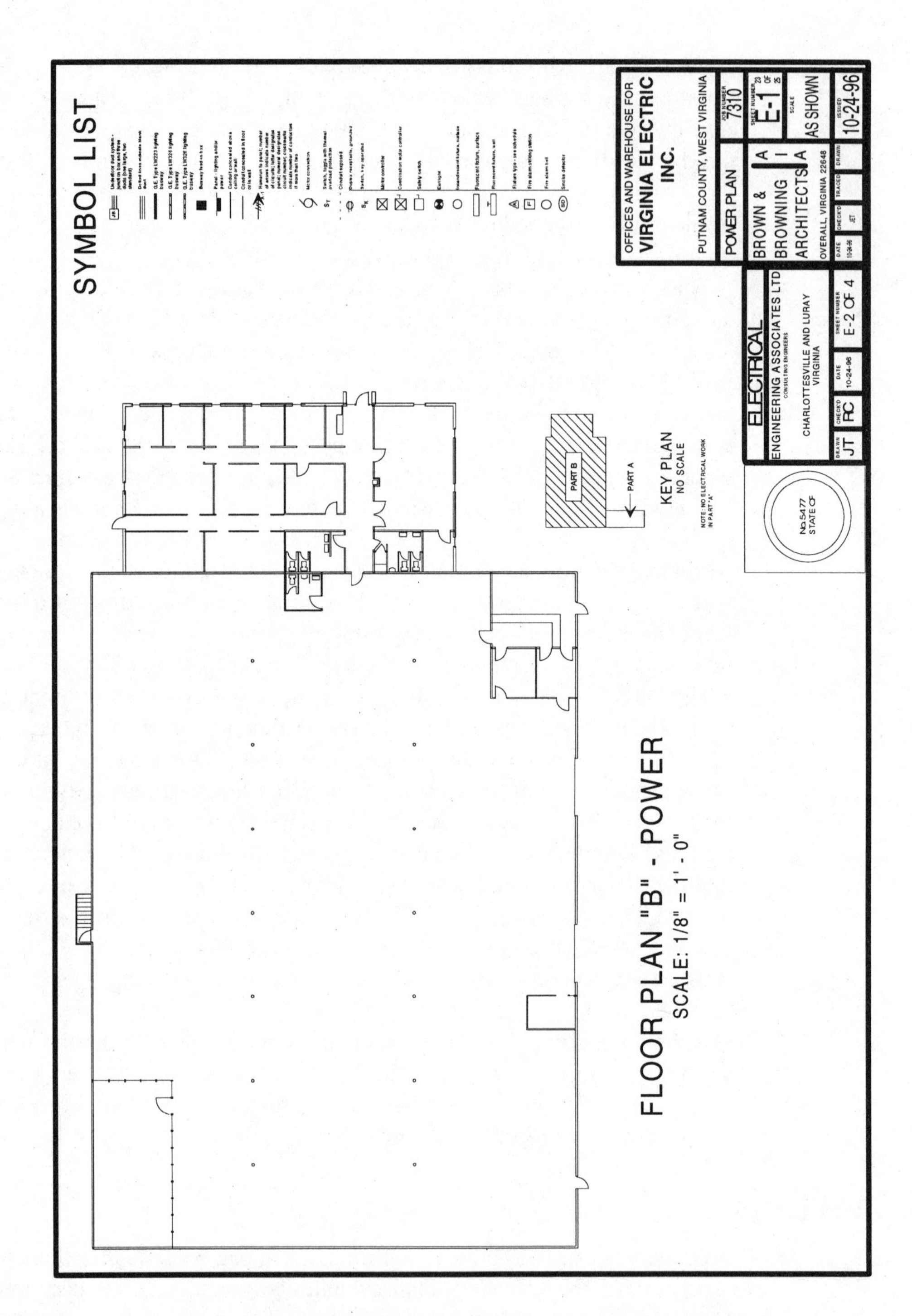

Figure 3-11: Drawing Sheet E-1 — Power Plan for the office/warehouse under discussion.

are shown on one drawing; that is, outlets for lighting, power, signal and communication, special electrical systems, and related equipment. However, on complex installations, the drawing would become cluttered if both lighting and power were shown on the same floor plan. Therefore, most projects will have a separate drawing for power and another for lighting. Riser diagrams and details may be shown on yet another drawing sheet, or if room permits, they may be shown on the lighting or power floor plan sheets.

A closer look at the drawing in Figure 3-11 reveals the title blocks in the lower right corner of the drawing sheet. These blocks list both the architectural and engineering firms, along with information to identify the project and drawing sheet. Also note that the floor plan is titled, "Floor Plan B - Power" and is drawn to a scale of $\frac{1}{8}$ = 1 - 0 . There are no revisions shown on this drawing sheet.

Key Plan

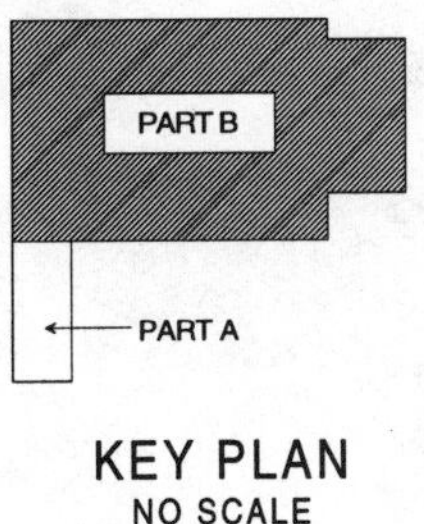

NOTE: NO ELECTRICAL WORK IN PART "A"

Figure 3-12: Key plan that appears on sheet E-2.

A "Key Plan" appears on the drawing sheet immediately above the engineer's title block (Figure 3-12). The purpose of this key plan is to identify that part of the project to which the drawing sheet applies. In this case, the project involves two buildings: Building "A" and Building "B." Since the outline of Building "B" is cross-hatched in the key plan, this is the building to which drawing sheet E-2 applies. Note that this key plan is not drawn to scale — only its approximate shape.

Although Building "A" is also shown on this key plan, a note below the key plan title states, "Note: No electrical work in Part A."

On some larger installations, the overall project may involve several buildings requiring appropriate key plans on each drawing to help the workers orient the drawings to the appropriate building. In some cases, separate drawing sheets may be used for each room or department in an industrial project — again requiring key plans on each drawing sheet to identify applicable drawings for each room.

Symbol List

A "Symbol List" appears on drawing sheet E-2 (immediately above the architect's title block) to identify the various symbols used for both power and lighting on this project. In most cases, only symbols are listed that apply to the particular project. In other cases, however, a standard list of symbols is used for all projects with the following note.

These are standard symbols and may not all appear on the project drawing; however, wherever the symbol on the project drawings occurs, the item shall be provided and installed.

Only electrical symbols that are actually used for the office/warehouse drawings are shown in this list. A close-up look at these symbols appears in Figure 3-13.

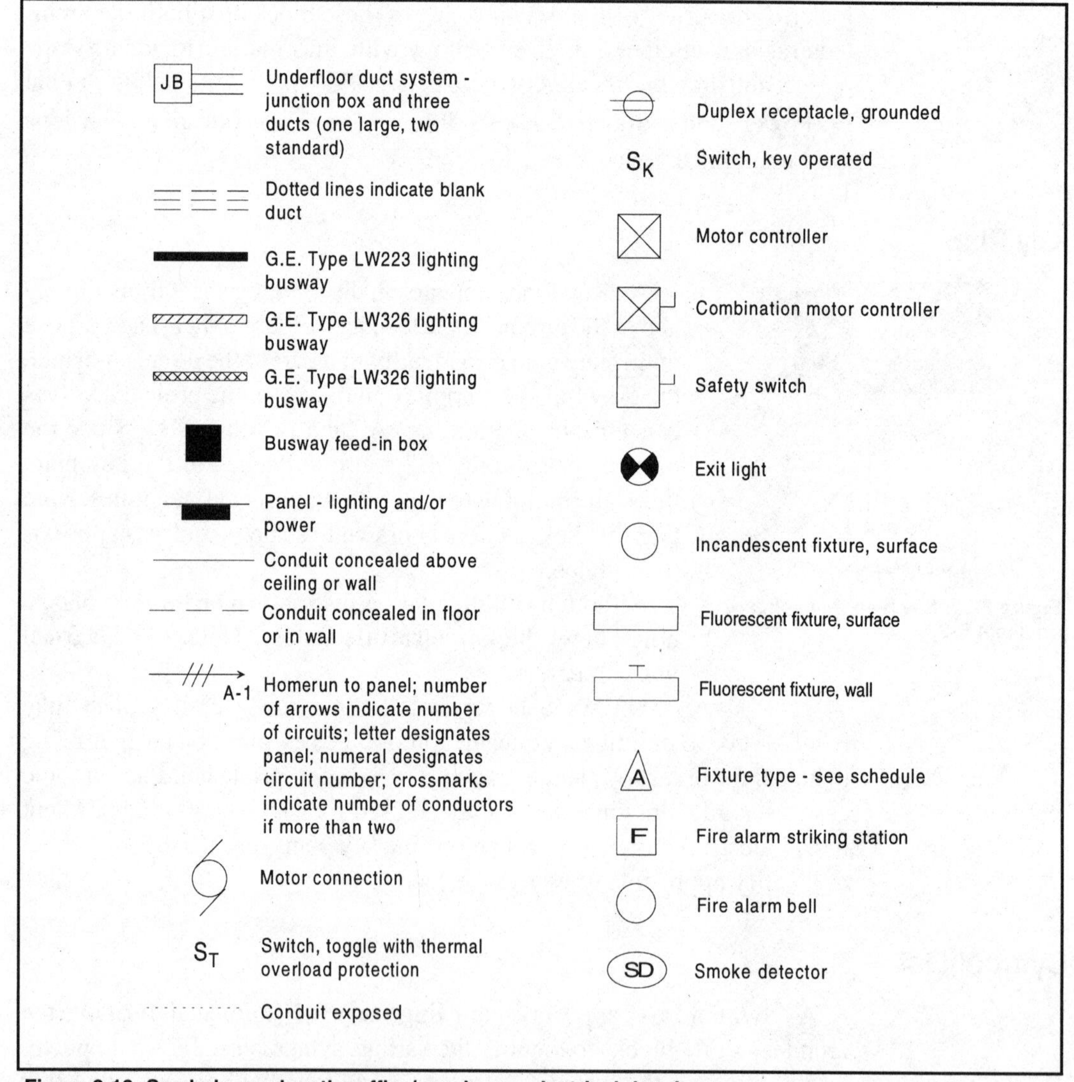

Figure 3-13: Symbols used on the office/warehouse electrical drawings.

Floor Plan

A somewhat enlarged view of the electrical floor-plan drawing (sheet E-1) is shown in Figure 3-14 on the next page. However, due to the size of the drawing in comparison with the size of the pages in this book, it is still difficult to see very much detail. This illustration is meant to show the overall layout of the floor plan and how the symbols and notes are arranged.

In general, this plan shows the service equipment (in plan view), receptacles, the underfloor duct system, motor connections, motor controllers, electric heat, busway, and similar details. The electric panels and other service equipment are drawn close to scale. The locations of other electrical outlets and similar components are only approximated on the drawings because they have to be exaggerated to show up on the prints. To illustrate, a common duplex receptacle is only about three inches wide. If such a receptacle were to be located on the floor plan of this building (drawn to a scale of $\frac{1}{8}$″ = 1′ - 0″), even a small dot on the drawing would be too large to draw the receptacle exactly to scale. Therefore, the receptacle symbol is exaggerated. When such receptacles are scaled on the drawings to determine the proper location, a measurement is usually taken to the center of the symbol to determine the distance between outlets. Junction boxes, switches, and other electrical connections shown on the floor plan will be exaggerated in a similar manner.

A partial floor-plan drawing appears in Figure 3-15 which allows a somewhat better view of the drawing details.

Notes and Building Symbols

Referring again to Figures 3-14 and 3-15, you will notice numbers placed inside of an oval symbol in each room. These numbered ovals represent the room name or type and correspond to a room schedule in the architectural drawings. For example, room number 112 is designated "lobby" in the room schedule, room number 113 is designated "office No. 1," etc. On some drawings, however, these room symbols are omitted and the room names are written out on the drawings.

There are also several notes appearing at various places on the floor plan. These notes offer additional information to clarify certain aspects of the drawings. For example, only one electric heater is to be installed by the electrical contractor; this heater is located in the building's vestibule. Rather than have a symbol in the symbol list for this one heater, a note is used to identify it on the drawing.

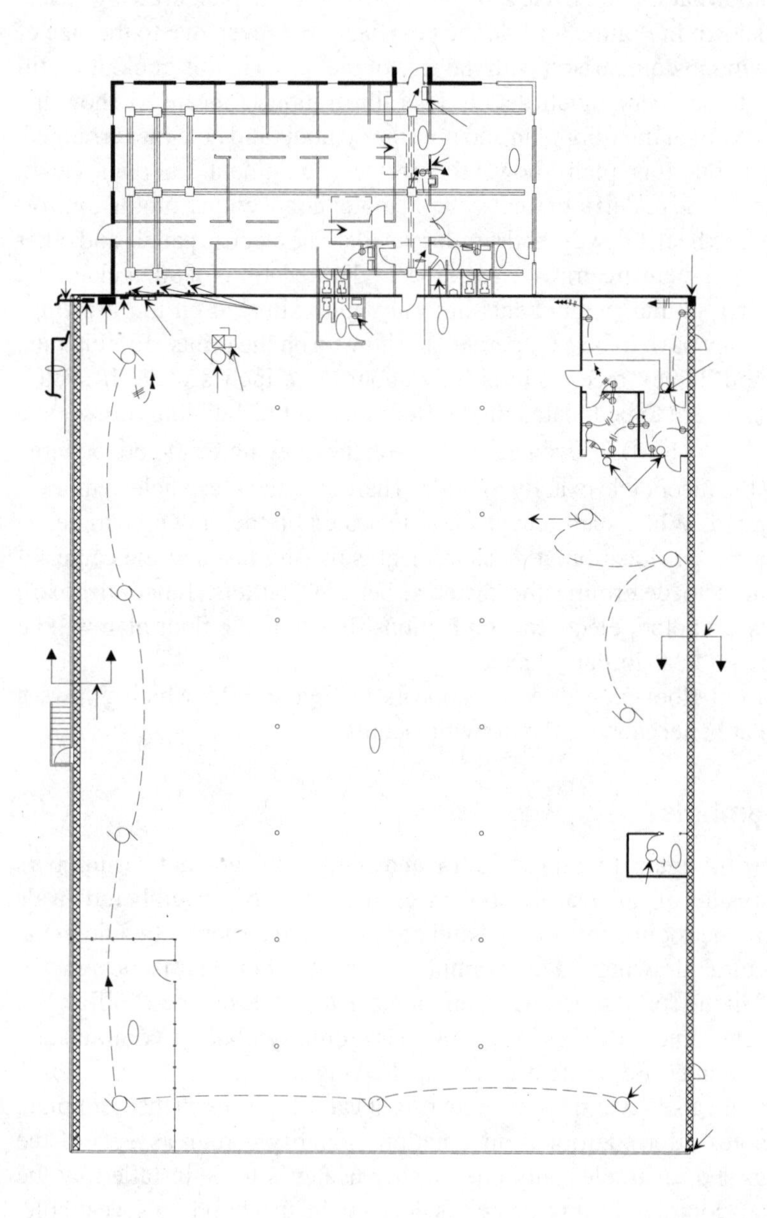

Figure 3-14: Power plan for office/warehouse building.

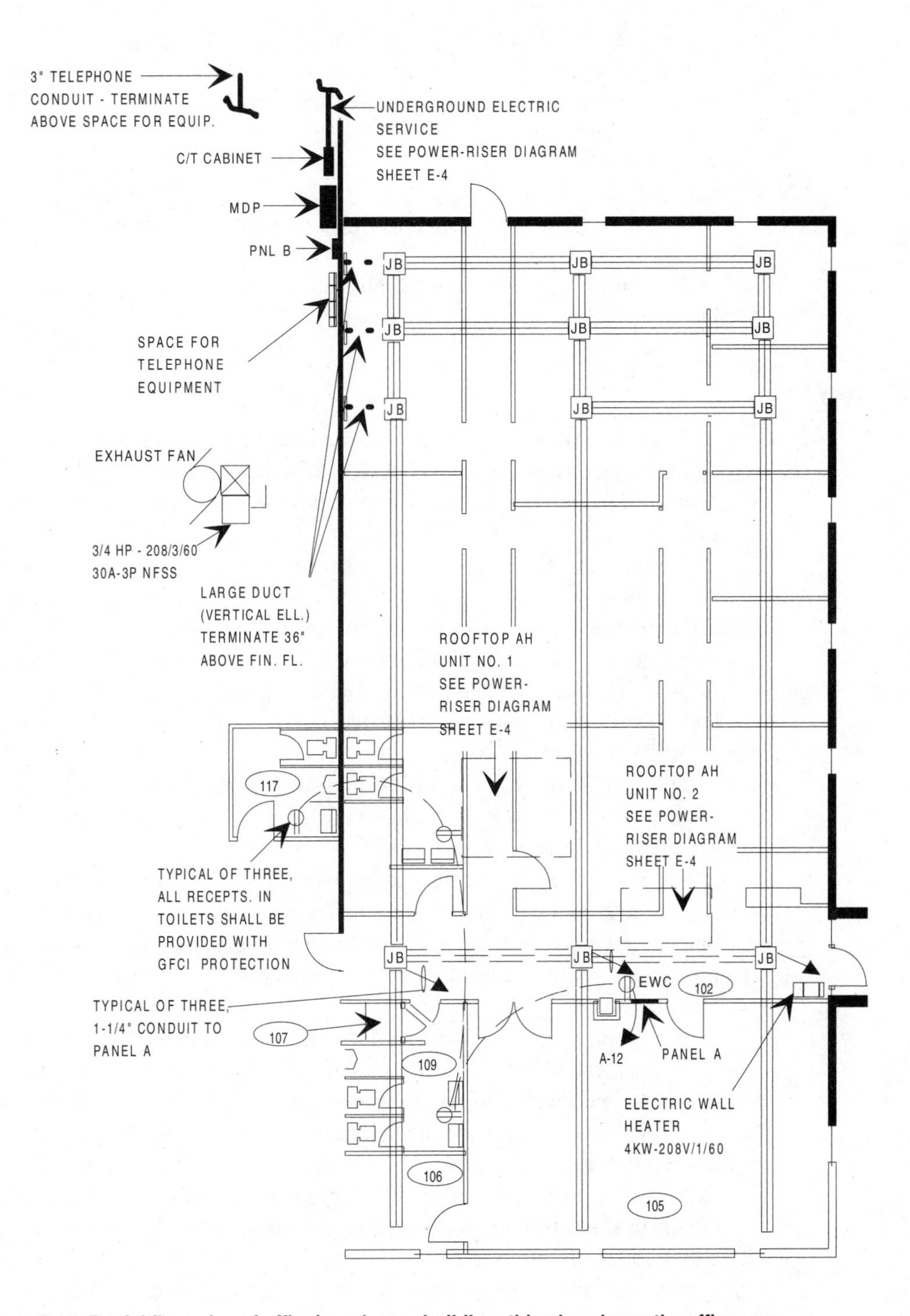

Figure 3-15: Partial floor plan of office/warehouse building; this plan shows the office area.

Other notes on this drawing describe how certain parts of the system are to be installed. For example, in the office area (rooms 112, 113, and 114), you will see the following note:

Conduit turned up and stubbed out above ceiling

This empty conduit is for telephone/communication cables that will be installed later by the telephone company.

Busway

The office/warehouse project utilizes three types of busways: two types of lighting busways, and one power busway. Only the power busway is shown on the power plan; the lighting busways will appear on the lighting plan.

Figure 3-14 shows two runs of busways — one running the length of the building on the south end (top wall on drawing), and one the length of the north wall. The symbol list in Figure 3-13 shows this busway to be designated by two parallel lines with a series of Xs inside. The symbol list further describes the busway as General Electric Type LW326. These busways are fed from the main distribution panel (circuits MDP-1 and MDP-2) through tap boxes.

The *NEC* defines busway to be a grounded metal enclosure containing factory-mounted, bare or insulated conductors, which are usually copper or aluminum bars, rods, or tubes.

The relationship of the busway and hangers to the building construction should be checked prior to commencing the installation and any problems due to space conflicts, inadequate or inappropriate supporting structure, openings through walls, etc., are worked out in advance so as not to incur lost time.

For example, the drawings and specifications may call for the busway to be suspended from brackets clamped or welded to steel columns. However, the spacing of the columns may be such that additional supplementary hanger rods suspended from the ceiling or roof structure may be necessary for the adequate support of the busway. To offer more assistance to workers on the office/warehouse project, the engineer has provided a detail on sheet E-4 of the drawings that shows how the busway is to be mounted. *See* Figure 3-16.

Other details that appear on the floor plan in Figure 3-15 include the general arrangement of the underfloor duct system, junction boxes, and

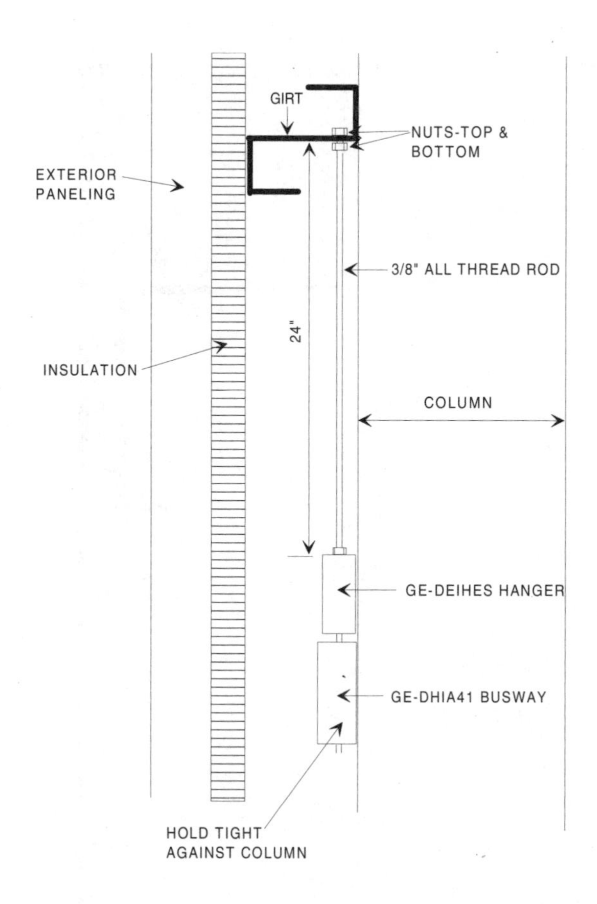

Figure 3-16: Section A-A on the drawing in Figure 3-14.

feeder conduit for the underfloor duct system, plan views of the service and telephone equipment, along with duplex receptacle outlets. A note on the drawing requires all receptacles in the toilets to be provided with ground-fault circuit-interruption protection. The letters "EWC" next to the receptacle in the vestibule designates this receptacle to feed an electric water cooler.

LIGHTING FLOOR PLAN

A skeleton view of drawing sheet E-3 — Lighting Floor Plan is shown in Figure 3-17 on the next page. Again, the architect/engineer's title blocks appear in the lower right corner of the drawing. A "Key Plan," as discussed previously appears above the engineer's title block. This drawing is titled "Lighting Plan" and is drawn to the same scale as the power plan; that is, ⅛′′ = 1′ - 0′′. A lighting-fixture schedule appears in the upper right corner of the drawing and some installation notes appear below this schedule.

The lighting-outlet symbols found on the office/warehouse building represent both incandescent and fluorescent types; a circle on most electrical drawings usually represents an incandescent fixture and a rectangle represents a fluorescent one. All of these symbols are designed to indicate the physical shape of a particular fixture and are usually drawn to scale.

The type of mounting used for all lighting fixtures is usually indicated in a lighting-fixture schedule, which is shown on the drawings in this case, but the schedule may be found only in the written specifications on some projects.

The type of lighting fixture is identified by a numeral placed inside a triangle near each lighting fixture. If one type of fixture is used exclusively

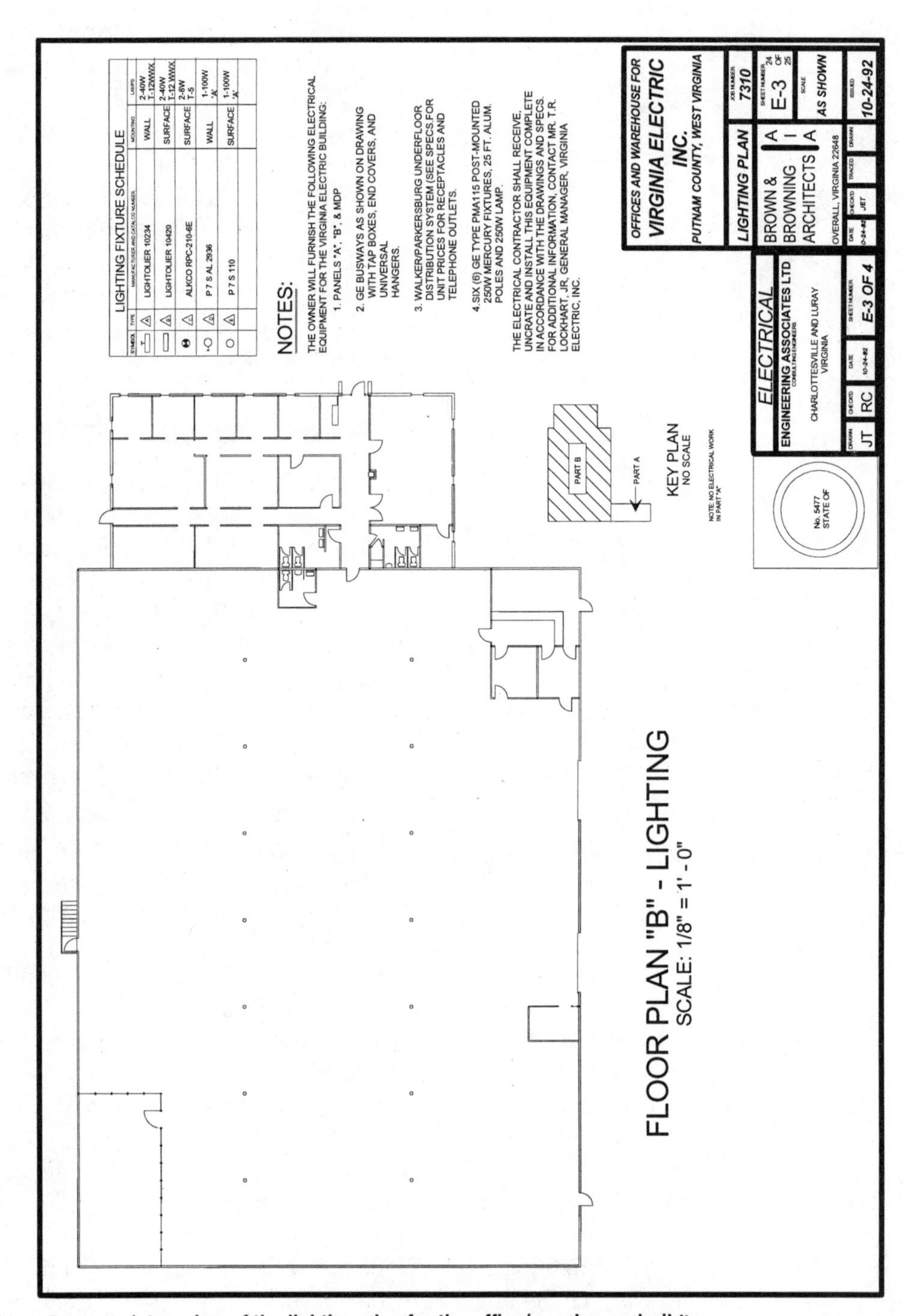

Figure 3-17: Skeleton view of the lighting plan for the office/warehouse building.

in one room or area, the triangular indicator need only appear once with the word "all" lettered at the bottom of the triangle.

Drawing Schedules

A schedule is a systematic method of presenting notes or lists of equipment on a drawing in tabular form. When properly organized and thoroughly understood, schedules are not only powerful time-saving devices for those preparing the drawings, but they can also save the workers on the job much valuable time.

For example, the lighting-fixture schedule shown in Figure 3-18 lists the fixture type and identifies each fixture type on the drawing by number. The manufacturer and catalog number of each type are given along with the number, size, and type of lamp for each.

Sometimes all of the same information found in schedules will be duplicated in the written specifications, but combing through page after page of written specifications can be time consuming and workers do not always have access to the specifications while working, whereas they usually do have access to the working drawings. Therefore, the schedule is an excellent means of providing essential information in a clear and accurate manner, allowing the workers to carry out their assignments in the least amount of time.

Other schedules that are frequently found on electrical working drawings include:

LIGHTING FIXTURE SCHEDULE

SYMBOL	TYPE	MANUFACTURER AND CATALOG NUMBER	MOUNTING	LAMPS
	A	LIGHTOLIER 10234	WALL	2-40W T-12WWX
	B	LIGHTOLIER 10420	SURFACE	2-40W T-12 WWX
	C	ALKCO RPC-210-6E	SURFACE	2-8W T-5
	D	P 7 S AL 2936	WALL	1-100W 'A'
	E	P 7 S 110	SURFACE	1-100W 'A'

Figure 3-18: Lighting-fixture schedule.

- Connected load schedule
- Panelboard schedule
- Electric-heat schedule
- Kitchen-equipment schedule
- Schedule of receptacle types

There are also other schedules found on electrical drawings, depending upon the type of project. Most, however, will deal with lists of equipment; that is, such items as motors, motor controllers, and the like.

ELECTRICAL DETAILS AND DIAGRAMS

Electrical diagrams are drawings that are intended to show, in diagrammatic form, electrical components and their related connections. They are

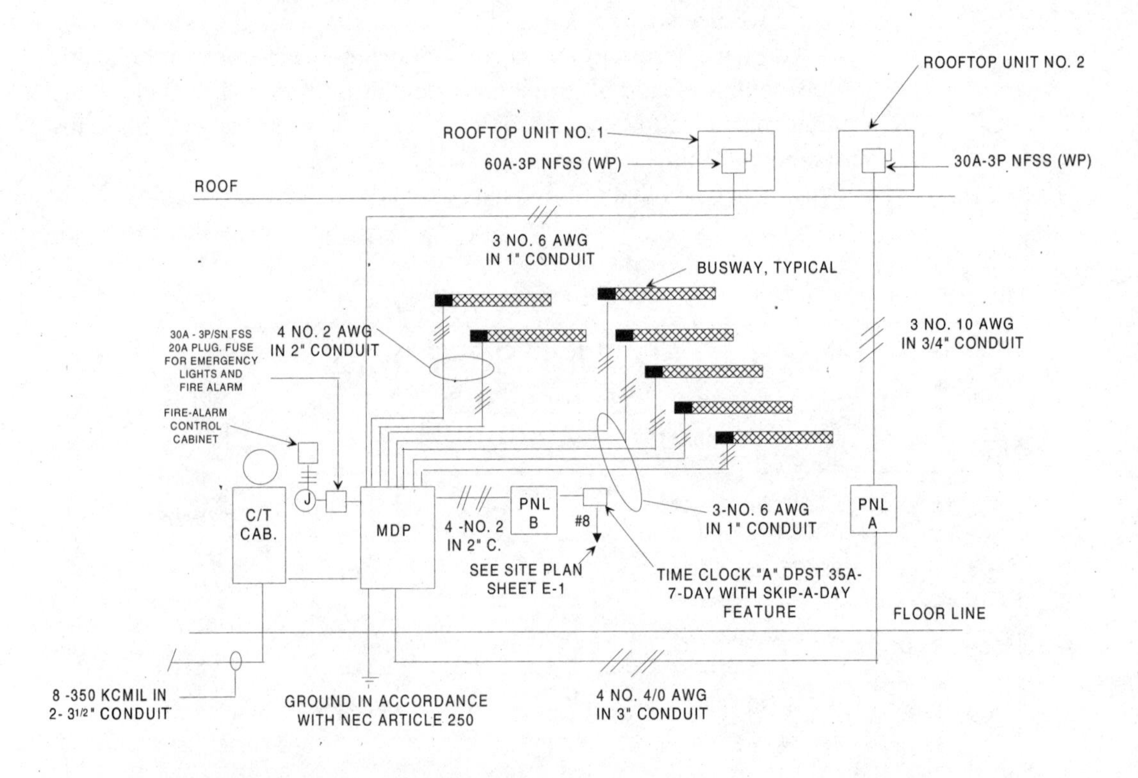

Figure 3-19: Power-riser diagram for office/warehouse building.

PANELBOARD SCHEDULE										
PANEL No.	CABINET TYPE	PANEL MAINS			BRANCHES					ITEMS FED OR REMARKS
		AMPS	VOLTS	PHASE	1P	2P	3P	PROT.	FRAME	
MDP	SURFACE	600A	120/208	3 ϕ, 4-W	-	-	1	225A	25,000	PANEL "A"
					-	-	1	100A	18,000	PANEL "B"
					-	-	1	100A		POWER BUSWAY
					-	-	1	60A		LIGHTING BUSWAY
					-	-	1	70A		ROOFTOP UNIT #1
					-	-	1	70A	▼	SPARE
					-	-	1	600A	42,000	MAIN CIRCUIT BRKR

Figure 3-20: Corresponding panelboard schedule for the power-riser diagram in Figure 3-19.

seldom, if ever, drawn to scale, and show on the electrical association of the different components.

Power-Riser Diagrams

Single-line block diagrams are used extensively to show the arrangement of electric service equipment. The power-riser diagram in Figure 3-19, for example, was used on the office/warehouse building under discussion and is typical of such drawings. The drawing shows all pieces of electrical equipment as well as the connecting lines used to indicate service-entrance conductors and feeders. Notes are used to identify the equipment, indicate the size of conduit necessary for each feeder, and the number, size, and type of conductors in each conduit.

A panelboard schedule (Figure 3-20) is included with the power-riser diagram to indicate the exact components contained in each panelboard. This panelboard schedule is for the main distribution panel. Schedules will also be shown for the remaining schedules.

In general, panelboard schedules usually indicate the panel number, the type of cabinet (either flush- or surface-mounted), the panel mains (ampere and voltage rating), the phase (single- or three-phase), and the number of wires. A 4-wire panel, for example, indicates that a solid neutral exists in the panel. Branches indicate the type of overcurrent protection; that is, the

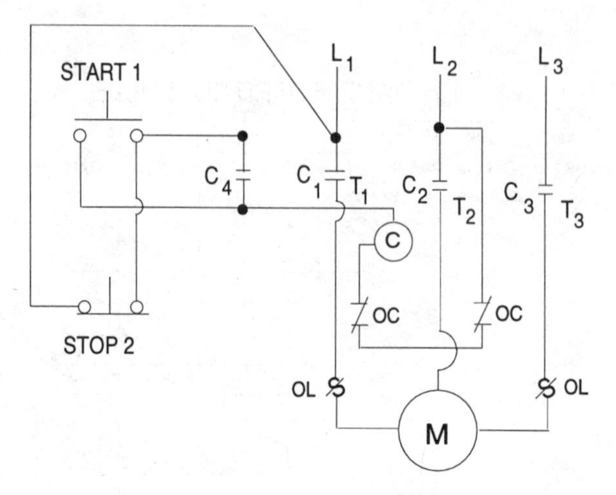

Figure 3-21: Schematic wiring diagram for a three-phase, ac magnetic nonreversing motor starter.

number of "poles," the trip rating, and the frame size. The items that each overcurrent device feeds is also indicated in one of the columns.

Schematic Diagrams

Complete schematic wiring diagrams are normally used only in highly unique and complicated electrical systems, such as control circuits. Components are represented by symbols, and every wire is either shown by itself or included in an assembly of several wires which appear as one line on the drawing. Each wire should be numbered when it enters an assembly and should keep the same number when it comes out again to be connected to some electrical component in the system. Figure 3-21 shows a complete schematic wiring diagram for a three-phase, ac magnetic non-reversing motor starter.

Note that this diagram shows the various devices in symbol form and indicates the actual connections of all wires between the devices. The three-wire supply lines are indicated by L_1, L_2, and L_3; the motor terminals of motor M are indicated by T_1, T_2, and T_3. Each line has a thermal overload-protection device (OL) connected in series with normally open line contactors C_1, C_2, and C_3, which are controlled by the magnetic starter coil, C. Each contactor has a pair of contacts that close or open during operation. The control station, consisting of start pushbutton 1 and stop pushbutton 2, is connected across lines L_1 and L_2. An auxiliary contactor (C_4) is connected in series with the stop pushbutton and in parallel with the start pushbutton. The control circuit also has normally closed overload contactors (OC) connected in series with the magnetic starter coil (C).

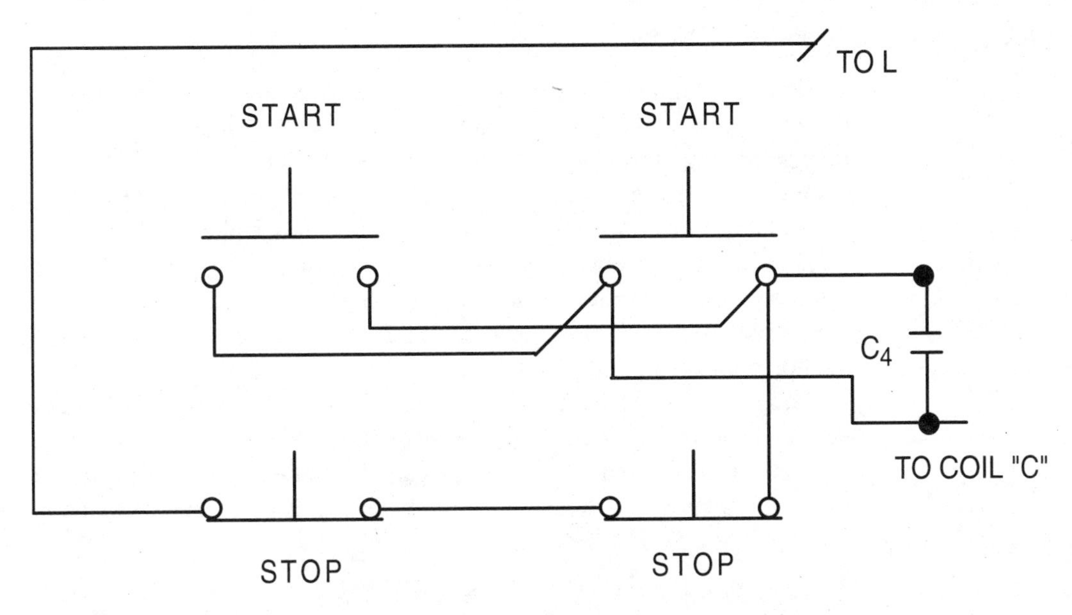

Figure 3-22: Schematic wiring diagram for a three-phase, ac magnetic nonreversing motor starter, controlled by two sets of start-stop pushbutton stations.

Any number of additional pushbutton stations may be added to this control circuit similarly to the way three- and four-way switches are added to control a lighting circuit. In adding pushbutton stations, the stop buttons are always connected in series and the start buttons are always connected in parallel. Figure 3-22 shows the same motor-starter circuit in Figure 3-21, but this time it is controlled by two sets of start-stop buttons.

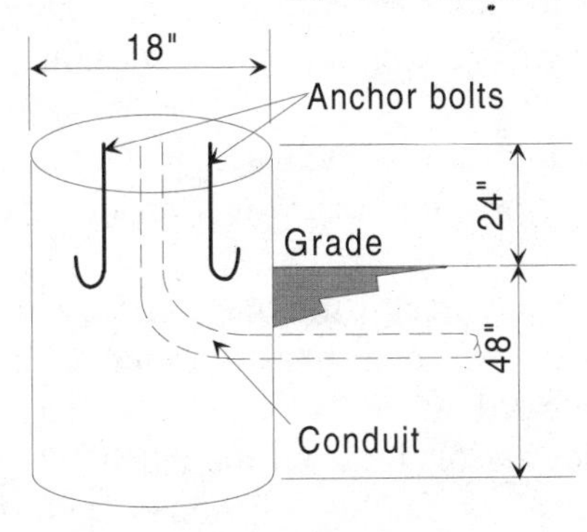

Figure 3-23: Pole-mounting detail for exterior lighting fixtures.

Drawing Details

A set of electrical drawings will sometimes require large-scale drawings of certain areas that are not indicated with sufficient clarity on the small-scale drawings. For example, the site plan in Figure 3-10 shows exterior pole-mounted lighting fixtures that are to be installed by the contractor. The detail in Figure 3-23 shows how the concrete base is constructed.

SINGLE-LINE DRAWING SYMBOLS

Single-line drawing symbols are used in conjunction with lines to show the component and equipment connections to an electrical system and not necessarily the physical location of such equipment.

A list of the most commonly-used symbols is shown in Figure 3-24. Note that some of the symbols are abbreviated idioms, like "CB" for circuit breaker or "R" for relay. Other symbols are simplified pictographs, like a stator and brushes for motor or a triangle for a pothead. In some cases there are combinations of idioms and pictographs, as in electric motors where the horsepower is indicated by a numeral.

Single-line electrical drawing symbols have evolved over the years to their present state after much discussion with electrical engineers, electrical drafters, electrical estimators, electricians, and others who are required to interpret electrical drawings. It is felt that the current list represents a good set of symbols in that they are:

- Easy to draw
- Easily interpreted by workers
- Sufficient for most applications

The use of "standard" symbols for single-line diagrams has been attempted for almost 100 years, but the standard symbols are frequently modified to suit a particular need. Consequently, if a deviation is made from the standard, a legend or symbol list normally appears on the working drawings or in the written specifications. When encountering a new set of drawings, always look for, and refer to, this symbol list. Then, continue referring to the list each time the drawings are used until the meaning of each symbol is memorized.

Refer again to Figure 3-24. Let's take each individual symbol and see how some of them might be modified on drawings that will be encountered on the job.

Electric motor: The symbol shown for an electric motor is a circle which represents the motor's stator. Two short diagonal lines represent motor brushes. The rated horsepower is indicated by numerals inside the circle. On some drawings, the diagonal lines are omitted and the letter "M" is inserted inside the circle while the horsepower rating is in numerical form (outside the circle) or else coded and the type and horsepower of the motor indicated in a motor schedule on the drawings or in the written specifications.

Power transformer: The symbol shown differs slightly from conventional transformer symbols in that the cooling tubes or radiator are depicted

MEANING	SYMBOL	NEC REFERENCE
Electric Motor (HP as Indicated)	1/4	NEC Articles 422, 424, 430, and 440
Power Transformer		NEC Article 450
Pothead (Cable Termination)		NEC Section 230-220
Circuit Element e.g. Circuit Breaker	CB	NEC Article 240 and Section 710-21
Circuit Breaker		NEC Article 240
Fusible Element		NEC Section 230-208 and Article 240
Single-Throw Knife Switch		NEC Articles 380 and 384
Double-Throw Knife Switch		NEC Articles 380 and 384
Ground		NEC Article 250
Battery		NEC Article 480
Contactor	C	
Photoelectric Cell	PE	NEC Article 690
Voltage Cycles, Phase	EX: 480/60/3	
Relay	R	NEC Sections 430-40 and 450-5
Equipment Connection (as noted)		

Figure 3-24: Single-line drawing symbols.

in a power-transformer symbol; they are omitted for other types of transformers. In many cases, only two coils will be shown, representing the primary and secondary of the transformer. Other symbols will also show core lines between the primary and secondary coils.

Pothead: The symbol for a pothead is a triangle which is the general shape of a pothead. The symbol is rarely modified from the one shown.

Circuit element: When a box is used in a single-line electrical drawing, some identifying letters or numerals are normally used inside the box. For example, "CB" stands for circuit breaker. This type of circuit breaker, however, represents the huge outdoor oil-immersed circuit breaker used on high-voltage systems rather than the plug-in type of circuit breaker used in panelboards and load centers.

Circuit breaker: This symbol is normally used for lower-voltage thermal circuit breakers of 600 V and below. Other symbols found on electrical drawings representing circuit breakers are also described in this and other chapters in this book.

Fusible element: The symbol shown is one of many that are used to represent overcurrent protection in an electrical system..

Single-throw knife switch: This is the standard symbol for a disconnecting switch, regardless of the voltage, and few modifications will be found on electrical drawings.

Double-throw knife switch: There are many variations of this symbol and the more common modifications are shown in this chapter.

Ground: This is the standard symbol for ground and is used in all types of drawings from wiring diagrams, ladder diagrams, schematic diagrams, as well as single-line electrical diagrams.

Battery: This is also the standard symbol for a battery. Sometimes the symbol is modified to indicate the number of battery cells; that is, the long and short lines are repeated for the number of cells in the battery.

Contactor: The symbol shown is frequently used as a circuit element in high-voltage systems, although two separated short lines are also common.

Photoelectric cell: There are many modifications for this symbol and most are described later in this book.

Voltage, cycle, phase: These electrical characteristics are most often represented by the numerals separated by slash marks. Sometimes letters are used in conjunction with the numerals to further clarify the intent; that is, 480 V/60 Hz/3-Phase.

Relay: The symbol shown is frequently used on one-line power diagrams. A circle is sometimes used on schematic or ladder diagrams.

Equipment connections (as noted): This symbol is used to describe a wide variety of electrical connections. Notes usually accompany the symbol or else a legend or symbol list is used to denote certain connections. For example, this symbol with the letter "W" next to it could specify an outlet for an electric welding machine. Always check the symbol list or legend for an explanation of this type of symbol.

INTERPRETING SINGLE-LINE DIAGRAMS

The single-line diagram shown in Figure 3-25 (pages 54 and 55) is typical of those used to show workers how an electrical system is to be installed. In general, a one-line diagram is never drawn to scale. Such drawings show the major components in an electrical system and then utilize only one drawing line to indicate the connections between these components. Even though only one line is used between components, this single line may indicate a raceway of two, three, four, or more conductors. Notes, symbols, tables, and detailed drawings are used to supplement and clarify a one-line diagram.

Referring again to Figure 3-25, this drawing was prepared by an electrical manufacturing company to give workers at the job site an overview of a 2000 kVA substation utilizing a 13.8 kV primary and a 4.16 kV, 3-phase, 3-wire, 60 Hz secondary. Note that this drawing sheet is divided into the following sections:

- Service order numbers
- Unit numbers
- One-line diagram
- Title block
- Revision block

Service order numbers: These numbers are arranged at the top of the drawing sheet in a time-sequence, bar-chart type arrangement. For example, S.O. #58454 deals with the primary side of the 2000 kVA transfromer, including the transformer itself. This section includes a high-voltage switchgear, with an in-outdoor enclosure. The switchgear itself consists of two HPL-C interrupter switches, each rated at 15 kV, 600-A, and 150E current-limiting fuses (CLF).

Service order #58455 deals with the wiring and related components on the secondary side of the transformer and begins with a low-voltage switchgear with an in-outdoor enclosure. Details of this combined switch-

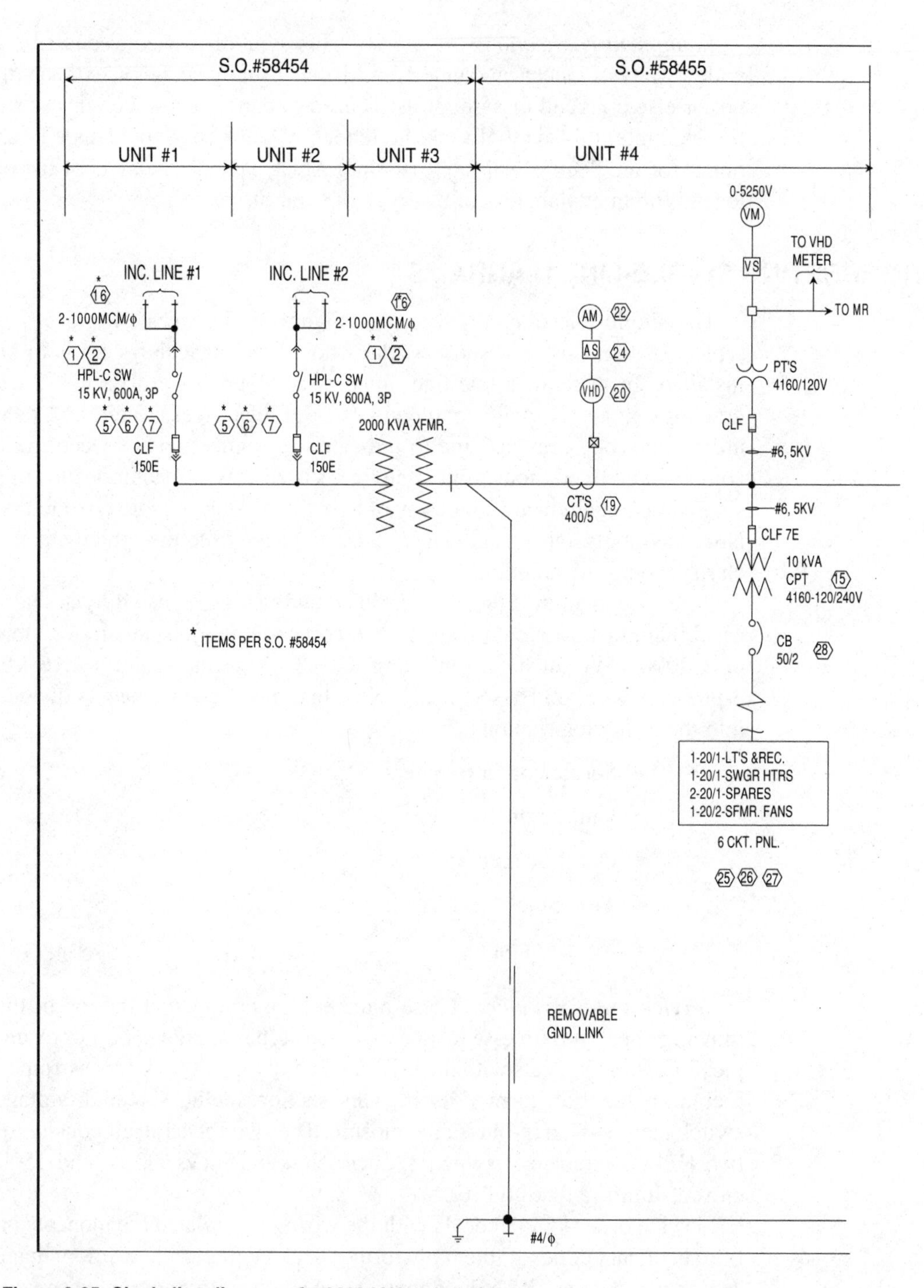

Figure 3-25: Single-line diagram of a 2000 kVA substation.

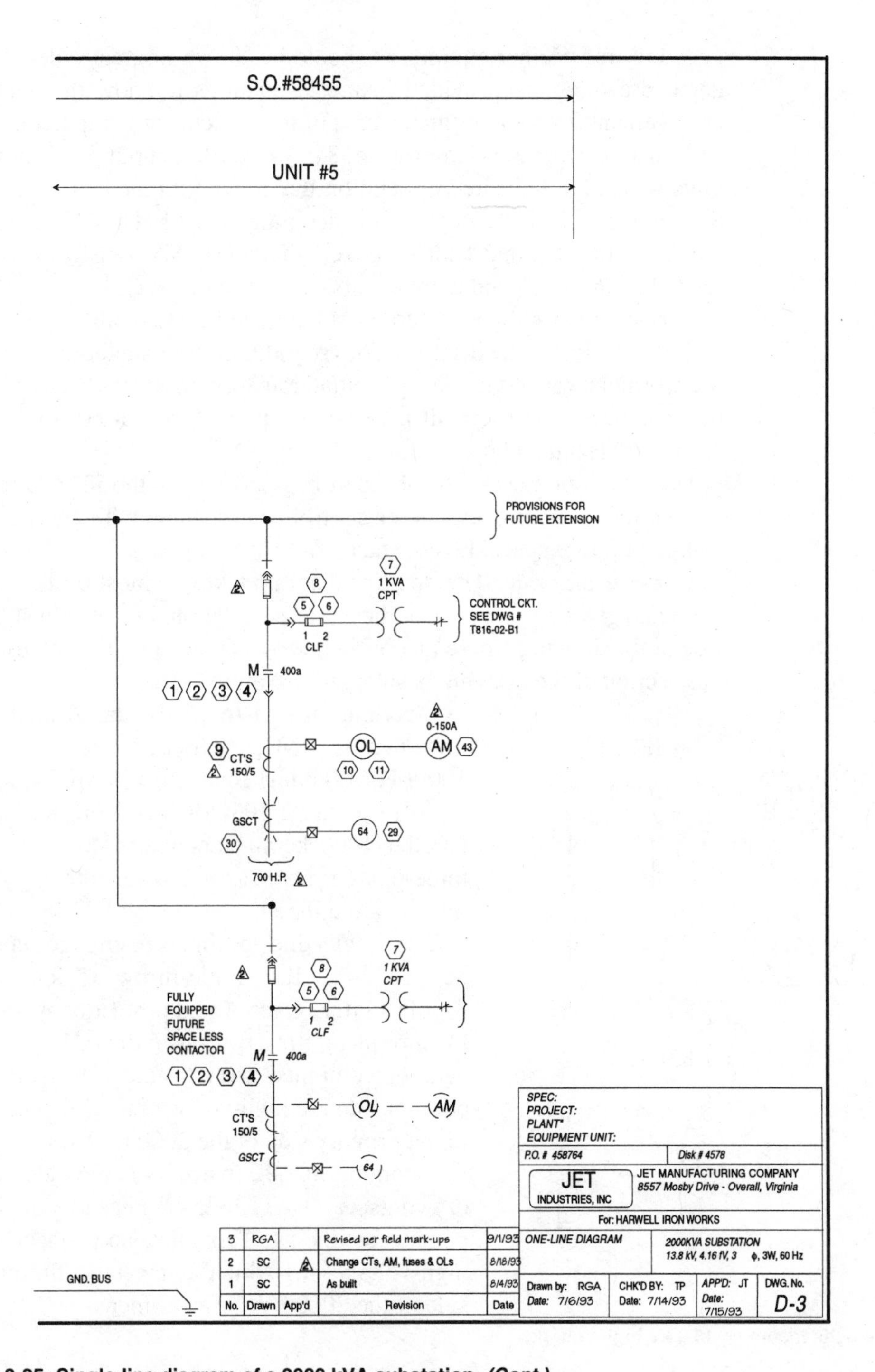

Figure 3-25: Single-line diagram of a 2000 kVA substation. ***(Cont.)***

gear and transformer equipment should be shown in greater detail. Such detail drawings are provided in most situations to show the anchoring arrangement (including dimensions) of the switchgear equipment.

Unit numbers: Service order #58454 is further subdivided into three units which are indicated as such on the drawing immediately under the S.O. number. Unit #1 deals with incoming line #1; Unit #2 deals with incoming line #2, and Unit #3 covers the 2000 kVA transformer and its related connections and components.

Service order #58455 is further subdivided into two units — Units # 4 and #5. Basically, Unit #4 covers grounding, the installation of current transformers, various meters, potential transformers, a 10 kVA 4160/240V transformer, and a 6-circuit panel — all derived from a 600-A, 4160-V, 3-wire, 60 Hz main bus.

Unit #5 continues with the main bus and covers the installation and connection of a complete motor-control center, along with another fully-equipped future space, less contactors.

One-line diagram: The one-line diagram takes up most of the drawing sheet and gives an overview of the entire installation. Let's begin at the left side of the drawing where incoming line #1 is indicated. For clarification, this section of the drawing is enlarged in Figure 3-26.

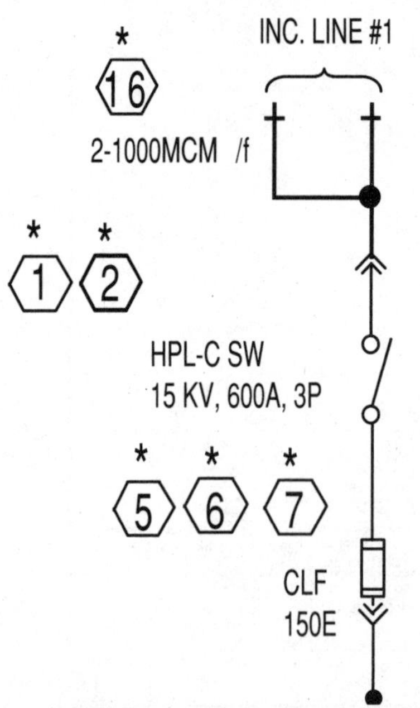

Figure 3-26: Incoming 13.8 kV high-voltage line.

Incoming line #1 (partially abbreviated on the drawings as "INC. LINE #1") consists of two, 1000MCM (kcmil) conductors per phase as indicated by note "2-1000MCM/f." In other words, parallel 1000 kcmil conductors. Since this is a three-phase system, a total of six 1000 kcmil conductors are utilized.

The single-line continues to engage separable connectors at the single-throw 15 kV, 600-A, 3-pole switch. Overcurrent protection is provided by current-limiting fuses as indicated by the fuse symbol combined with a note. The single-line continues to the high-voltage bus which connects to the primary side of the 2000 kVA transformer.

Incoming line #2 (partially abbreviated on the drawings as "INC. LINE #2") has identical components as line #1. This line also connects to the high-voltage bus which connects to the primary side of the 2000 kVA transformer.

Notice the numerals, each enclosed by a hexagon, placed near various components in these two high-voltage primaries. Note also that an asterisk is placed above each of these marks. A note on the drawing indicates the following:

* ITEMS PER S.O. #58454

These marks appear in a supplemental schedule titled "Bill of Materials" which describes the marked items, lists the number required, manufacturer, catalog number, and a brief description of each. Such schedules are extremely useful to estimators, job superintendents, and workers on the job to ensure that each required item is accounted for and installed.

Title block: Every electrical drawing should have a title block, and it is normally located in the lower right-hand corner of the drawing sheet; the size of the block varies with the size of the drawing and also with the information required.

In general, the title block for an electrical drawing should contain the following:

- Name of the project
- Address of the project
- Name of the owner or client
- Name of the person or firm who prepared the drawing
- Date drawing was made
- Scale(s), if any
- Initials of the drafter, checker, designer, and engineer, with dates under each
- Job number
- Drawing sheet number
- General description of the drawing

The title block for the project in question is shown in Figure 3-27.

Revison block: Sometimes electrical drawings will have to be partially redrawn or modified during the planning or construction of a project. It is extremely important that such modifications are noted and dated on the drawings to ensure that the workers have an up-to-date set of drawings to work from. In some situations, sufficient space is left near the title block for dates and description of revisions as shown in Figure 3-28.

SPEC: PROJECT: PLANT: EQUIPMENT UNIT:			
P.O. # 458764		Disk # 4578	
JET INDUSTRIES, INC	JET MANUFACTURING COMPANY 8557 Mosby Drive - Overall, Virginia		
For: HARWELL IRON WORKS			
ONE-LINE DIAGRAM	2000KVA SUBSTATION 13.8 kV, 4.16 kV, 3 ϕ, 3W, 60 Hz		
Drawn by: RGA Date: 7/6/93	CHK'D BY: TP *Date:* 7/14/93	APP'D: JT Date: 7/15/93	DWG. No. D-3

Figure 3-27: Typical drawing title block.

CAUTION!

When a set of electrical drawings has been revised, always make certain that the most up-to-date set is used for all future layout work. Either destroy the obsolete set of drawings, or else clearly mark on the affected drawing sheets, "Obsolete Drawing — Do Not Use." Also, when working with a set of working drawings and written specifications for the first time, thoroughly check each page to see if any revisions or modifications have been made to the originals. Doing so can save much time and expense to all concerned with the project.

No.	Drawn	App'd	Revision	Date
3	RGA		Revised per field mark-ups	9/1/93
2	SC		Change CTs, AM, fuses & OLs	8/18/93
1	SC		As built	8/4/93

Figure 3-28: Typical drawing revision block.

Interpreting Secondary Single-Line Diagrams

Referring again to Figure 3-25, note that a 600-A, 4160-V, 3-phase, 3-wire, 60 Hz aluminum main bus is used to feed the remaining secondary elements. The drawing in Figure 3-25 shows this bus to consist of busbars with dimensions of $1\frac{5}{16}'' \times 2''$ for each phase.

A grounding conductor is shown immediately at the secondary side of the 2000 kVA transformer. This grounding conductor angles off the main line to a vertical line that proceeds toward the bottom of the drawing where it connects to the ground bus of the system which, in turn, is bonded to all qualifying grounding electrodes on the premises. Note the removable ground link between the transformer and the grounding bus connection. The drawing shows this conductor to be No. 4/0 AWG.

Looking back at the 2000 kVA transformer, note that the main bus continues in a horizontal line to the right of the transformer symbol. The first group of equipment encountered is the metering section. Note the current transformers (CTs) which are designated by both symbol and note. The "400/5" note indicates that the CTs have a ratio of 400 to 5; that is, if 400 A are flowing in the main bus, only 5 amperes will flow to the meters. Again, numerals enclosed in hexagons are placed at each component in this section. Referring to the "Bill of Materials" schedule in Figure 3-29, we see a description of Item #19 to be "CT's 400/5 Type JAF-0;" two are required; catalog number is 750X10G304 and is manufactured by GE (General Electric). Continuing from the CTs up to Item #20, the schedule describes this as a three-phase, three-wire, watthour meter with a 15-minute demand and is designed to register with CTs with a ratio of 400/5 and PTs with primary/secondary at 4160/120 V. Locate the remaining numerals in this group and find their description in the schedule in Figure 3-29.

The next group to the right of this first group of metering equipment is a second group of metering equipment, connected to the vertical line above the main bus. Notes on the drawing indicate #6 AWG, 5 kV conductors are connected to the main bus and are protected with current-limiting fuses. These conductors continue to two 4160/120-V PTs and then the 120-V conductors continue to a junction box, a voltage-sensing device (voltage synchroscope), and finally to a voltmeter. Also note the branches "To VHD Meter" (varhours demand meter) and "To MR" (meter recorder). This latter device is normally referred to as a "recording demand meter."

At this point on the main bus, note also that a vertical line extends below the main bus. Again, #6 AWG, 5 kV conductors tap onto the main bus and are protected with current-limiting fuses. These high-voltage conductors terminate at a 10 kVA 4160-120/240-V transformer. The secondary side

Mark	Req'd	Cat. No.	Mfg.	Description
1	2	IC2957B103C	GE	Disc. Handle & Elec. Interlock ASM.
				(400A) (CAT#116C9928G1)
2	2	IC2957B108E	GE	Vert. Bus (CAT#195B4010G1)(400A)
				Shutter ASM. (CAT#116C9927G1) (400A)
3	2	1C2957B10BF	GE	Coil Finger ASM. (CAT#194A6949G1) (400A)
				Safety Catch (CAT#194A6994G1) (400A)
				Stab Fingers (CAT#232A6635G) (400A)
4	1		Toshiba	5kV, 300A, 3P, Vacuum Contactor
				120VAC Rectified Control
				Type CV461J-GAT2
5	4	2033A73G03	W	5kV Fuse MTG (2/CPT)
6	4	677C592G09	W	5kV, CLF, 2E Fuses Type CLE-PT
7	2	HN1K0EG15	Micron	1kVA, 4160-120 CPT
8	3	9F60LJD809	GE	CLF Size 9R (170A) Type EJ-2 (600HP)
9	3	615X3	GE	CT'S 150/5A Type JCH-0
10	1	CR224C610A	GE	200 Line Block O.L. Rly. 3 Elements
				Ambient Compensated W/INC. Contact
11	0	CR123C3.56A	GE	O.L. HTR (600HP)

Figure 3-29: Bill of Materials Schedule for the 2000 kV substation under discussion.

Mark	Req'd	Cat. No.	Mfg.	Description
11A	3	CR123C3.26A	GE	O.L. HTR. (2.79A) (700HP)
12	1	7022AB	AG	Off Delay R.Y .5-5 SEC.
13	0	CR2810A14A	GE	Machine Tool RLY. INO&INC 120VAC (MR)
14	1	CR294OUM301	GE	Emergency Stop PB (Push to Stop Pull to Reset) W/NP
15	1	9T28Y5611	GE	10kVA CPT. 4160-120/240V
16	2	643X92	GE	PT'S 4160/120V Type JVM-3/2FU
17	2	9F60CED007	GE	CLF 7E, 4.8kV Type EJ-1
18	2	9F61BNW451	GE	Fuse Clips Size C
19	2	750X10G304	GE	CT'S 400/5 Type JAF-0
20	1	700X64G885	GE	DWH-Meter 3ϕ, 3W, 60HZ, Type DSM-63 W/15MIN. Demand Register CT'S Ratio 400/5 & PT'S 4160-120V
21	1	50-103021P	GE	VM Scale 0-5250V Type AB-40
22	1	50-103131L	GE	AM Scale 0-400A Type AB-40
23	1	10AA004	GE	VS Type SBM
24	1	10AA012	GE	AS Type SBM
25	1	TL612FL	GE	6 CKT. PNL.
26	4	TQL1120	GE	20/1 C/B Type TQL.
27	1	TQL2120	GE	20/2 C/B Type TQL.

Figure 3-29: Bill of Materials Schedule for the 2000 kV substation under discussion. ***(Cont.)***

Mark	Req'd	Cat. No.	Mfg.	Description
28	1	TEB12050WL	GE	50/2 C/B Type TEB
29	1	3512C12H02	W	Type GR Groundgard RLY. Solid State
30	1	3512C13H03	W	GRD. Sensor
31	2	H	Smout Hollman	1/2 LT. REC.
32	2	7604-1	GE	LT. SW. & Receptacle
33	2	4D846G20	GE	120VAC, 250W HTR
34	1		Econo	Econo Lift for Contactor
35	11	Lot	Cook	NP/Schedule DWG. 58455-A1
36	3	Hold	T & B	Lug
37	0	50250440LSPK	GE	AM Scale 0-100A PNL. Type 2%
				ACC. Type 250 4-1/2 Case
38	1	NON10	Bus	10A, 250V Fuse
39	1	CP232	AH	2P, 250V Pull-Apart Fuse Block
40	1		Cook	SWGR NP S.O.#58455

Figure 3-29: Bill of Materials Schedule for the 2000 kV substation under discussion. ***(Cont.)***

of this transformer has its conductors protected by means of a 50-ampere, 2-pole circuit breaker which feeds a 6-circuit panel. Note that this panel contains four, 20-ampere, 1-pole circuit breakers and one 20-ampere, 2-pole circuit breaker. The circuits for the 1-pole breakers are also indicated on the drawing as follows:

- 1-20/1-LTs & REC.
- 1-20/1-SWGR HTRS

- 2-20/1-SPARES
- 1-20/2-XFMR. FANS

The interpretation of the abbreviations are as follows:

LTS	=	Lights
REC	=	Receptacles
SWGR HRTS	=	Switchgear heaters
XFMR FANS	=	Transformer fans

The remaining two taps from the main bus in the drawing in question are for feeding two motor-control centers (MCC); one to be put into use immediately while the other is a fully-equipped MCC, less contactors, for future use. Let's look at the complete MCC first. This is the last tap from the main bus in Figure 3-25. An enlarged view of this section is shown in Figure 3-30 on the next page for clarification.

This feeder is provided with overcurrent protection by means of current-limiting fuses (CLF 9R), which are fuse type EJ-2, rated at 170 A. Immediately beneath this device, note that a tap is taken from the main line, fused with 5 kV MTG fuses and also 5 kV, CLF, 2E fuses before terminating at a 1 kVA, 4160/120-V CPT transformer. This transformer is provided to accommodate the 120-volt control circuit shown in Figure 3-31.

Now let's backtrack to the main feeder and continue downward to a contactor before another group of current transformers (CTs) are installed in the circuit. These CTs are accompanied by notes and Mark No. 9. Referring to the schedule in Figure 3-29 for a description of Mark #9, we see that these three CTs have a ratio of 150/5; that is, when the circuit is drawing 150 A, the metering devices will receive only 5 A, but the meter itself will indicate 150 A. This circuit continues to a 200 line block overload relay with three elements, and then on to an ammeter with a range of 0-150 A.

The next item on this main vertical feeder is a ground sensor which is connected to a solid-state groundguard relay. The feeder then enters, and connects to the busbars, in a motor-control center (MCC) enclosure. The remaining feeder in the one-line wiring diagram under consideration is for future use and is similar to the circuit just described.

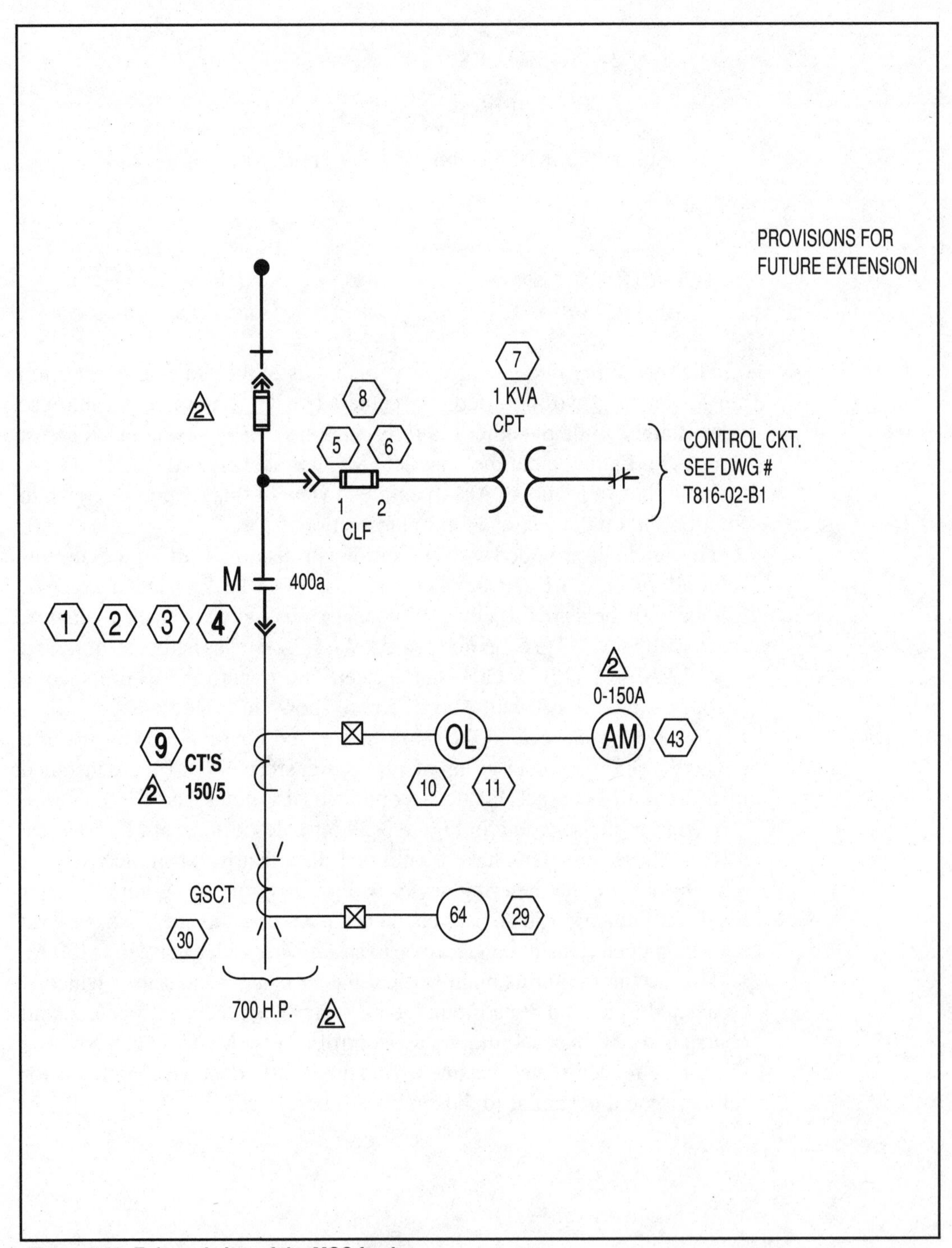

Figure 3-30: Enlarged view of the MCC feeder.

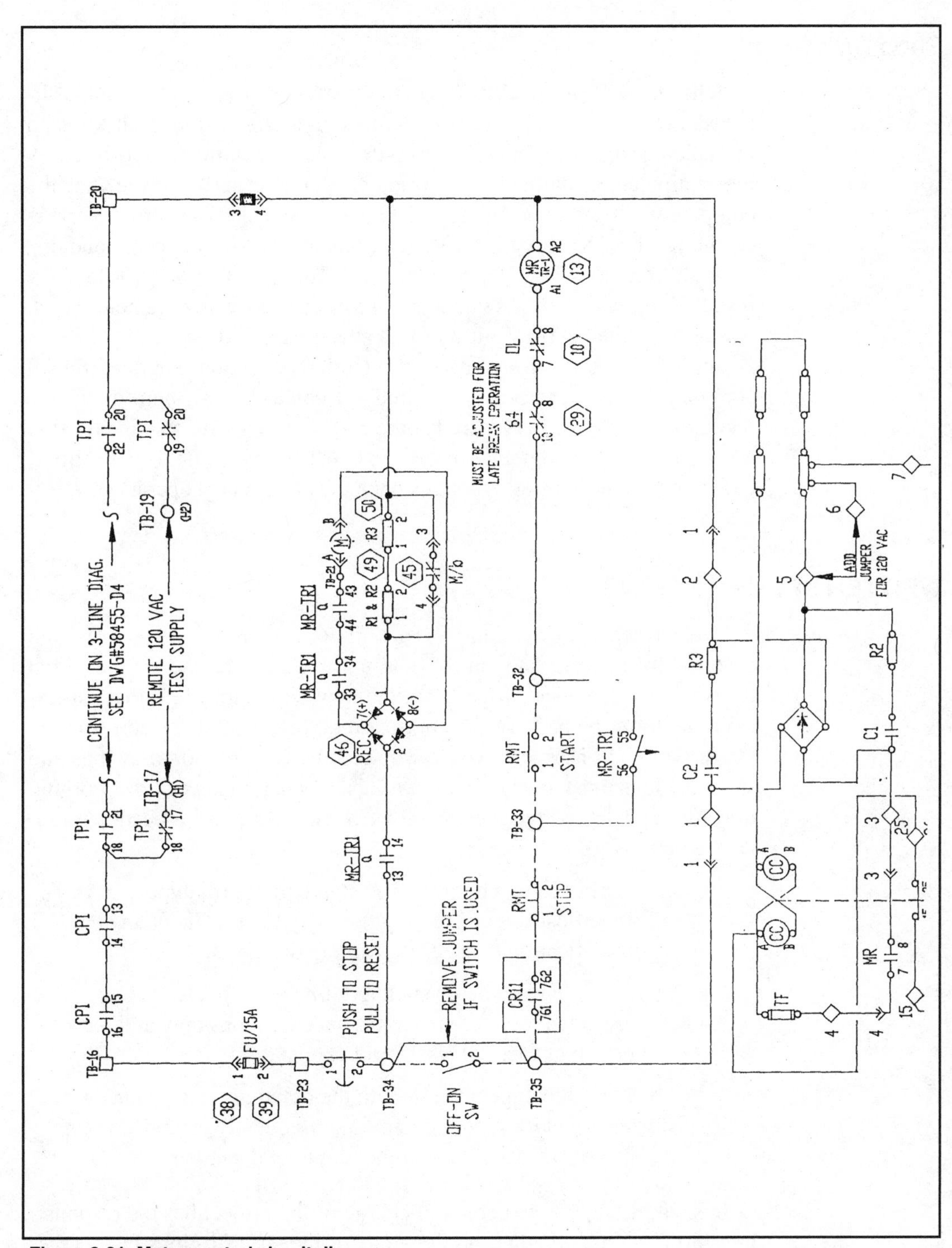

Figure 3-31: Motor-control circuit diagram.

Shop Drawings

When large pieces of electrical equipment are needed, such as high-voltage switchgear and motor control centers, most are custom built for each individual project. In doing so, shop drawings are normally furnished by the equipment manufacturer — prior to shipment — to ensure that the equipment will fit the location at the job site, and also to instruct workers on the job how to prepare for the equipment; that is, rough-in conduits, cabletray, and the like. The drawings in Figures 3-31 through 3-36 show installation details for a switchgear and transformer arrangement which was used on the 2000 kV substation that is under discussion.

Shop drawings will also usually include connection diagrams for all components that must be "field wired" or connected. As-built drawings, including detailed factory-wired connection diagrams are also included to assist workers and maintenance personnel in making the final connections, and then troubleshooting problems once the system is in operation.

WRITTEN SPECIFICATIONS

The written specifications for a building or project are the written descriptions of work and duties required of the owner, the architect, and the consulting engineer. Together with the working drawings, these specifications form the basis of the contract requirements for the construction of the building or project. Those who use the construction drawings and specifications must always be alert to discrepancies between the working drawings and the written specifications. Such discrepancies occur particularly when:

- Architects or engineers use standard or prototype specifications and attempt to apply them without any modification to specific working drawings.
- Previously prepared standard drawings are changed or amended by reference in the specifications only and the drawings themselves are not changed.
- Items are duplicated in both the drawings and specifications, but an item is subsequently amended in one and overlooked on the other contract document.

In such instances, the person in charge of the project has the responsibility to ascertain whether the drawings or the specifications take precedence. Such questions must be resolved, preferably before the work is

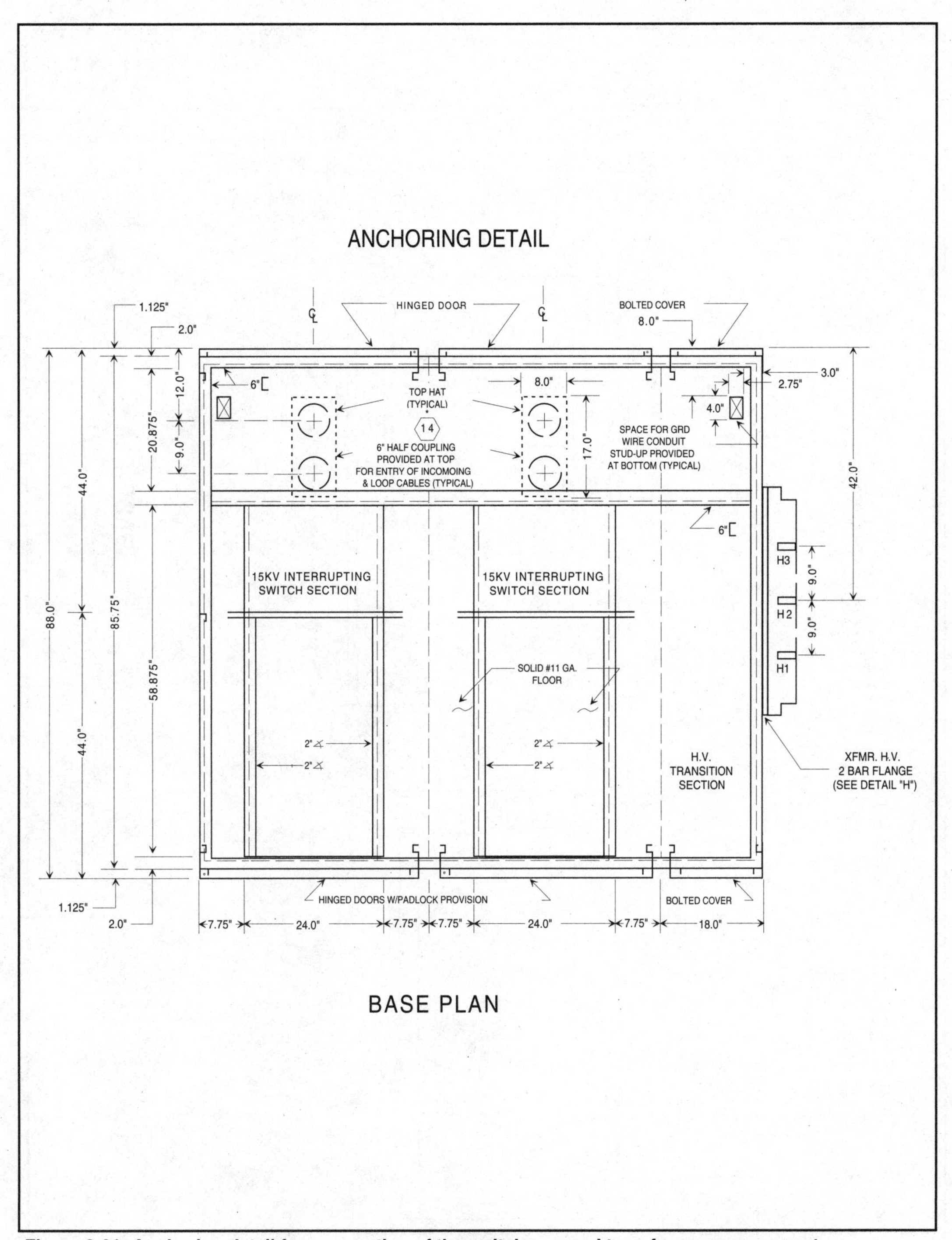

Figure 3-31: Anchoring detail for one section of the switchgear and transformer arrangement.

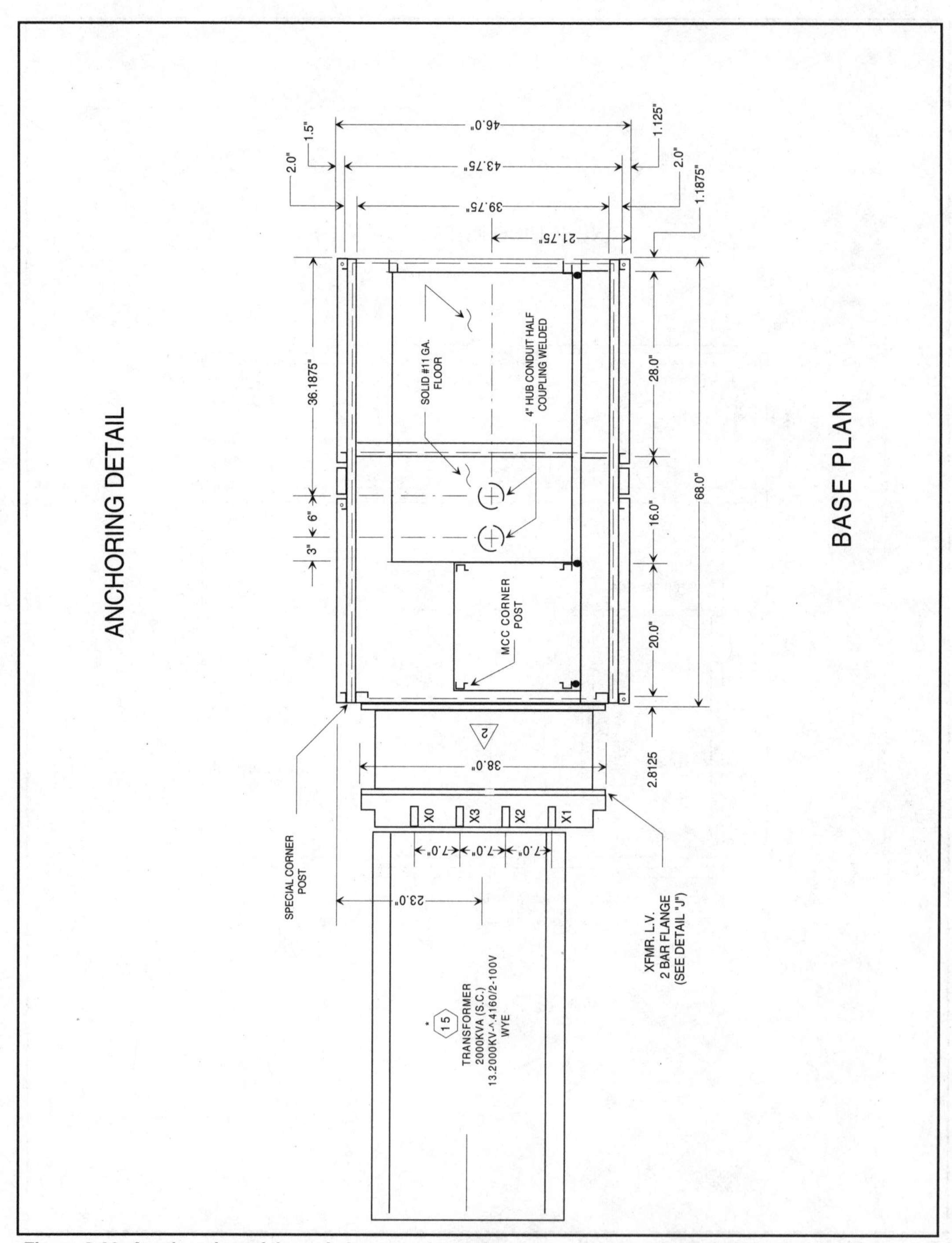

Figure 3-32: Another view of the switchgear and transformer arrangement for the 2000 kV substation.

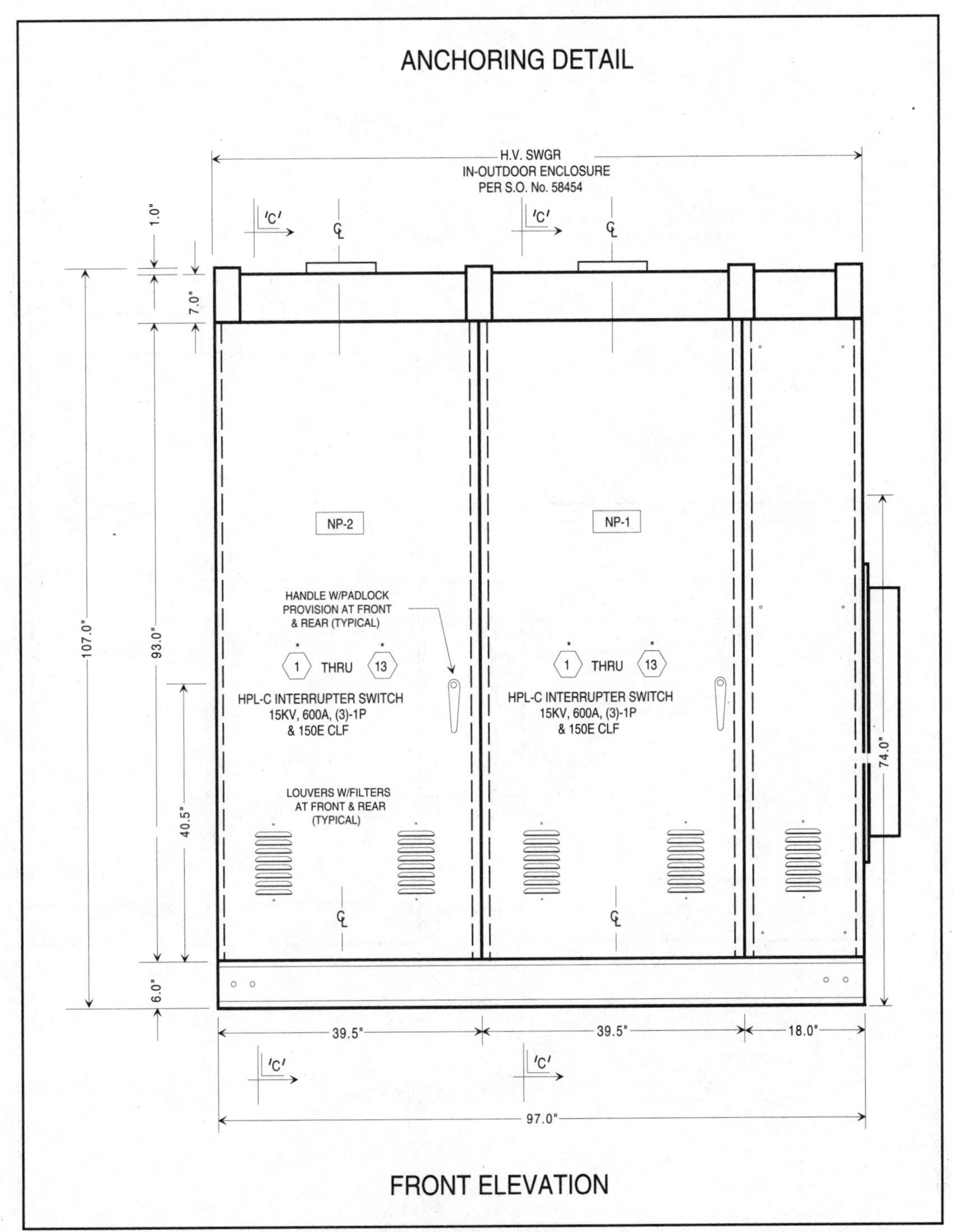

Figure 3-33: Front elevation of a high-voltage switchgear section.

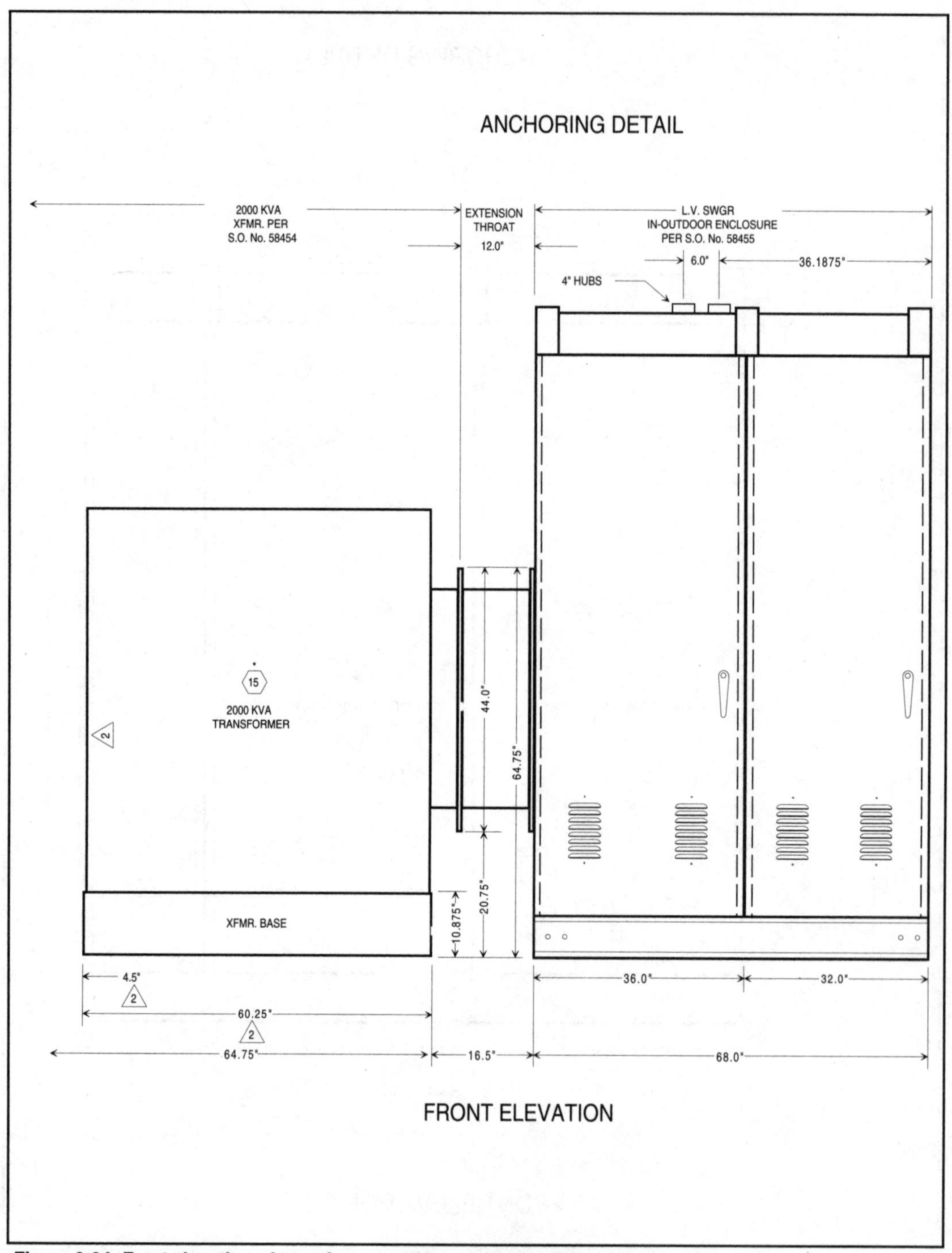

Figure 3-34: Front elevation of transformer and low-voltage switchgear.

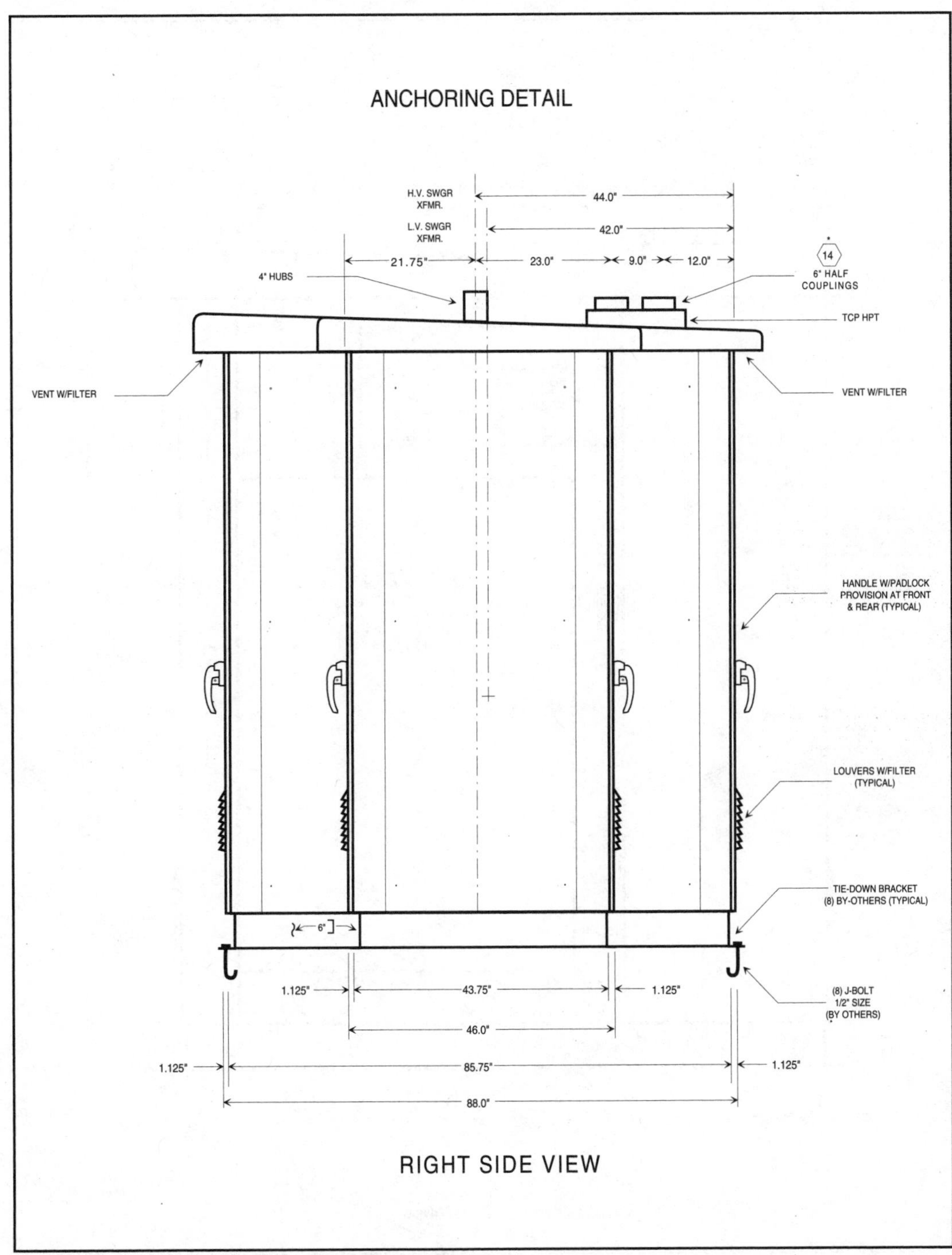

Figure 3-35: Right elevation of switchgear/transformer arrangement.

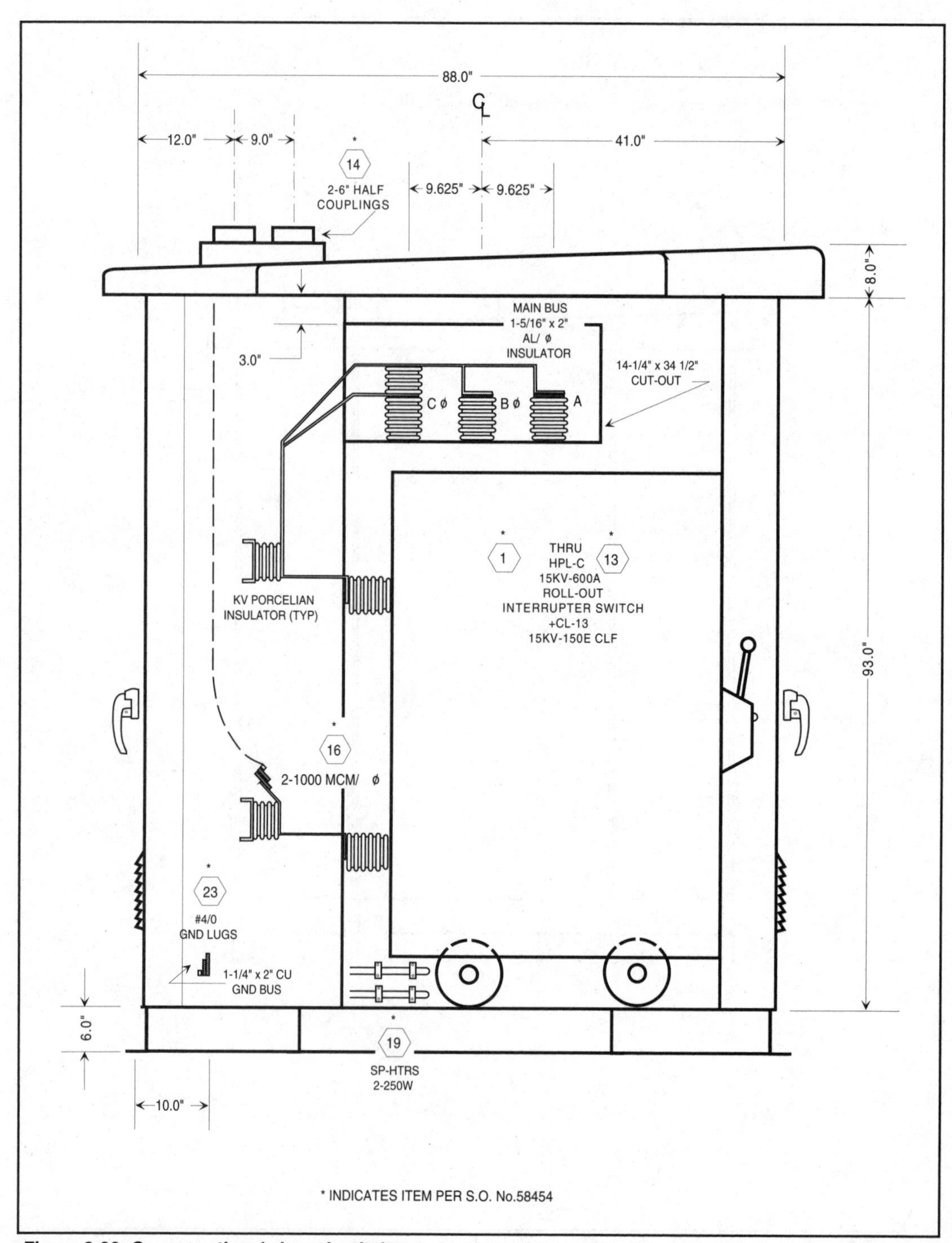

Figure 3-36: Cross-sectional view of switchgear.

installed, to avoid added cost to either the owner, the architect/engineer, or the contractor.

How Specifications Are Written

Writing accurate and complete specifications for building construction is a serious responsibility for those who design the buildings because the specifications, combined with the working drawings, govern practically all important decisions made during the construction span of every project. Compiling and writing these specifications is not a simple task, even for those who have had considerable experience in preparing such documents. A set of written specifications for a single project usually will contain thousands of products, parts and components, and methods of installing them, all of which must be covered in either the drawings and/or specifications. No one can memorize all of the necessary items required to accurately describe the various areas of construction. One must rely upon reference materials — manufacturer's data, catalogs, checklists, and, best of all, a high-quality master specification.

The CSI Format

The Construction Specification Institute (CSI) developed the Uniform Construction Index some years ago that allowed all specifications, product information, and cost data to be arranged into a uniform system. This format is now followed on most large construction projects in North America. All construction is divided into 16 Divisions, and each division has several sections and subsections. The following outline describes the various divisions normally included in a set of specifications for building construction.

Division 1—General Requirements. This division summarizes the work, alternatives, project meetings, submissions, quality control, temporary facilities and controls, products, and the project closeout. Every responsible person involved with the project should become familiar with this division.

Division 2—Site Work. This division outlines work involving such items as paving, sidewalks, outside utility lines (electrical, plumbing, gas, telephone, etc.), landscaping, grading, and other items pertaining to the outside of the building.

Division 3—Concrete. This division covers work involving footings, concrete formwork, expansion and contraction joints, cast-in-place concrete, specially finished concrete, precast concrete, concrete slabs, and the like.

Division 4—Masonry. This division covers concrete, mortar, stone, masonry accessories, and the like.

Division 5—Metals. Metal roofs, structural metal framing, metal joists, metal decking, ornamental metal, and expansion control normally fall under this division.

Division 6—Carpentry. Items falling under this division include: rough carpentry, heavy timber construction, trestles, prefabricated structural wood, finish carpentry, wood treatment, architectural woodwork, and the like. Plastic fabrications may also be included in this division of the specifications.

Division 7—Thermal and Moisture Protection. Waterproofing is the main topic discussed under this division. Other related items such as dampproofing, building insulation, shingles and roofing tiles, preformed roofing and siding, membrane roofing, sheet metal work, wall flashing, roof accessories, and sealants are also included.

Division 8—Doors and Windows. All types of doors and frames are included under this division: metal, plastic, wood, etc. Windows and framing are also included along with hardware and other window and door accessories.

Division 9—Finishes. Included in this division are the types, quality, and workmanship of lath and plaster, gypsum wallboard, tile, terrazzo, acoustical treatment, ceiling suspension systems, wood flooring, floor treatment, special coatings, painting, and wallcovering.

Division 10—Specialties. Specialty items such as chalkboards and tackboards; compartments and cubicles, louvers and vents that are not connected with the heating, ventilating, and air conditioning system; wall and corner guards; access flooring; specialty modules; pest control; fireplaces; flagpoles; identifying devices; lockers; protective covers; postal specialties; partitions; scales; storage shelving; wardrobe specialties; and the like are covered in this division of the specifications.

Division 11—Equipment. The equipment included in this division could include central vacuum cleaning systems, bank vaults, darkrooms, food service, vending machines, laundry equipment, and many similar items.

Division 12—Furnishing. Items such as cabinets and storage, fabrics, furniture, rugs and mats, seating, and other similar furnishing accessories are included under this division.

Division 13—Special Construction. Such items as air-supported structures, incinerators, and other special items will fall under this division.

Division 14—Conveying Systems. This division covers conveying apparatus such as dumbwaiters, elevators, hoists and cranes, lifts, material-handling systems, turntables, moving stairs and walks, pneumatic tube systems, and powered scaffolding.

Division 15—Mechanical. This division includes plumbing, heating, ventilating, and air conditioning and related work. Electric heat is sometimes covered under Division 16, especially if individual baseboard heating units are used in each room or area of the building.

Division 16—Electrical. This division covers all electrical requirements for the building including lighting, power, alarm and communication systems, special electrical systems, and related electrical equipment. This is the Division that electricians will use the most. Division 16 contains the following sections:

DIVISION 16—ELECTRICAL

16050 Electrical Contractors
16200 Power Generation
16300 Power Transmission
16400 Service and Distribution
16500 Lighting
16600 Special Systems
16700 Communications
16850 Heating and Cooling
16900 Controls and Instrumentation

The above sections are further subdivided into many subsections. For example, items covered under Section 16400 — Service and Distribution — will usually include the project's service entrance, metering, grounding, service-entrance conductors, and similar details.

Chapter 4

Service and Distribution

An electric power system consists of several systems, all of which contribute to the power available for utilization. The main parts of the power system are the generating system, the transmission system, and the distribution system. The flow of power from generation to utilization can be illustrated in a simple way by a single-line diagram, such as that shown in Figure 4-1.

The voltage is generated in one or several generators, such as the generator located in the generating station. Then the voltage is stepped up by the transmission transformer to a very high value suitable for transmission. The generator circuit breaker is used with the transformer. The transmission line carries the very high voltage from the generator bus in the station to the high-voltage bus in a substation. The line is protected by the circuit breakers. A substation which steps the voltage down for a subtransmission system is a bulk-power substation. The substation is protected by the circuit breakers, and has a step-down transformer which provides by its secondary a lower voltage for the substation bus. Subtransmission circuits branch off to the distribution substations, such as the substation with the distribution transformer. Primary distribution feeders take the still lower distribution voltage through the circuit breakers to the utilization substations, which contain the transformers. From these transformers, secondary feeders provide utilization voltage to lighting and power circuits at the customer's premises.

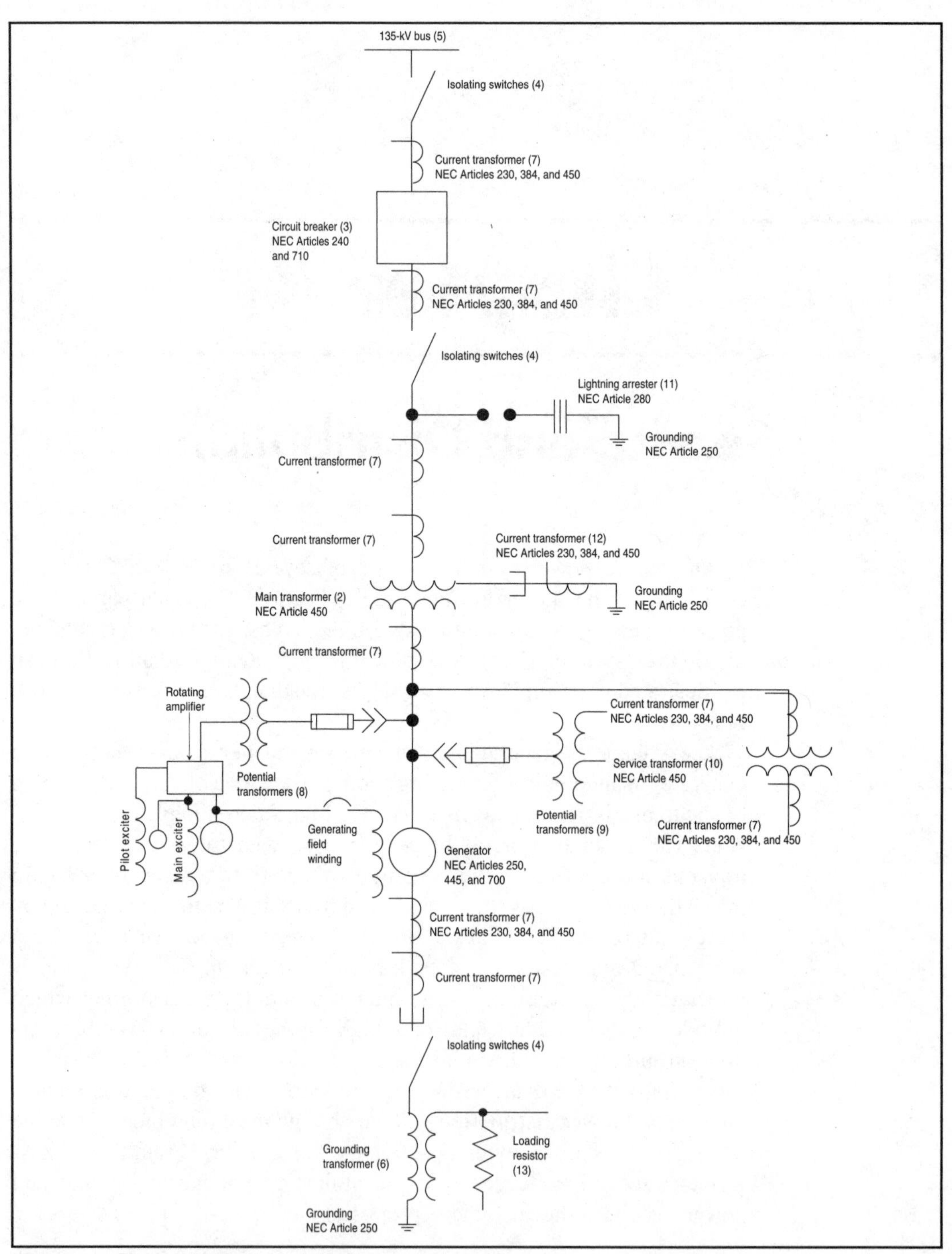

Figure 4-1: Single-line diagram of an electrical generating unit.

PARTS OF AN ELECTRICAL DISTRIBUTION SYSTEM

The simple single-line diagram in Figure 4-1 points out only the essential parts of the power system. In each system there are many devices necessary for protection, regulation, control, and measuring. There are also deviations in connections and various combinations of generated-transmission, and distribution voltage, as well as deviations in the length of the transmission line and in the area covered by the system of substations and distribution systems.

Generating Station

A generating station contains one or more ac generators, or alternators, which generate ac power at a certain voltage. The main power system in a generating station starts with the water, steam, or nuclear energy input into a prime mover, which may be a steam or water turbine or a steam engine. The prime mover turns the alternator shaft and the voltage is generated in the armature windings, which cut across the lines of force produced by the field. The field electromagnets are excited by a dc generator called an exciter. The main power system continues to the low-voltage switching section, then to a step-up power transformer, and high-voltage switching. The transmission line conveys the high-voltage power from the station. Within the generating station, however, there are many other devices which make station operation possible. There is, first, the control-feedback system, which controls the prime mover through the governor, and the excitation system, which controls the generator voltage by controlling the field excitation. Another important system is the power-feedback system, which provides the power for the equipment used for station-service lighting, heating, cooling, and communication within the generating station.

Electrical Components of a Generating Unit

The single-line diagram shown in Figure 4-1 indicates all the essential electrical components of one generator, or generating unit. The voltage is generated at 13,800 V in the generator and stepped up to 138 kV (kilovolts) in the main transformer, and through the 138-kV circuit breaker and the isolating switches it is impressed on the 138-kV bus. The generator is grounded through the grounding transformer. The current transformers provide power for the measuring instruments and relays, and the potential transformer provides voltage for the voltage regulation. Another potential transformer serves for relays and measuring. The station-service transformer serves as a source of voltage for the station equipment. Generator

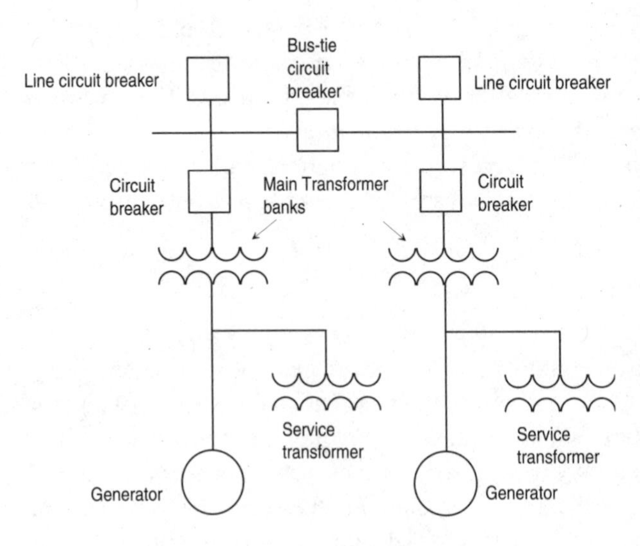

Figure 4-2: Two-unit generating station.

protection is obtained by the lightning arrester and the main-transformer neutral point is grounded through the current transformer. The grounding transformer is provided with a loading resistor. The generator field is energized by the main exciter and the exciter field is energized by the pilot exciter. The field is controlled by a rotating amplifier, which is energized by an auxiliary power source.

When several generating units are used in a generating station, they may be connected in a unit system as shown in Figure 4-2, or in a multiple-supply system, shown in Figure 4-3. In the unit system, each unit has its own main transformer, and in the multiple-supply system, one main transformer receives power from all units. The main components are identified by the callouts.

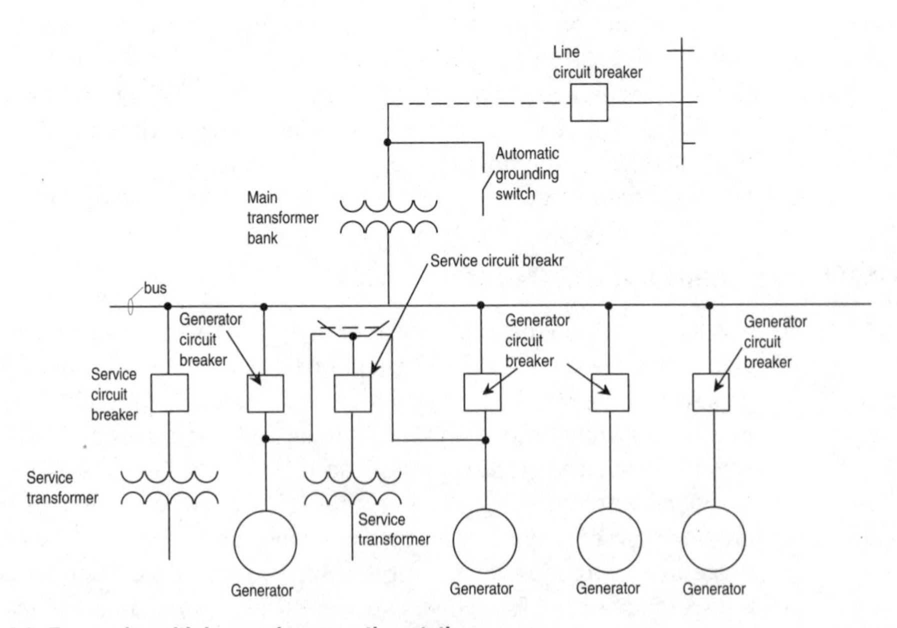

Figure 4-3: Four-unit multiple-supply generating station.

Synchronization of High-Voltage Alternator

When an alternator is connected to the bus of an ac system, it is of the utmost importance that the incoming generator run in synchronism with the frequency of the ac voltage in the bus. A synchronizer, or synchroscope, indicates whether the generator and the bus are in synchronism. A vertical position of the indicator means that the generator and the bus are running in phase and at the same frequency. The connections of a synchroscope are shown in Figure 4-4 on the next page. The incoming high-voltage three-phase alternator is connected to the high-voltage three-phase bus through the disconnecting switches and the circuit breaker. One potential transformer is connected to one phase of the alternator and another one to the same phase of the bus. The secondaries of the potential transformers are connected to the synchronizing bus. A voltmeter is connected through the double receptacle to the synchronizing bus and then through the receptacle to the synchronizing lamps, along with the synchroscope. The voltmeter can indicate the voltage of the bus or that of the generator, depending on the way it is plugged into the double receptacle. Similarly, the generator voltage is applied through plugging to one element of the synchroscope through the reactor and the resistor. The bus voltage is applied directly to the other element of the synchroscope. If the two voltages are not in phase, the indicator of the synchroscope moves and indicates the need for synchronization.

Station-Service Supply

The electric power necessary for proper operation of station-service equipment is up to 5 to 10 percent of the power generated by the generators. Transformer banks of 10,000 kVA or more are in common use to supply power for lighting and equipment at the generating station.

TRANSMISSION SYSTEMS

From the generating station, electric power is transmitted by high-voltage transmission lines to the areas of distribution. The system of transmission lines covers great distances and provides large areas with electric power. The required reliability, security, and stability of transmission systems are insured by transmission substations which supply the switching, voltage-transformation, and control facilities. High-voltage switchgear gives protection against line faults, contains disconnecting means for maintenance purposes, and may have equipment for tying two transmission lines and synchronizing their voltages.

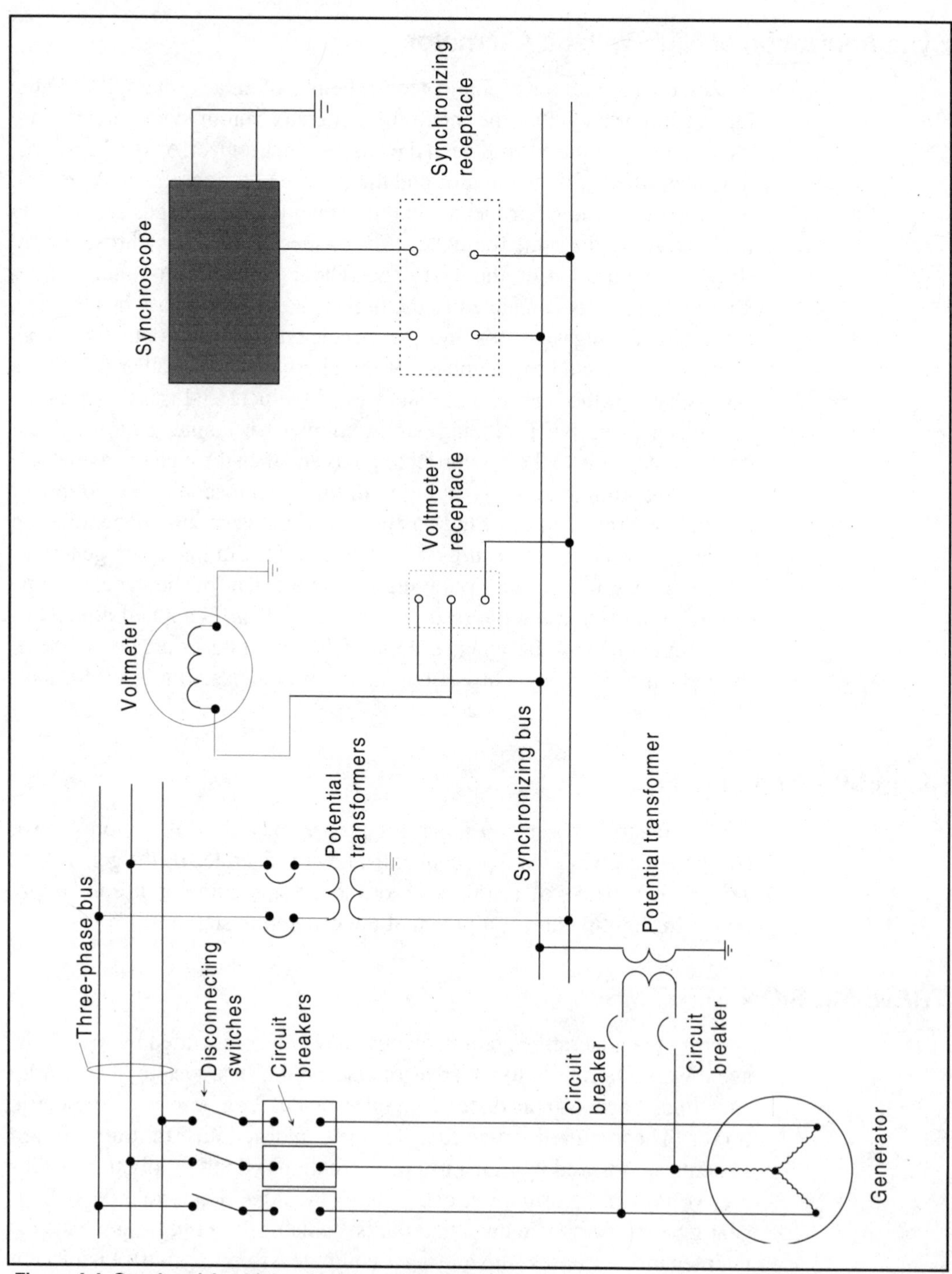

Figure 4-4: Synchronizing high-voltage generating system.

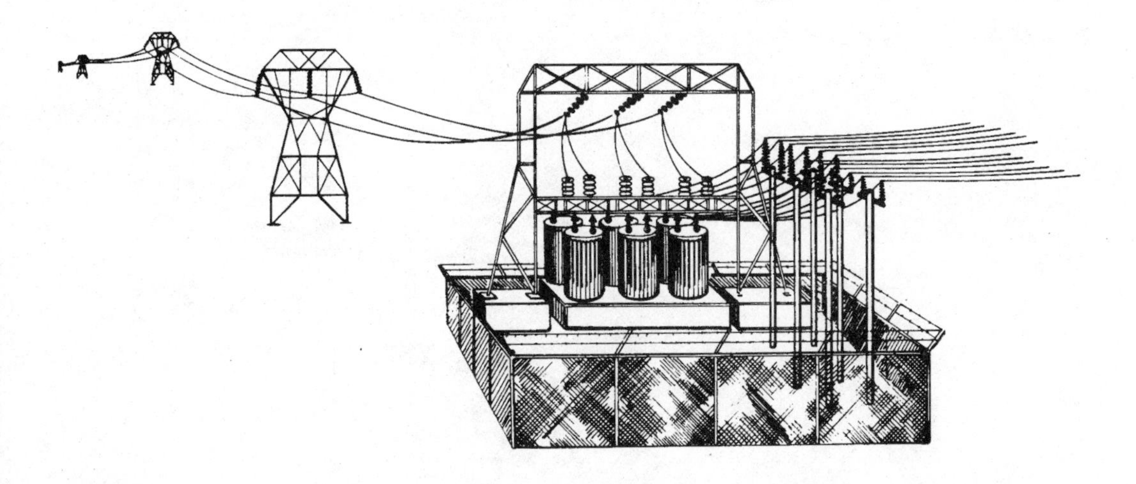

Figure 4-5: Typical automatically-operated substation.

Large power transformers, mostly three-phase, and rated up to more than 100,000 kVA, and autotransformers rated as high as 215,000 kVA are located in most transmission substations in order to step down the high transmission voltage for subtransmission and distribution systems.

Voltage control in transmission lines may be obtained by line-drop compensation or by improving the power factor with synchronous capacitors.

The equipment of substations is rated first according to the nominal voltage. Some common nominal voltages are, for example, 240 or 120 kV for transmission voltage and 13.8 or 4.16 kV for distribution voltage. The substation equipment is further rated according to the high-potential test voltage and according to the basic installation level (BIL), which is the resistance to sudden stresses, such as caused by lightning. The BIL rating is normally 4 to 15 times the normal voltage rating.

Substations in the transmission system may he operated automatically, manually, or by remote control from a distance by an attendant at a supervising control substation. *See* Figure 4-5.

Substation Switching System

Circuit breakers and other switching equipment in a substation may be arranged to separate a bus, a part of a transformer, or a control device from

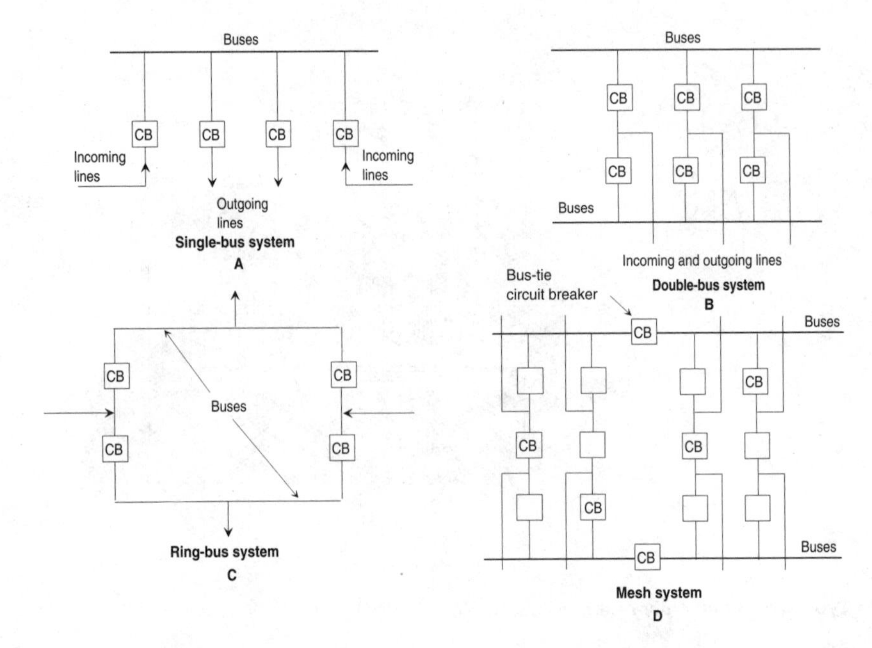

Figure 4-6: Switching systems in substations.

other equipment. The common arrangements are single-bus, double-bus, transfer-bus, ring-bus, or mesh systems of switching.

Switching systems are shown in Figure 4-6. The simple single-bus switching system in (a) has the bus protected by the circuit breakers on the incoming and outgoing lines. The double-bus switching system in (b) has two main buses, but only one is normally in operation; the other is a reserve bus. The ring-bus system in (c) has the bus arranged in a loop with breakers placed so that the opening of one breaker does not interrupt the power through the substation. The greatest security and flexibility of switching is obtained by the mesh system shown in (d). Bus-tie breakers are added and placed between pairs of main buses. Assuming any circuit breaker opens, you may verify that the circuits will not be interrupted by tracing the circuits.

Voltage Regulation in Substations

One way of obtaining voltage control uses on-load changers in transformers. The taps may be changed automatically and in many small steps according to the need.

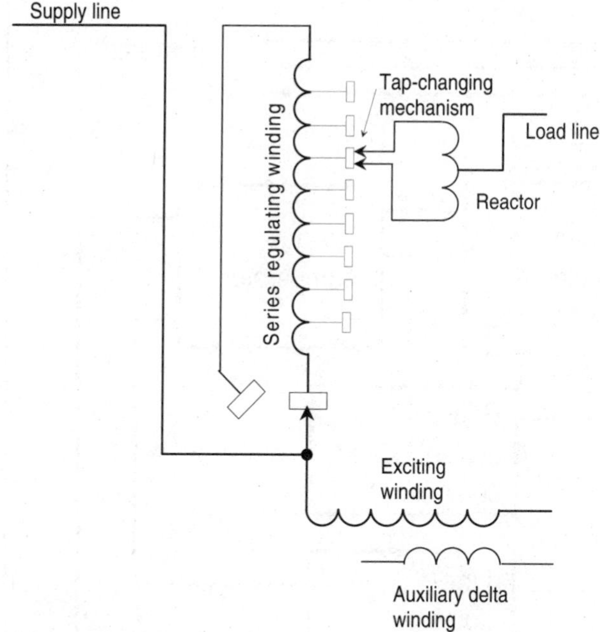

Figure 4-7: One phase of a single-core, step-voltage regulator.

Another method of voltage control is possible with a step-voltage regulator. One phase of such a regulator is shown in Figure 4-7. The supply line is connected to the exciting winding of the regulator. The auxiliary delta winding provides a path for circulating currents. The regulating winding in series with the exciting winding has taps which are connected to the tap-changing mechanism, which in turn is connected through the reactor to the load line.

A system of voltage regulation in a transmission line which uses a synchronous capacitor is shown by the block diagram in Figure 4-8 on the next page. The synchronous capacitor provides reactive power for the line and so improves the power factor of the power system. The regulation is obtained automatically by changing the strength of the exciter field of the synchronous capacitor. The changes in voltage are detected by potential transformers on the three-phase lines and are transferred to the static control devices which consist of the voltage-adjusting device, the voltage comparator, and the reactive-current compensator. The compensator receives the current from the current transformer on one phase.

The voltage comparator sends a boosting or a bucking signal to the stabilizer and to the rotating amplifier, which in turn increase or decrease the exciter field. These changes in the exciter field are reflected in the synchronous-condenser field and finally in the output of the condenser. The line-current limiter is used in automatic operation to guard against overload of the condenser. The negative-excitation limiter prevents too high a leading current in the condenser field, which might increase the line voltage dangerously.

The power factor of the system may be improved by installation of series and shunt banks of capacitors in the transmission system. One single-phase capacitor unit is shown in Figure 4-9. The capacitor is connected in series with the line through two disconnecting switches. A normally open bypass switch is in the main line. The holding coil and the operating coil are

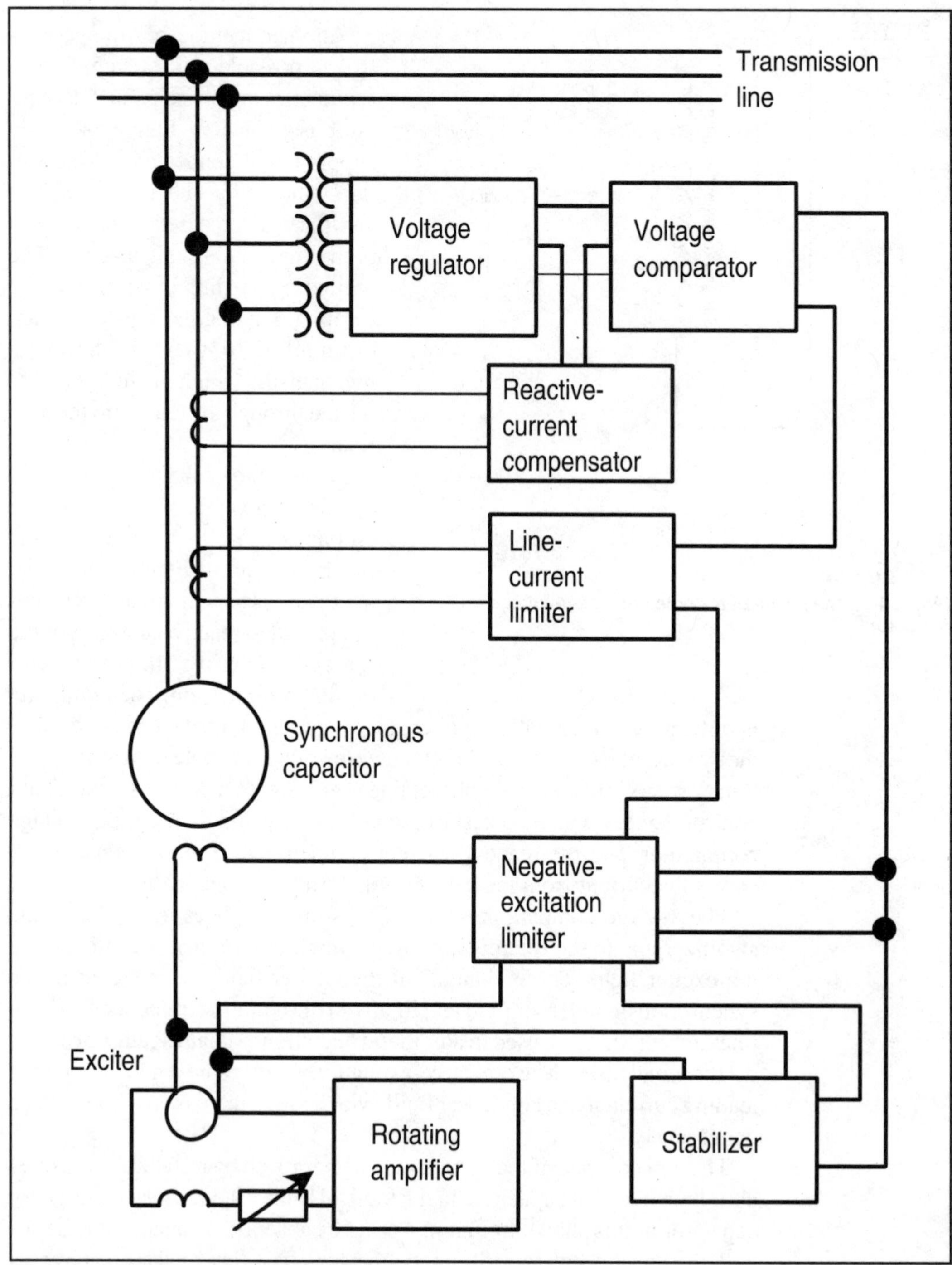

Figure 4-8: Synchronous capacitor in transmission substation.

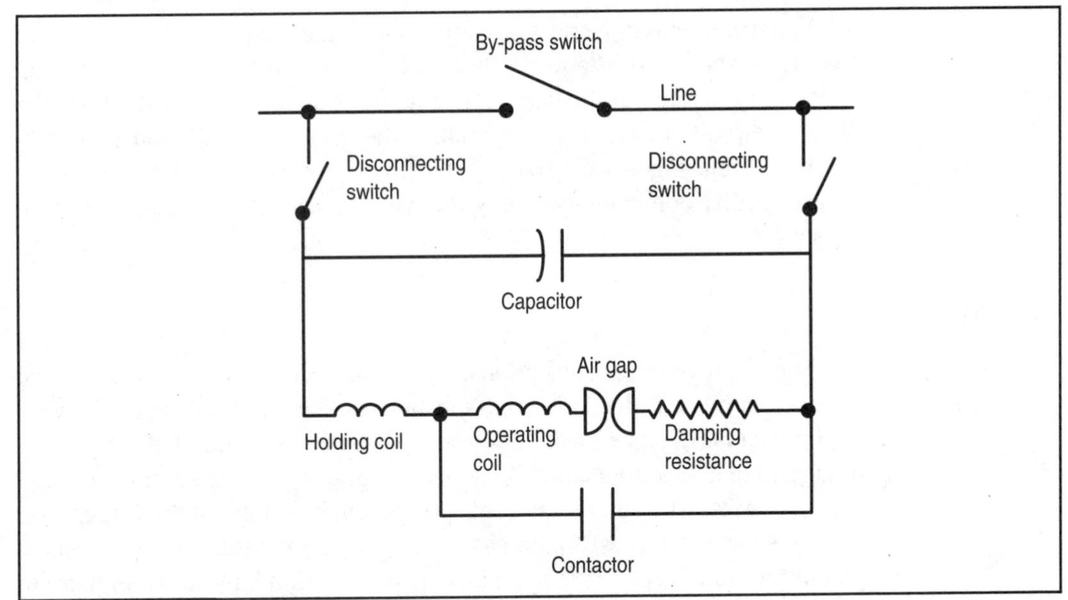

Figure 4-9: Capacitor unit in series with line.

connected in series with an air gap and a damping resistance. When the voltage across the capacitor rises over a certain value, a spark jumps across the gap and causes a current in the circuit parallel with the capacitor. This current closes the contactor and so short-circuits the capacitor. The holding coil keeps the contacts closed as long as the current is higher than normal.

Distribution Systems

In the flow of power from the generating station to the utilization point, distribution systems provide the link between the high-voltage transmission or subtransmission systems and the consumer of electric power. Key points of distribution systems are the distribution substations, which contain transformers and the switchgear necessary to deliver low-voltage power to the distribution-system primary feeders. A common voltage transformation is from 33 kV to the standard distribution voltage of 4.16 kV.

Voltage control is obtained economically with on-load tap changers on the transformers in distribution substations, but some installations require separate voltage-regulating transformers, individual feeder-voltage regulators, or induction-voltage regulators.

Metering, relaying, and automatic controlling are functions included in most distribution substations. There is rarely a distribution substation that does not operate automatically. The main items to be controlled are the feeder circuit breakers, which makes the control installation relatively simple. Metering is required for statistical purposes or for billing if the power-utility company providing the power does not own the distribution system.

Rural Substations

The main types of substations, classified according to their purposes, are the rural and the industrial substation. The rural substation is very simple and is designed with the lowest possible cost in mind. A single-line diagram of a rural substation is shown in Figure 4-10. The subtransmission line connects through the disconnecting switch and the high-voltage fuse to the substation transformer. No circuit breaker is used. The low voltage from the transformer is supplied to the distribution bus and from there to several feeders. Each feeder has a load-disconnecting switch and a fuse. Meters are connected to the current transformer and the potential transformer, which has its own disconnecting switch.

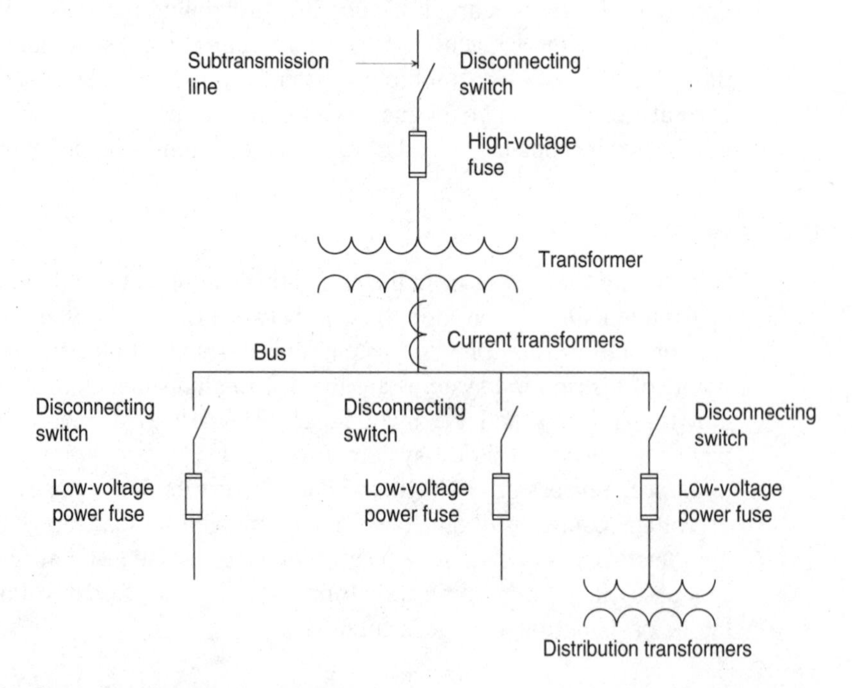

Figure 4-10: Single-line diagram of a substation.

Industrial Substation

An industrial substation could be located near an industrial plant or in the plant building. It must supply a low utilization voltage of 120 V for lighting and small motors, and voltages of 6.9 or 13.8 kV for large motors, and should provide a high security of service. Industrial substations very often include capacitors or a synchronous condenser to control the power factor. An industrial substation built as a unit is shown diagrammatically in Figure 4-11. It is a structure containing five sections labeled 1 to 5. Section 1 is the supply section, which houses the pothead and the disconnecting switch. The transformer section 2 houses the 4160/600-V transformer and the 600-V bus. The feeder sections 3 to 5 each contain a bus, an air circuit breaker, and a feeder cable which connects to the motor load. One section includes the 600-120/208-V lighting transformer and the secondary feeder, which connects to the distribution panel for lighting circuits.

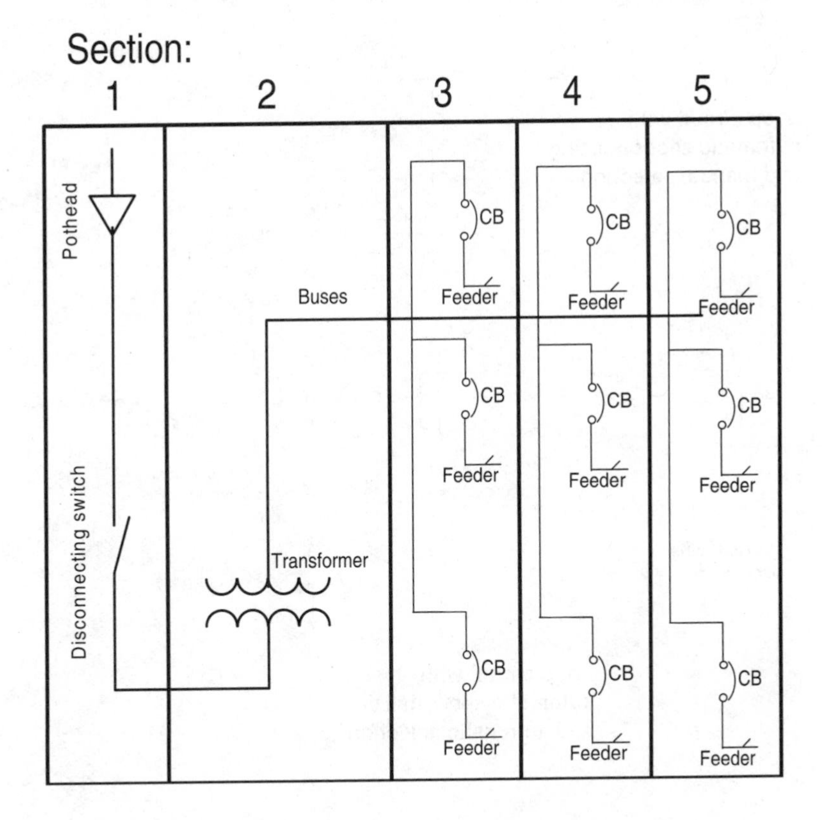

Figure 4-11: Industrial substation unit.

High-Voltage Lines in Substations

The incoming transmission or subtransmission lines and the outgoing distribution lines of a distribution substation are most frequently arranged in a radial system or in a network system. The system selected for either incoming or outgoing lines depends on the purpose of the substation.

According to the degree of security required of the substation, the incoming, or the high-voltage, lines may be arranged in a radial, double radial, or loop system. The most commonly-used high-voltage switchgear arrangement is shown in Figure 4-12.

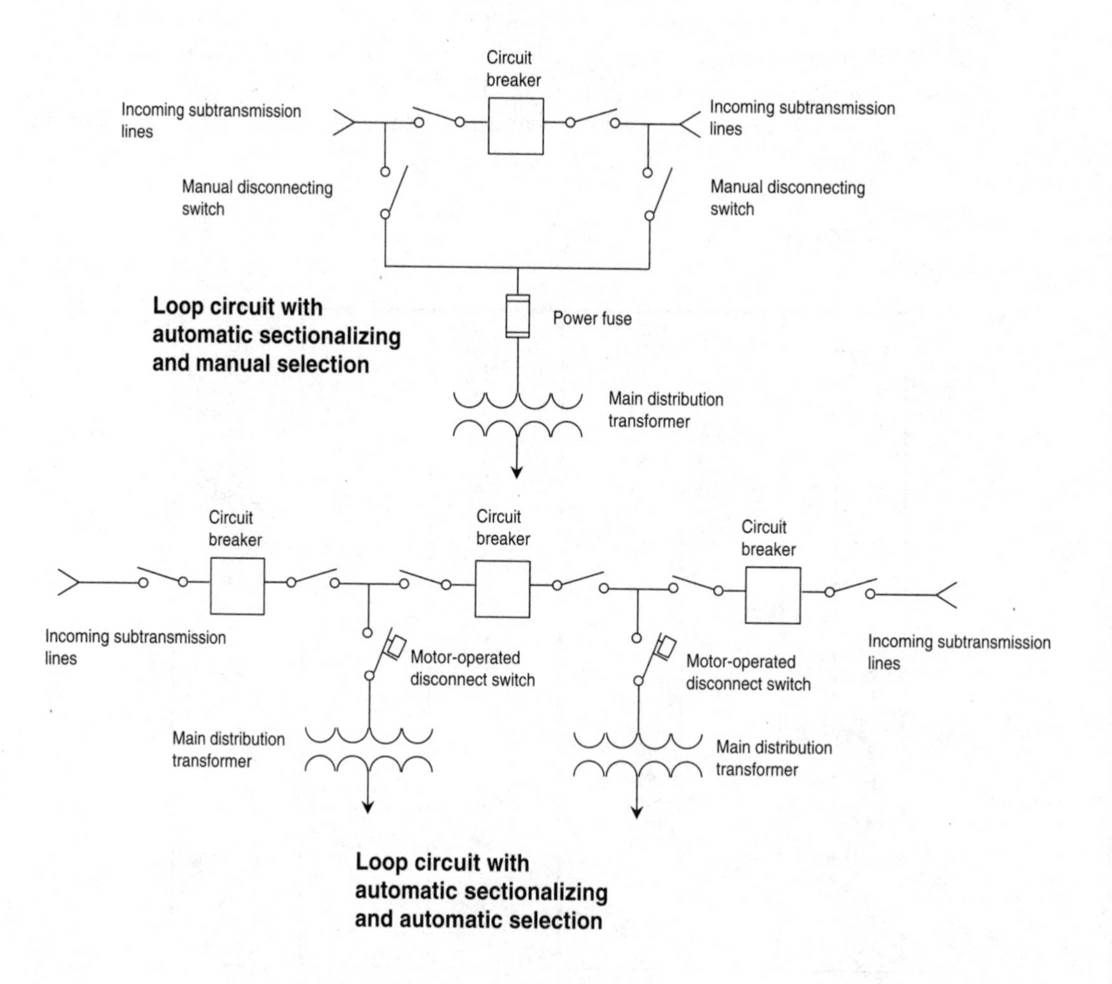

Figure 4-12: High-voltage arrangements of distribution substations.

The incoming subtransmission line in each arrangement connects through a switch and a power fuse to the transformers. Normally open switches close and connect the second line to the transformer if the normally closed switch in the first line opens because of a fault.

The single radial circuit in Figure 4-12 depends on only one incoming line; the double radial circuit in this same illustration has the second line for emergency; the loop circuit has a manual switch between the two lines for sectionalizing; the other loop circuit has automatic sectionalizing by a circuit breaker between the two lines, but manual selection of lines by the switches; the remaining loop circuit has automatic sectionalizing and automatic selection because it uses motor-operated disconnecting switches in the lines.

Distribution substations use outdoor switching or metal enclosed switchgear for high-voltage switching, and low-voltage metal-enclosed switchgear for switching of low-voltage out-going primary feeders.

Low-Voltage Lines in Substations

The primary feeders, or the low-voltage lines, in a distribution substation are arranged as a single feeder, double feeder, or a network. There are various types of networks. An ordinary network arrangement of primary distribution feeders is shown in Figure 4-13 on the next page. Three substations receive the power from the subtransmission lines and deliver it to the network through the transformers. The transformer circuit breakers protect the substation low-voltage buses. The primary tie feeders with their circuit breakers interconnect the substations in the network. Radial primary feeders and their circuit breakers feed the secondary distribution transformers. From these transformers, the secondary distribution feeders deliver power to the customers. A primary fuse protects the secondary transformer in each primary feeder.

ELECTRIC SERVICES

Electric services can range in size from a small 120-V, single-phase, 15-A installation (the minimum allowed by *NEC* Section 230-79a) to huge industrial installations involving substations dealing with thousands of volts and amperes. Regardless of the size, all electric services are provided for the same purpose: for delivering electrical energy from the supply system to the wiring system on the premises served. Consequently, all establishments containing equipment that utilizes electricity require an electric service.

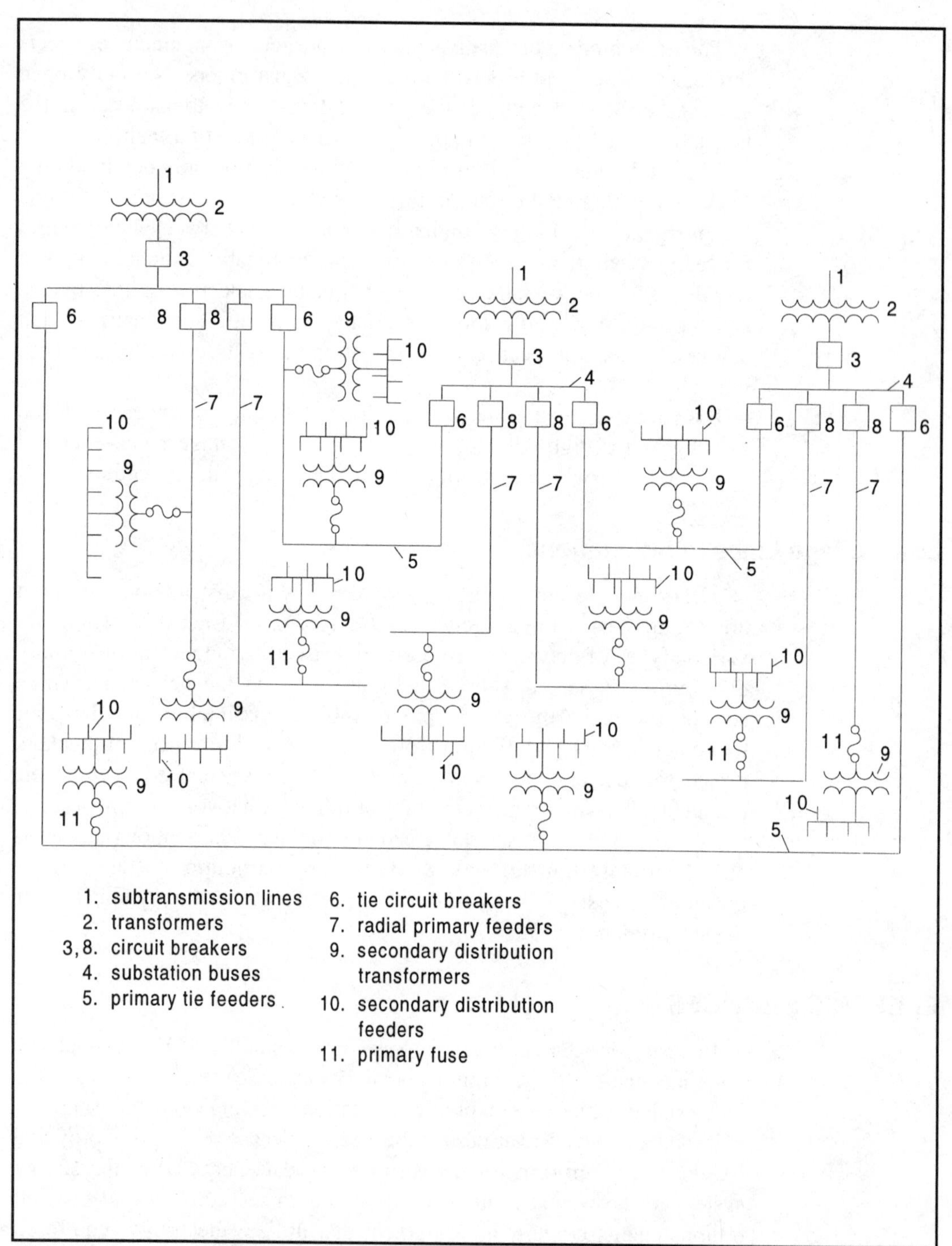

Figure 4-13: Primary network system.

The essential elements of an ac electrical system capable of producing useful power include generating stations, transformers, substations, transmission lines, and distribution lines. Figure 4-14 on the next page shows these elements and their relationships. The remaining portions of electric services include the following:

- *Service drop:* The overhead conductors, through which electrical service is supplied, between the last power company pole and the point of their connection to the service facilities located at the building or other support used for the purpose.
- *Service entrance:* All components between the point of termination of the overhead service drop or underground service lateral and the building's main disconnecting device, except for metering equipment.
- *Service-entrance conductors:* The conductors between the point of termination of the overhead service drop or underground service lateral and the main disconnecting device in the building or on the premises.
- *Service-entrance equipment:* Provides overcurrent protection to the feeder and service conductors, a means of disconnecting the feeders from energized service conductors, and a means of measuring the energy used by the use of metering equipment.

When the service conductors to the building are routed underground, these conductors are known as the service lateral, defined as follows:

- *Service lateral:* The underground conductors through which service is supplied between the power company's distribution facilities and the first point of their connection to the building or area service facilities located at the building or other support used for the purpose.

UNDERGROUND SYSTEMS

There are several methods used to install underground wiring, but the most common include direct-burial cables and the use of duct lines or duct banks.

The method used depends on the type of wiring, soil conditions, allotted budget for the work, etc.

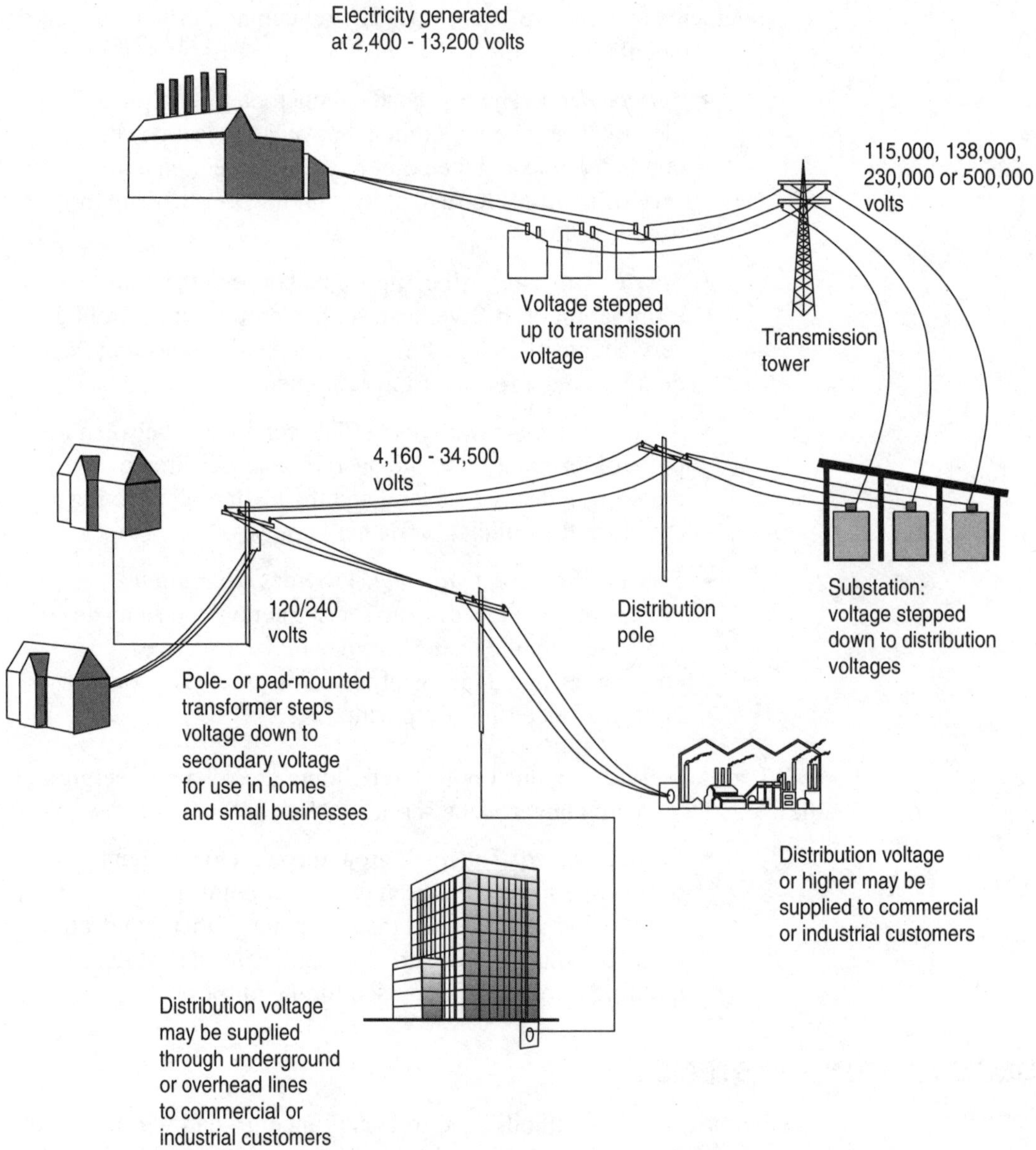

Figure 4-14: Parts of a typical electrical distribution system.

Direct-burial installations will range from small, single-conductor wires to multiconductor cables for power or communications or alarm systems. In any case, the conductors are installed in the ground either by placing them in an excavated trench, which is later backfilled, or by burying them directly by means of some form of cable plow, which opens a furrow, feeds the conductors into the furrow, and closes the furrow over the conductor.

Sometimes it becomes necessary to use lengths of conduit in conjunction with direct-burial installations, especially where the cables emerge on the surface of the ground or terminate at an outlet or junction box. Also, where the cables cross a roadway or concrete pavement, it is best to install a length of conduit under these areas in case the cable must be removed at a later date. By doing so, the road or concrete pad will not have to be disturbed.

Figure 4-15 shows a cross section of a trench with direct-burial cable installed. Note the sand base on which the conductors lie to protect them from sharp stones and such. A treated board is placed over the conductors in the trench to offer protection during any digging that might occur in the future. Also, a continuous warning ribbon is laid in the trench, some distance above the board, to warn future diggers that electrical conductors are present in the area.

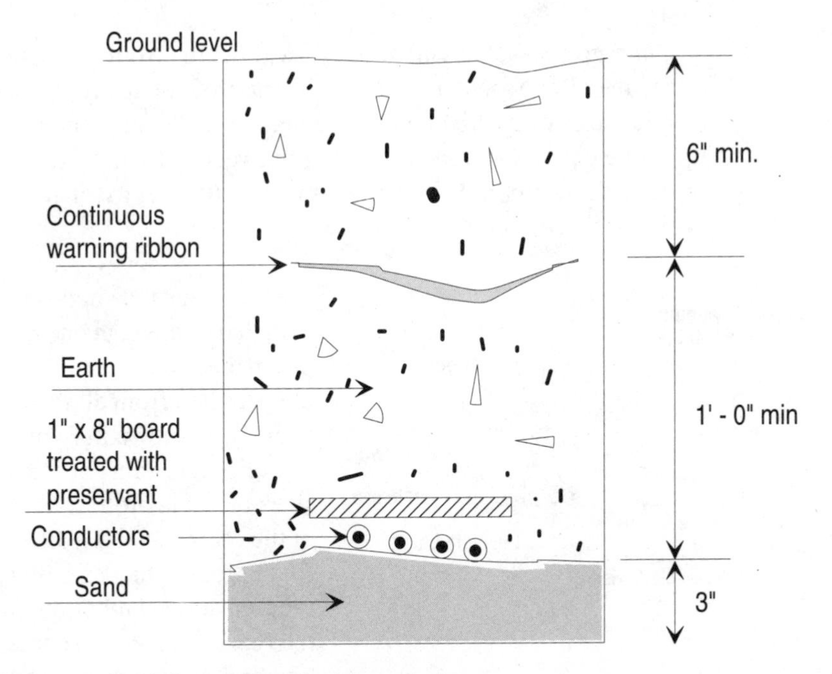

Figure 4-15: Cross-sectional view of direct-burial conductors.

Minimum Cover Requirements

The *NEC* specifies minimum cover requirements for direct-buried cable. Furthermore, all underground installations must be grounded and bonded in accordance with *NEC* Article 250.

Where direct buried cables emerge from the ground, they must be protected by enclosures or raceways extending from the minimum cover distance specified. However, in no cases will the protection be required to exceed 18 in below the finished grade.

Practical Application

Methods of installing direct-burial underground cable vary according to the length of the installation, the size of cable being installed, and the soil conditions.

In general, the trench is opened to the correct depth with an entrenching tractor or backhoe. All sharp rocks, roots, and similar items are then removed from the trench to prevent these objects from damaging the direct-burial cable. If soil conditions dictate, a 3-in layer of clean sand is poured into the bottom of the trench to further protect the direct-burial cable.

Underground cable will almost always come from the manufacturer on either metal or wooden reels. In the case of relatively short runs of the smaller sizes of multiconductor cable, the reel containing the cable is set up at one end of the trench — using two reel jacks and a length of conduit through the center hole in the reel to allow the reel to rotate as the cable is "paid off." *See* Figure 4-16.

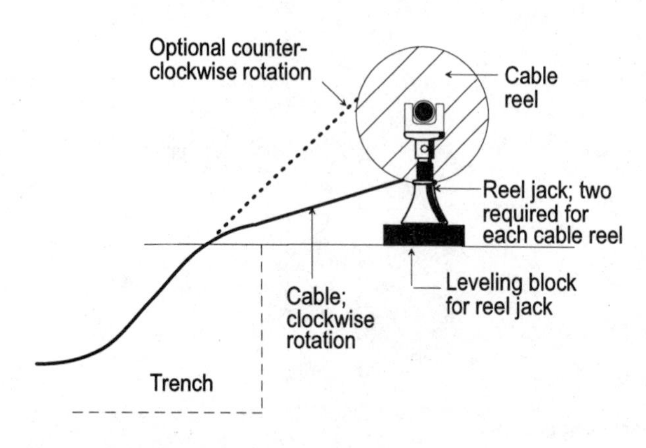

Figure 4-16: Method of setting up cable reel for underground pull.

Where several single conductors are installed, more than one reel is set up in the manner described above. Then cable is pulled from all the reels simultaneously as shown in Figure 4-17.

For longer runs, or where the larger sizes of cable are installed, the weight of the cable dictates a different method. In this case, the end of the cable is secured at one end of the trench, while the reel is attached to a tractor or backhoe and arranged so that the reel will rotate freely. The tractor or backhoe then runs along the trench — either at

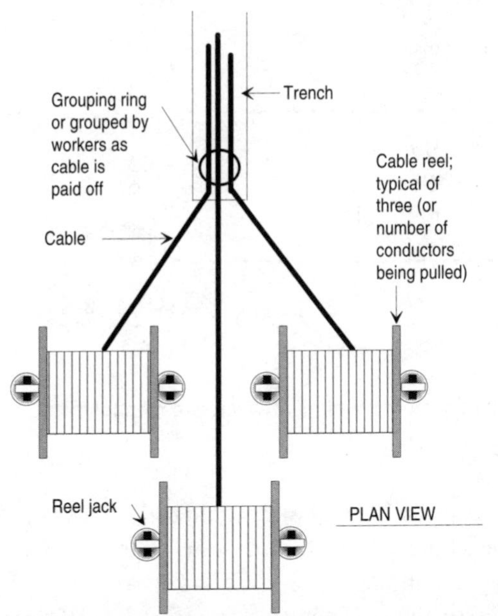

Figure 4-17: Method used to pull three conductors at once for direct burial in an open trench. In actual practice, the two outside reels should be turned so that the cable feeds off the reel exactly perpendicular to the reel.

the trench edge or straddles it — and the cable is paid out by workers and allowed to fall into the trench.

Once installed, another layer of sand may be placed over the cable for protection against sharp rocks; the trench is then backfilled. A treated wooden plank may also be used for cable protection; a yellow warning ribbon — designed for the purpose — is also a good idea; both are shown in Figure 4-15.

Duct Systems

A duct is a single enclosed raceway through which conductors or cables are pulled. One or more ducts in a single trench are usually referred to as a duct bank. A duct system provides a safe passageway for power lines, communication cables, or both.

Depending upon the wiring system and the soil conditions, a duct bank may be placed in a trench and covered with earth or enclosed in concrete. Underground duct systems also include manholes, handholes, transformer vaults, and risers.

Manholes are set at various intervals in an underground duct system to facilitate pulling conductors or cables when first installed, and to allow for testing and maintenance later on. Access to manholes are provided through *throats* extending from the manhole compartment to the surface (ground level). At ground level, a manhole cover closes off the manhole area tightly.

In general, underground cable runs normally terminate at a manhole, where they are spliced to another length of cable. Manholes are sometimes constructed of brick and concrete. Most, however, are prefabricated, reinforced concrete, made in two parts — the base and the throat — for quicker installation. Their design provides room for workers to carry out all appropriate activities inside them, and they are also provided with a means for drainage. *See* Figure 4-18.

There are three basic designs of manholes: two-way, three-way, and four-way. In a two-way manhole, ducts and cables enter and leave in two directions. A three-way manhole is similar to a two-way manhole, except

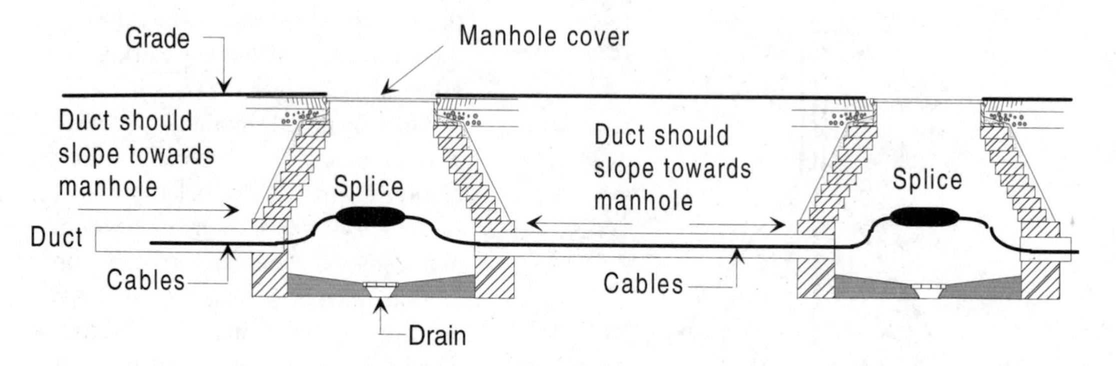

Figure 4-18: Cross section of typical underground duct system running between two manholes.

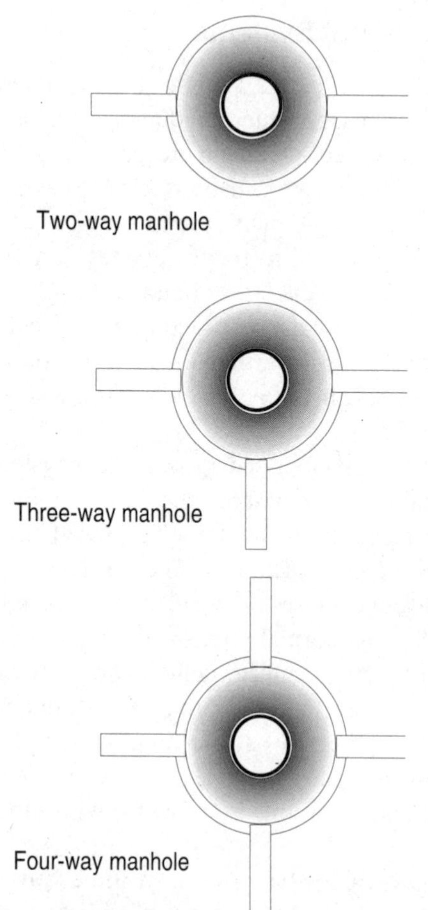

Figure 4-19: Plan view of two-, three-, and four-way manholes.

that one additional duct/cable run leaves the manhole. Four duct/cable runs are installed in a four-way manhole. *See* Figure 4-19. Also *see* Figure 4-20 on the next page for specifications of a typical manhole.

Transformer vaults house power transformers, voltage regulators, network protectors, meters and circuit breakers. Other cables end at a substation or terminate as risers — connecting to overhead lines by means of a pothead. *See* Chapter 10 of this book.

Types of Ducts

Ducts for use in underground electrical systems are made of fiber, vitrified tile, metal conduit, plastic or poured concrete. In some existing installations, the worker may find that asbestos/cement ducts have been used. In most areas, a contractor must be certified before removing or disturbing asbestos ductwork, and then extreme caution must be practiced at all times.

The inside diameter of ducts for specific installations is determined by the size of the cable that the ducts will house. Sizes from 2 to 6 in are common.

Fiber duct: Fiber duct is made with wood pulp and various chemicals to provide a lightweight

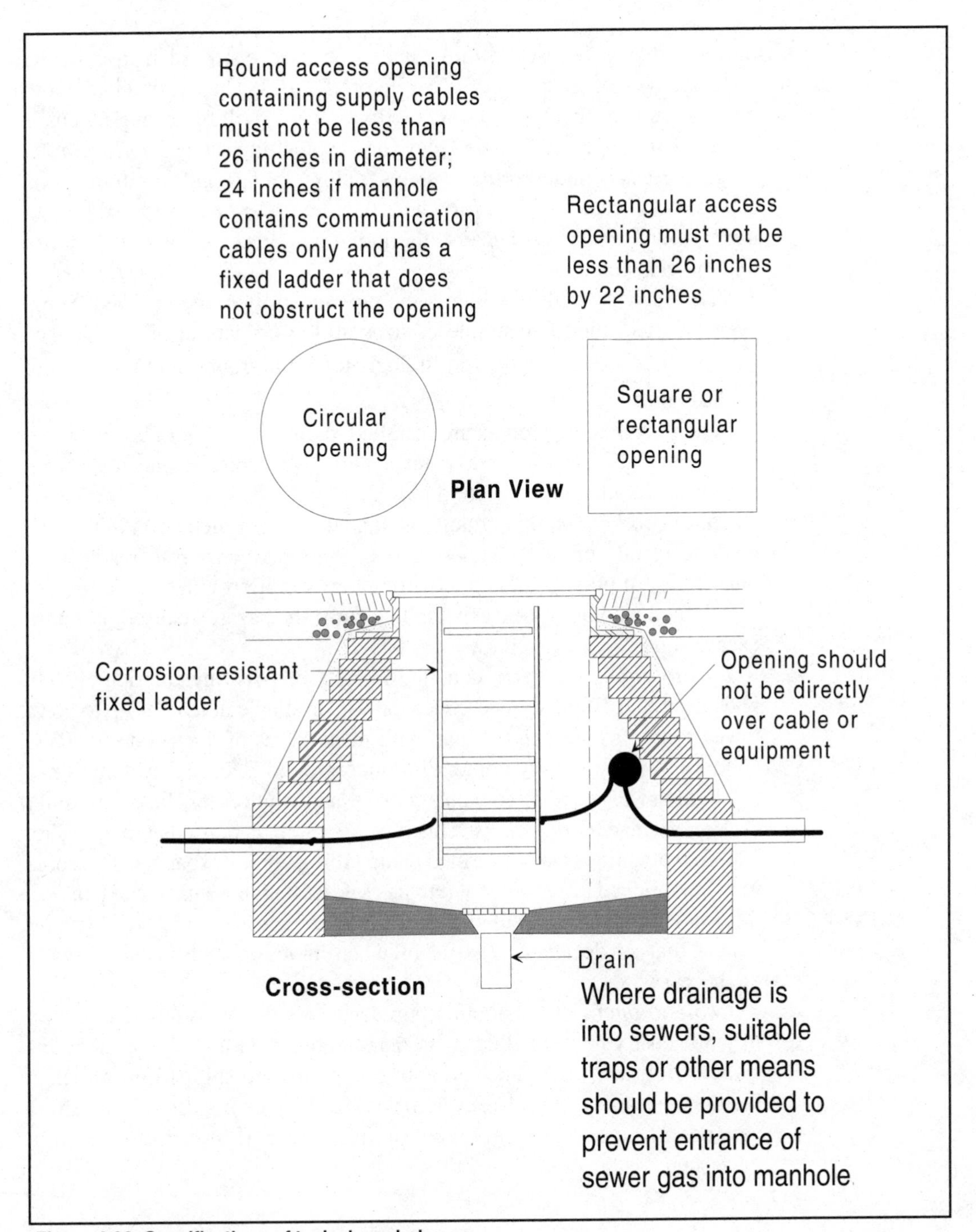

Figure 4-20: Specifications of typical manhole.

raceway that will resist rotting. It can be used enclosed in a concrete envelope with at least 3 in of concrete on all sides. The extremely smooth interior walls of this type of duct facilitates cable pulling through them.

Vitrified clay duct: Vitrified clay tile is sometimes called *hollow brick*. Its main use is in underground systems for low-voltage and communication cables and is especially useful where the duct run must be routed around underground obstacles, because the individual pieces of duct are shorter than other types.

The four-way multiple duct is the type most often encountered. However, vitrified duct is available in sizes up to 16 ducts in one bank. The square ducts are usually 3½ in in diameter, while round ducts vary from 3½ to 4½ in.

When vitrified clay ducts are installed, their joints should be staggered to prevent a flame or spark from a defective cable in one duct from damaging cable in an adjacent duct.

Metal conduit: Metal conduit, such as iron, rigid metal conduit, intermediate metal conduit, etc., is relatively more expensive to install than other kinds of underground ductwork. However, it provides better protection than most other types, especially against the hazards caused by future excavation.

Plastic conduit: Plastic conduit is made of polyvinylchloride (PVC), polyethylene (PE), or styrene. Since they are available in lengths up to 30 ft, fewer couplings are needed than with many types of duct systems. PVC conduit is currently very popular for underground electrical systems since it is light in weight, relatively inexpensive, and requires less labor to install.

Monolithic concrete ducts: This type of system is poured at the job site. Multiple duct lines can be formed using tubing cores or spacers. The cores may be removed after the concrete has set. Although relatively expensive, this system has the advantage of creating a very clean duct interior with no residue that can decay. It is also useful when curves or bends in duct systems are necessary.

Cable-in-duct: This is another popular duct type that offers a reduction in labor cost when installed. It is manufactured with cables already installed. Both the duct and the cable it contains are shipped on a reel to facilitate installing the entire system with ease. Once installed, the cables can be withdrawn or replaced in the future if it should become necessary.

Installing Underground Duct

The selection of high-voltage cables and their installation in underground ductwork is not part of the requirements of the *NEC*. However, Appendix B of the *NEC* provides installation criteria "for information purposes only." Furthermore, the National Electrical Safety Code (NESC) covers regulations governing high-voltage underground installations in detail. Section 33 of the NESC covers supply cable including detailed requirements of conductors, insulation, sheaths, jackets, and shielding. Section 34 also covers underground installations. Both of these sections should be studied by anyone involved in underground duct systems — either design or actual installations.

In practical applications of underground duct systems, present needs and potential growth are both considered. Construction aspects such as trenching and pouring concrete are cost factors, so it is economically sound to provide for future growth as part of an original installation. Often the number of conduits laid is twice the number needed for present usage.

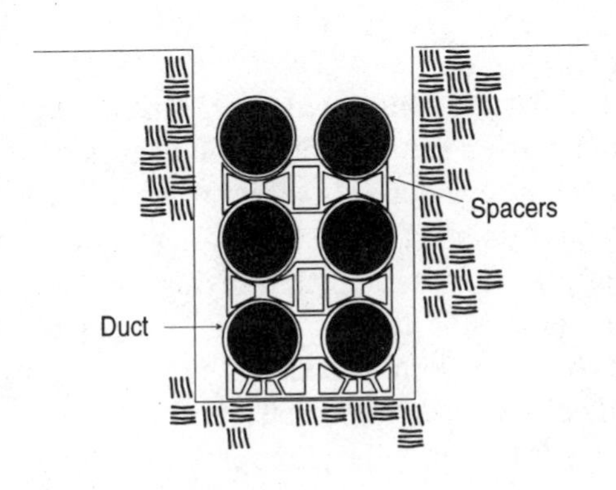

Figure 4-21: Duct bank with spacer separation.

In general, there should be at least 1 to 3 in of earth or concrete between adjacent conduits containing power cables. This insulation will ensure that the heat radiated by one line will not affect the surrounding ducts. Heat will cause insulation to deteriorate faster and, in general, the hotter the cable, the least amount of load it will carry. Consequently, duct banks should be designed so that heat from conductors is dissipated into either the surrounding concrete envelope or earth.

During the installation of a duct system, spacers should be used to hold the ducts in place while concrete is being poured. Figures 4-21 and 4-22 show how spacers are arranged prior to a concrete pour.

Duct Installation

Excavation is the first order of business; that is, a backhoe or other digging apparatus is employed to dig the trenches and ground openings for manholes. Manholes are then constructed at various intervals throughout the "run." In the case of two-piece, prefabricated manholes, only the bases are installed prior to installing the ductwork.

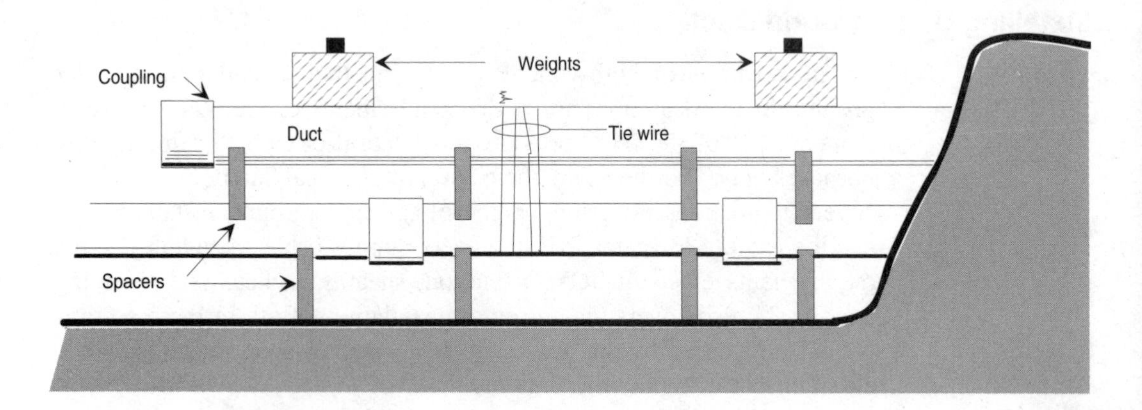

Figure 4-22: Duct bank, prior to pouring concrete, using spacers, weights, and tie wire to keep duct in place during the pouring operation.

The bottom of the trenches must be flat and compacted prior to installing the ducts. This is to prevent the trench from settling and putting stress on the duct banks. In most cases, if a concrete envelope is to be used, the trench is first filled with 3 in of concrete and finished at the appropriate grade. Once hardened, duct lines are then placed in the trench utilizing fiber or plastic spacers at various intervals along the duct run. This process is repeated until the final row of duct is laid and imbedded.

During the pouring process, and before the concrete sets, it is important to make absolutely certain that the ducts line up at the joints and couplings. All duct installations must join in a manner sufficient to prevent solid matter from entering the raceway. Furthermore, the joints must form a sufficiently continuous smooth interior surface between joining duct sections so that supply cable will not be damaged when pulled past the joint.

To ensure alignment, various dowels, mandrels, and scrapers may be used. The mandrel, for example, must be long enough to reach back two joints so that at least three sections of duct will be aligned. A leather or rubber washer is attached to the mandrel which serves to clean out the conduit as the mandrel is pulled through.

A wire brush — slightly larger than the interior duct diameter — or a scraper should be pulled through the ducts after the concrete has been poured. This will eliminate any cement or dirt that has penetrated the duct at the joints.

One method used to prevent concrete from entering the joints during pouring is to wrap each joint with coarse muslin or some similar material prior to pouring the concrete envelope. The muslin is dampened to help it stick to the cut and then coated with cement. Although time consuming, this procedure prevents concrete from entering into the raceway. It is also best to stagger the joints in a multiple duct run.

Soil conditions will dictate whether concrete encasement is required. If the soil is not firm, concrete encasement is mandatory. Concrete is also required with certain types of duct lines that are not able to withstand the pressure of an earth covering. If the soil is firm, and concrete encasement is still desired (or specified), the trench need only be wide enough for the ducts and concrete encasement. The concrete is then poured between the conduit and the earth wall. If the soil is not very firm and concrete is required, 3 additional inches should be allowed on each side of the duct banks to permit the use of concrete forms.

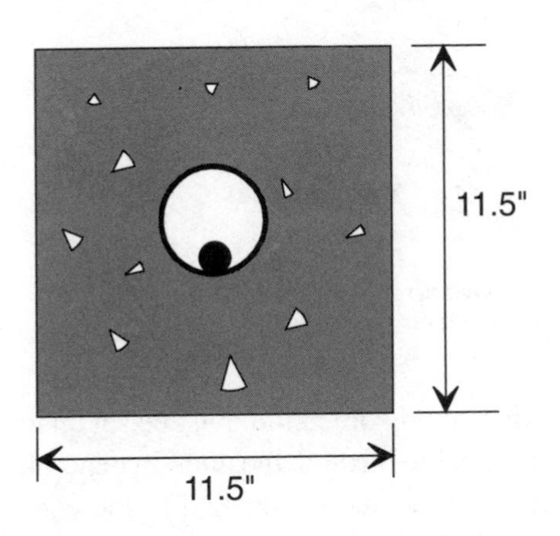

Figure 4-23: Electrical duct bank for one electrical duct.

Ducts may be grouped in any of several different ways, but for power distribution, each duct should have at least one side exposed to the earth or the outside of the concrete envelope. This means that the pattern of ducts for power distribution should be restricted to either a two-conduit width or a two-conduit depth. This permits the heat generated by power transmission to dissipate into the surrounding earth. In other words, ducts for power distribution should not be completely surrounded by other ducts. When this type of situation exists, the inner ducts may be referred to as *dead ducts*. The heat that these ducts radiate is not dissipated as fast as from the ducts surrounding them. While not suited for power cable, these dead ducts may be used for street lighting, control cable, or communication cable. The heat generated by these types of cables is relatively low, so the ducts can be arranged in any convenient configuration.

Figures 4-23 through 4-26 show duct-bank configurations and dimensions given in the *NEC*, and should serve as a guide.

Pulling Cable

There are several preliminary operations prior to pulling cable through a duct bank. *Rodding* involves the use of many short wooden rods or dowels which are joined together on a long, flexible steel rod, stiff enough to be pushed or pulled through the duct. When the rod reaches the far end of the duct, steel "fish" wire is attached to the near end. Then the rod is pulled through, followed by the wire. This wire is then used to pull the cables through the ductwork.

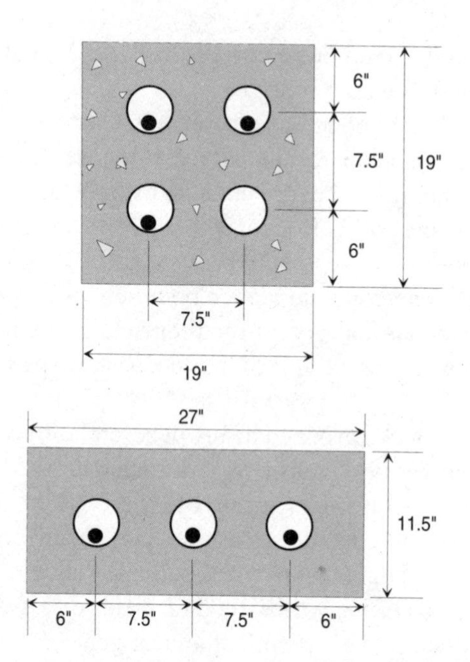

Figure 4-24: Two arrangements for three electrical ducts.

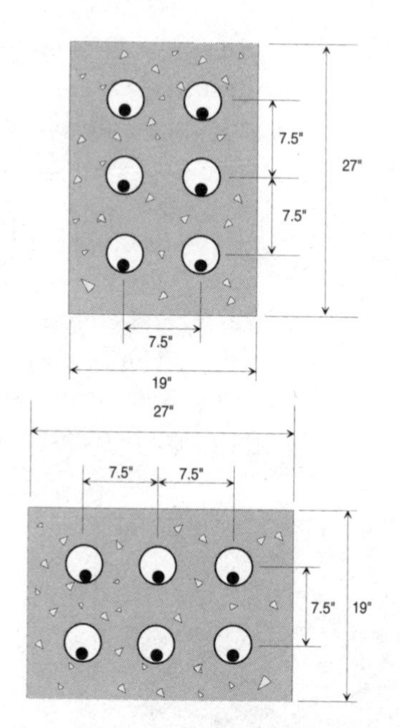

Figure 4-25: Two arrangements for duct bank containing six electrical ducts.

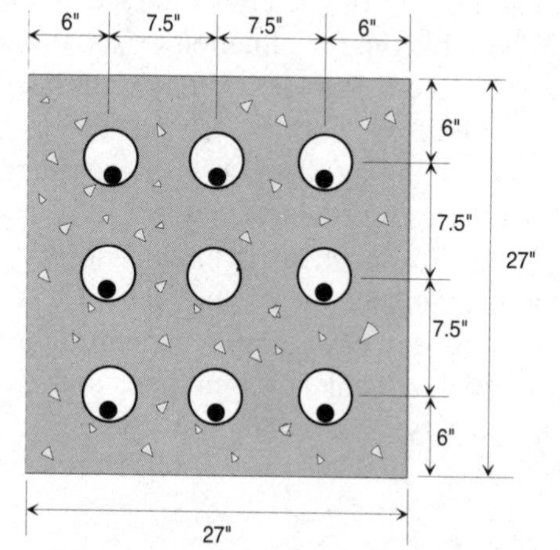

Figure 4-26: Electrical duct bank with nine ducts. The center duct, however, should not be used for power transmission; only control or communication cables.

Rigging of some kind must be set up to pull the cable through the duct. In general, a gripping device is attached to the ends of the cable. This device consists of flexible steel mesh so that the harder the pull, the tighter the device grips the cable. Before the pulling operation is started, however, a wire lubrication known as "soap" in the trade is freely applied to the ends of the cable to reduce friction during the pull. A winch or other pulling mechanism is used to move the cable through the duct. More soap or wire lubricant is applied at regular intervals as the cable enters the ducts.

Part 341A of the National Electric Safety Code covers control of bending, pulling, tensions, and sidewall pressures during the installation of cable. This section also covers cleaning foreign material from ducts, selection of cable lubricants that will not damage any part of the installation, and restraint of cables in sloping or vertical runs to prevent them from moving downhill. It further specifies that power supply cables, control cables, and communication cables should not be installed in the same duct unless the same utility maintains or operates them.

SERVICE EQUIPMENT

Wye-connected, 480/277-V services or delta-connected 480/240-V services are both a common low-voltage power source in industrial establishments. Such installations frequently utilize switchgear enclosures such as the one shown in Figure 4-27. Service equipment of this type is made up of vertical sections that connect to a common bus system within the enclosure. These sections contain fusible switches or circuit breakers, metering equipment, or other devices related to the electric service.

NEC Section 230-71 allows a maximum of six service disconnecting means per service grouped in any one location. If there are more than six switches or circuit breakers in the switchboard, then a main switch or circuit breaker must be provided to disconnect all service conductors in the building or structure from the power supply.

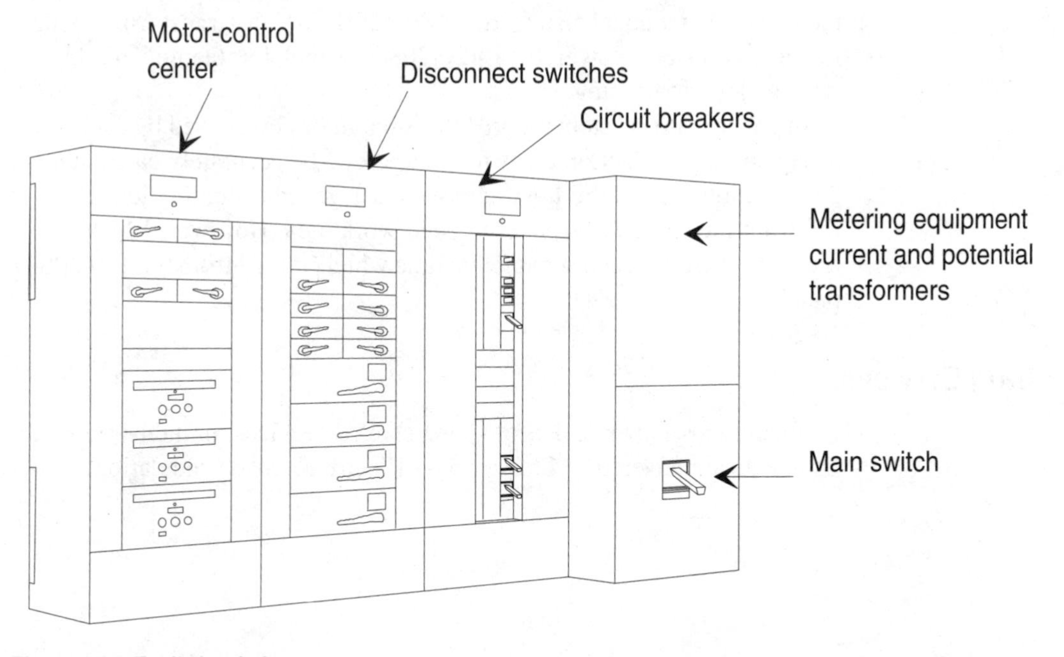

Figure 4-27: Typical switchgear.

Typically, the metering equipment (CTs, potential transformers, etc.) are installed in the same enclosure as the main disconnecting means; additional space is usually provided in the switchgear for this equipment. Furthermore, taps are normally provided in the main bus in a barriered section for connection to emergency switches, such as for fire-alarm systems, emergency lighting, etc., as allowed by *NEC* Section 230-82, Exception 5.

Practical Application

Figure 4-28 shows a plan view of a garment factory that utilizes a 480/277-V, wye-connected, three-phase service to supply numerous sewing areas within the building. In general, a pad-mounted transformer installed on the property perimeter reduces the distribution voltage to 480/277 V. An underground service is installed from this pad-mounted transformer to a switchgear room in one section of the factory. A single-line diagram of the electrical system for this project is shown in Figure 4-29. Note that only six fusible switches exist in the main switchgear so no main disconnecting means is necessary. Each of these six feeders supply a bussed-gutter system, each of which contains six meter bases, a fused safety switch that feeds a 480/277-V panel for lighting and HVAC equipment. This latter panel also feeds a 480-120/208-V dry transformer which, in turn, furnishes power to a 120/208-V panel for feeding work area receptacles and lighting.

In general, this system records the amount of power used by each work group so that production costs for each group be accurately calculated. A 277-V fluorescent lighting system provides general illumination which is fed from the 480/277-V panel in each work area. However, 120-V lights are also utilized at each sewing machine which are fed from the 120/208-V panels.

Sizing Services

Techniques and examples of sizing electric services for industrial establishments are covered in Chapter 5 — Electrical Load Calculations.

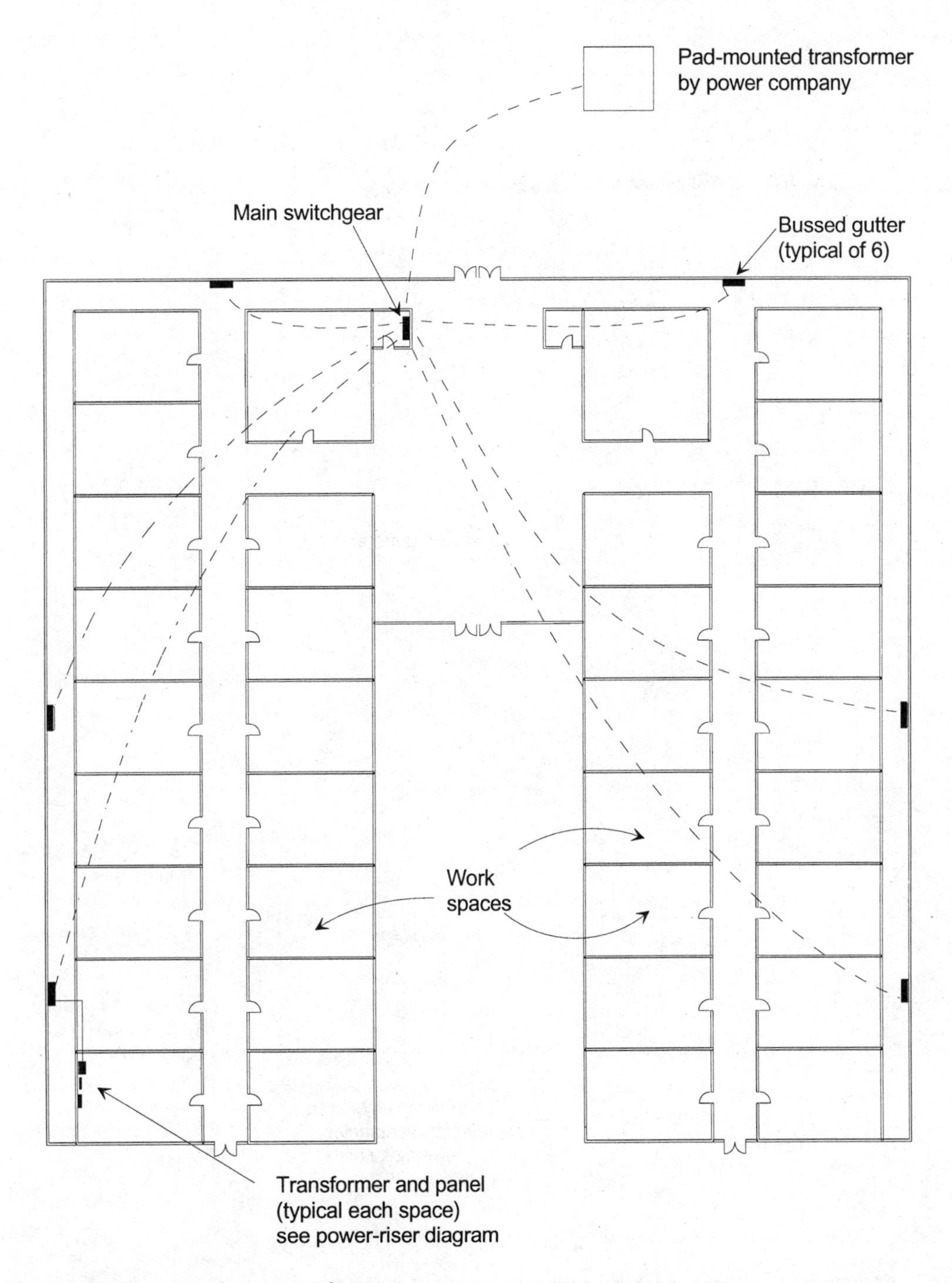

Figure 4-28: Plan view of a garment factory.

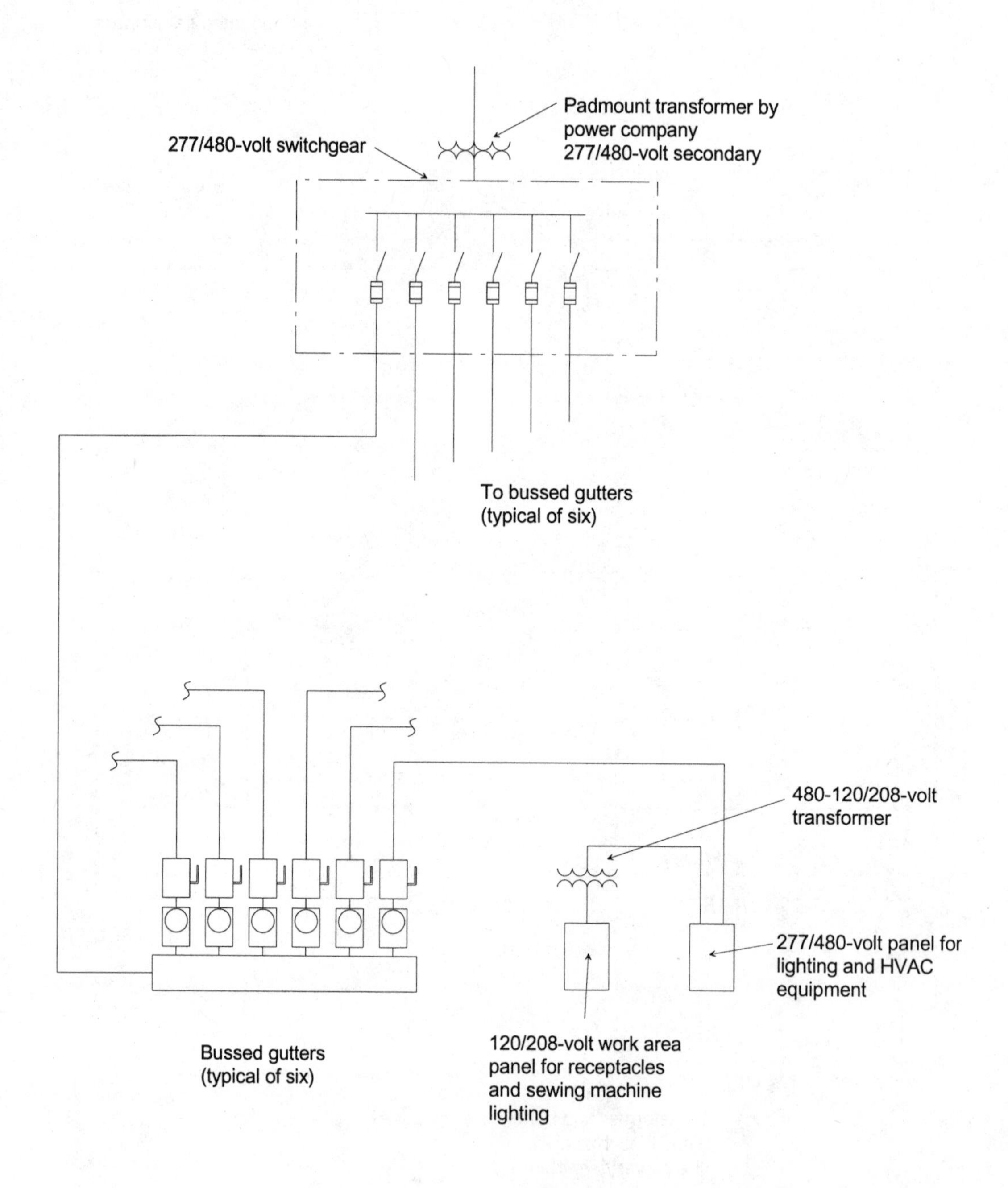

Figure 4-29: Power-riser diagram for the garment factory in Figure 4-28.

Chapter 5

Electrical Load Calculations

Sometimes it is confusing as to which comes first, the layout of the outlets, or the sizing of the electric service for a building or project. In many cases, the service size (size of main disconnect, panelboard, service conductors, etc.) can be sized using *NEC* procedures before the outlets are actually laid out. In other cases, the outlets and equipment connections will have to be laid out first. However, in either case, the service-entrance and panelboard size will have to be calculated and located before the circuits may be designed or installed.

Traditionally, consulting engineers and electrical designers locate all outlets on the working drawings prior to sizing the service entrance. Once the total connected load has been determined, demand factors and continuous-load factors are applied to size the branch-circuit requirements. Subpanels are then placed at strategic locations throughout the building, and feeders are then sized according to *NEC* requirements. The sum of these feeder circuits (allowing appropriate demand factors) determines the size of the electric service. The main distribution panel, number and size of disconnect switches, and overcurrent protection is then determined to finalize the load calculation. Such procedures for determining the load of any project normally surpass *NEC* requirements, but this is quite acceptable since the *NEC* specifies minimum requirements. In fact, in *NEC* Section 90-1(b), the *NEC* itself states:

> "This Code contains provisions considered necessary for safety. Compliance therewith and proper maintenance will result in an

> installation essentially free from hazard but not necessarily efficient, convenient, or adequate for good service or future expansion of electrical use."

Consequently, many electrical installations are designed for conditions that surpass *NEC* requirements to obtain a more efficient and convenient electrical system.

Steps for Calculating Service Loads

In general, the types of areas in an industrial establishment (office area, warehouse, machine shop, etc.) are first determined and categorized before the load calculations for the electric service begin. The basic steps proceed as follows:

Step 1. Determine the area of the building using outside dimensions, less any areas such as garages, porches, or any unfinished spaces not adaptable for future use.

Step 2. Multiply the resulting area by the load per ft^2 amount listed in *NEC* Table 220-3(b) for the type of occupancy.

Step 3. Determine the volt-amperes (va) of continuous loads (if any) and multiply the va for these loads by a factor of 1.25 (125 percent) as per *NEC* Section 220-3(a).

Step 4. Apply demand factors to any qualifying loads.

Step 5. Calculate the total adjusted general lighting load.

Step 6. Determine the type and va ratings of any other loads, such as motor circuits, machine tools, testing equipment and the like.

Step 7. Add these "other loads" to the total adjusted general lighting load to obtain the total load (in va) for the project.

In actual situations, plans and specifications that provide ample space in raceways, spare raceways, and additional spaces in panelboards will allow for future increases in the use of electricity. Distribution centers located in readily accessible locations will provide convenience and safety of operation.

Although the *NEC* does not specifically state the exact amount of additional space to allow for future expansion, electrical inspectors normally require a minimum of 20 percent. Therefore, if the required service for an electrical installation is, say, exactly 1000 A, electrical inspectors may require that the service size be increased to 1200 A — the next higher standard overcurrent protective-device rating.

Furthermore, if a lighting panelboard requires 20 spaces for circuit breakers to protect branch-circuit loads, a panelboard or load center that has at least 24 spaces (excluding any main circuit breaker or disconnect) should be considered minimum; a 30-space panelboard would be better.

The flowchart in Figure 5-1 gives an overview of the basic steps for determining electrical loads for all types of occupancies. Details concerning the use of this chart should become apparent as the following practical examples are reviewed.

Industrial Motor-Control Center

Electricians working on industrial projects are frequently required to size the feeder load for motor-control centers, as well as the motor branch-circuit conductors and overcurrent devices. In general, when sizing feeders for motor-control centers — serving several motors — the feeder conductors must have an ampacity at least equal to the sum of the full-load current rating of all the motors, plus 25 percent of the highest rated motor in the group, plus the ampere rating of any other loads determined in accordance with *NEC* Article 220 and other *NEC* applicable sections.

A typical motor-control center (MCC) is shown in Figure 5-2, while its related motor-control schedule is shown in Figure 5-3. Note that this MCC contains spaces for 29 controllers ranging in size from NEMA size 00 to NEMA size 3. A total of 19 motors are currently connected with 10 spaces remaining as "spares" for future extensions.

Assuming a power factor of 90 percent, and using connected loads only, one method of sizing the feeder for this motor-control center is explained in the following steps.

Step 1. Find the full-load current rating of all motors in the group: Since the *NEC* full-load amperes have already been determined in the motor-control schedule shown in Figure 5-3, the sum of all motor FLA is first determined. The total FLA of all motors = 484.8 A.

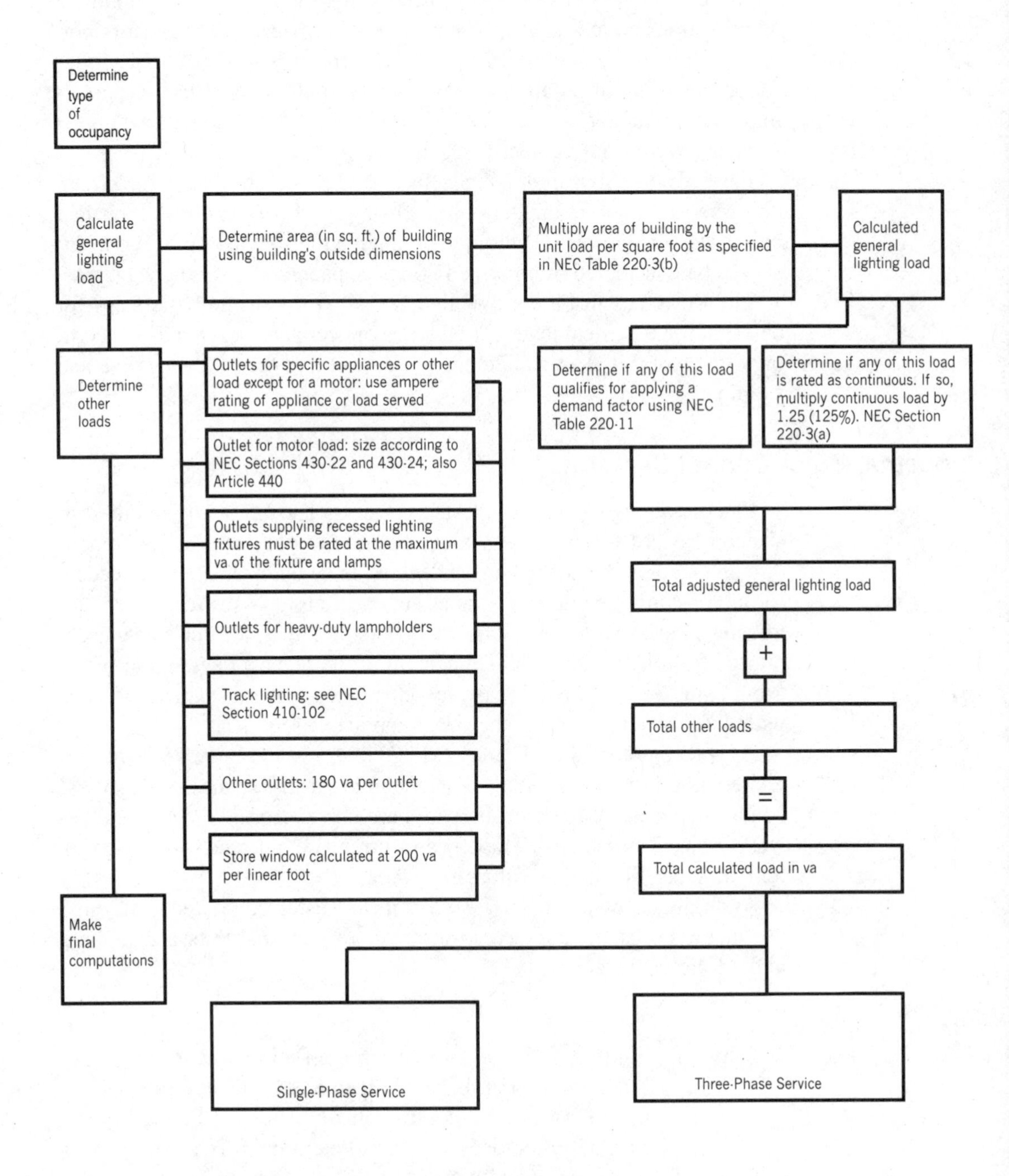

Figure 5-1: Basic steps required for electrical load calculations.

Step 2. Calculate 25 percent of the largest motor's FLA: The 50 hp is the largest motor in the group with an FLA of 130 A. Therefore,

$$130 \times .25 = 32.5 \text{ A}$$

Step 3. Add the results obtained from the last calculation to the sum of all motor FLA obtained previously; that is,

$$484.8 + 32.5 = 517.3 \text{ A}$$

Step 4. The results obtained thus far would normally be the number of amperes to determine the feeder size. However, since the power factor of this electrical system is 90 percent, the footnotes to *NEC* Table 430-150 require that the load determined previously must be multiplied by a factor of 1.1. Therefore, the following equation must be used to determine the demand amperes for the MCC with a 90 percent power factor:

$$517.3 \times 1.1 = 569.03 \text{ A}$$

Step 5. Refer to *NEC* Table 310-16 to find a conductor, say, THHN, that will carry no less than 569.03 A. In doing so, it is found that 900 kcmil is the closest size obtainable to the load; that is, this conductor is rated at 585 A.

Industrial Office Building

A 20,000 ft^2 office/reception building is a part of an industrial complex and is to be fed from the main service by a 480Y/277-V, three-phase feeder. The building contains the following loads:

- 10,000 va, 208-V, three-phase sign
- 100 duplex receptacles supplying continuous loads
- 40-ft long show window
- 12 kVA, 208/120-V, three-phase electric cooking unit

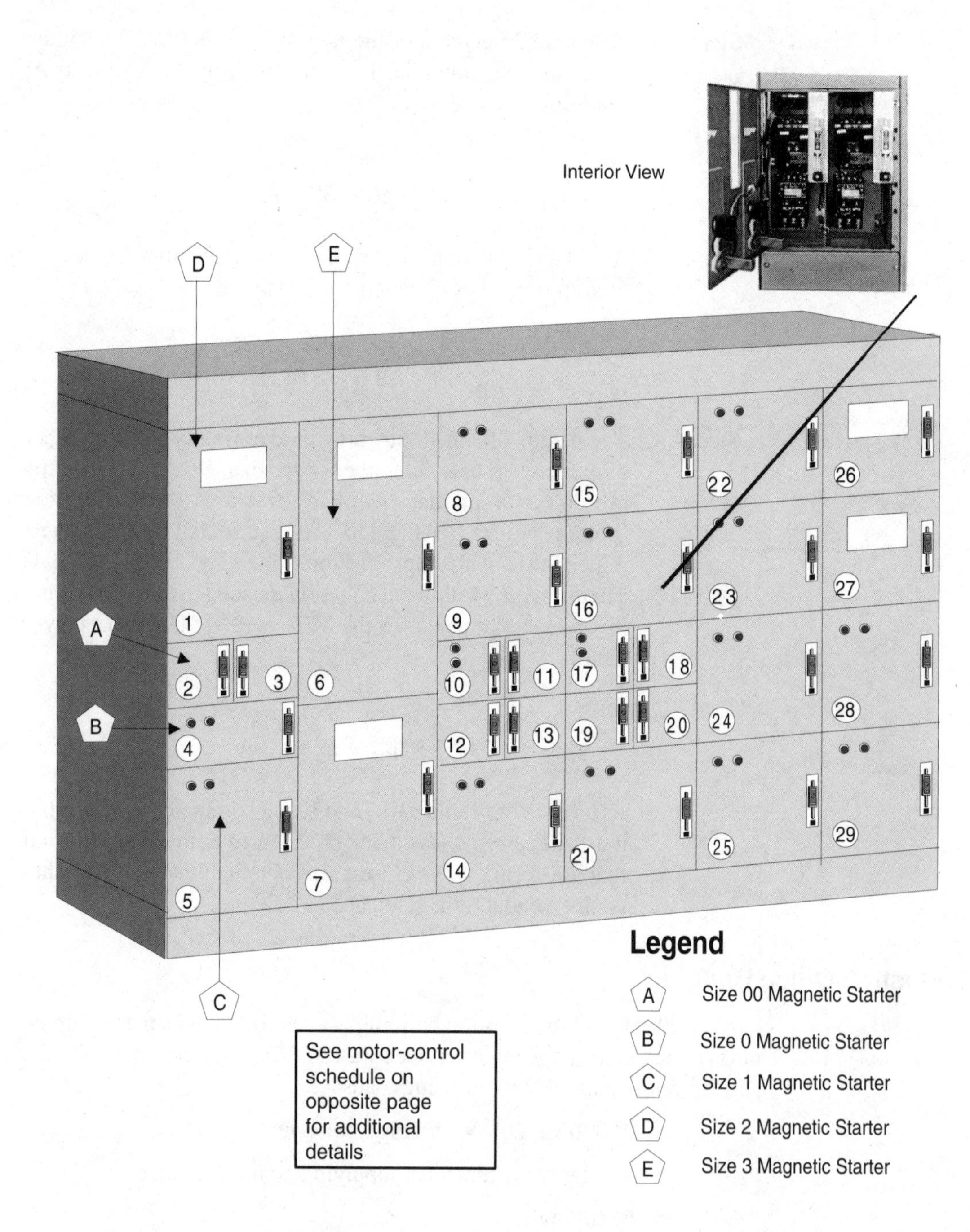

Figure 5-2: Typical motor-control center.

MOTOR-CONTROL CENTER SCHEDULE						
Circuit Number	Motor Horsepower	Temp. °C	Code Letter	Service Factor	NEC FLA	NEMA Size
1	20	40	B	1.15	54	2
2	1½				5.2	00
3	1½				5.2	00
4	5				15.2	0
5	7½				22	1
6	50				130	3
7	20				54	2
8	Spare				—	1
9	Spare				—	1
10	2				6.8	00
11	Spare				—	00
12	1½				5.2	00
13	Spare				—	00
14	10				28	1
15	7½				22	1
16	Spare				—	1
17	1				3.6	00
18	¾				2.8	00
19	¾				2.8	00
20	Spare				—	00
21	10				28	1
22	10				28	1
23	7½				22	1
24	7½				22	1
25	10				28	1
26	Spare				—	1
27	Spare				—	1
28	Spare				—	1
29	Spare				—	1

Figure 5-3: Motor-control center schedule.

- 10 kVA, 208/120-V, three-phase electric oven
- 20 kVA, 480-V, three-phase water heater
- Seventy-five 150-W, 120-volt incandescent lighting fixtures
- Two hundred 200-W, 277-V fluorescent lighting fixtures
- 7.5 hp, 480-V, three-phase motor for fan-coil unit
- 40 kVA, 480-V, three-phase electric heating unit
- 60-A, 480-V, three-phase air-conditioning unit

The ratings of the service equipment, transformers, feeders, and branch circuits are to be determined, along with the required size of service grounding conductor. Circuit breakers are used to protect each circuit and THHN copper conductors are used throughout the electrical system.

A one-line diagram of the electrical system is shown in Figure 5-4. Note that the incoming 3-phase, 4-wire, 480Y/277-V main service terminates into a main distribution panel containing six overcurrent protective devices. Since there are only six circuit breakers in this enclosure, no main circuit breaker or disconnect is required as allowed by *NEC* Section 230-71. Five of these circuit breakers protect feeders and branch circuits to 480/277-V equipment, while the sixth circuit breaker protects the feeder to a 480 - 208Y/120-V transformer. The secondary side of this transformer feeds a 208/120-V lighting panel with all 120-V loads balanced and all loads on this panel are continuous. Now let's start at the loads connected to the 208/120-V panel and perform the required calculations.

Step 1. Calculate the load for the 100 receptacles, remembering that all of these are rated as a continuous load.

$$100 \times 180 \times 1.25 = 22{,}500 \text{ va}$$

Step 2. Calculate the load for the show window using 200 va per linear foot.

$$200 \text{ va} \times 40 \text{ feet} = 8{,}000 \text{ va}$$

Step 3. Calculate the load for the outside lighting.

$$75 \text{ lamps} \times 150 \text{ va} \times 1.25 = 14{,}062.5 \text{ or } 14{,}063 \text{ va}$$

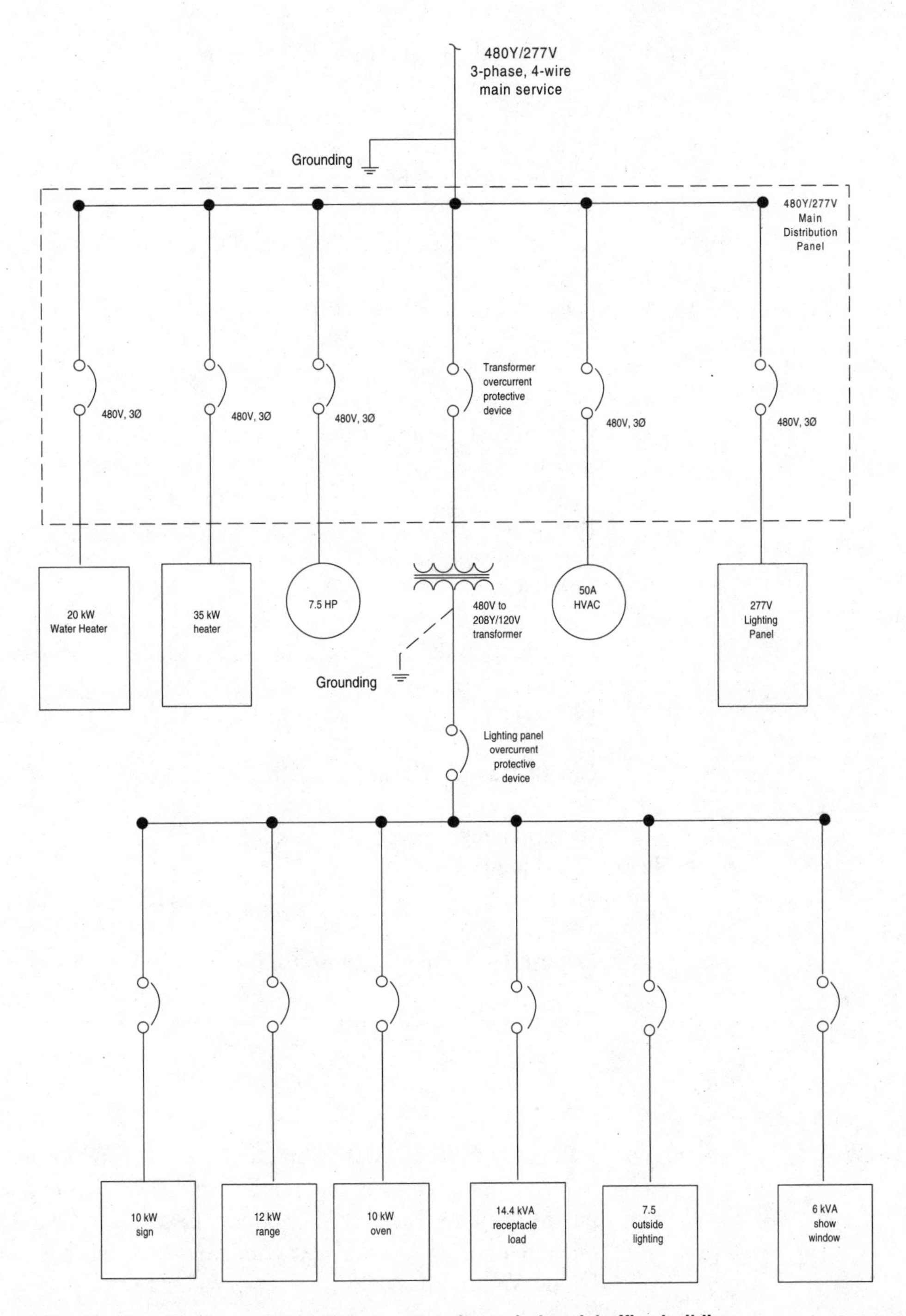

Figure 5-4: One-line diagram of the electrical system for an industrial office building.

Step 4. Calculate the load for the 10 kW sign.

$$10{,}000 \times 1.25 = 12{,}500 \text{ va}$$

Step 5. Calculate the load for the 12 kW range.

$$12{,}000 \times 1.25 = 15{,}000 \text{ va}$$

Step 6. Calculate the load for the 10 kW oven.

$$10{,}000 \times 1.25 = 12{,}500 \text{ va}$$

Step 7. Determine the sum of the loads on the 208/120-V lighting panel.

Receptacle load.................................. 22,500

Show window.................................... 8,000

Outside lighting...................................14,063

Electric sign...12,500

Electric range.......................................15,000

Electric oven...12,500

Total connected load on subpanel **84,563**

Step 8. Determine the feeder rating for the subpanel.

$$\frac{84563}{\sqrt{3} \times 208V} = 235\ A$$

Step 9. Refer to *NEC* Table 310-16 and find that 4/0 THHN conductor (rated at 260 A) is the closest conductor size that will handle the load. Since a 235-A circuit breaker is not standard, one rated at 250-A is the next standard size, and is permitted by the *NEC*.

The 208Y/120-V circuit is a separately derived system from the transformer and is grounded by means of a grounding electrode conductor that must be at least a No. 2 copper conductor based on the No. 4/0 copper feeder conductors.

Step 10. Size the transformer and the transformer overcurrent protective device. Since the transformer supplying the subpanel is rated for a continuous load, the kVA rating of the transformer need only be the actual connected load. The 1.25 continuous-rating factor does not have to be applied. Therefore, the kVA rating of the transformer need only be 69.3 kVA. A commercially available 75-kVA transformer would be selected.

Step 11. Select the overcurrent protective device for the transformer.

$$\frac{75{,}000\,va}{\sqrt{3} \times 480\,V} = 90.32\ A$$

The maximum setting of the transformer overcurrent protective device is then 1.25 × 90.32 = 112.89 A; that is, it may not exceed this rating. Since 110 A is a standard fuse and circuit-breaker rating, this would be the maximum size normally used. No. 4 THHN copper conductors may be used for the transformer feeder.

Calculations for the Primary Feeder

Calculations for the primary feeder and other 480/277-V circuits are based on the assumption that all of these loads are continuous, including the water heater for the building. Let's take a look at the lighting first. *NEC* Table 220-3(b) requires a minimum general lighting load for office buildings to be based on 3.5 va per ft^2. Since the building has already been sized at 20,000 ft^2, the minimum va for lighting may be determined as follows:

$$20{,}000 \times 3.5 = 70{,}000\ va$$

Since this load is continuous, the 70 kVA figure will have to be multiplied by a factor of 1.25 to obtain the total calculated load:

$$70{,}000 \times 1.25 = 87{,}500$$

The actual connected load, however, is as follows:

$$200 \text{ light fixtures} \times 200 \text{ va each} \times 1.25 = 50{,}000 \text{ va}$$

Since this connected load is less than the *NEC* requirements, it is neglected in the calculation and the 3.5 va per-ft^2 figure is used. Therefore, the total load on the 277-V lighting panel may be determined as follows:

$$\frac{1.25 \times 70000}{\sqrt{3} \times 480V} = 105.37\,A$$

No. 3 THHN conductors will be used for the feeder supplying the 277-V lighting panel, and the overcurrent protective device will be rated for 110 A — the closest standard fuse or circuit breaker size.

NEC Section 422-14 requires that the feeder or branch circuit supplying a fixed storage-type water heater having a capacity of 120 gal or less be sized not less than 125 percent of the nameplate rating of the water heater. Consequently, the load for the water heater may be calculated as follows:

$$1.25 \times \frac{20{,}000\,va}{\sqrt{3} \times 480V} = 30.1\,A$$

The feeder for the water heater will therefore require a No. 8 AWG THHN conductor with an overcurrent protective device of either 35 or 40 A.

The electric-heating load is sized in a similar way. *NEC* Section 424-3(b) requires that this load be sized at not less than 125 percent of the total load of the motors and the heaters. Therefore, the load for the electric heating may be calculated as follows:

$$1.25 \times \frac{40{,}000}{\sqrt{3} \times 480} = 60.2\,A$$

This load requires a conductor size of No. 6 AWG THHN with an overcurrent protective device rated at 70 A.

An instantaneous trip circuit breaker is selected to protect the 7.5 hp motor circuit. This arrangement is allowed if the breaker is adjustable and is part of an approved controller. A full-load current of 11 A requires No.

14 conductors protected by a breaker set as high as $(13 \times 11) = 143$ A. However, in this case, the circuit breaker will probably be set at approximately 70 to 80 A.

The air-conditioning circuit must have an ampacity of

$$1.25 \times 60 \text{ A} = 75 \text{ A}$$

for which No. 6 THHN conductors are selected from *NEC* Table 310-16. An overcurrent protective device for this circuit cannot be set higher than $(1.75 \times 60) = 105$ A.

Main Feeder Calculations

When performing the calculations for the main feeder, assume that all loads are balanced and may be computed in terms of va, but are more than likely computed in terms of amperes. Calculation of the loads in amperes also simplifies the selection of the main overcurrent protective device and service conductors. A summary of this calculation is shown in Figure 5-5.

MAIN SERVICE LOAD CALCULATION			
Type of Load	**Load in Amperes**	**Neutral**	**NEC Reference**
208Y/120-V system	101.83	0	
277-V lighting panel	105.37	104	
Water heater	30.1	0	
Electric heat (neglected)	0	0	Section 220-21
7.5 hp motor	11.0	0	
Air conditioner	75	0	
25% of largest motor = .25 x 60 A	15	0	Section 530-25
Service Load	**338.3**	**104**	

Figure 5-5: Summary of main feeder load calculation.

The 84,563 va, 208Y/120-V lighting panel load represents a line load of 101.83 A at 480 V. The 480/277-V loads are then added to this for a total of 338.3 A. The conductors are then selected from *NEC* Table 310-16 which indicates that 350 kcmil THHN conductors have a current-carrying capacity of 350 A. An overcurrent protective device of the same amperage may also be selected. The load on the neutral is only 104 A, but the neutral conductor cannot be smaller than the grounding electrode conductor. Referring to *NEC* Table 250-94, the grounding electrode conductor must be 1/0 copper. Consequently, the neutral cannot be smaller than 1/0 THHN copper conductor.

In actual practice, the service or feeder size for this office building would probably be increased to 400 A to allow for some future development.

MOTOR CIRCUIT CONDUCTORS

The basic elements that must be accounted for in any motor circuit are shown in Figure 5-6. Although these elements are shown separately in this illustration, there are certain cases where the *NEC* permits a single device to serve more than one function. For example, in some cases, one switch can serve as both the disconnecting means and controller. In other instances, short-circuit protection and overload protection can be combined in a single circuit breaker or set of fuses.

NOTES: *The basic NEC rule for sizing conductors supplying a single-speed motor used for continuous duty specifies that conductors must have a current-carrying capacity of not less than 125 percent of the motor's full-load current rating.*

Conductors on the line side of the controller supplying multi-speed motors must be based on the highest of the full-load current ratings shown on the motor nameplate.

Conductors between the controller and the motor must have a current-carrying rating based on the current rating for the speed of the motor each set of conductors is feeding.

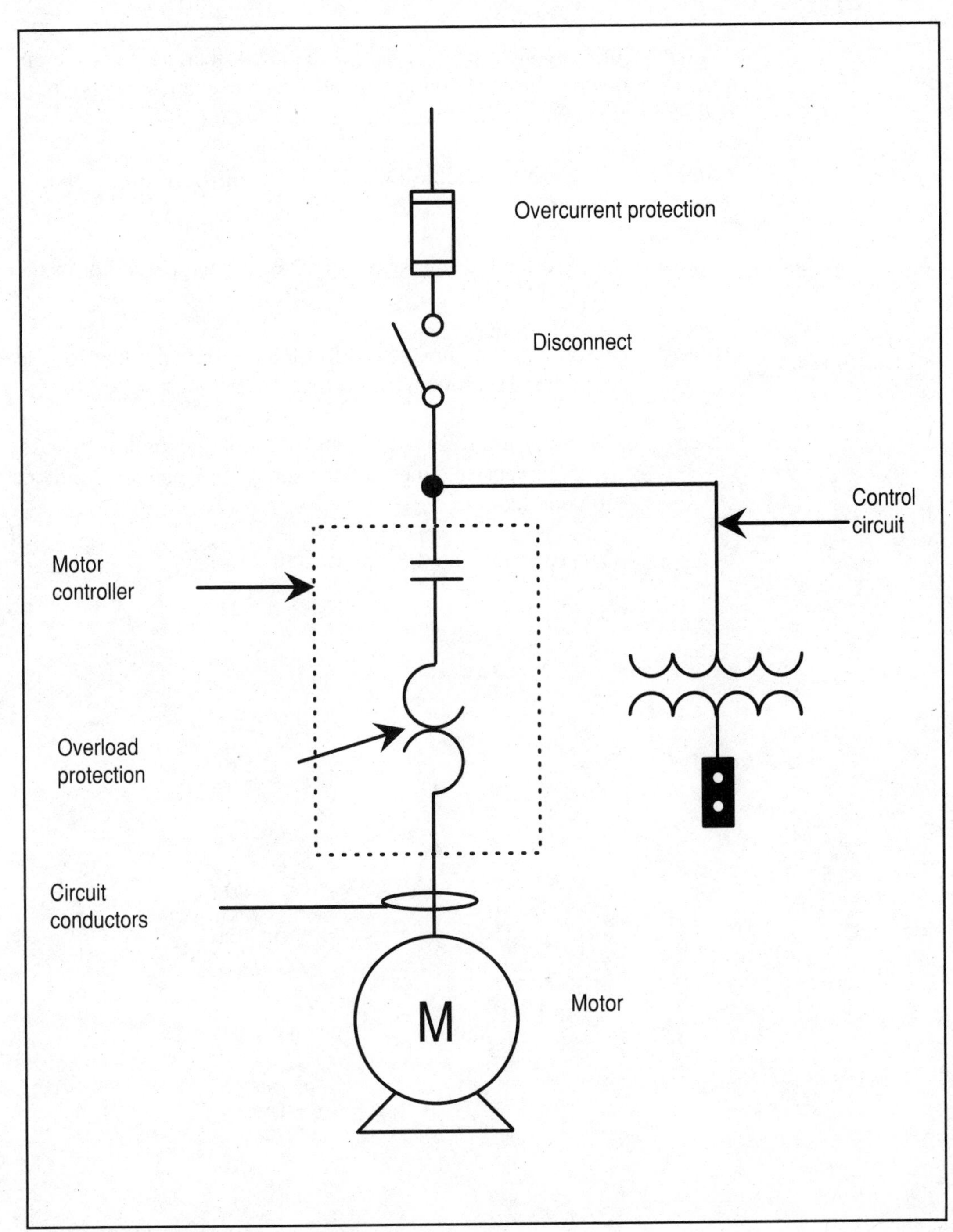

Figure 5-6: Basic elements of any motor circuit.

A typical motor-control center and branch circuits feeding four different motors are shown in Figure 5-7. Let's see how the feeder and branch-circuit conductors are sized for these motors.

Step 1. Refer to *NEC* Table 430-150 for the full-load current of each motor.

Step 2. Determine the full-load current of the largest motor in the group.

Step 3. Calculate the sum of the full-load current ratings for the remaining motors in the group.

Step 4. Multiply the full-load current of the largest motor by 1.25 (125 percent) and then add the sum of the remaining motors to your answer.

Step 5. The combined total of Step 4 will give the *minimum* feeder size.

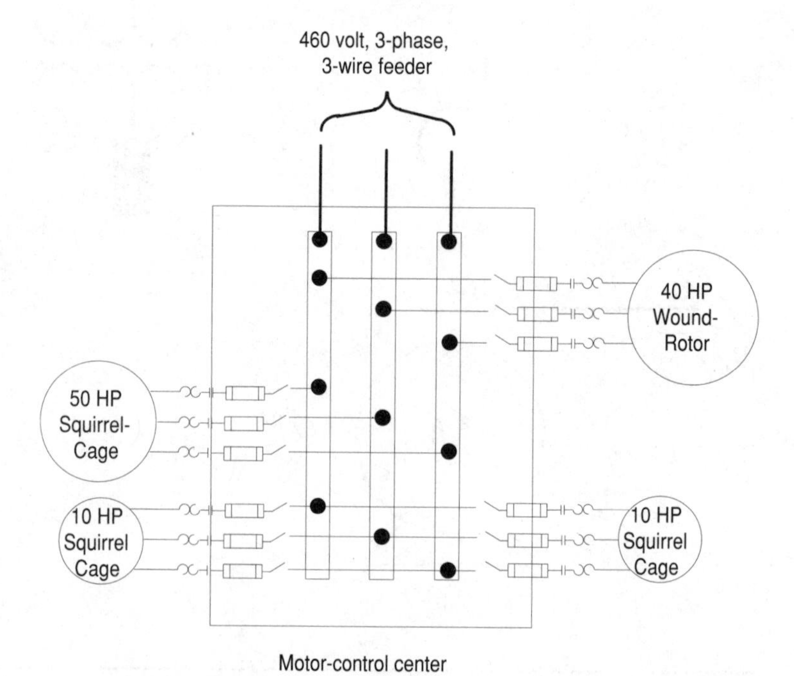

Figure 5-7: Motor branch circuits from motor-control center.

When sizing feeder conductors for motors, be aware that the procedure described in the previous five steps will give the *minimum* conductor rating based on temperature rise only. Consequently, it is often necessary to increase the size of conductors to compensate for voltage drop and power loss in the circuit.

Now let's complete the conductor calculations for the motor circuits in Figure 5-7.

Step 1. A partial list of the motors listed in *NEC* Table 430-150 is shown in Figure 5-8. Referring to this table, the motor horsepower is shown in the very left-hand column. Follow

INDUCTION TYPE SQUIRREL-CAGE AND WOUND-ROTOR AMPERES					
HP	**115 V**	**200 V**	**208 V**	**230 V**	**460 V**
½	4	2.3	2.2	2	I
¾	5.6	3.2	3.1	2.8	1.4
1	7.2	4.1	4.0	3.6	1.8
1½	10.4	6.0	5.7	5.2	2.6
2	13.6	7.8	7.5	6.8	3.4
3	-	11.0	10.6	9.6	4.8
5	-	17.5	16.7	15.2	7.6
7½	-	25.3	24.2	22	11
10	-	32.2	30.8	28	14
15	-	48.3	46.2	42	21
20	-	62.1	59.4	54	27
25	-	78.2	74.8	68	34
30	-	92	88	80	40
40	-	119.6	114.4	104	52
50		149.5	143.0	130	65

Figure 5-8: NEC Table 430-150 (partial).

across the appropriate row until you come to the column titled " 460 V "— the voltage of the motor circuits in question. In doing so, we find that the ampere ratings for the motors in question are as follows:

50 hp = 65 A
40 hp = 52 A
10 hp = 14 A

Step 2. The largest motor in this group is the 50 hp squirrel-cage motor which has a full-load current of 65 A.

Step 3. The sum of the remaining motors is as follows:

40 hp = 52 A
10 hp = 14 A
10 hp = 14 A

80 A

Step 4. Multiplying the full-load current of the largest motor and then adding the total amperage of the remaining motors results in the following:

$$(1.25)(65) + 80 = 161.25 \text{ A}$$

Step 5. Therefore, the minimum feeder size for the 460 V, 3-phase, 3-wire motor control center will be 161.25 A. Referring to *NEC* Table 310-16, under the column headed "90°C," the closest conductor size is 1/0 copper (rated at 170 A) or 3/0 aluminum (rated at 175 A).

The branch-circuit conductors feeding the individual motors are calculated somewhat differently. *NEC* Section 430-22(a) requires that the ampacity of branch-circuit conductors supplying a single continuous-duty motor must not be less than 125 percent of the motor's full-load current rating. Therefore, the current-carrying capacity of the branch-circuit conductors feeding the four motors in question are calculated as follows:

- 50 hp motor = 65 A × 1.25 = 81.25 A
- 40 hp motor = 52 A × 1.25 = 65.00 A
- 10 hp motor = 14 A × 1.25 = 17.5 A

Referring to *NEC* Table 310-16, the closest size THHN copper conductors that will be permitted to be used on these various branch circuits are as follows:

- 50 hp motor = 81.25 A requires No. 4 AWG THHN conductors.
- 40 hp motor = 65 A requires No. 6 AWG THHN conductors.
- 10 hp motor = 17.5 A requires No. 12 AWG THHN conductors.

Refer to Figure 5-9 on the next page for a summary of the conductors used to feed the motor-control center in question, along with the branch-circuits supplying the individual motors.

For motors with other voltages (up to 2300 volts) or for synchronous motors, refer to *NEC* Table 430-150.

Branch-circuit conductors serving motors used for short-time, intermittent, or other varying duty, must have an ampacity not less than the percentage of the motor nameplate current rating shown in *NEC* Table 430-22(a), Exception. However, to qualify as a short-time, intermittent motor, the nature of the apparatus that the motor drives must be arranged so that the motor cannot operate continuously with load under any condition of use. Otherwise, the motor must be considered continuous duty. Consequently, the majority of motors encountered in the electrical trade must be rated for continuous duty, and the branch-circuit conductors sized accordingly.

Wound-Rotor Motors

The primary full-load current of wound-rotor motors is listed in NEC Table 430-150 and is the same as squirrel-cage motors. Conductors connecting the secondary leads of wound-rotor induction motors to their controllers must have a current-carrying capacity at least equal to 125 percent of the motor's full-load secondary current if the motor is used for continuous duty. If the motor is used for less than continuous duty, the conductors must

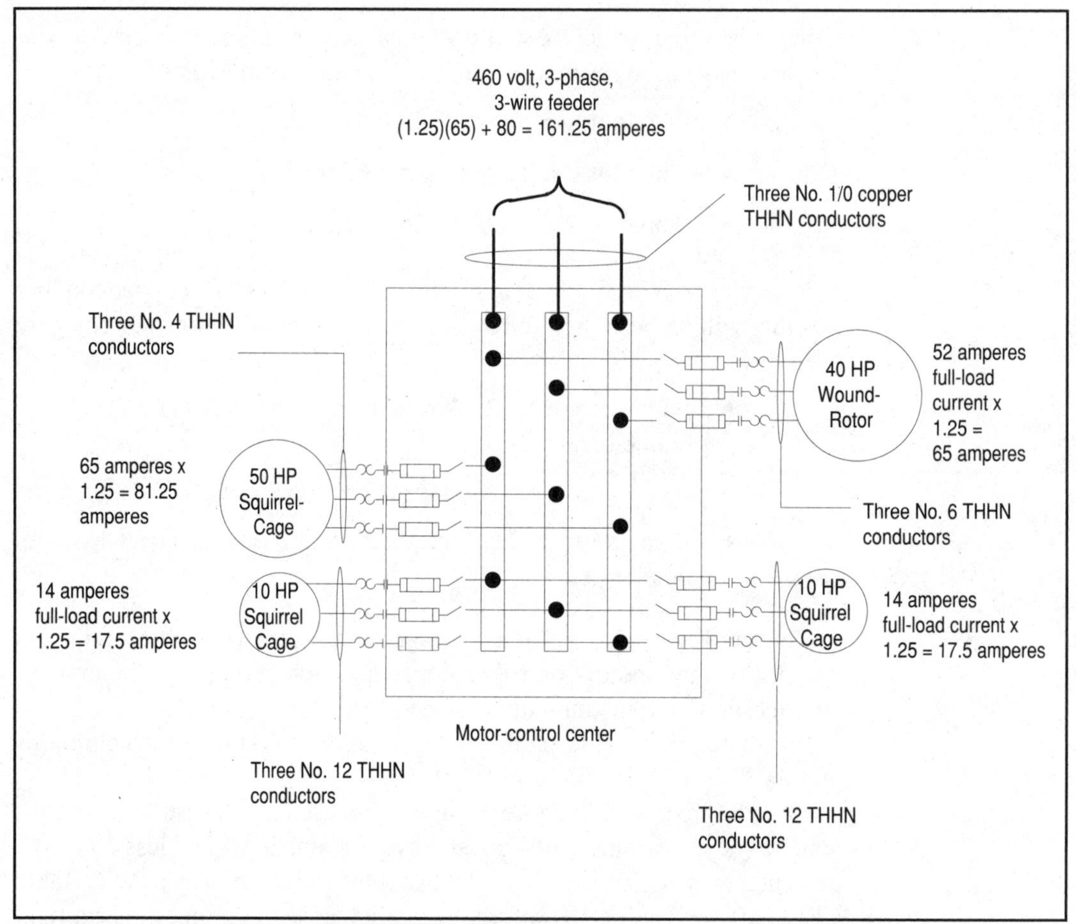

Figure 5-9: Sized branch-circuit and feeder conductors.

have a current-carrying capacity of not less than the percentage of the full-load secondary nameplate current given in *NEC* Table 430-22(a). Conductors from the controller of a wound-rotor induction motor to its starting resistors must have an ampacity in accordance with *NEC* Table 430-22(c).

Note: *NEC Section 430-6 specifies that for general motor applications (excluding applications of torque motors and sealed hermetic-type refrigeration compressor motors), the values given in NEC Tables 430-147, 430-148, 430-149, and 430-150 should be used instead of the actual current rating marked on the motor nameplate when sizing conductors, switches, and overcurrent protection. Overload protection, however, is based on the marked motor nameplate.*

Conductors for DC Motors

NEC Section 430-29 covers the rules governing the sizing of conductors from a dc motor controller to separate resistors for power accelerating and dynamic braking. This section, with its table of conductor ampacity percentages, assures proper application of dc constant-potential motor controls and power resistors. However, when selecting overload protection, the actual motor nameplate current rating must be used.

Conductors for Miscellaneous Motor Applications

NEC Section 430-6 should be referred to for torque motors, shaded-pole motors, permanent split-capacitor motors, and ac adjustable-voltage motors.

NEC Section 430-6(b) specifically states that the motor's nameplate full-load current rating is used to size ground-fault protection for a torque motor. However, branch-circuit conductors and overcurrent protection is sized by the provisions listed in *NEC* Section 430-52(b) and the full-load current rating listed in *NEC* Tables 430-147 through 430-150 are used instead of the motor's nameplate rating.

For sealed (hermetic-type) refrigeration compressor motors, the actual nameplate full-load running current of the motor must be used in determining the current rating of the disconnecting means, the controller, branch-circuit conductor, overcurrent-protective devices, and motor overload protection.

Multimotor Branch Circuits

NEC Section 430-53(a) and 430-53(b) permits the use of more than one motor on a branch circuit provided the following conditions are met:

- Two or more motors, each rated not more than 1 hp, and each drawing a full-load current not exceeding 6 A, may be used on a branch circuit protected at not more than 20 A at 125 V or less, or 15 A at 600 V or less. The rating of the branch circuit protective device marked on any of the controllers must not be exceeded. Individual overload protection is necessary in such circuits unless the motor is not permanently installed, or is manually started and is within sight from the controller location, or has sufficient winding impedance to prevent overheating due to locked-rotor current, or is part of an approved assembly which does not subject the motor to overloads and which incorporates protection for the motor against locked-rotor, or the motor cannot operate continuously under load.

- Two or more motors of any rating, each having individual overload protection, may be connected to a single branch circuit that is protected by a short-circuit protective device (MSCP). The protective device must be selected in accordance with the maximum rating or setting that could protect an individual circuit to the motor of the smallest rating. This may be done only where it can be determined that the branch-circuit device so selected will not open under the most severe normal conditions of service that might be encountered. The permission of this *NEC* section offers wide application of more than one motor on a single circuit, particularly in the use of small integral-horsepower motors installed on 208-V, 240-V, and 480-V, 3-phase industrial and commercial systems. Only such 3-phase motors have full-load operating currents low enough to permit more than one motor on circuits fed from 15-A protective devices.

Using these *NEC* rules, let's take a typical branch circuit (Figure 5-10) with more than one motor connected and see how the calculations are made.

Step 1. The full-load current of each motor is taken from *NEC* Table 430-150 as required by *NEC* Section 430-6(a).

Step 2. A circuit breaker must be chosen that does not exceed the maximum value of short-circuit protection (250 percent) required by *NEC* Section 430-52 and *NEC* Table 430-152 for the smallest motor in the group. In this case: 1.5 hp. Since the listed full-load current for the smallest motor (1.5 hp) is 2.6 A, the calculation is made as follows:

$$2.6\text{ A} \times 2.5\ (250\%) = 6.5\text{A}$$

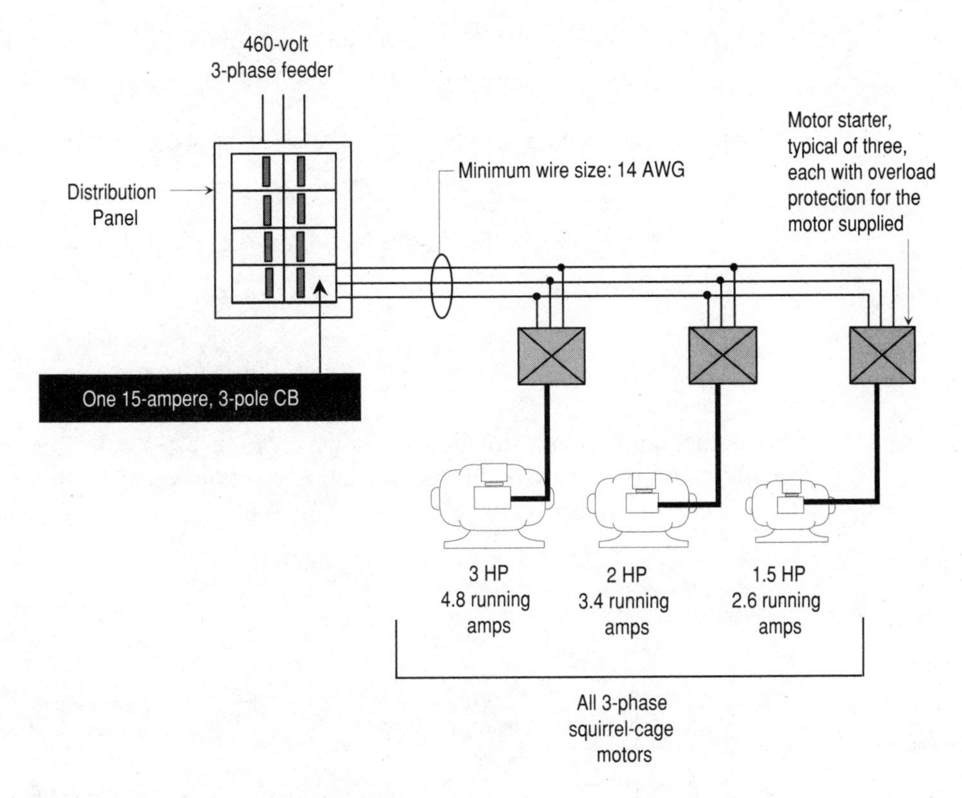

Figure 5-10: Several motors on one branch circuit.

Note: *NEC Section 430-52, Exception No. 1, allows the next higher size, rating or setting for a standard circuit breaker. Since a 15-A circuit breaker is the smallest standard rating recognized by NEC Section 240-6, a 15-A, 3-pole circuit breaker may be used.*

Step 3. The total load of the motor currents must be calculated as follows:

$$4.8 + 3.4 + 2.6 = 10.8 \text{ A}$$

The total full-load current for the three motors (10.8 A) is well within the 15-A circuit breaker rating, which has sufficient time delay in its operation to permit starting of any one of these motors with the other two already operating. Torque characteristics of the loads on starting are not high. Therefore, the circuit breaker will not open under the most severe normal service.

Step 4. Make certain that each motor is provided with the properly rated individual overload protection in the motor starter.

Step 5. Branch-circuit conductors are sized in accordance with *NEC* Section 430-24. In this case:

$$4.8 + 3.4 + 2.6 + (25\% \text{ the largest motor — } 4.8 \text{ A}) = 12 \text{ A}.$$

No. 14 AWG conductors rated at 75°C will fully satisfy this application.

Another multimotor situation is shown in Figure 5-11. In this case, smaller motors are used. In general, *NEC* Section 430-53(b) requires branch-circuit protection to be no greater than the maximum A permitted

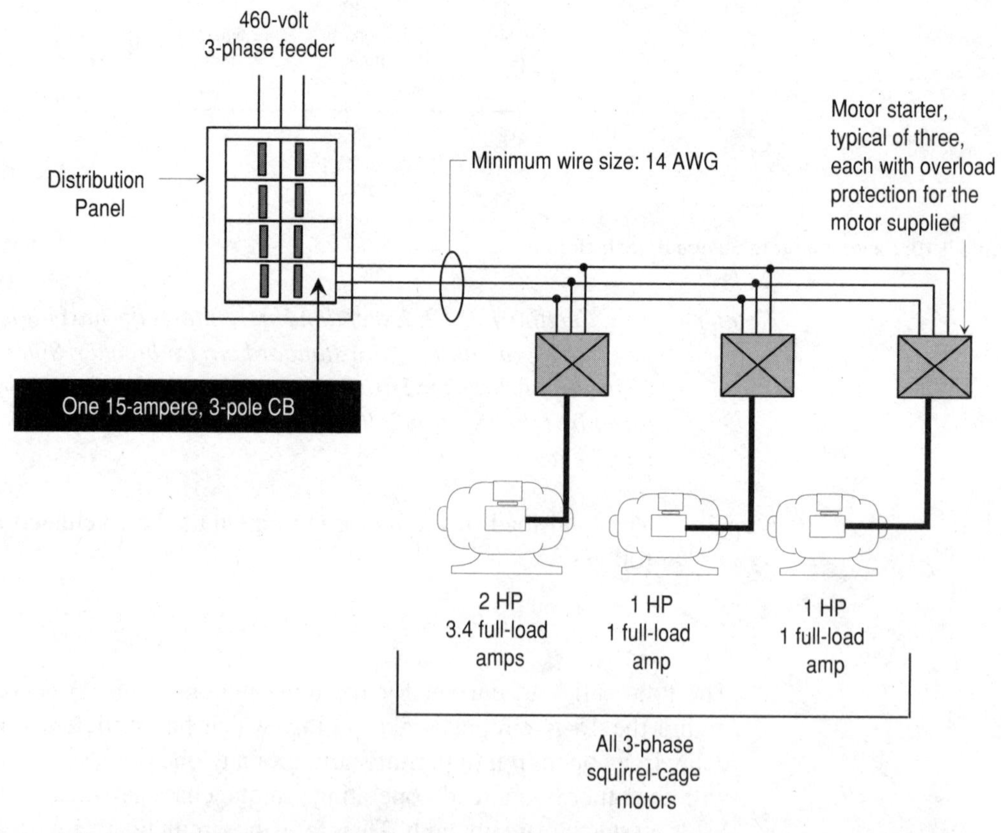

Figure 5-11: Several smaller motors supplied by one branch circuit.

by *NEC* Section 430-52 for the lowest rated motor of the group, which, in our case, is 1 A for the 0.5 hp motors. With this information in mind, let's size the circuit components for this application.

Step 1. From *NEC* Section 430-52 and *NEC* Table 430-152, the maximum protection rating for a circuit breaker is 250% of the lowest rated motor. Since this rating is 1 ampere, the calculation is performed as follows:

$$2.5 \times 1 = 2.5 \text{ A}$$

Note: *Since 2.5 A is not a standard rating for a circuit breaker, according to NEC Section 240-6, NEC Section 430-52 (Exception 1) permits the use of the next higher rating. Because 15 A is the lowest standard rating of circuit breakers, it is the next higher device rating* above 2.5 *A and satisfies NEC rules governing the rating of the branch-circuit protection.*

These two previous applications permit the use of several motors up to the circuit capacity, based on *NEC* Sections 430-24 and 430-53(b) and on starting torque characteristics, operating duty cycles of the motors and their loads, and the time-delay of the circuit breaker. Such applications greatly reduce the number of circuit breakers, number of panels, and the amount of wire used in the total system. One limitation, however, is placed on this practice in *NEC* Section 430-52(b):

- Where maximum branch-circuit, short-circuit, and ground-fault protective device ratings are shown in the manufacturer's overload relay table for use with a motor controller or are otherwise marked on the equipment, they shall not be exceeded even if higher values are allowed as shown in the preceding examples.

POWER-FACTOR CORRECTION AT MOTOR TERMINALS

Generally, the most effective method of power-factor correction is the installation of capacitors at the source of poor power fact — the induction motor. This not only increases power factor, but also releases system capacity, improves voltage stability and reduces power losses.

When power factor correction capacitors are used, the total corrective Kvar on the load side of the motor controller should not exceed the value required to raise the no-load power factor to unity. Corrective Kvar in excess of this value may cause over excitation that results in high transient voltages, currents and torques that can increase safety hazards to personnel and possibly damage the motor or driven equipment.

Do not connect power factor correction capacitors at motor terminals on elevator motors, multispeed motors, plugging or jogging applications or open transition, wye-delta, autotransformer starting and some part-winding start motors.

If possible, capacitors should be located at position No. 2 in Figure 5-12. This does not change the current flowing through motor overload protectors.

Connection of capacitors at position No. 3 requires a change of overload protectors. Capacitors should be located at position No. 1 for the following:

- Elevator motors
- Multispeed motors
- Plugging or jogging applications
- Open transition, wye-delta, autotransformer starting
- Some part-winding motors

Note: *Make sure that bus power factor is not increased above 95 percent under all loading conditions to avoid over excitation.*

The table in Figure 5-13 allows the determination of corrective Kvar required where capacitors are individually connected at motor leads. These values should be considered the maximum capacitor rating when the motor and capacitor are switched as a unit. The figures given are for 3-phase, 60 Hz, NEMA Class B motors to raise full-load power factor to 95 percent.

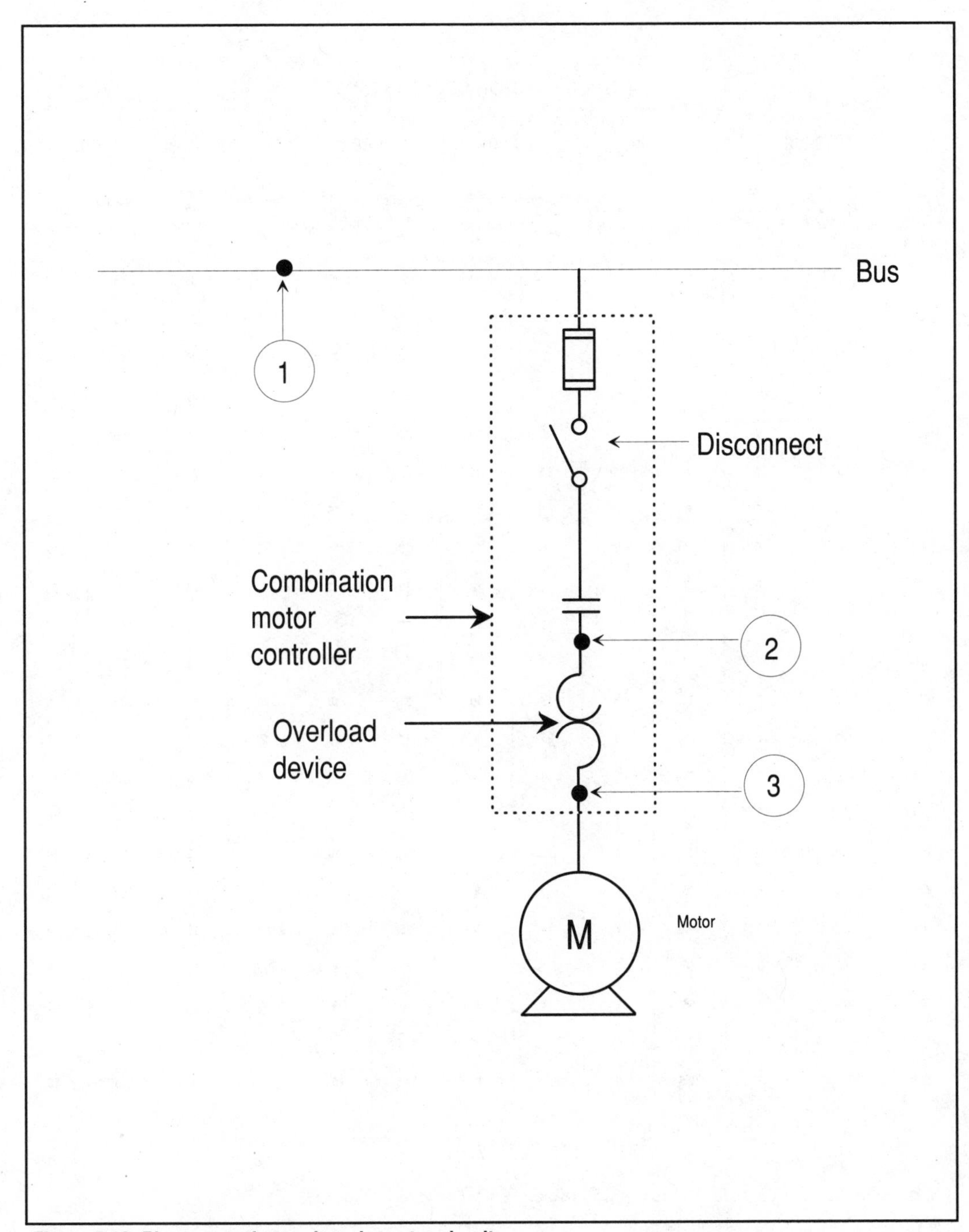

Figure 5-12: Placement of capacitors in motor circuits.

	NOMINAL MOTOR SPEED IN RPM											
	3600		1800		1200		900		720		600	
Induction Motor Horsepower Rating	Capacitor Rating KVAR	Line Current Reduction %	Capacitor Rating KVAR	Line Current Reduction %	Capacitor Rating KVAR	Line Current Reduction %	Capacitor Rating KVAR	Line Current Reduction %	Capacitor Rating KVAR	Line Current Reduction %	Capacitor Rating KVAR	Line Current Reduction %
3	1.5	14	1.5	15	1.5	20	2	27	2.5	35	3.5	41
5	2	12	2	13	2	17	3	25	4	32	4.5	37
7½	2.5	11	2.5	12	3	15	4	22	5.5	30	6	34
10	3	10	3	11	3.5	14	5	21	6.5	27	7.5	31
15	4	9	4	10	5	13	6.5	18	8	23	9.5	27
20	5	9	5	10	6.5	12	7.5	16	9	21	12	25
25	6	9	6	10	7.5	11	9	15	11	20	14	23
30	7	8	7	9	9	11	10	14	12	18	16	22
40	9	8	9	9	11	10	12	13	15	16	20	20
50	12	8	11	9	13	10	15	12	19	15	24	19
60	14	8	14	8	15	10	18	11	22	15	27	19
75	17	8	16	8	18	10	21	10	26	14	32.5	18
100	22	8	21	8	25	9	27	10	32.5	13	40	17
125	27	8	26	8	30	9	32.5	10	40	13	47.5	16
150	32.5	8	30	8	35	9	37.5	10	47.5	12	52.5	15
200	40	8	37.5	8	42.5	9	47.5	10	60	12	65	14
250	50	8	45	7	52.5	8	57.5	9	70	11	77.5	13
300	57.5	8	52.5	7	60	8	65	9	80	11	87.5	12
350	65	8	60	7	67.5	8	75	9	87.5	10	95	11
400	70	8	65	6	75	8	85	9	95	10	105	11
450	75	8	67.5	6	80	8	92.5	9	100	9	110	11
500	77.5	8	72.5	6	82.5	8	97.5	9	107.5	9	115	10

Figure 5-13: Motor power-factor correction table.

Chapter 6

Overcurrent Protection

Electrical distribution systems are often quite complicated. They cannot be absolutely fail-safe. Circuits are subject to destructive overcurrents. Harsh environments, general deterioration, accidental damage or damage from natural causes, excessive expansion or overloading of the electrical distribution system are factors which contribute to the occurrence of such overcurrents. Reliable protective devices prevent or minimize costly damage to transformers, conductors, motors, and the many other components and loads that make up the complete distribution system. Reliable circuit protection is essential to avoid the severe monetary losses which can result from power blackouts and prolonged downtime of facilities. It is the need for reliable protection, safety, and freedom from fire hazards that has made overcurrent protective devices absolutely necessary in all electrical systems — both large and small.

Overcurrent protection of electrical circuits is so important that the *National Electrical Code* devotes an entire Article to this subject. *NEC* Article 240 — Overcurrent Protection — provides the general requirements for overcurrent protection and overcurrent protective devices; that is, *NEC* Article 240, Parts A through G cover systems 600 *V*, nominal and under, while Part H covers overcurrent protection over 600 *V*, nominal. This entire *NEC* Article will be thoroughly covered in this chapter with practical examples.

All conductors must be protected against overcurrents in accordance with their ampacities as set forth in *NEC* Section 240-3. They must also be protected against short-circuit current damage as required by *NEC* Sections

110-10 and 240-1. Two basic types of overcurrent protective devices that are in common use include:

- Fuses
- Circuit breakers

Overcurrents

An overcurrent is either an overload current or a short-circuit current. The overload current is an excessive current relative to normal operating current but one which is confined to the normal conductive paths provided by the conductor and other components and loads of the distribution system. As the name implies, a short-circuit current is one which flows outside the normal conducting paths.

Overloads

Overloads are most often between one and six times the normal current level. Usually, they are caused by harmless temporary surge currents that occur when motors are started-up or transformers are energized. Such overload currents or transients are normal occurrences. Since they are of brief duration, any temperature rise is trivial and has no harmful effect on the circuit components.

Continuous overloads can result from defective motors (such as worn motor bearings), overloaded equipment, or too many loads on one circuit. Such sustained overloads are destructive and must be cut-off by protective devices before they damage the distribution system or system loads. However, since they are of relatively low magnitude compared to short-circuit currents, removal of the overload current within a few seconds will generally prevent equipment damage. A sustained overload current results in overheating of conductors and other components and will cause deterioration of insulation which may eventually result in severe damage and short circuits if not interrupted.

Short Circuits

The ampere interrupting capacity (AIC) rating of a circuit breaker or fuse is the maximum short-circuit current which the breaker will interrupt safely. This AIC rating is at rated voltage and frequency.

Whereas overload currents occur at rather modest levels, the short-circuit or fault current can be many hundreds of times larger than the normal operating current. A high level fault may be 50,000 A (or larger). If not cut off within a matter of a few thousands of a second, damage and destruction can become rampant — there can be severe insulation damage, melting of conductors, vaporization of metal, ionization of gases, arcing, and fires. Simultaneously, high-level short-circuit currents can develop huge magnetic-field stresses. The magnetic forces between bus bars and other conductors can be many hundreds of pounds per lineal foot; even heavy bracing may not be adequate to keep them from being warped or distorted beyond repair.

NEC Section 110-9 clearly states:

> *". . . equipment intended to break current at fault levels (fuses and circuit breakers) must have an interrupting rating sufficient for the nominal circuit* voltage *and the current that is available at the line terminals of the equipment."*

Equipment intended to break current at other than fault levels must have an interrupting rating at nominal circuit voltage sufficient for the current that must be interrupted.

These *NEC* statements mean that fuses and circuit breakers (and their related components) designed to break fault or operating currents (open the circuit) must have a rating sufficient to withstand such currents. This *NEC* section emphasizes the difference between clearing fault level currents and clearing operating currents. Protective devices such as fuses and circuit breakers are designed to clear fault currents and therefore must have short-circuit interrupting ratings sufficient for fault levels. Equipment such as contactors and safety switches have interrupting ratings for currents at other than fault levels. Thus, the interrupting rating of electrical equipment is now divided into two parts:

- Current at fault (short-circuit) levels.
- Current at operating levels.

Most people are familiar with the normal current-carrying A rating of fuses and circuit breakers. For example, if an overcurrent protective device is designed to open a circuit when the circuit load exceeds 20 A for a given time period, as the current approaches 20 A, the overcurrent protective device begins to overheat. If the current barely exceeds 20 A, the circuit breaker will open normally or a fuse link will melt after a given period of time with little, if any, arcing. If, say, 40 A of current were instantaneously applied to the circuit, the overcurrent protective device will open instantly, but again with very little arcing. However, if a ground fault occurs on the circuit that ran the amperage up to, say, 5000 A, an explosion effect would occur within the protective device. One simple indication of this "explosion effect" is blackened windows of plug fuses.

If this fault current exceeds the interrupting rating of a fuse or circuit breaker, the protective device can be damaged or destroyed; such current can also cause severe damage to equipment and injure personnel. Therefore, selecting overcurrent protective devices with the proper interrupting capacity is extremely important in all electrical systems.

There are several factors that must be considered when calculating the required interrupting capacity of an overcurrent protective device. *NEC* Section 110-10 states the following:

> *". . . The overcurrent protective devices, the total impedance, the component short-circuit withstand ratings, and other characteristics of the circuit to be protected shall be so selected and coordinated as to permit the circuit protective devices that are used to clear a fault without the occurrence of extensive damage to the electrical components of the circuit. This fault shall be assumed to be either between two or more of the circuit conductors, or between any circuit conductor or the grounding conductor or enclosing metal raceway."*

Component short-circuit rating is a current rating given to conductors, switches, circuit breakers, and other electrical components, which, if exceeded by fault currents, will result in "extensive" damage to the component. The rating is expressed in terms of time intervals and/or current values. Short-circuit damage can be the result of heat generated or the electro-mechanical force of high-intensity, magnetic field.

The *NEC*'s intent is that the design of a system must be such that short-circuit currents cannot exceed the withstand ratings of the components selected as part of the system. Given specific system components, and level of "available" short-circuit currents that could occur, overcurrent protective devices (mainly fuses and/or circuit breakers) must be used which will limit the energy let-through of fault currents to levels within the withstand ratings of the system components.

FUSEOLOGY

The fuse is a reliable overcurrent protective device. A "fusible" link or links encapsulated in a tube and connected to contact terminals comprise the fundamental elements of the basic fuse. Electrical resistance of the link is so low that it simply acts as a conductor. However, when destructive currents occur, the link very quickly melts and opens the circuit to protect conductors and other circuit components and loads. Fuse characteristics are stable. Fuses do not require periodic maintenance or testing. Fuses have three unique performance characteristics.

Voltage Rating

Most low-voltage power distribution fuses have 250-V or 600-V ratings. The voltage rating of a fuse must be at least equal to or greater than the circuit voltage. It can be higher, but never lower. For example, a 600-V fuse can be used in a 240-V circuit.

The voltage rating of a fuse is a function of or depends upon its capability to open a circuit under an overcurrent condition. Specifically, the voltage rating determines the ability of the fuse to suppress the internal arcing that occurs after a fuse link melts and an arc is produced. If a fuse is used with a voltage rating lower than the circuit voltage, arc suppression will be impaired and, under some fault-current conditions, the fuse may not safely clear the overcurrent. Special consideration is necessary for semiconductor fuse application, where a fuse of a certain voltage rating is used on a lower-voltage circuit.

Ampere Rating

Every fuse has a specific ampere rating. In selecting the ampere rating of a fuse, consideration must be given to the type of load and *NEC* requirements. The ampere rating of a fuse should normally not exceed current-carrying capacity of the circuit. For instance, if a conductor is rated to carry 20 A, a 20-A fuse is the largest that should be used. However, there are specific circumstances where the A rating is permitted to be greater than the current-carrying capacity of the circuit. A typical example is the motor circuit; dual-element fuses generally are permitted to be sized up to 175 percent and non-time delay fuses up to 300 percent of the motor full-load amperes. Generally, the ampere rating of a fuse and switch combination should be selected at 125 percent of the continuous load current (this usually corresponds to the circuit capacity which is also selected at 125 percent of the load current). There are exceptions such as when the fuse-switch combination is approved for continuous operation at 100 percent of its rating.

Interrupting Rating-Safe Operation

A protective device must be able to withstand the destructive energy of short-circuit currents. If a fault current exceeds a level beyond the capability of the protective device, the device may actually rupture, causing additional damage. Therefore, it is important in applying a fuse or circuit breaker to use one which can sustain the largest potential short-circuit currents. The rating which defines the capacity of a protective device to maintain its integrity when reacting to fault currents is termed its "interrupting rating."

NEC Section 110-9 requires equipment intended to break current at fault levels to have an interrupting rating sufficient for the current that must be interrupted.

Selective Coordination

The coordination of protective devices prevents system power outages or blackouts caused by overcurrent conditions. When only the protective device nearest a faulted circuit opens and larger upstream fuses remain closed, the protective devices are "selectively" coordinated (they discriminate). The word "selective" is used to denote total coordination . . . isolation of a faulted circuit by the opening of only the localized protective device.

The diagram in Figure 6-1 shows the minimum ratios of ampere ratings of low-peak fuses that are required to provide "selective coordination" of upstream and downstream fuses.

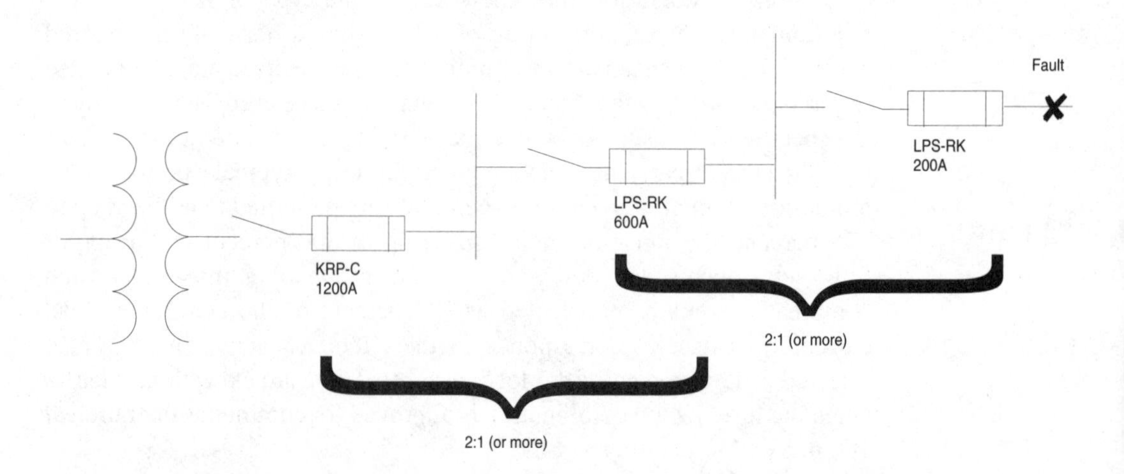

Figure 6-1: Minimum ratios of ampere ratings of fuses that will provide selective coordination.

Current Limitation

If a protective device cuts off a short-circuit current in less than one-half cycle, before it reaches its total available (and highly destructive) value, the device is a "current limiting" device. Most modern fuses are current limiting. They restrict fault currents to such low values that a high degree of protection is given to circuit components against even very high short-circuit currents. They permit breakers with lower interrupting ratings to be used. They can reduce bracing of bus structures. They minimize the need of other components to have high short-circuit current "withstand" ratings. If not limited, short-circuit currents can reach levels of 30,000 or 40,000 A or higher in the first half cycle (.008 seconds at 60 Hz) after the start of a short circuit. The heat that can be produced in circuit components by the immense energy of short-circuit currents can cause severe insulation damage or even explosion. At the same time, huge magnetic forces developed between conductors can crack insulators and distort and destroy bracing structures. Thus, it is important that a protective device limit fault currents before they can reach their full potential level.

A noncurrent-limiting protective device, by permitting a short-circuit current to build up to its full value, can let an immense amount of destructive short-circuit heat energy through before opening the circuit as shown in Figure 6-2 on the next page. On the other hand, a current-limiting fuse has such a high speed of response that it cuts off a short circuit long before it can build up to its full peak value as shown in Figure 6-3.

OPERATING PRINCIPLES OF FUSES

There are several different types of fuses, and although all operate in a similar fashion, all have slightly different characteristics. Each is described in the paragraphs that follow.

Non-Time-Delay Fuses

The basic component of a fuse is the link. Depending upon the A rating of the fuse, the single-element fuse may have one or more links. They are electrically connected to the end blades (or *ferrules*) and enclosed in a tube or cartridge surrounded by an arc quenching filler material.

Under normal operation, when the fuse is operating at or near its A rating, it simply functions as a conductor. However, as illustrated in Figure 6-4, if an overload current occurs and persists for more than a short interval of time, the temperature of the link eventually reaches a level which causes a restricted segment of the link to melt; as a result, a gap is formed and an

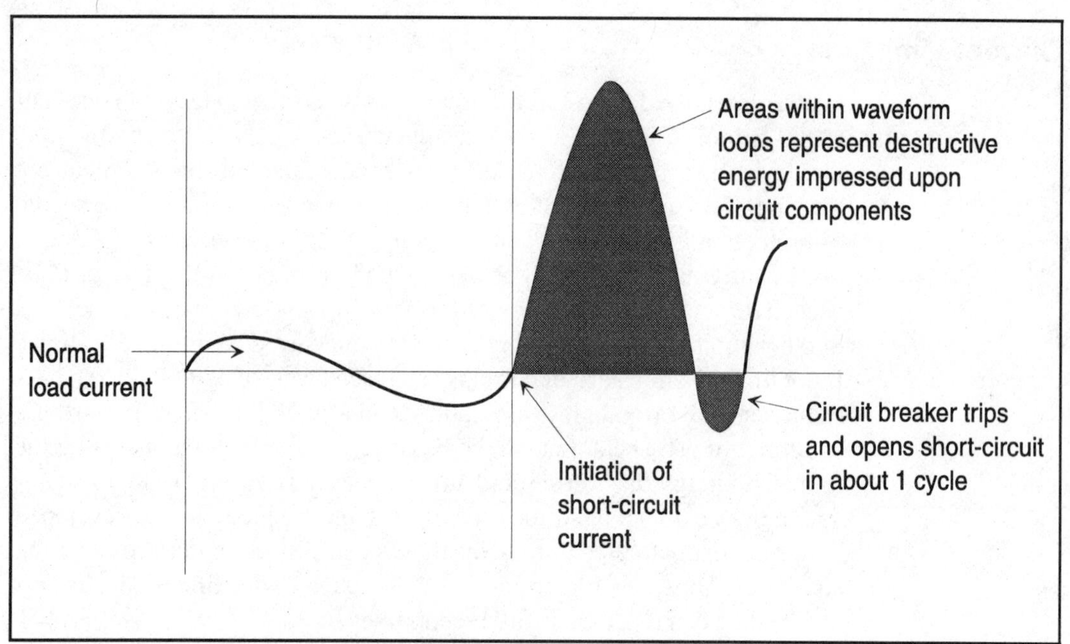

Figure 6-2: Characteristics of non-current-limiting protective device.

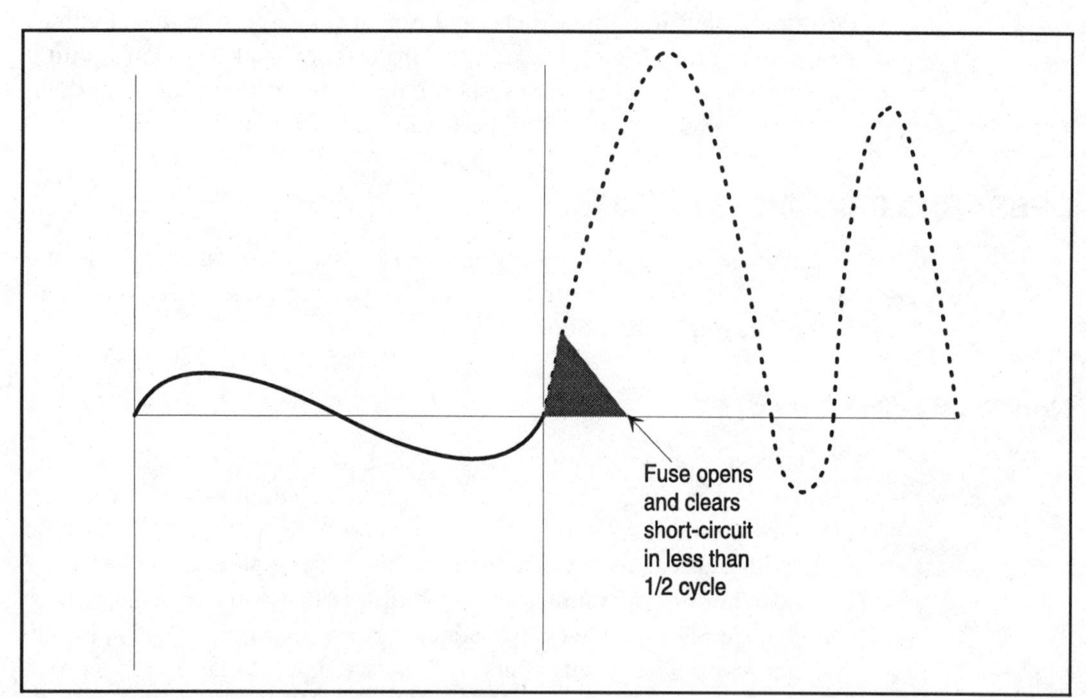

Figure 6-3: Characteristics of current-limiting fuse.

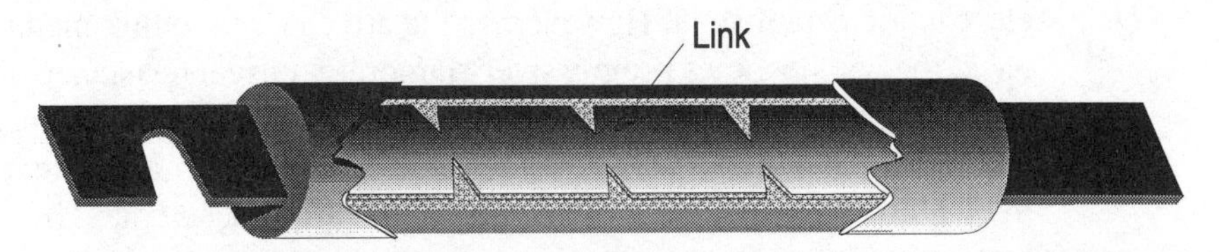

Cut-away view of single-element fuse.

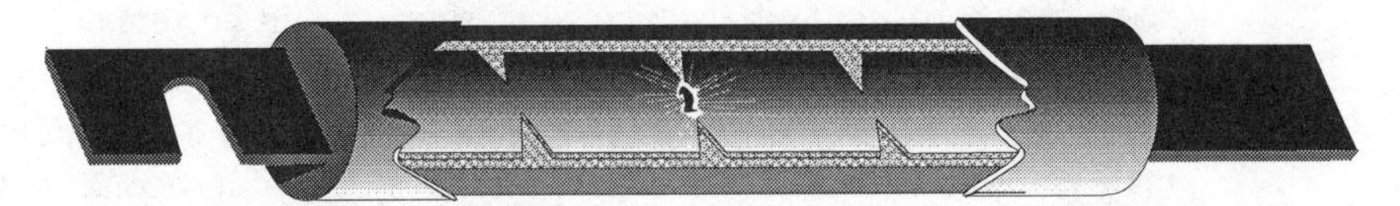

Under sustained overload a section of the link melts and an arc is established.

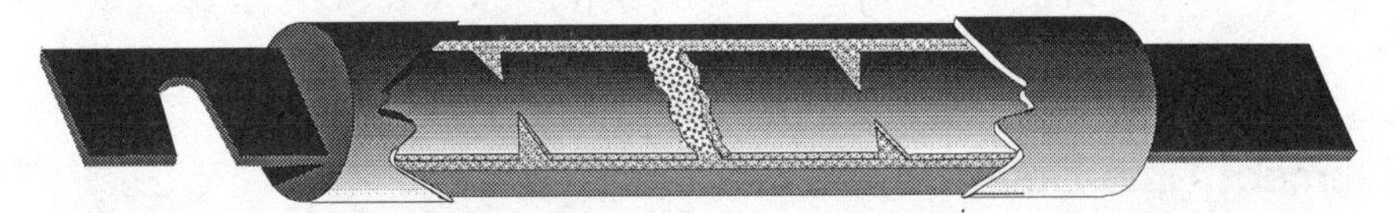

The "open" single-element fuse after opening a circuit overload.

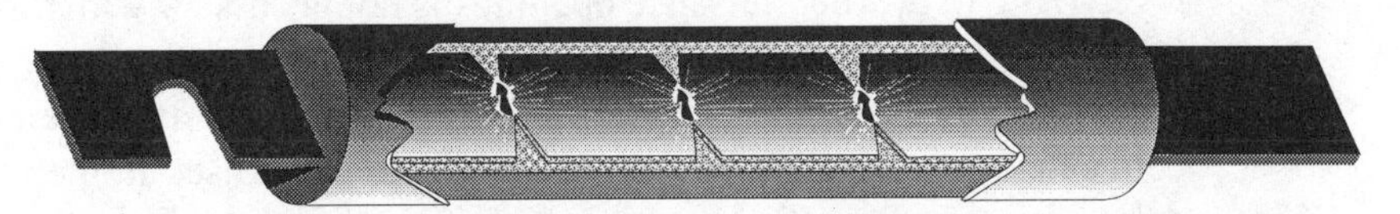

When subjected to a short-circuit, several sections of the fuse link melt almost instantly.

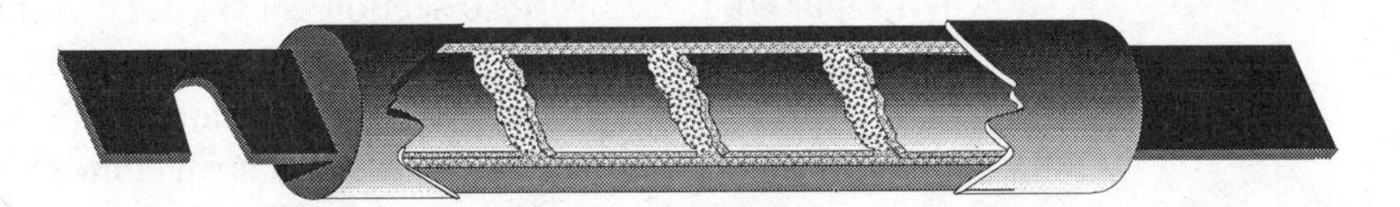

The appearance of an "open" single-element fuse after opening a short-circuit.

Figure 6-4: Characteristics of a single-element fuse.

electric arc established. However, as the arc causes the link metal to burn back, the gap becomes progressively larger. Electrical resistance of the arc eventually reaches such a high level that the arc cannot be sustained and is extinguished; the fuse will have then completely cut off all current flow in the circuit. Suppression or quenching of the arc is accelerated by the filler material.

Overload current normally falls within the region of between one and six times normal current — resulting in currents that are quite high. Consequently, a fuse may be subjected to short-circuit currents of 30,000 or 40,000 A or higher. Response of current-limiting fuses to such currents is extremely fast. The restricted sections of the fuse link will simultaneously melt within a matter of two or three-thousandths of a second in the event of a high-level fault current.

The high resistance of the multiple arcs, together with the quenching effects of the filler particles, results in rapid arc suppression and clearing of the circuit. Again, refer to Figure 6-4. Short-circuit current is cut-off in less than a half-cycle — long before the short-circuit current can reach its full value.

Dual-Element Time-Delay Fuses

Unlike single-element fuses, the dual-element time-delay fuse can be applied in circuits subject to temporary motor overloads and surge currents to provide both high performance short-circuit and overload protection. Oversizing to prevent nuisance openings is not necessary with this type of fuse. The dual-element time-delay fuse contains two distinctly separate types of elements. *See* Figure 6-5. Electrically, the two elements are connected in series. The fuse links similar to those used in the non-time-delay fuse perform the short-circuit protection function; the overload element provides protection against low-level overcurrents or overloads and will hold an overload which is five times greater than the ampere rating of the fuse for a minimum time of 10 seconds.

As shown in Figure 6-5, the overload section consists of a copper heat absorber and a spring-operated trigger assembly. The heat absorber bar is permanently connected to the heat absorber extension and to the short-circuit link on the opposite end of the fuse by the S-shaped connector of the trigger assembly. The connector electrically joins the short-circuit link to the heat absorber in the overload section of the fuse. These elements are joined by a "calibrated" fusing alloy. An overload current causes heating of the short-circuit link connected to the trigger assembly. Transfer of heat from the short-circuit link to the heat-absorbing bar in the mid-section of

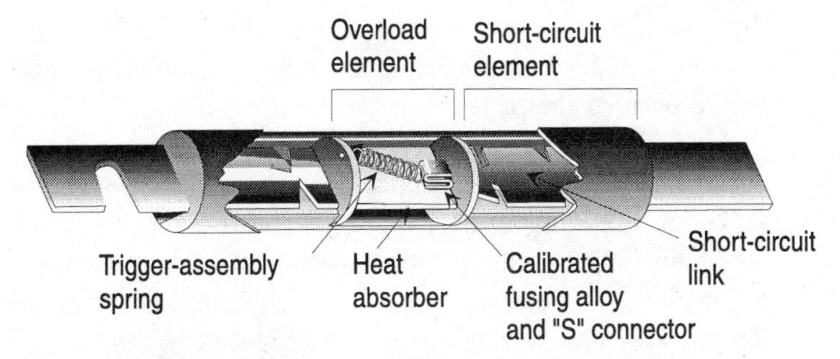

The true dual-element fuse has distinct and separate overload and short-circuit elements.

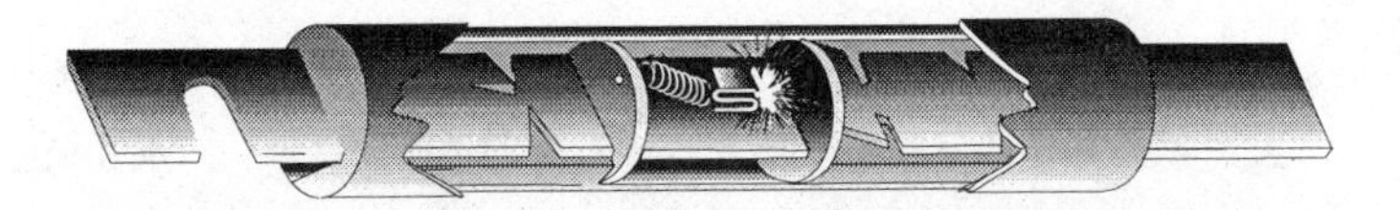

Under sustained overload conditions, the trigger spring fractures the calibrated fusing alloy and releases the "connector."

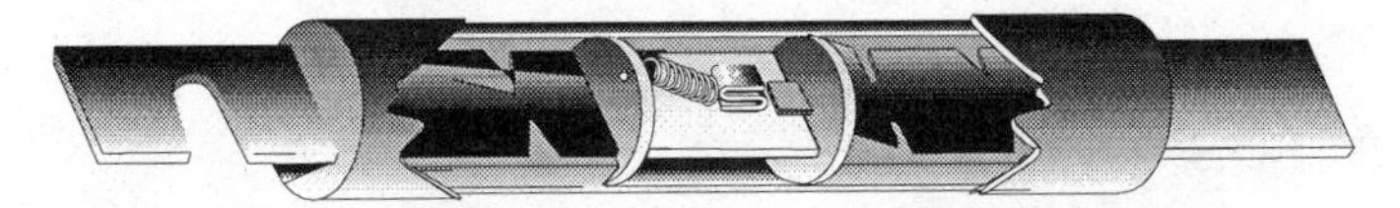

The "open" dual-element fuse after opening under an overload

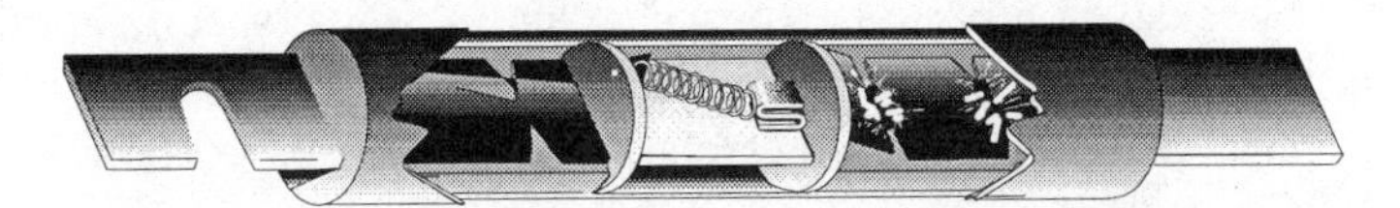

Like the single-element fuse, a short-circuit current causes the restricted portions of the short-circuit elements to melt and arcing to burn back the resulting gaps until the arcs are suppressed by the arc-quenching material and increased are resistance.

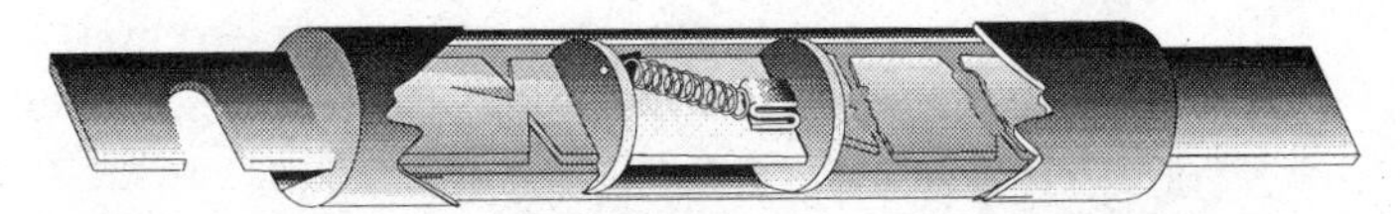

The "open" dual-element fuse after opening under a short-circuit condition.

Figure 6-5: Characteristics of dual-element, time-delay fuses.

the fuse begins to raise the temperature of the heat absorber. If the overload is sustained, the temperature of the heat absorber eventually reaches a level which permits the trigger spring to "fracture" the calibrated fusing alloy and pull the connector free of the short-circuit link and the heat absorber. As a result, the short-circuit link is electrically disconnected from the heat absorber, the conducting path through the fuse is opened, and overload current is interrupted. A critical aspect of the fusing alloy is that it retains its original characteristic after repeated temporary overloads with degradation. The main purpose of dual-element fuses are as follows:

- Provide motor overload, ground-fault and short-circuit protection.
- Permit the use of smaller and less costly switches.
- Give a higher degree of short-circuit protection (greater current limitation) in circuits in which surge currents or temporary overloads occur.
- Simplify and improve blackout prevention (selective coordination).

U.L. FUSE CLASSES

Safety is the U.L. mandate. However, proper selection, overall functional performance and reliability of a product are factors that are not within the basic scope of U.L. activities. To develop its safety test procedures, U.L. does develop basic performance and physical specifications of standards of a product. In the case of fuses, these standards have culminated in the establishment of distinct classes of low-voltage (600 V or less) fuses, Classes FK 1, RK 5, G, L, T, J. H, and CC being the more important.

Class R fuses: U.L. Class R (rejection) fuses are high performance $^1/_{10}$ to 600 A units, 250 V and 600 V, having a high degree of current limitation and a short-circuit interrupting rating of up to 200,000 A (rms symmetrical). This type of fuse is designed to be mounted in rejection type fuseclips to prevent older type Class H fuses from being installed. Since Class H fuses are not current limiting and are recognized by U.L. as having only a 10,000 A interrupting rating, serious damage could result if a Class H fuse were inserted in a system designed for Class R fuses. Consequently, *NEC* Section 240-60(b) requires fuseholders for current-limiting fuses to reject non-current-limiting type fuses.

Figure 6-6 shows a standard Class H fuse (left) and a Class R fuse (right). A grooved ring in one ferrule of the Class R fuse provides the rejection

Figure 6-6: Comparison of Class H and Class R fuses.

feature of the Class R fuse in contrast to the lower interrupting capacity, nonrejection type. Figure 6-7 shows Class R type fuse rejection clips that accept only the Class R rejection type fuses.

Class CC fuses: 600-V, 200,000-A interrupting rating, branch circuit fuses with overall dimensions of $^{15}/_{32}''$ × $1^{1}/_{2}''$. Their design incorporates rejection features that allow them to be inserted into rejection fuse holders and fuse blocks that reject all lower *V*age, lower interrupting rating $^{15}/_{32}'' \times 1^{1}/_{2}''$ fuses. They are available from $^{1}/_{10}$ A through 30 A.

Class G fuses: 300-V, 100,000-A interrupting rating branch circuit fuses that are size rejecting to eliminate overfusing. The fuse diameter is $^{13}/_{32}''$ while the length varies from $1^{5}/_{16}''$ to $2^{1}/_{4}''$. These are available in ratings from 1 amp through 60 amps.

Class H fuses: 250-V and 600-V, 10,000-A interrupting rating branch circuit fuses that may be renewable or non-renewable. These are available in A ratings of 1 A through 600 A.

Class J fuses: These fuses are rated to interrupt 200,000 A ac. They are U.L. labeled as "current limiting", rated for 600 V ac, and are not interchangeable with other classes.

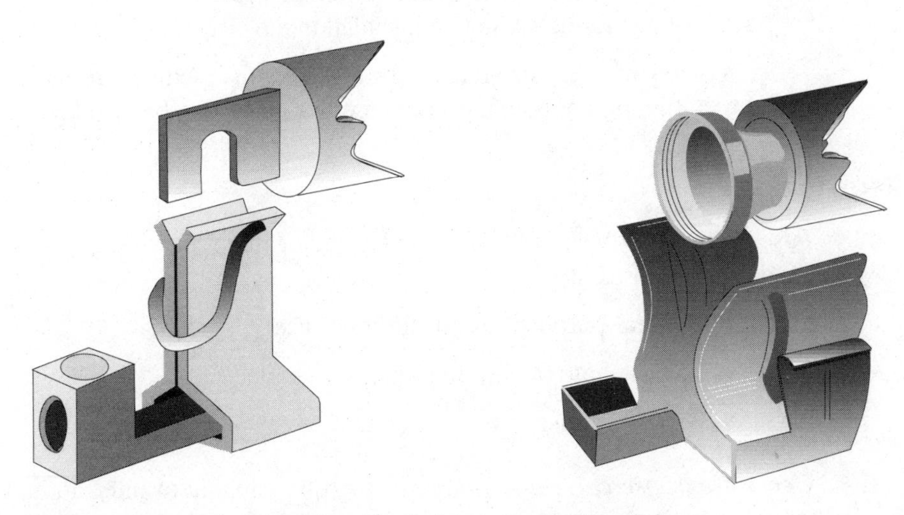

Figure 6-7: Class R fuse rejection clips that accept only Class R fuses.

Class K fuses: These are fuses listed by U.L. as K-1, K-5, or K-9 fuses. Each subclass has designated I^2t and Ip maximums. These are dimensionally the same as Class H fuses, (*NEC* dimensions) and they can have interrupting ratings of 50,000, 100,000, or 200,000 A. These fuses are current limiting, however, they are not marked "current limiting" on their label since they do not have a rejection feature.

Class L fuses: These fuses are rated for 601 through 6000 A, and are rated to interrupt 200,000 A ac. They are labeled "current limiting" and are rated for 600 V ac. They are intended to be bolted into their mountings and are not normally used in clips. Some Class L fuses have designed in time-delay features for all-purpose use.

Class T fuses: A U.L. classification of fuses in 300 V and 600 V ratings from 1 A through 1200 A. They are physically very small and can be applied where space is at a premium. They are fast acting fuses, with an interrupting rating of 200,000 amps RMS.

Branch-Circuit Listed Fuses

Branch-circuit listed fuses are designed to prevent the installation of fuses that cannot provide a comparable level of protection to equipment. The characteristics of branch-circuit fuses are as follows:

- They just have a minimum interrupting rating of 10,000 A.
- They must have a minimum voltage rating of 125 V.
- They must be size rejecting such that a fuse of a lower voltage rating cannot be installed in the circuit.
- They must be size rejecting such that a fuse with a current rating higher than the fuseholder rating cannot be installed.

Medium-Voltage Fuses

Fuses above 600 V are classified under one of three classifications as defined in ANSI/IEEE 40-1981:

- General-purpose current-limiting fuse
- Back-up current-limiting fuse
- Expulsion fuse

General-purpose current-limiting fuse: A fuse capable of interrupting all currents from the rated interrupting current down to the current that causes melting of the fusible element in one hour.

Back-up current-limiting fuse: A fuse capable of interrupting all currents from the maximum rated interrupting current down to the rated minimum interrupting current.

Expulsion fuse: A vented fuse in which the expulsion effect of gasses produced by the arc and lining of the fuseholder, either alone or aided by a spring, extinguishes the arc.

One should note that in the definitions just given, the fuses are defined as either expulsion or current limiting. A current-limiting fuse is a sealed, non-venting fuse that, when melted by a current within its interrupting rating, produces arc *V*ages exceeding the system *V*age which in turn forces the current to zero. The arc *V*ages are produced by introducing a series of high resistance arcs within the fuse. The result is a fuse that typically interrupts high fault currents with the first half-cycle of the fault.

In contrast an expulsion fuse depends on one arc to initiate the interruption process. The arc acts as a catalyst causing the generation of de-ionizing gas from its housing. The arc is then elongated either by the force of the gasses created or a spring. At some point the arc elongates far enough to prevent a restrike after passing through a current zero. Therefore, it is not atypical for an expulsion fuse to take many cycles to clear.

Application of Medium-Voltage Fuses

Many of the rules for applying expulsion fuses and current-limiting fuses are the same, but because the current-limiting fuse operates much faster on highfault, some additional rules must be applied.

Three basic factors must be considered when applying any fuse:

- Voltage
- Continuous current-carrying capacity
- Interrupting rating

Voltage: The fuse must have a voltage rating equal to or greater than the normal frequency recovery voltage which will be seen across the fuse under all conditions. On three-phase systems it is a good rule-of-thumb that the voltage rating of the fuse be greater than or equal to the line-to-line voltage of the system.

Continuous current-carrying capacity: Continuous current values that are shown on the fuse represent the level of current the fuse can carry continuously without exceeding the temperature rises as specified in ANSI C37.46. An application that exposes the fuse to a current slightly above its continuous rating but below its minimum interrupting rating may damage the fuse due to excessive heat. This is the main reason overload relays are used in series with back-up current-limiting fuses for motor protection.

Interrupting rating: All fuses are given a maximum interrupting rating. This rating is the maximum level of fault current that the fuse can safely interrupt. Back-up current-limiting fuses are also given a minimum interrupting rating. When using back-up current-limiting fuses, it is important that other protective devices are used to interrupt currents below this level.

When choosing a fuse, it is important that the fuse be properly coordinated with other protective devices located upstream and downstream. To accomplish this, one must consider the melting and clearing characteristics of the devices. Two curves, the minimum melting curve and the total clearing curve, provide this information. To insure proper coordination, the following rules should be used:

- The total clearing curve of any downstream protective device must be below a curve representing 75 percent of the minimum melting curve of the fuse being applied.
- The total clearing curve, of the fuse being applied, must lie below a curve representing 75 percent of the minimum melting curve for any upstream protective device.

Current-Limiting Fuses

To insure proper application of a current-limiting fuse, it is important that the following additional rules be applied.

1. Current-limiting fuses produce arc voltages that exceed the system voltage. Care must be taken to make sure that the peak voltages do not exceed the insulation level of the system. If the fuse voltage rating is not permitted to exceed 140 percent of the system voltage, there should not be a problem. This does not mean that a higher rated fuse cannot be used, but points out that one must be assured that the system insulation level (BIL) will handle the peak arc voltage produced. BIL stands for basic impulse level which is the reference impulse insulation strength of an electrical system.

2. As with the expulsion fuse, current-limiting fuses must be properly coordinated with other protective devices on the system. For this to happen, the rules for applying an expulsion fuse must be used at all currents that cause the fuse to interrupt in 0.01 seconds or greater.

When other current-limiting protective devices are on the system, it becomes necessary to use I^2t values for coordination at currents causing the fuse to interrupt in less than 0.01 seconds. These values may be supplied as minimum and maximum values or minimum melting and total clearing I^2t curves. In either case, the following rules should be followed.

1. The minimum melting I^2t of the fuse should be greater than the total clearing I^2t of the downstream current-limiting device.

2. The total clearing I^2t of the fuse should be less than the minimum melting I^2t of the upstream current-limiting device.

Valuable information may be found in catalogs furnished by manufacturers of overcurrent protective devices. These are usually obtainable from electrical supply houses or from manufacturers' reps. You may also write the various manufacturers for a complete list (and price, if any) for all reference materials offered by them.

Fuses for Selective Coordination

The larger the upstream fuse is relative to a downstream fuse (feeder to branch, etc.), the less possibility there is of an overcurrent in the downstream circuit causing both fuses to open. Fast action, non-time-delay fuses require at least a 3:1 ratio between the ampere rating of a large upstream, line-side time-delay fuse to that of the downstream, load-side fuse in order to be selectively coordinated. In contrast, the minimum selective coordination ratio necessary for dual-element fuses is only 2:1 when used with low peak load-side fuses. *See* Figure 6-8.

The use of time-delay, dual-element fuses affords easy selective coordination — coordination hardly requires anything more than a routine check of a tabulation of required selective ratios. As shown in Figure 6-9, close sizing of dual-element fuses in the branch circuit for motor overload protection provides a large difference (ratio) in the ampere ratings between the feeder fuse and the branch fuse compared to the single-element, non-time-delay fuse.

CIRCUIT BREAKERS

Basically, a circuit breaker is a device for closing and interrupting a circuit between separable contacts under both normal and abnormal conditions. This is done manually (normal condition) by use of its "handle"

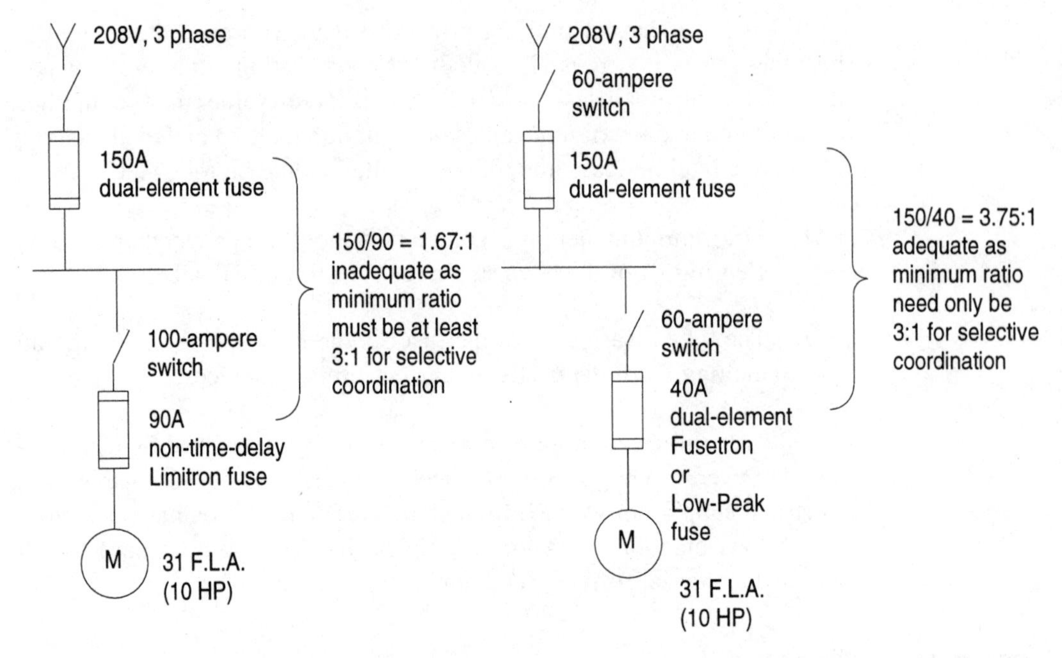

Figure 6-8: Fuses used for selective coordination.

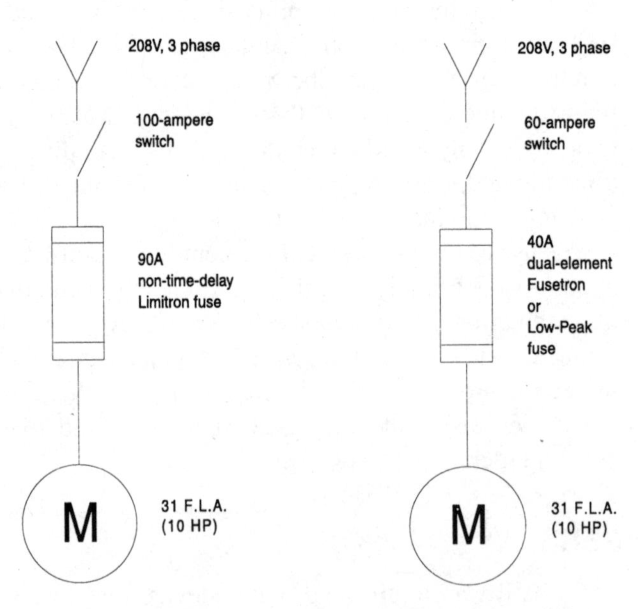

Figure 6-9: Dual-element fuses permit the use of smaller and less costly switches.

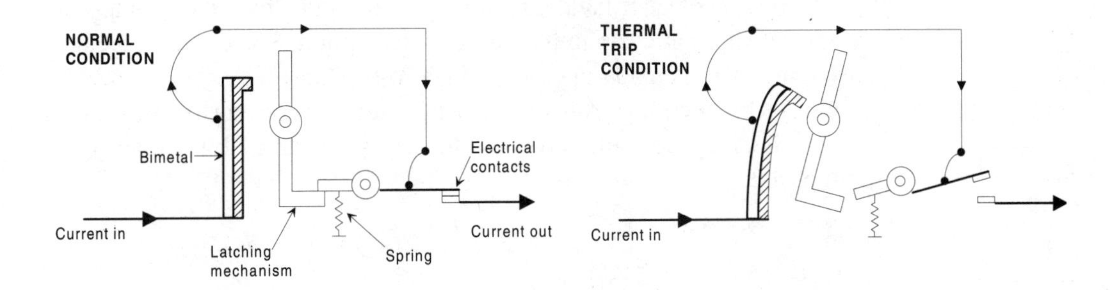

Figure 6-10: Characteristics of a thermal-trip circuit breaker.

by switching to the ON or OFF positions. However, the circuit breaker also is designed to open a circuit automatically on a predetermined overload or ground-fault current without damage to itself or its associated equipment. As long as a circuit breaker is applied within its rating, it will automatically interrupt any "fault" and therefore must be classified as an inherently safe, overcurrent, protective device.

The internal arrangement of a circuit breaker is shown in Figure 6-10 while its external operating characteristics are shown in Figure 6-11. Note that the handle on a circuit breaker resembles an ordinary toggle switch. On an overload, the circuit breaker opens itself or *trips.* In a tripped position, the handle jumps to the middle position (Figure 6-11). To reset, turn the handle to the OFF position and then turn it as far as it will go beyond this position; finally, turn it to the ON position.

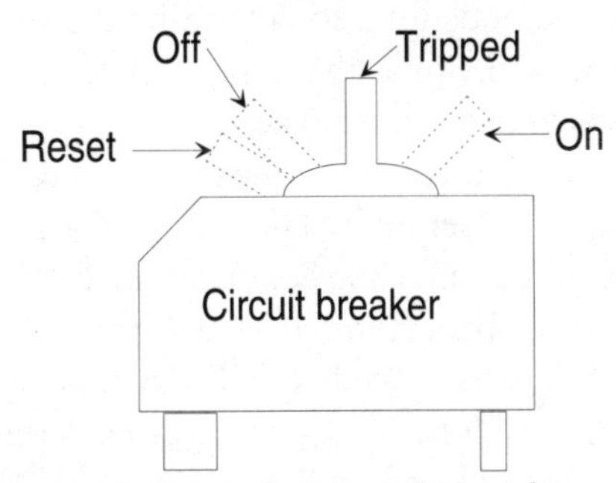

Figure 6-11: External characteristics of a circuit breaker.

A standard molded case circuit breaker usually contains:

- A set of contacts
- A magnetic trip element
- A thermal trip element
- Line and load terminals
- Bussing used to connect these individual parts
- An enclosing housing of insulating material

The circuit-breaker handle manually opens and closes the contacts and resets the automatic trip units after an interruption. Some circuit breakers also contain a manually operated "push-to-trip" testing mechanism.

Circuit breakers are grouped for identification according to given current ranges. Each group is classified by the largest A rating of its range. These groups are:

- 15-100 A
- 125-225 A
- 250-400 A
- 500-1000 A
- 1200-2000 A

Therefore, they are classified as 100-, 225-, 400-, 1000- and 2000-A frames. These numbers are commonly referred to as "frame classification" or "frame sizes" and are terms applied to groups of molded case circuit breakers which are physically interchangeable with each other.

Interrupting Capacity Rating

In most large commercial and industrial installations, it is necessary to calculate available short-circuit currents at various points in a system to determine if the equipment meets the requirements of *NEC* Sections 110-9 and 110-10. There are a number of methods used to determine the short-circuit requirements in an electrical system. Some give approximate values; some require extensive computations and are quite exacting.

The breaker interrupting capacity is based on tests to which the breaker is subjected. There are two such tests; one set up by U.L. and the other by NEMA. The NEMA tests are self-certification while U.L. tests are certified by unbiased witnesses. U.L. tests have been limited to a maximum of 10,000 A in the past, so the emphasis was placed on NEMA tests with higher ratings. U.L. tests now include the NEMA tests plus other ratings. Consequently, the emphasis is now being placed on U.L. tests.

The interrupting capacity of a circuit breaker is based on its rated voltage. Where the circuit breaker can be used on more than one voltage, the interrupting capacity will be shown for each voltage level. For example, the LA type circuit breaker has 42,000 A, symmetrical interrupting capacity at 240 V, 30,000 A symmetrical at 480 V, and 22,000 A symmetrical at 600 V.

CONDUCTOR PROTECTION

All conductors are to be protected against overcurrents in accordance with their ampacities as set forth in *NEC* Section 240-3. They must also be protected against short-circuit current damage as required by *NEC* Sections 240-1 and 110-10.

A ratings of overcurrent-protective devices must not be greater than the ampacity of the conductor. There is, however, an exception. *NEC* Section 240-3 states that if such conductor rating does not correspond to a standard size overcurrent-protective device, the next larger size overcurrent-protective device may be used provided its rating does not exceed 800 A and when the conductor is not part of a multioutlet branch circuit supplying receptacles for cord-and-plug connected portable loads. When the ampacity of busway or cablebus does not correspond to a standard overcurrent-protective device, the next larger stand rating may be used even though this rating may be greater than 800 A (*NEC* Sections 364-10 and 365-5).

Standard fuse sizes stipulated in *NEC* Section 240-6 are: 1, 3, 6, 10, 15, 20, 25, 30, 35, 40, 45, 50, 60, 70, 80, 90, 100, 110, 125, 150, 175, 200, 225, 250, 300, 350, 400, 450, 500, 600, 700, 800, 1000, 1200, 1600, 2000, 2500, 3000, 4000, 5000, and 6000 A.

Note: *The small fuse A ratings of 1, 3, 6, and 10 have recently been added to the* NEC *to provide more effective short-circuit and ground-fault protection for motor circuits in accordance with Sections 430-40 and 430-52 and U.L. requirements for protecting the overload relays in controllers for very small motors.*

Protection of conductors under short-circuit conditions is accomplished by obtaining the maximum short-circuit current available at the supply end of the conductor, the short-circuit withstand rating of the conductor, and the short-circuit let-through characteristics of the overcurrent device.

When a non-current-limiting device is used for short-circuit protection, the conductor's short-circuit withstand rating must be properly selected based on the overcurrent protective device's ability to protect. *See* Figure 6-12 on the next page.

It is necessary to check the energy let-through of the overcurrent device under short-circuit conditions and select a wire size of sufficient short-circuit withstand ability.

In contrast, the use of a current-limiting fuse permits a fuse to be selected which limits short-circuit current to a level less than that of the conductor's

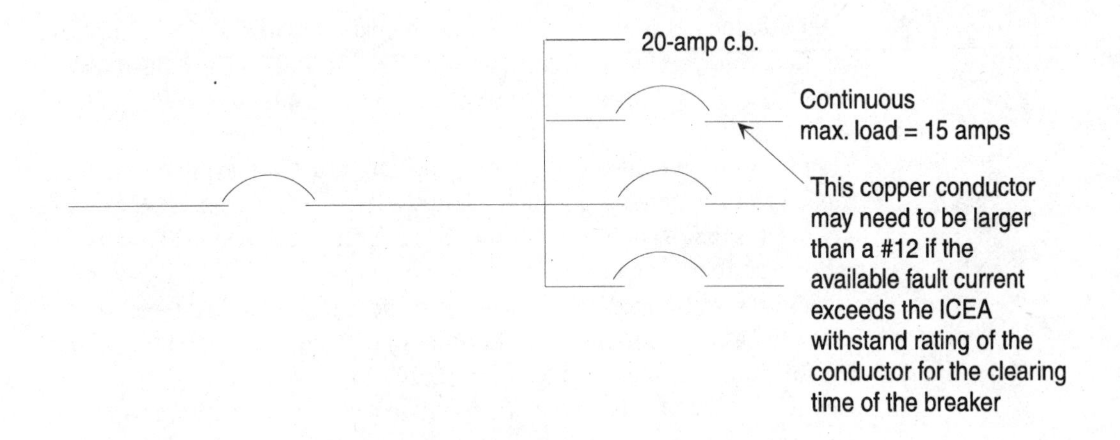

Figure 6-12: Conductor protection with noncurrent-limiting device.

short-circuit withstand rating — doing away with the need of oversized ampacity conductors. *See* Figue 6-13.

In many applications, it is desirable to use the convenience of a circuit breaker for a disconnecting means and general overcurrent protection, supplemented by current-limiting fuses at strategic points in the circuits.

Flexible cord, including tinsel cords and extension cords, must be protected against overcurrent in accordance with their ampacities.

Location of Protective Devices in Circuits

In general, overcurrent-protective devices must be installed at points where the conductors receive their supply; that is, at the beginning or line-side of a branch circuit or feeder. Exceptions to this rule are given in *NEC* Section 240-21.

Exception No. 1: Overcurrent-protective devices are not required at the conductor supply if the devices protecting one conductor are small enough to protect a small conductor connected thereto.

Exception No. 2: Overcurrent-protective devices are not required at the conductor supply if a feed tap conductor is not over ten feet long; is enclosed in raceway; does not extend beyond the switchboard, panelboard, or control device which it supplies, and has an ampacity not less than the combined computed loads supplied and not less than the rating of the device supplied unless the tap conductors are terminated in a fuse not exceeding

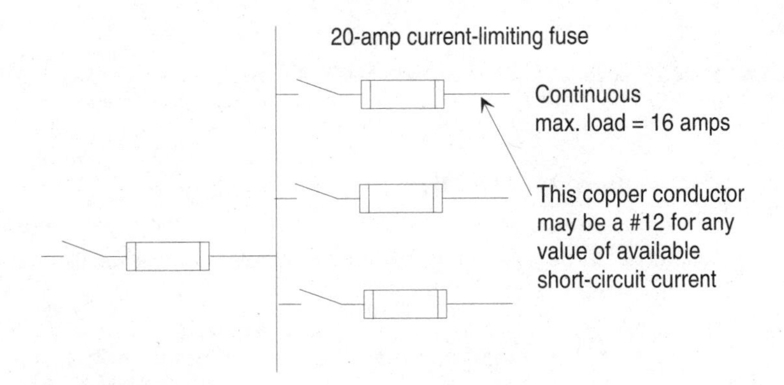

Figure 6-13: Circuits protected by current-limiting devices.

the tap conductors' ampacities. For field installed taps, the ampacity of the tap conductor must be at least 10 percent of the overcurrent device rating.

Exception No. 3: Overcurrent-protective devices are not required at the conductor supply if a feeder tap conductor is not over 25 ft long; is suitably protected from physical damage; has an ampacity not less than ⅓ that of the feeder conductors or fuses from which the tap conductors receive their supply; and terminate in a single set of fuses sized not more than the tap conductor ampacity.

Exception No. 8: Overcurrent-protective devices are not required at the conductor supply if a transformer feeder tap has primary conductors at least ⅓ ampacity and/or secondary conductors at least ⅓ ampacity when multiplied by the approximate transformer turns ratio of the fuse or conductors from which they are tapped; the total length of one primary plus one secondary conductor (excluding any portion of the primary conductor that is protected at its ampacity) is not over 25 ft in length; the secondary conductors terminate in a set of fuses rated at the ampacity of the tap conductors; and if the primary and secondary conductors are suitably protected from physical damage.

Exception No. 10: Overcurrent-protective devices are not required at the conductor supply if a feeder tap is not over 25 ft long horizontally and not over 100 ft total length in high bay manufacturing buildings when only qualified persons will service such a system, and the ampacity of the tap conductors is not less than ⅓ of the fuse rating from which they are supplied, that will limit the load to the ampacity of the tap are at least No. 6 AWG copper or No. 4 AWG aluminum, do not penetrate walls, floors, or ceilings, and are made no less than 30 ft from the floor.

WARNING!

Smaller conductors tapped to larger conductors can be a serious hazard. If not protected against short-circuit conditions, these unprotected conductors can vaporize or incur severe insulation damage.

Exception No. 11: Transformer secondary conductors of separately derived systems do not require overcurrent-protective devices at the transformer terminals when all of the following conditions are met:

- Must be an industrial location
- Secondary conductors must be less than 25 ft long
- Secondary conductor ampacity must be at least equal to secondary full-load current of transformer and sum of terminating, grouped, overcurrent devices
- Secondary conductors must be protected from physical damage

Note: *Switchboard and panelboard protection (*NEC *Section 384-16) and transformer protection (*NEC *Section 450-3) must still be observed.*

Lighting/Appliance Loads

The branch-circuit rating must be classified in accordance with the rating of the overcurrent protective device. Classifications for those branch circuits other than individual loads must be: 15, 20, 30, 40, and 50 A as specified in *NEC* Section 210-3.

Branch-circuit conductors must have an ampacity of the rating of the branch circuit and not less than the load to be served (*NEC* Section 210-19). The minimum size branch-circuit conductor that can be used is No. 14 (*NEC* Section 210-19). However, there are some exceptions as specified in *NEC* Section 210-19.

Branch-circuit conductors and equipment must be protected by a fuse whose A rating conforms to *NEC* Section 210-20. Basically, the branch circuit conductor and fuse must be sized for the actual noncontinuous load and 125 percent for all continuous loads. The fuse size must not be greater than the conductor ampacity. Branch circuits rated 15 through 50 A with two or more outlets (other than receptacle circuits) must be fused at their rating and the branch-circuit conductor sized according to *NEC* Table 210-24.

Feeder Circuits With No Motor Load

The feeder fuse ampere rating and feeder conductor ampacity must be at least 100 percent of the non continuous load plus, 125 percent of the continuous load as calculated per *NEC* Article 220. The feeder conductor must be protected by a fuse not greater than the conductor ampacity. Motor loads shall be computed in accordance with Article 430.

Service Equipment

Each ungrounded service-entrance conductor must have a fuse in series with a rating not higher than the ampacity of the conductor. There service fuses shall be part of the service-disconnecting means or be located immediately adjacent thereto (*NEC* Section 230-91).

Service disconnecting means can consist of one to six switches or circuit breakers for each service or for each set of service-entrance conductors permitted in *NEC* Section 230-2. When more than one switch is used, the switches must be grouped together (*NEC* Section 230-71).

Transformer Secondary Conductors

Field installations indicate nearly 50 percent of transformers installed do not have secondary protection. The *NEC* recommends that secondary conductors be protected from damage by the proper overcurrent-protective device. For example, the primary overcurrent device protecting a 3-wire transformer cannot offer protection of the secondary conductors. Also see *NEC* exception in Section 240-3 for 2-wire primary and secondary circuits.

Motor Circuit Protection

Motors and motor circuits have unique operating characteristics and circuit components. Therefore, these circuits must be dealt with differently from other types of loads. Generally, two levels of overcurrent protection are required for motor branch circuits:

- Overload protection — Motor running overload protection is intended to protect the system components and motor from damaging overload currents.

- Short-circuit protection (includes ground-fault protection) — Short-circuit protection is intended to protect the motor circuit components such as the conductors, switches, controllers, overload relays, motor, etc. against short-circuit currents or grounds. This level of protection is commonly referred to as motor branch-circuit protection applications. Dual-element fuses are designed to provide this protection provided they are sized correctly.

There are a variety of ways to protect a motor circuit — depending upon the user's objective. The A rating of a fuse selected for motor protection depends on whether the fuse is of the dual-element time-delay type or the non-time-delay type.

In general, non-time-delay fuses can be sized at 300 percent of the motor full-load current for ordinary motors so that the normal motor starting current does not affect the fuse. Dual-element, time-delay fuses are able to withstand normal motor-starting current and can be sized closer to the actual motor rating than can nontime-delay fuses.

Summary

Reliable overcurrent-protective devices prevent or minimize costly damage to transformers, conductors, motors, and the other many components and electrical loads that make up the complete electrical distribution system. Consequently, reliable circuit protection is essential to avoid the severe monetary losses which can result from power blackouts and prolonged downtime of various types of facilities. Knowing these facts, the NFPA — via the *NEC* —has set forth various minimum requirements dealing with overcurrent devices, and how they should be installed in various types of electrical circuits.

Knowing how to select and size the type of overcurrent devices for specific applications is one of the basic requirements of every electrical engineer, contractor, and worker.

Chapter 7

Grounding

The grounding system is a major part of the electrical system. Its purpose is to protect life and equipment against the various electrical faults that can occur. It is sometimes possible for higher-than-normal voltages to appear at certain points in an electrical system or in the electrical equipment connected to the system. Proper grounding ensures that the high electrical charges that cause these high voltages are channeled to earth or ground before damaging equipment or causing danger to human life. Therefore, circuits are grounded to limit the voltage on the circuit, improve overall operation of the electrical system, and the continuity of service. Grounding also provides the following:

- Rapid operation of fuses, circuit breakers, and circuit interrupters.
- Minimizes the magnitude and duration of step-and-touch potentials in substations.
- Decreases the duration and magnitude of lightning current effects on the electrical system and transfers fault current into the earth.
- Voltage surges higher than that for which the circuit is designed are transferred into the earth.
- Transfers fault current into the earth.
- Helps prevent low-voltage problems when a grounded (neutral) conductor opens up.

- Minimizes heat damage to the electrical system and related equipment.
- Increases the load capacity of the primary line of electrical equipment such as motor starters.
- Substantially lowers the neutral-to-earth voltages.

When we refer to *ground*, we are talking about ground potential or earth ground. If a conductor is connected to the earth or to some conducting body that serves in place of the earth, such as a driven ground rod (electrode) or cold-water pipe, the conductor is said to be *grounded*. The neutral conductor in a three- or four-wire service, for example, is intentionally grounded and therefore becomes a *grounded conductor*. However, a wire used to connect this neutral conductor to a grounding electrode or electrodes is referred to as a *grounding conductor*. Note the difference in the two meanings; one is grounded, while the other is grounding. *See* Figure 7-1.

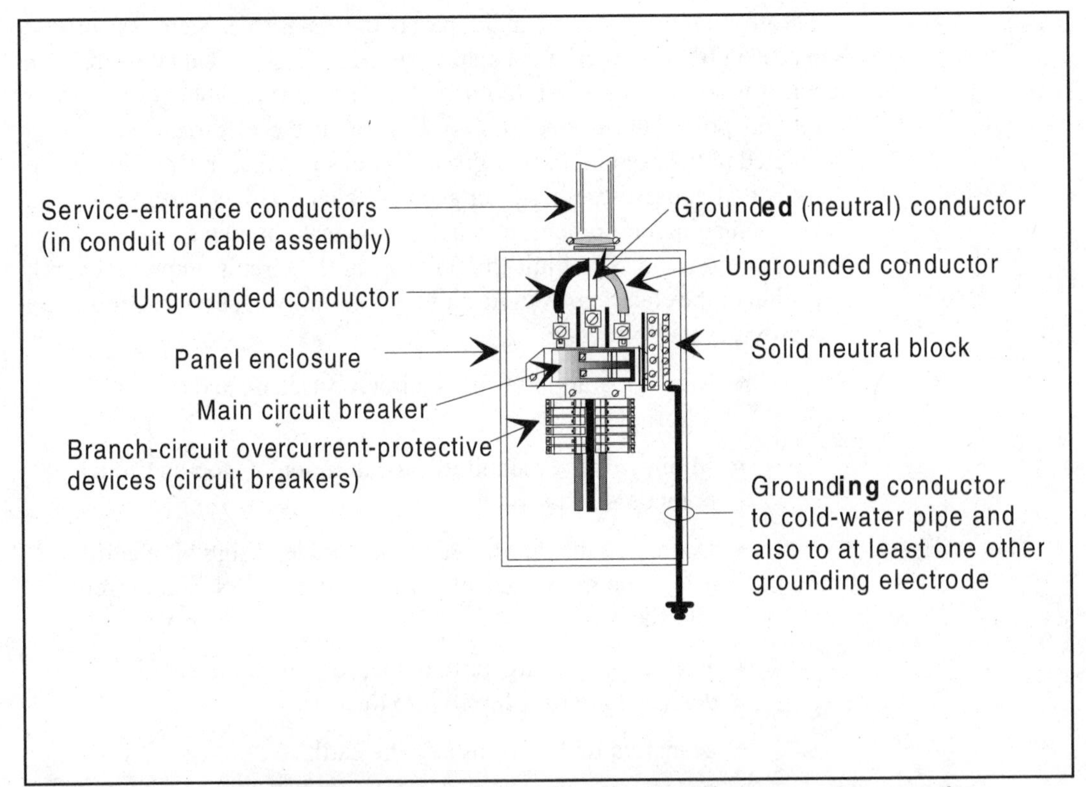

Figure 7-1: Simple single-phase panelboard used to show the difference between the connection grounded and grounding conductors.

TYPES OF GROUNDING SYSTEMS

There are two general classifications of protective grounding:

- System grounding
- Equipment grounding

The system ground relates to the service-entrance equipment and its interrelated and bonded components. That is, system and circuit conductors are grounded to limit voltages due to lighting, line surges, or unintentional contact with higher voltage lines, and to stabilize the voltage to ground during normal operation.

Equipment grounding conductors are used to connect the noncurrent-carrying metal parts of equipment, conduit, outlet boxes, and other enclosures to the system grounded conductor, the grounding electrode conductor, or both, at the service equipment or at the source of a separately derived system. Equipment grounding conductors are bonded to the system grounded conductor to provide a low impedance path for fault current that will facilitate the operation of overcurrent devices under ground-fault conditions. Article 250 of the *NEC* covers general requirements for grounding and bonding.

Single-Phase Systems

To better understand a complete grounding system, let's take a look at a conventional residential or small commercial system beginning at the power company's high-voltage lines and transformer as shown in Figure 7-2 on the next page. The transformer is fed with a two-wire 7200-V system which is transformed and stepped down to a 3-wire, 120/240-V, single-phase electric service suitable for residential use. Note that the voltage between phase A and phase B is 240 V. However, by connecting a third wire (neutral) on the secondary winding of the transformer — between the other two — the 240 V are split in half, giving 120 V between either phase A or phase B and the neutral conductor. Consequently, 240 V are available for household appliances such as ranges, hot-water heaters, clothes dryers, and the like, while 120 V are available for lights, small appliances, TVs, and similar electrical appliances.

Referring again to the diagram in Figure 7-2, conductors A and B are ungrounded conductors, while the neutral is a grounded conductor. If only 240-V loads were connected, the neutral (grounded conductor) would carry no current. However, since 120-V loads are present, the neutral will carry the unbalanced load and becomes a current-carrying conductor. For exam-

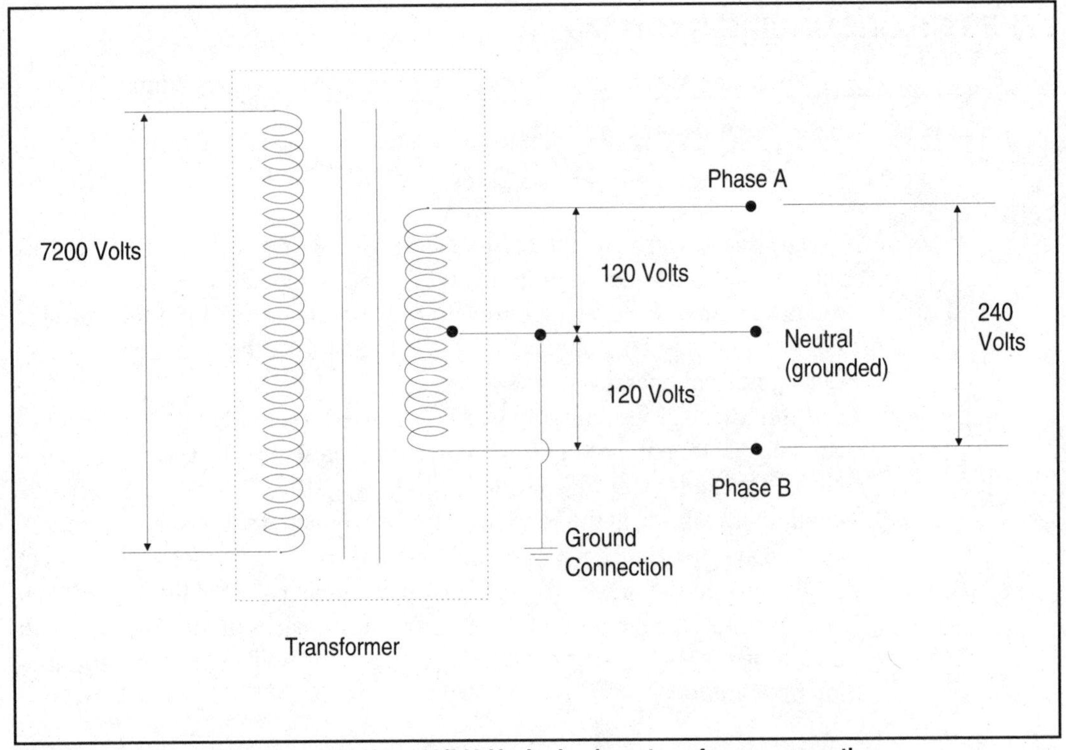

Figure 7-2: Wiring diagram of 7200-V to 120/240-V, single-phase transformer connection.

ple, if phase A carries 60 amperes and phase B carries 50 amperes, the neutral conductor would carry only (60 - 50 =) 10 amperes. This is why the *NEC* allows the neutral conductor in an electric service to be smaller than the ungrounded conductors.

CAUTION!

Exercise extreme caution when lifting a ground. Never grab a disconnected ground wire with one hand and the grounding electrode with the other. Your body will act as a conductor for any fault current; the results could be fatal.

OSHA AND NEC REQUIREMENTS

The grounding equipment requirements established by Underwriters' Laboratories, Inc., has served as the basis for approval for grounding of the *NEC*. The *NEC*, in turn, provides the grounding premises of the Occupational Safety and Health Act (OSHA).

All electrical systems must be grounded in a manner prescribed by the *NEC* to protect personnel and valuable equipment. To be totally effective, a grounding system must limit the voltage on the electrical system and protect it from:

- Exposure to lightning.
- Voltage surges higher than that for which the circuit is designed.
- An increase in the maximum potential to ground due to abnormal voltages.

Grounding Methods

Methods of grounding an electric service are covered in *NEC* Section 250-81. In general, all of the following (if available) and any made electrodes must be bonded together to form the grounding electrode system:

- An underground water pipe in direct contact with the earth for no less than 10 ft.
- The metal frame of a building where effectively grounded.
- An electrode encased by at least 2 in of concrete, located within and near the bottom of a concrete foundation or footing that is in direct contact with the earth. Furthermore, this electrode must be at least 20 ft long and must be made of electrically conductive coated steel reinforcing bars or rods of not less than ½-in diameter, or consisting of at least 20 ft of bare copper conductor not smaller than No. 2 AWG wire size.
- A ground ring encircling the building or structure, in direct contact with the earth at a depth below grade not less than 2½ ft. This ring must consist of at least 20 ft of bare copper conductor not smaller than No. 2 AWG wire size. *See* Figure 7-3.

Grounding systems used in industrial buildings will frequently use all of the methods shown in Figure 7-3, and the methods used will often surpass the *NEC*, depending upon the manufacturing process, and the calculated requirements made by plant engineers. Figure 7-4 shows a floor plan of a typical industrial grounding system.

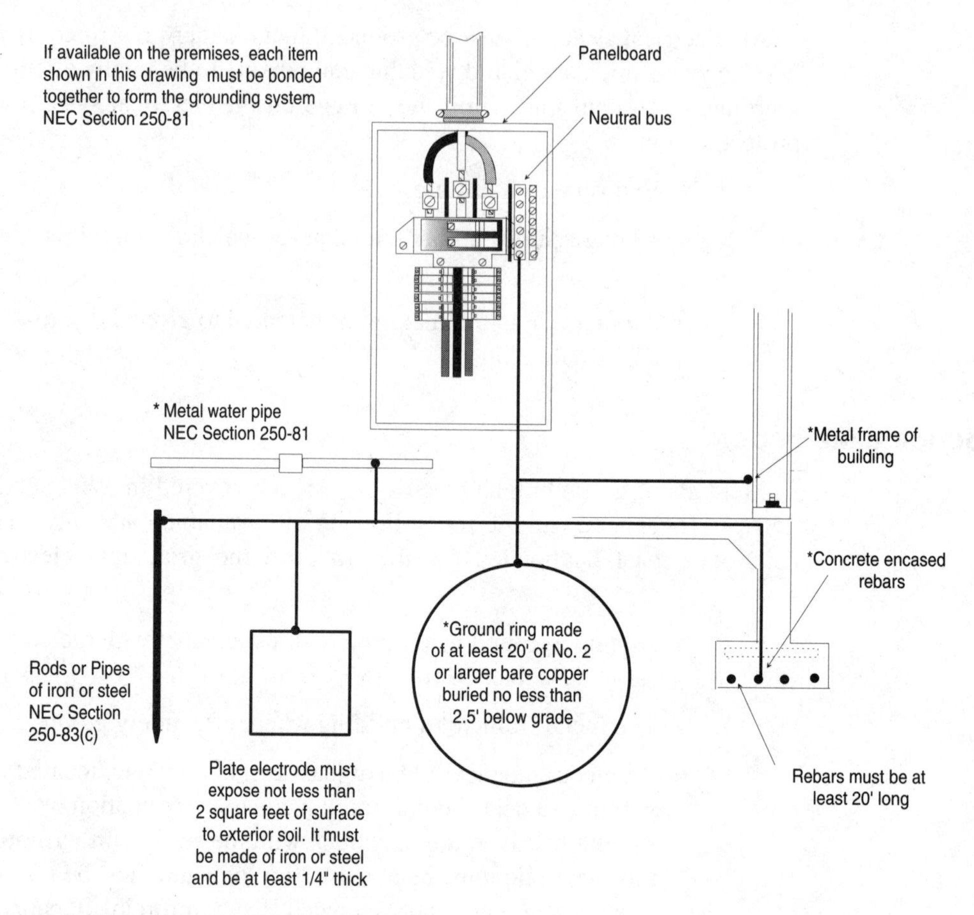

Figure 7-3: NEC approved grounding electrodes.

In some structures, only the water pipe will be available, and this water pipe must be supplemented by an additional electrode as specified in *NEC* Sections 250-81(a) and 250-83.

Earth Electrodes

The area of contact between the earth and ground rod must be sufficient so that the resistance of the current path into and through the earth will be within the allowable limits of the particular application. The resistance of this earth path must be relatively low and must remain reasonably constant through the seasons of the year.

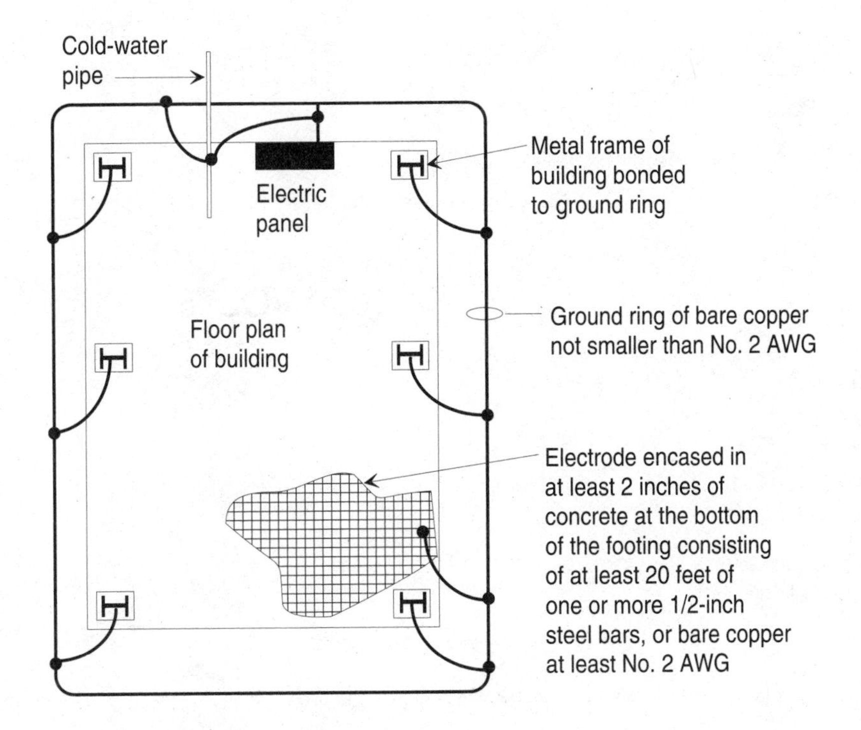

Figure 7-4: Floor plan of the grounding system for a typical industrial building.

To understand why earth resistance must be low (*see* Figure 7-5 on the next page) and then apply Ohm's law, $E = I \times R$. (E is volts, I is the current in amperes, and R is the resistance in ohms.) For example, assume a 4000-V supply (2300 V to "ground"), with a resistance of 13 ohms. Now assume an exposed ungrounded wire in this system touches a motor frame that is connected to a grounding system that has a 10-ohm resistance to earth.

According to Ohm's law, there will be a current of 100 A through the fault, from the motor frame to the earth. If a person touches the motor frame and is solidly grounded to earth, the person could be subjected to 1000 V (10 ohms times 100 A). This is more than enough for a fatality.

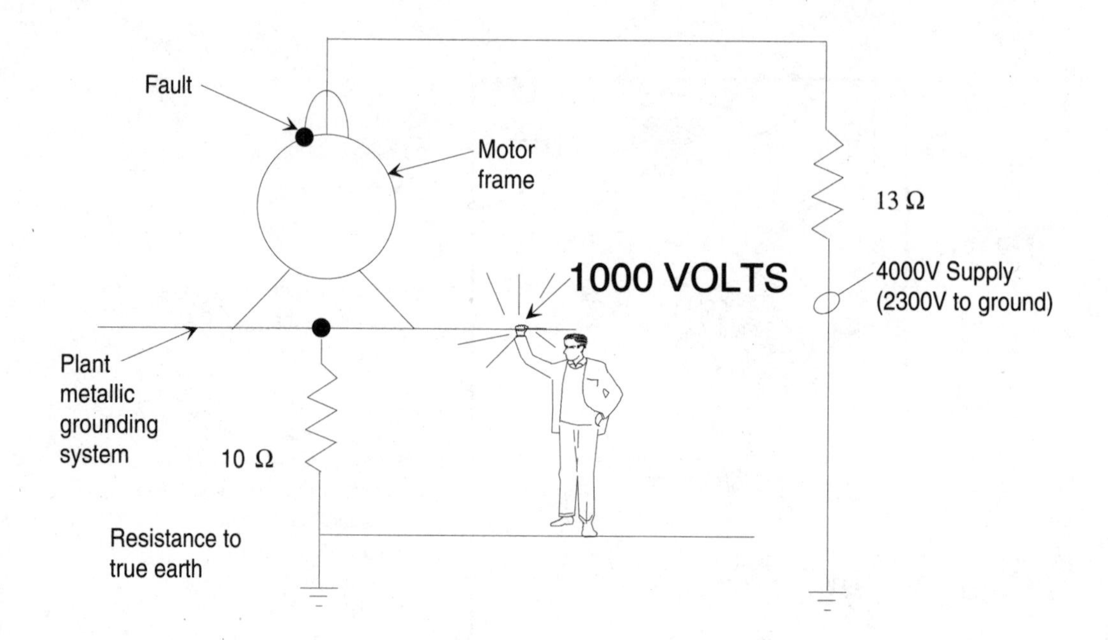

Figure 7-5: An electric circuit with an earth resistance that is too high.

Summary

No other phase of the electrical industry is more important than grounding. It is the chief means of protecting life and property from electrical hazards and also ensures proper operation of the system as well as helping other protective devices to function properly.

The term *grounded* means connecting to earth by a conductor or to some conducting body that serves in place of the earth. The earth as a whole is properly classed as a conductor. For convenience, its electric potential is assumed to be zero. When a metal object is grounded, it too is thereby forced to take the same zero potential as the earth. Therefore, the main purpose of grounding is to ensure that the grounded object cannot take on a potential differing sufficiently from earth potential to be hazardous.

Chapter 8

Transformers

The electric power produced by alternators in a generating station is transmitted to locations where it is utilized and distributed to users. Many different types of transformers play an important role in the distribution of electricity. Power transformers are located at generating stations to step up the voltage for more economical transmission. Substations with additional power transformers and distribution equipment are installed along the transmission line. Finally, distribution transformers are used to step down the voltage to a level suitable for utilization.

Transformers are also used quite extensively in all types of control work, to raise and lower ac voltage on control circuits. They are also used in 480Y/277-V systems to reduce the voltage for operating 208Y/120-V lighting and other electrically-operated equipment. Buck-and-boost transformers are used for maintaining appropriate voltage levels in certain electrical systems.

It is important for anyone working with electricity to become familiar with transformer operation; that is, how they work, how they are connected into circuits, their practical applications and precautions to take during the installation or while working on them. This chapter is designed to cover these items as well as overcurrent protection and grounding procedures for transforms.

TRANSFORMER TAPS

If the exact rated voltage could be delivered at every transformer location, transformer taps would be unnecessary. However, this is not possible, so taps are provided on the secondary windings to provide a means of either increasing or decreasing the secondary voltage.

Generally, if a load is very close to a substation or power plant, the voltage will consistently be above normal. Near the end of the line the voltage may be below normal.

In large transformers, it would naturally be very inconvenient to move the thick, well-insulated primary leads to different tap positions when changes in source-voltage levels make this necessary. Therefore, taps are used, such as those shown in the wiring diagram in Figure 8-1. In this transformer, the permanent high-voltage leads would be connected to H_1 and H_2, and the secondary leads, in their normal fashion, to X_1 and X_2, X_3, and X_4. Note, however, the tap arrangements available at taps 2 through 7. Until a pair of these taps is interconnected with a jumper wire, the primary circuit is not completed. If this were, say, a typical 7200-V primary, the transformer would have a normal 1620 turns. Assume 810 of these turns are between H_1 and H_6 and another 810 between H_3 and H_2. Then, if taps

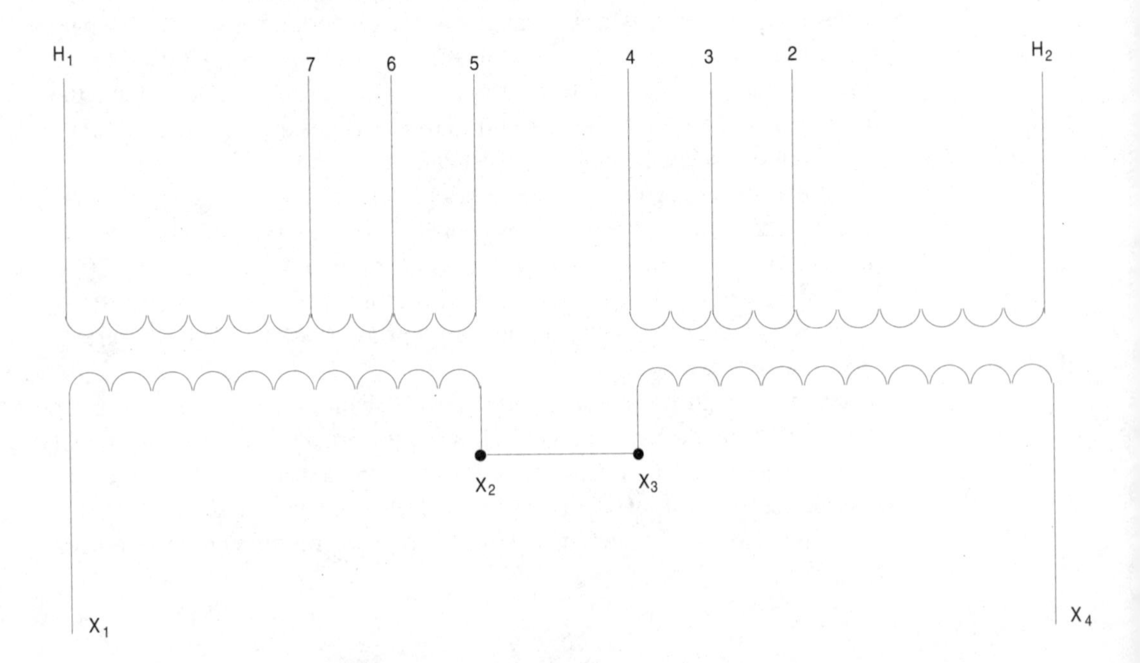

Figure 8-1: Transformer taps used to adjust secondary voltage.

6 and 3 were connected together with a flexible jumper on which lugs have already been installed, the primary circuit is completed, and we have a normal ratio transformer that could deliver 120/240 V from the secondary.

Between taps 6 and either 5 or 7, 40 turns of wire exist. Similarly, between taps 3 and either 2 or 4, 40 turns are present. Changing the jumper from 3 to 6 to 3 to 7 removes 40 turns from the left half of the primary. The same condition would apply on the right half of the winding if the jumper were between taps 6 and 2. Either connection would boost secondary voltage by 2½ percent. Had taps 2 and 7 been connected, 80 turns would have been omitted and a 5 percent boost would result. Placing the jumper between taps 6 and 4 or 3 and 5 would reduce the output voltage by 5 percent.

BASIC TRANSFORMER CONNECTIONS

Transformer connections are many, and space does not permit the description of all of them here. However, an understanding of a few will give the basic requirements and make it possible to use manufacturer's data for others should the need arise.

Single-Phase Light And Power

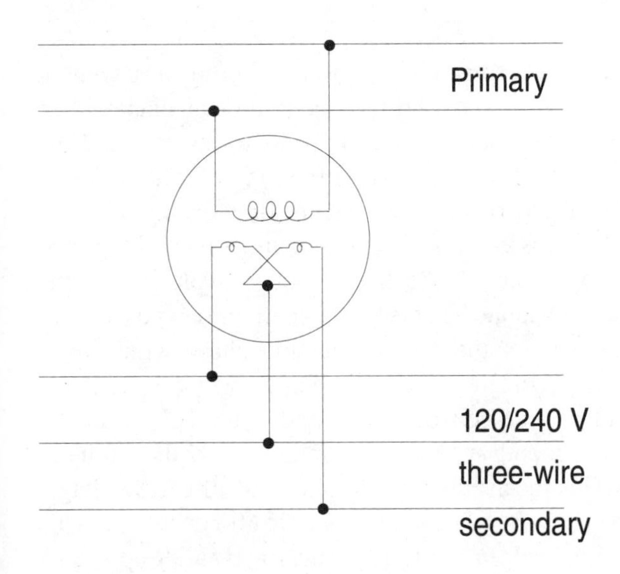

Figure 8-2: 120/240-V , three-wire, single-phase transformer connection.

The diagram in Figure 8-2 is a transformer connection used quite extensively for residential and small commercial applications. It is the most common single-phase distribution system in use today. It is known as the three-wire, 240/120-V single-phase system and is used where 120 and 240 V are used simultaneously.

Y-Y for Light and Power

The primaries of the transformer connection in Figure 8-3 are connected in wye — sometimes called *star* connection. When the primary system is 2400/4160Y volts, a 4160-V transformer is required when the system is connected in delta-Y. However, with a

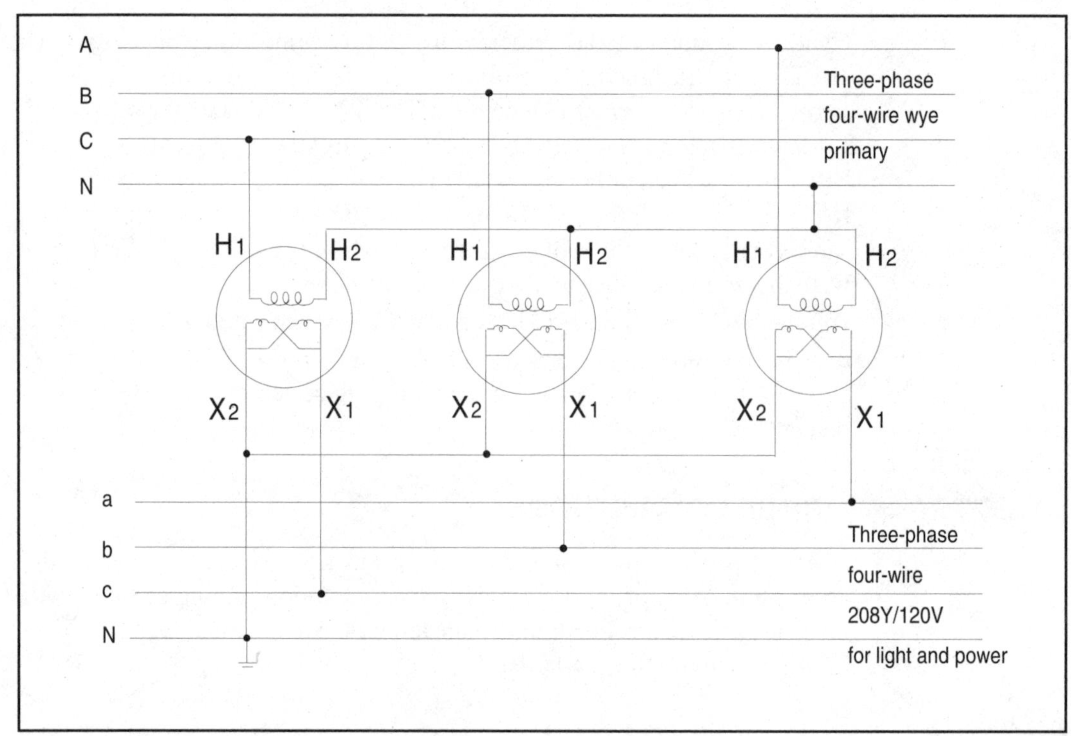

Figure 8-3: Three-phase, 4-wire, Y-Y connected transformer system.

Y-Y system, a 2400-V transformer can be used, offering a saving in transformer cost. It is necessary that a primary neutral be available when this connection is used, and the neutrals of the primary system and the transformer bank are tied together as shown in the diagram. If the three-phase load is unbalanced, part of the load current flows in the primary neutral. For these reasons, it is essential that the neutrals be tied together as shown. If this tie were omitted, the line-to-neutral voltages on the secondary would be very unstable. That is, if the load on one phase were heavier than on the other two, the voltage on this phase would drop excessively and the voltage on the other two phases would rise. Also, varying voltages would appear between lines and neutral, both in the transformers and in the secondary system, in addition to the 60-hertz component of voltage. This means that for a given value of RMS voltage, the peak voltage would be much higher than for a pure 60-Hz voltage. This overstresses the insulation both in the transformers and in all apparatus connected to the secondaries.

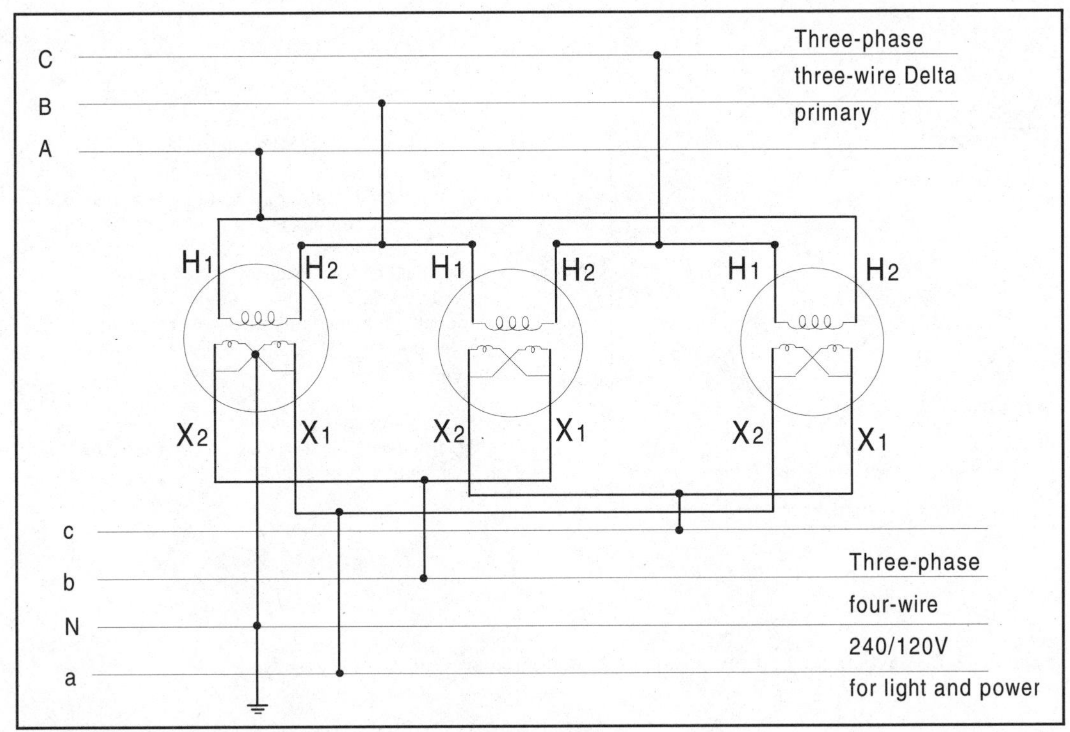

Figure 8-4: Three-phase, 4-wire, delta-connected secondary.

Delta-Connected Transformers

The delta-connected system in Figure 8-4 operates a little differently from the previously described wye-wye system. While the wye-connected system is formed by connecting one terminal from three equal voltage transformer windings together to make a common terminal, the delta-connected system has its windings connected in series, forming a triangle or the Greek delta symbol Δ. Note in Figure 8-5 that a center-tap terminal is used on one winding to ground the system. On a 240/120-V system, there are 120 V between the center-tap terminal and each ungrounded terminal on either side; that is, phases A and C. There are 240 V across the full winding of each phase.

Refer to Figure 8-5 and note that a high leg results at point "B." This is known in the trade as the "high leg," "red leg," or "wild leg." This high leg has a higher voltage to ground than the other two phases. The voltage of the high leg can be determined by multiplying the voltage to ground of either of the other two legs by the square root of 3. Therefore, if the voltage

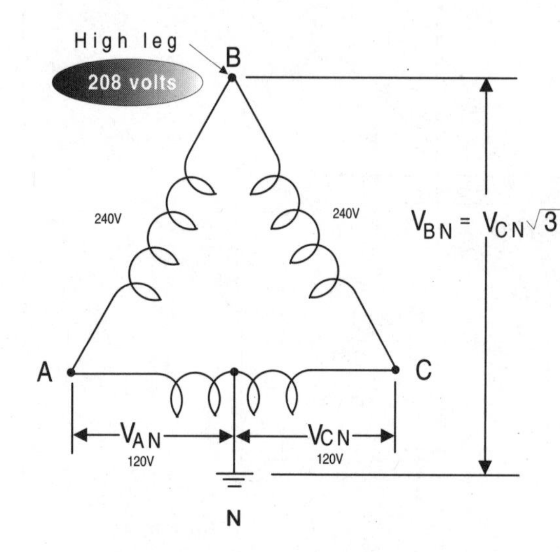

On a 3-phase, 4-wire 120/240V delta-connected system, the midpoint of one phase winding is grounded to provide 120V between phase A and ground; also between phase C and ground. Between phase B and ground, however, the voltage is higher and may be calculated by multiplying the voltage between C and ground (120V) by the square root of 3 or 1.73. Consequently, the voltage between phase B and ground is approximately 208 volts. Thus, the name "high leg."

The NEC requires that conductors connected to the high leg of a 4-wire delta system be color coded with orange insulation or tape.

Figure 8-5: Characteristics of a center-tap, delta-conneted system.

between phase A to ground is 120 V, the voltage between phase B to ground may be determined as follows:

$$120 \times \sqrt{3} = 207.84 = 208 \text{ volts}$$

From this, it should be obvious that no single-pole breakers should be connected to the high leg of a center-tapped, 4-wire delta-connected system. In fact, *NEC* Section 215-8 requires that the phase busbar or conductor having the higher voltage to ground to be permanently marked by an outer finish that is orange in color. By doing so, this will prevent future workers from connecting 120-V single-phase loads to this high leg which will probably result in damaging any equipment connected to the circuit. Remember the color ***orange***; no 120-V loads are to be connected to this phase.

WARNING!

Always use caution when working on a center-tapped, 4-wire, delta-connected system. Phase B has a higher voltage to ground than phases A and C. Never connect 120-V circuits to the high leg. Doing so will result in damage to the circuits and equipment.

Open Delta

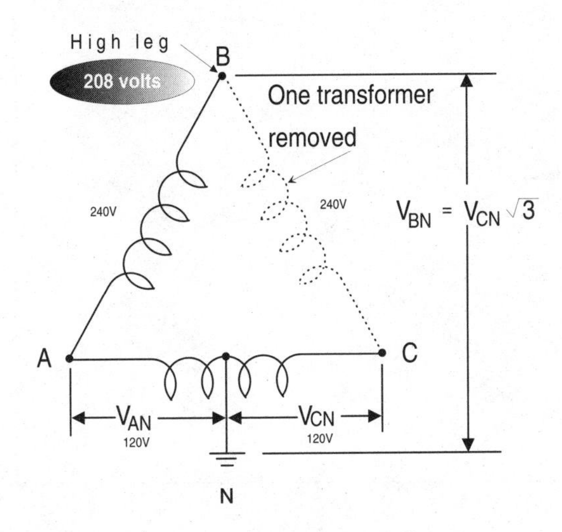

Figure 8-6: Open delta system.

Three-phase, delta-connected systems may be connected so that only two transformers are used; this arrangement is known as *open delta* as shown in Figure 8-6. This arrangement is frequently used on a delta system when one of the three transformers becomes damaged. The damaged transformer is disconnected from the circuit and the remaining two transformers carry the load. In doing so, the three-phase load carried by the open delta bank is only 86.6 percent of the combined rating of the remaining two equal sized units. It is only 57.7 percent of the normal full-load capability of a full bank of transformers. In an emergency, however, this capability permits single- and three-phase power at a location where one unit burned out and a replacement was not readily available. The total load must be curtailed to avoid another burnout.

Tee-Connected Transformers

When a delta-wye transformer is used, we would usually expect to find three primary and three secondary coils. However, in a tee-connected three-phase transformer, only two primary and two secondary windings are used as shown in Figure 8-7. If an equilateral triangle is drawn as indicated by the dotted lines in Figure 8-7 so that the distance between H_1 and H_3 is 4.8 in, you would find that the distance between H_2 to the midpoint of H_1 - H_3 measures 4.16 in. Therefore, if the voltage between outside phases is 480 V, the voltage between H_2 to the midpoint of H_1 - H_3 will equal 480 V $\times .866 = 415.68$ or 416 V. Also, if you were to place an imaginary dot exactly in the center of this triangle it would lay on the horizontal winding — the one containing 416 V. If you measured the distance from this dot to H_2, you would find it to be twice as long as the distance between the dot and the midpoint of H_1 to H_3. The measured distances would be 2.77 in and 1.385 in or the equivalent of 277 V and 138½ V respectively.

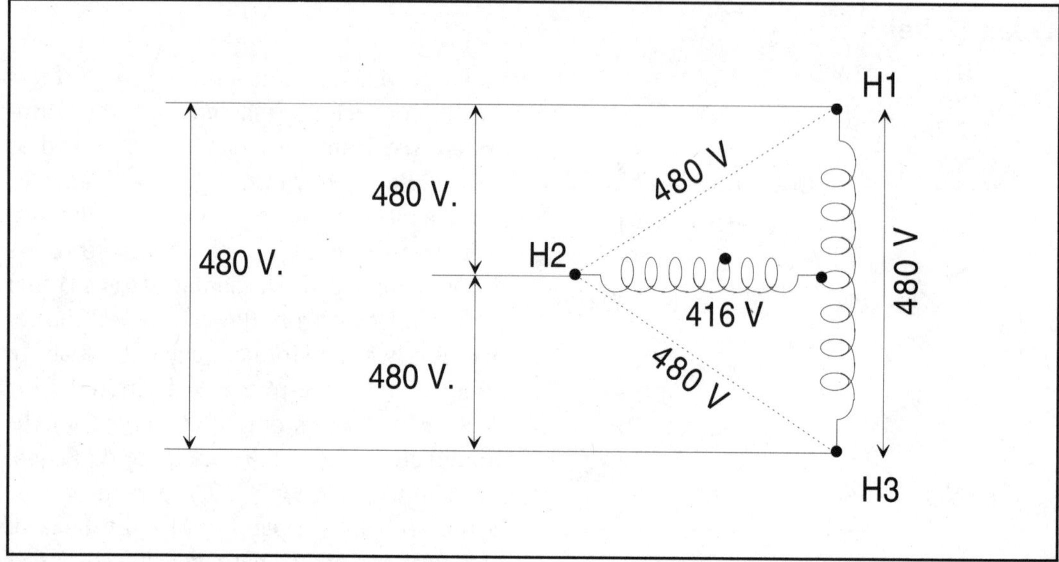

Figure 8-7: Typical Tee-connected transformer.

Now, let's look at the secondary winding in Figure 8-8. By placing a neutral tap X_0 so that one-third the number of turns exist between it and the midpoint of X_1 and X_3, as exist between it and X_2, we then can establish X_0 as a neutral point which may be grounded. This provides 120 V between X_0 and any of the three secondary terminals and the three-phase voltage between X_1, X_2, and X_3, will be 208 V.

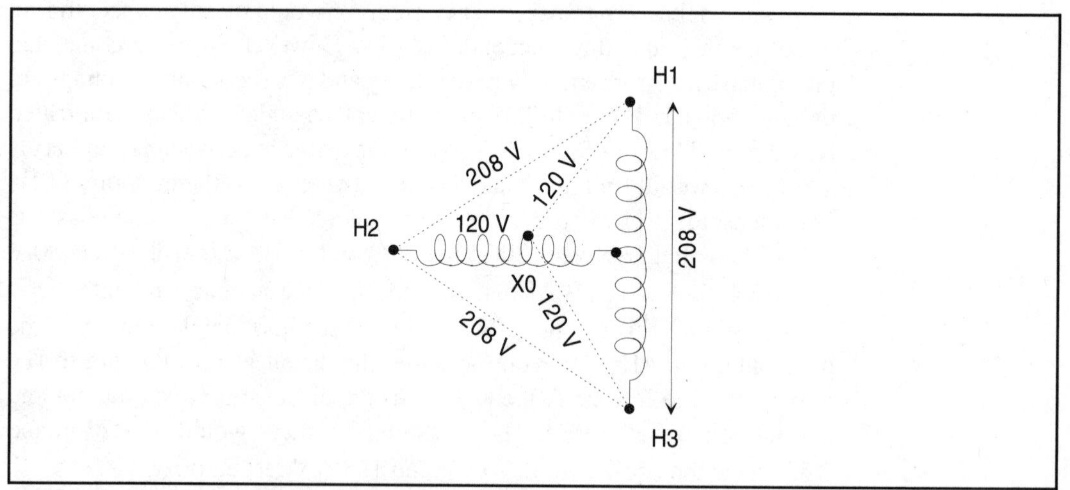

Figure 8-8: Secondary voltage on Tee-connected system.

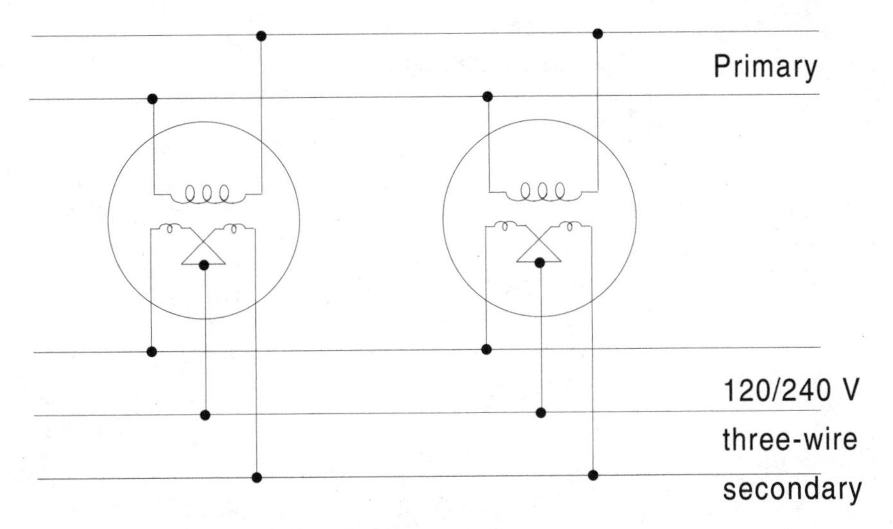

Figure 8-9: Parallel operation of single-phase transformers.

PARALLEL OPERATION OF TRANSFORMERS

Transformers will operate satisfactorily in parallel on a single-phase, three-wire system if the terminals with the same relative polarity are connected together. However, the practice is not very economical because the individual cost and losses of the smaller transformers are greater than one larger unit giving the same output. Therefore, paralleling of smaller transformers is usually done only in an emergency. In large transformers, however, it is often practical to operate units in parallel as a regular practice. *See* Figure 8-9.

In connecting large transformers in parallel, especially when one of the windings is for a comparatively low voltage, the resistance of the joints and interconnecting leads must not vary materially for the different transformers, or it will cause an unequal division of load.

Two three-phase transformers may also be connected in parallel provided they have the same winding arrangement, are connected with the same polarity, and have the same phase rotation. If two transformers — or two banks of transformers — have the same voltage ratings, the same turn ratios, the same impedances, and the same ratios of reactance to resistance, they will divide the load current in proportion to their kVA ratings, with no phase difference between the currents in the two transformers. However, if any of the preceding conditions are not met, then it is possible for the load current to divide between the two transformers in proportion to their

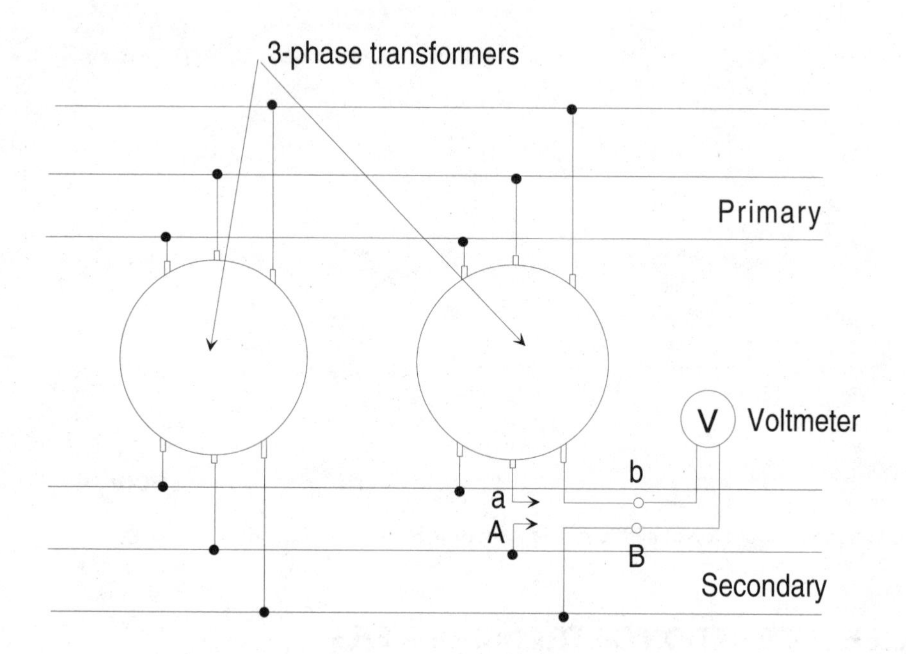

Figure 8-10: Testing three-phase transformers for parallel operation.

kVA ratings. There may also be a phase difference between currents in the two transformers or banks of transformers.

Some three-phase transformers cannot be operated properly in parallel. For example, a transformer having its coils connected in delta on both high-tension and low-tension sides cannot be made to parallel with one connected either in delta on the high-tension and in Y on the low-tension or in Y on the high-tension and in delta on the low-tension side and in Y on the low-tension side can be made to parallel with transformers having their coils joined in accordance with certain schemes, connected in star of Y on the high-tension side and in delta on the low-tension side.

To determine whether or not three-phase transformers will operate in parallel, connect them as shown in Figure 8-10, leaving two leads on one of the transformers unjoined. Test with a voltmeter across the unjoined leads. If there is no voltage between the points shown in the drawing, the polarities of the two transformers are the same, and the connections may then be made and put into service.

If a reading indicates a voltage between the points indicated in the drawing (either one of the two or both), the polarity of the two transformers are different. Should this occur, disconnect transformer lead A successively to mains 1, 2, and 3 as shown in Figure 8-10 and at each connection test with the voltmeter between b and B and the legs of the main to which lead A is connected. If with any trial connection the voltmeter readings between b and B and either of the two legs is found to be zero, the transformer will operate with leads b and B connected to those two legs. If no system of connections can be discovered that will satisfy this condition, the transformer will not operate in parallel without changes in its internal connections, and there is a possibility that it will not operate in parallel at all.

In parallel operation, the primaries of the two or more transformers involved are connected together, and the secondaries are also connected together. With the primaries so connected, the voltages in both primaries and secondaries will be in certain directions. It is necessary that the secondaries be so connected that the voltage from one secondary line to the other will be in the same direction through both transformers. Proper connections to obtain this condition for single-phase transformers of various polarities are shown in Figure 8-11 on the next page. In Figure 8-11(a), both transformers A and B have additive polarity; in Figure 8-11(b), both transformers have subtractive polarity; in Figure 8-11(c), transformer A has additive polarity and B has subtractive polarity.

Transformers, even when properly connected, will not operate satisfactorily in parallel unless their transformation ratios are very close to being equal and their impedance voltage drops are also approximately equal. A difference in transformation ratios will cause a circulating current to flow, even at no load, in each winding of both transformers. In a loaded parallel bank of two transformers of equal capacities, for example, if there is a difference in the transformation ratios, the load circuit will be superimposed on the circulating current. The result in such a case is that in one transformer the total circulating current will be added to the load current, whereas in the other transformer the actual current will be the difference between the load current and the circulating current. This may lead to unsatisfactory operation. Therefore, the transformation ratios of transformers for parallel operation must be definitely known.

When two transformers are connected in parallel, the circulating current caused by the difference in the ratios of the two is equal to the difference in open-circuit voltage divided by the sum of the transformer impedances, because the current is circulated through the windings of both transformers due to this voltage difference. To illustrate, let I represent the amount of circulating current — in percent of full-load current — and the equation will be as follows:

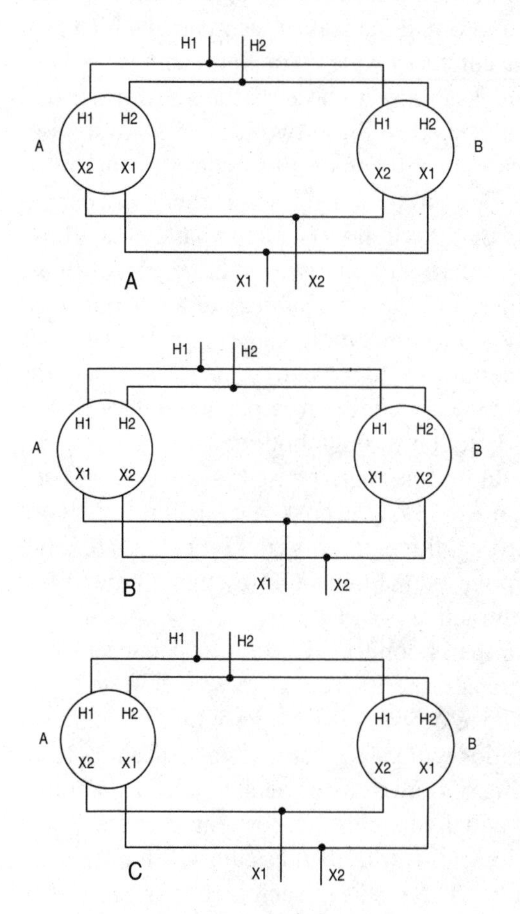

Figure 8-11: Transformers connected in parallel.

$$I = \frac{\textit{Percent voltage difference} \times 100}{\textit{Sum of percent impedances}}$$

Let's assume an open-circuit voltage difference of 3% between two transformers connected in parallel. If each transformer has an impedance of 5 percent, the circulating current, in percent of full-load current, is I = (3 × 100)/5 + 5) = 30 percent. A current equal to 30 percent full-load current therefore circulates in both the high-voltage and low-voltage windings. This current adds to the load current in the transformer having the higher induced voltage and subtracts from the load current of the other transformer. Therefore, one transformer will be overloaded, while the other may or may not be — depending on the phase-angle difference between the circulating current and the load current.

Impedance in Parallel-Operated Transformers

Impedance plays an important role in the successful operation of transformers connected in parallel. The impedance of the two or more transformers must be such that the voltage drop from no load to full load is the same in all transformer units in both magnitude and phase. In most applications, you will find that the total resistance drop is relatively small when compared with the reactance drop and that the total percent impedance drop can be taken as approximately equal to the percent reactance drop. If the percent impedances of the given transformers at full load are the same, they will, of course, divide the load equally.

The following equation may be used to obtain the division of loads between two transformer banks operating in parallel on single-phase systems. In this equation, it can be assumed that the ratio of resistance to reactance is the same in all units since the error introduced by differences in this ratio is usually so small as to be negligible:

$$\text{power} = \frac{(kVA-1)/(Z-1)}{[(kVA-1)/Z-1)]+[(kVA-2)/(Z-2)]} \times \text{total kVA load}$$

where:

kVA - 1 = kV A rating of transformer 1

kVA - 2 = kV A rating of transformer 2

Z - 1 = percent impedance of transformer 1

Z - 2 = percent impedance of transformer 2

The preceding equation may also be applied to more than two transformers operated in parallel by adding, to the denominator of the fraction, the kV A of each additional transformer divided by its percent impedance.

Parallel Operation of Three-Phase Transformers

Three-phase transformers, or banks of single-phase transformers, may be connected in parallel provided each of the three primary leads in one three-phase transformer is connected in parallel with a corresponding primary lead of the other transformer. The secondaries are then connected in the same way. The corresponding leads are the leads which have the same potential at all times and the same polarity. Furthermore, the transformers must have the same voltage ratio and the same impedance voltage drop.

When three-phase transformer banks operate in parallel and the three units in each bank are similar, the division of the load can be determined by the same method previously described for single-phase transformers connected in parallel on a single-phase system.

In addition to the requirements of polarity, ratio, and impedance, paralleling of three-phase transformers also requires that the angular displacement between the voltages in the windings be taken into consideration when they are connected together.

Phasor diagrams of three-phase transformers that are to be paralleled greatly simplify matters. With these, all that is required is to compare the two diagrams to make sure they consist of phasors that can be made to coincide; then connect together terminals corresponding to coinciding voltage phasors. If the diagram phasors can be made to coincide, leads that are connected together will have the same potential at all times. This is one of the fundamental requirements for paralleling.

AUTOTRANSFORMERS

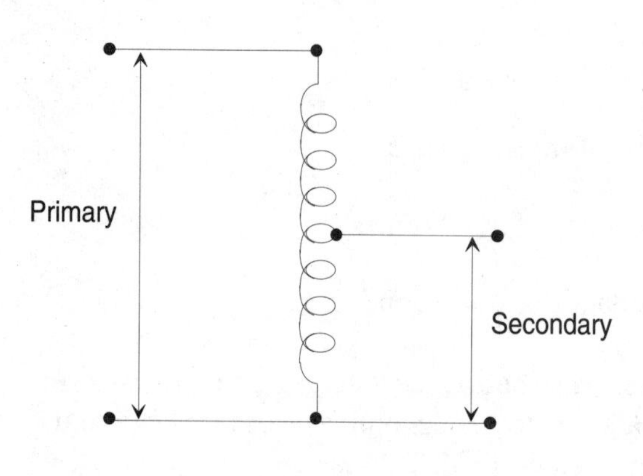

Figure 8-12: Step-down autotransformer.

An autotransformer is a transformer whose primary and secondary circuits have part of a winding in common and therefore the two circuits are not isolated from each other. *See* Figure 8-12. The application of an autotransformer is a good choice for some users where a 480Y/277- or 208Y/120-V, three-phase, four-wire distribution system is utilized. Some of the advantages are as follows:

- Lower purchase price
- Lower operating cost due to lower losses
- Smaller size; easier to install
- Better voltage regulation
- Lower sound levels

For example, when the ratio of transformation from the primary to secondary voltage is small, the most economical way of stepping down the voltage is by using autotransformers as shown in Figure 8-13. For this application, it is necessary that the neutral of the autotransformer bank be connected to the system neutral.

An autotransformer, however, cannot be used on a 480- or 240-V, three-phase, three-wire delta system. A grounded neutral phase conductor must be available in accordance with *NEC* Article 210-9, which states:

NEC Section 210-9: Circuits Derived from Autotransformers. Branch circuits shall not be supplied by autotransformers.

Exception No. 1: Where the system supplied has a grounded conductor that is electrically connected to a grounded conductor of the system supplying the autotransformer.

Exception No. 2: An autotransformer used to extend or add an individual branch circuit in an existing installation for an equipment load without the connection to a similar grounded conductor when transforming from a nominal 208 V to a nominal 240 V supply or similarly from 240 V to 208 volts.

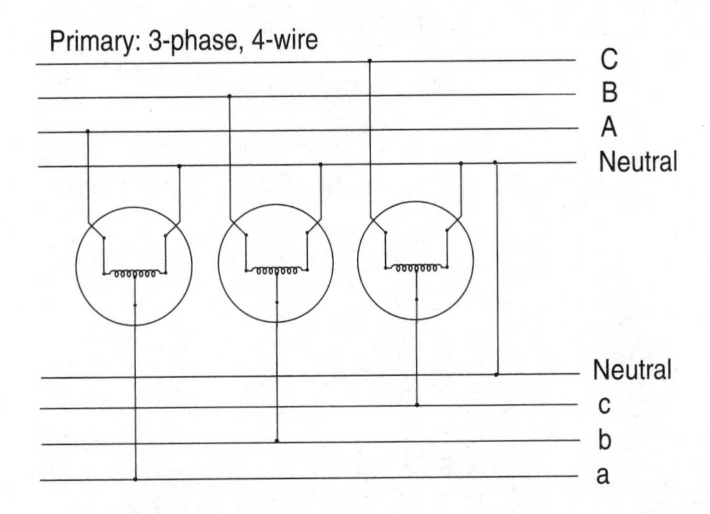

Figure 8-13: Autotransformers supplying power from a 3-phase, 4-wire system.

TRANSFORMER CONNECTIONS — DRY TYPE

Dry-type transformers are available in both single- and three-phase windings with a wide range of sizes from the small control transformers to those rated at 500 kVA or more. Such transformers have wide application in electrical systems of all types.

NEC Section 450-11 requires that each transformer must be provided with a nameplate giving the manufacturer; rated kVA; frequency; primary and secondary voltage; impedance of transformers 25 kVA and larger; required clearances for transformers with ventilating openings; and the amount and kind of insulating liquid where used. In addition, the nameplate of each dry-type transformer must include the temperature class for the insulation system. *See* Figure 8-14 on the next page.

In addition, most manufacturers include a wiring diagram and a connection chart as shown in Figure 8-15 for a 480-V delta primary to 208Y/120-V secondary. It is recommended that all transformers be connected as shown on the manufacturer's nameplate.

In general, this wiring diagram and accompanying table indicates that the 480-V, 3-phase, 3-wire primary conductors are connected to terminals H_1, H_2, and H_3, respectively — regardless of the desired voltage on the primary. A neutral conductor, if required, is carried from the primary through the transformer to the secondary. Two variations are possible on

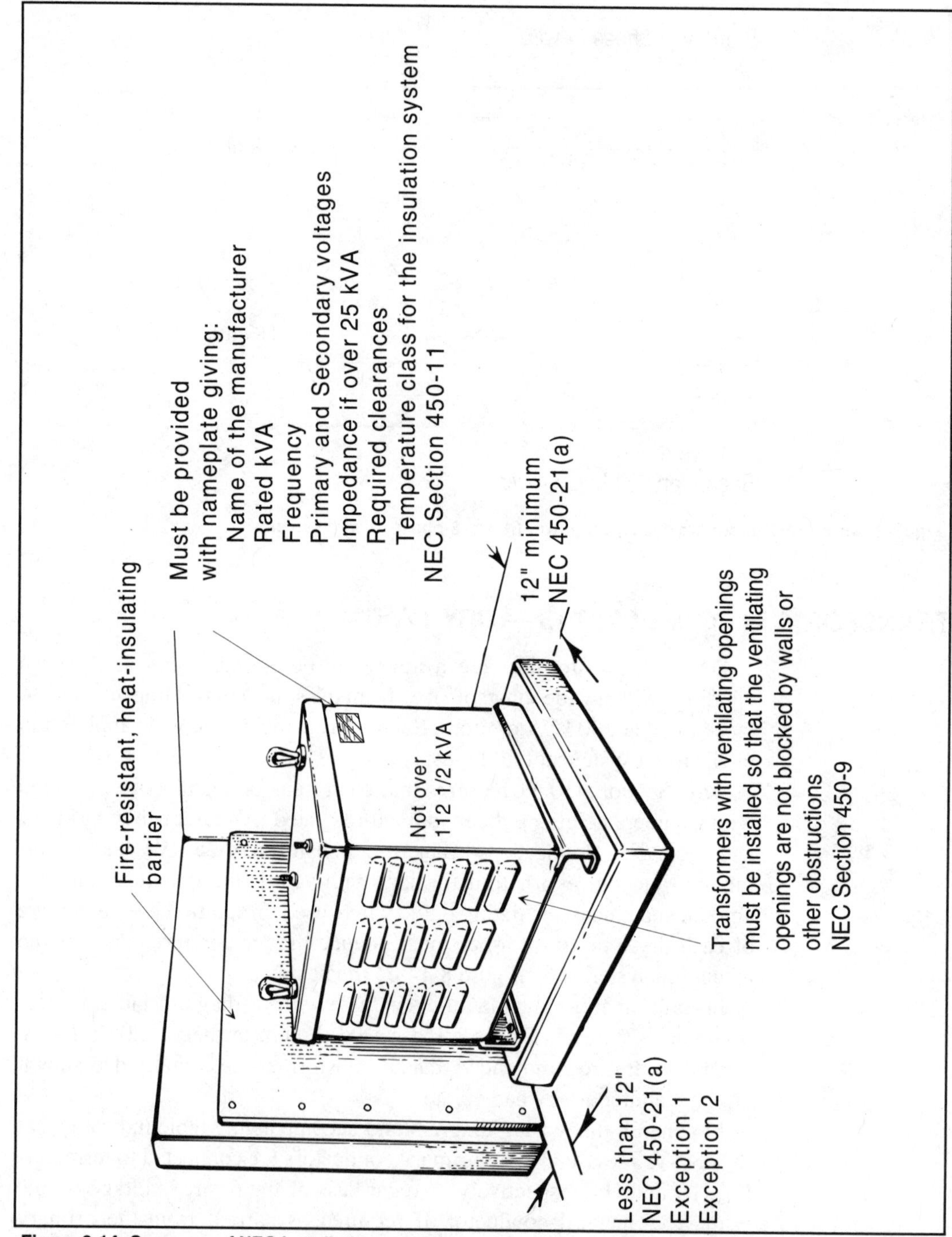

Figure 8-14: Summary of NEC installation requirements for dry-type transformers.

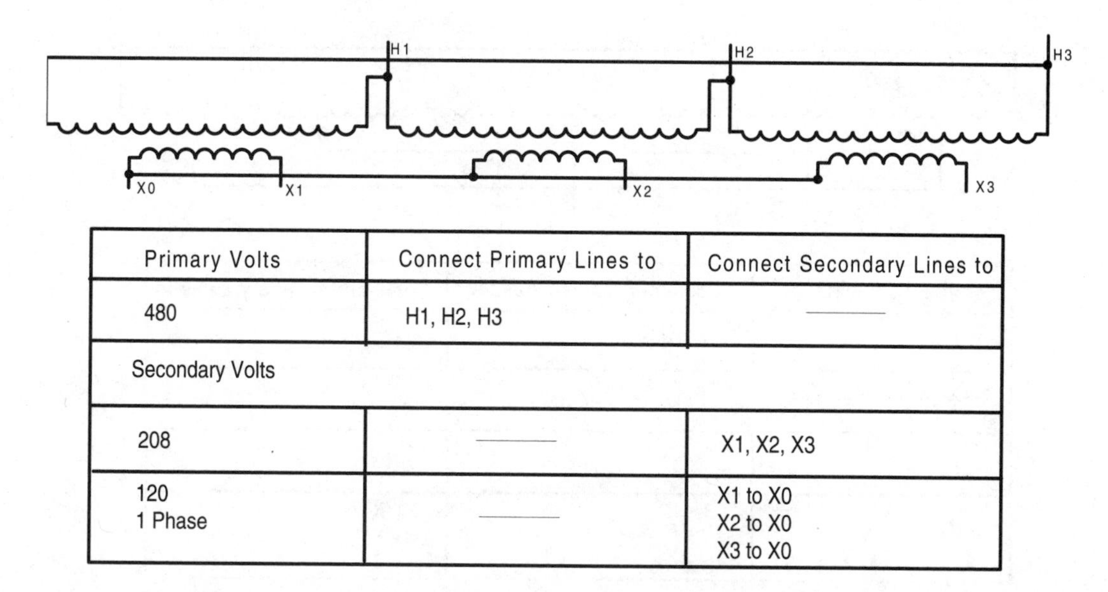

Primary Volts	Connect Primary Lines to	Connect Secondary Lines to
480	H1, H2, H3	———
Secondary Volts		
208	———	X1, X2, X3
120 1 Phase	———	X1 to X0 X2 to X0 X3 to X0

Figure 8-15: Typical transformer manufacturer's wiring diagram for a delta-wye dry-type transformer.

the secondary side of this transformer: 208-V, 3-phase, 3- or 4-wire or 120-V, 1-phase, 2-wire. To connect the secondary side of the transformer as a 208-V, 3-phase, 3-wire system, the secondary conductors are connected to terminals X_1, X_2, and X_3; the neutral is carried through with conductors usually terminating at a solid-neutral bus in the transformer.

Another popular dry-type transformer connection is the 480-V primary to 240-V delta/120 V secondary. This configuration is shown in Figure 8-16. Again, the primary conductors are connected to transformer terminals H_1, H_2, and H_3. The secondary connections for the desired voltages are made as indicated in the table.

Zig-Zag Connections

There are many occasions where it is desirable to upgrade a building's lighting system from 120-V fixtures to 277-V fluorescent lighting fixtures. Oftentimes these buildings have a 480/240-V, three-phase, four-wire delta system. One way to obtain 277 V from a 480/240-V system is to connect 480/240-V transformers in a zig-zag fashion as shown in Figure 8-17. In doing so, the secondary of one phase is connected in series with the primary of another phase, thus changing the phase angle.

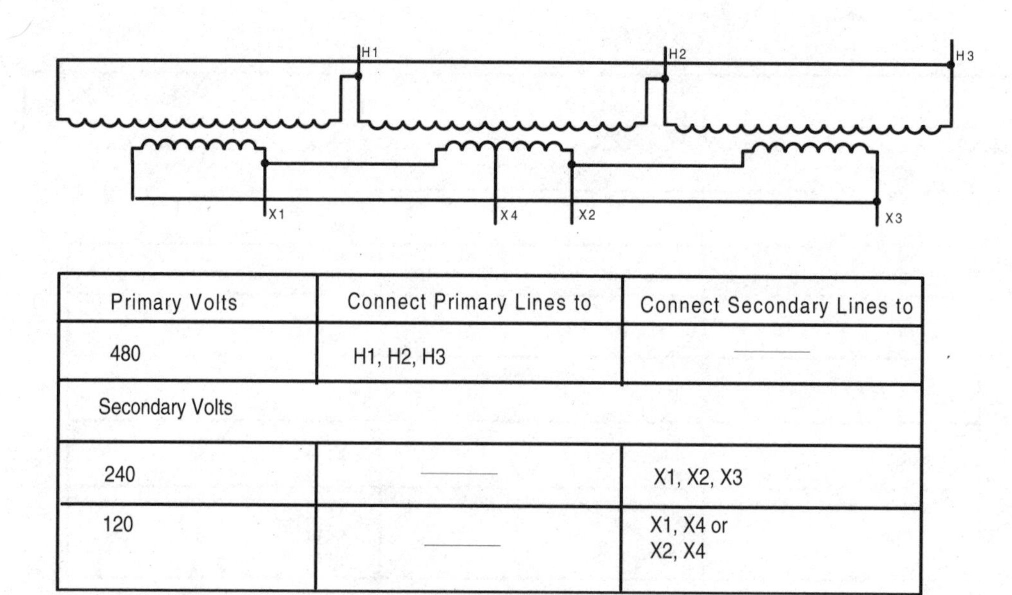

Primary Volts	Connect Primary Lines to	Connect Secondary Lines to
480	H1, H2, H3	——
Secondary Volts		
240	——	X1, X2, X3
120	——	X1, X4 or X2, X4

Figure 8-16: Typical transformer manufacturer's wiring diagram for a delta-delta dry-type transformer.

The zig-zag connection may also be used as a grounding transformer where its function is to obtain a neutral point from an ungrounded system. With a neutral being available, the system may then be grounded. When the system is grounded through the zig-zag transformer, its sole function is to pass ground current. A zig-zag transformer is essentially six impedances connected in a zig-zag configuration.

The operation of a zig-zag transformer is slightly different from that of the conventional transformer. We will consider current rather than voltage. While a voltage rating is necessary for the connection to function, this is actually line voltage and is not transformed. It provides only exciting current for the core. The dynamic portion of the zig-zag grounding system is the fault current. To understand its function, the system must also be viewed backward; that is, the fault current will flow into the transformer through the neutral as shown in Figure 8-18.

The zero sequence currents are all in phase in each line; that is, they all hit the peak at the same time. In reviewing Figure 8-18, we see that the current leaves the motor, goes to ground, flows up the neutral, and splits three ways. It then flows back down the line to the motor through the fuses which then open — shutting down the motor.

The neutral conductor will carry full fault current and must be sized accordingly. It is also time rated (0-60 seconds) and can therefore be

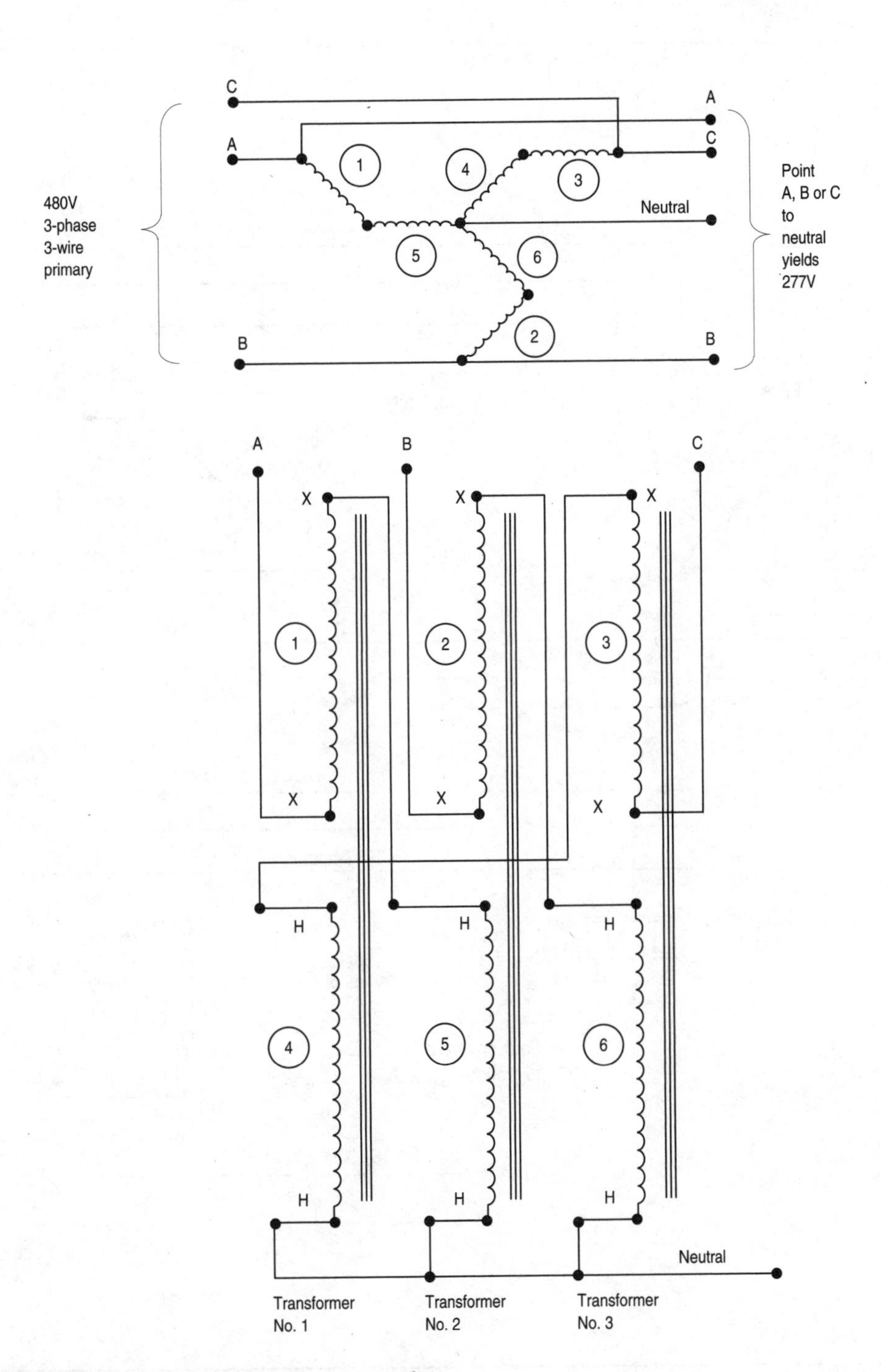

Figure 8-17: Typical zig-zag transformer connection.

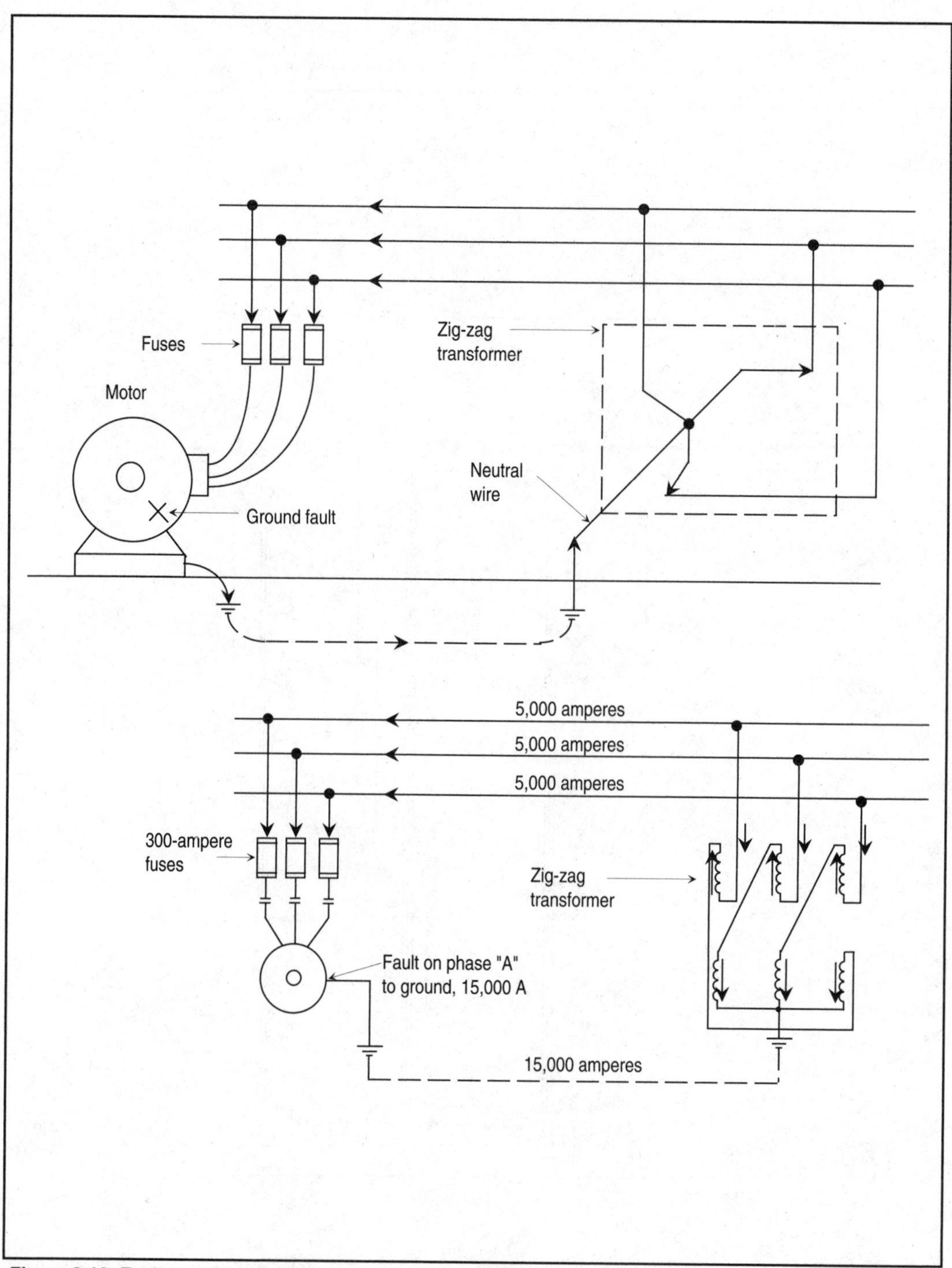

Figure 8-18: Fault-current paths for three-phase systems.

reduced in size. This should be coordinated with the manufacturer's time/current curves for the fuse.

To determine the size of a zig-zag grounding transformer, proceed as follows:

Step 1. Calculate the system line-to-ground asymmetrical fault current.

Step 2. If relaying is present, consider reducing the fault current by installing a resistor in the neutral.

Step 3. If fuses or circuit breakers are the protective device, you may need all the fault current to quickly open the overcurrent protective devices.

Step 4. Obtain time/current curves of relay, fuses, or circuit breakers.

Step 5. Select zig-zag transformer for:

a. Fault current — the line-to-ground

b. Line-to-line voltage

c. Duration of fault (determined from time/current curves)

d. Impedance per phase at 100 percent; for any other, contact manufacturer

Buck-and-Boost Transformers

The buck-and-boost transformer is a very versatile unit for which a multitude of applications exist. Buck-and-boost transformers, as the name implies, are designed to raise (boost) or lower (buck) the voltage in an electrical system or circuit. In their simplest form, these insulated units will deliver 12 or 24 V when the primaries are energized at 120 or 240 V respectively. Their prime use and value, however, lies in the fact that the primaries and the secondaries can be interconnected — permitting their use as an autotransformer.

Let's assume that an installation is supplied with 208Y/120-V service, but one piece of equipment in the installation is rated for 230 V. A buck-and-boost transformer may be used on the 230-V circuit to increase the voltage from 208 V to 230 V. *See* Figure 8-19. With this connection, the transformer is in the "boost" mode and delivers 228.8 V at the load.

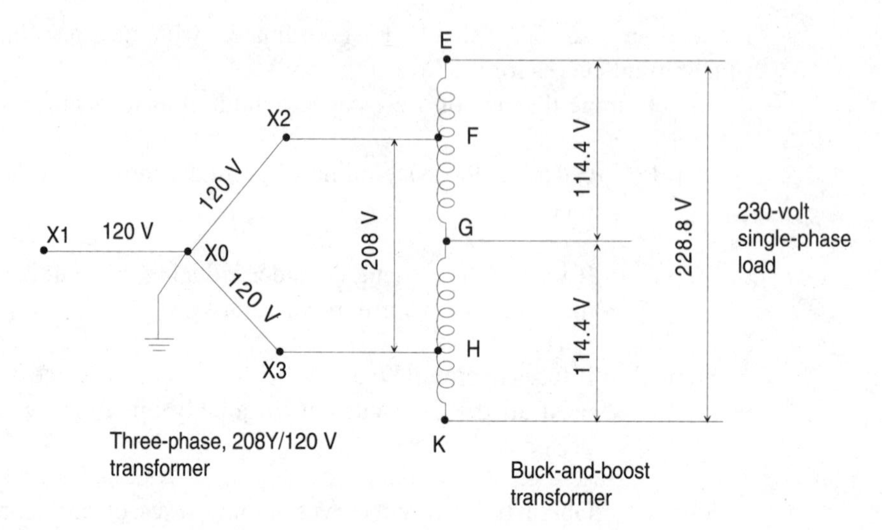

Figure 8-19: Buck-and-boost transformer connected to a 208-V system to raise the voltage to 230 V.

This is close enough to 230 V that the load equipment will function properly.

If the connections were reversed, this would also reverse the polarity of the secondary with the result that a voltage would be 208 V minus 20.8 V = 187.2 volts. The transformer is now operating in the "buck" mode.

Transformer connections for typical three-phase buck-and-boost open-delta transformers are shown in Figure 8-20. The connections shown are in the "boost" mode; to convert to "buck" mode, reverse the input and output.

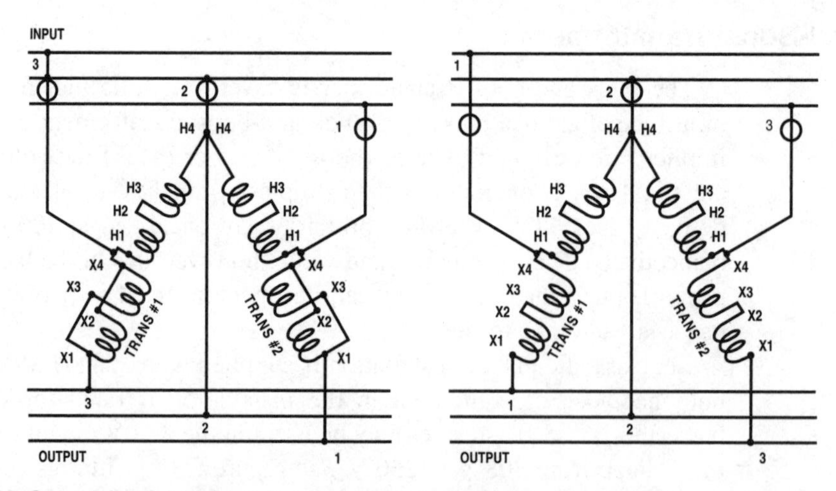

Figure 8-20: Open delta, three-phase, buck-and-boost transformer connections.

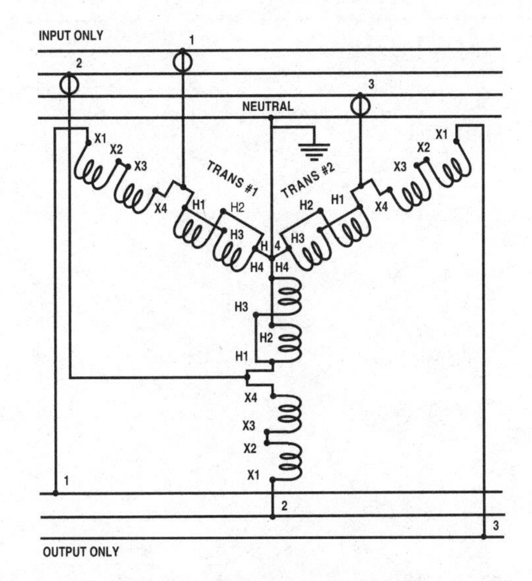

Figure 8-21: Three-phase, wye-connected buck-and-boost transformer in the "boost" mode.

Another three-phase buck-and-boost transformer connection is shown in Figure 8-21and is wye-connected. The connection shown is for the boost mode only.

Several typical single-phase buck-and-boost transformer connections are shown in Figure 8-22. Other diagrams may be found on the transformer's nameplate or with packing instructions that come with each new transformer.

CONTROL TRANSFORMERS

Control transformers are available in numerous types, but most control transformers are dry-type step-down units with the secondary control circuit isolated from the primary line circuit to assure maximum

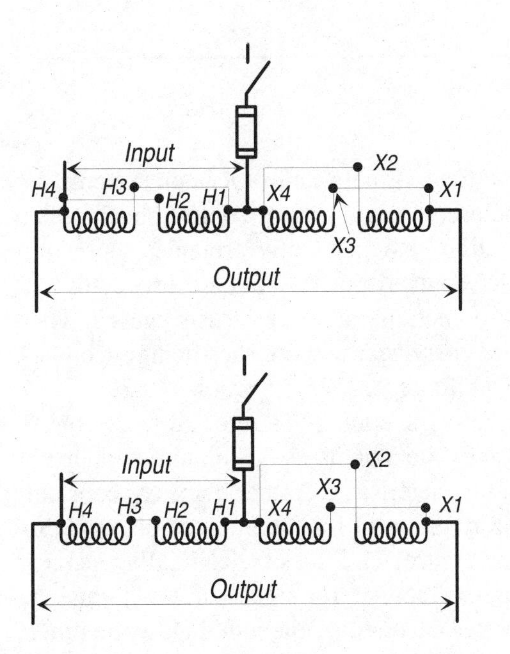

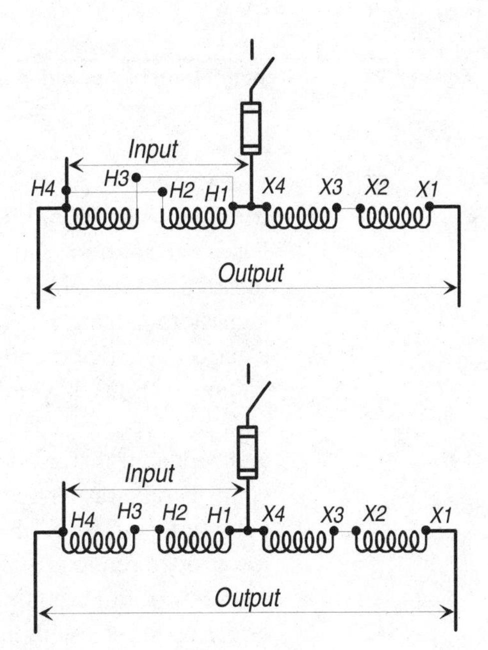

Figure 8-22: Typical single-phase buck-and-boost transformer connections.

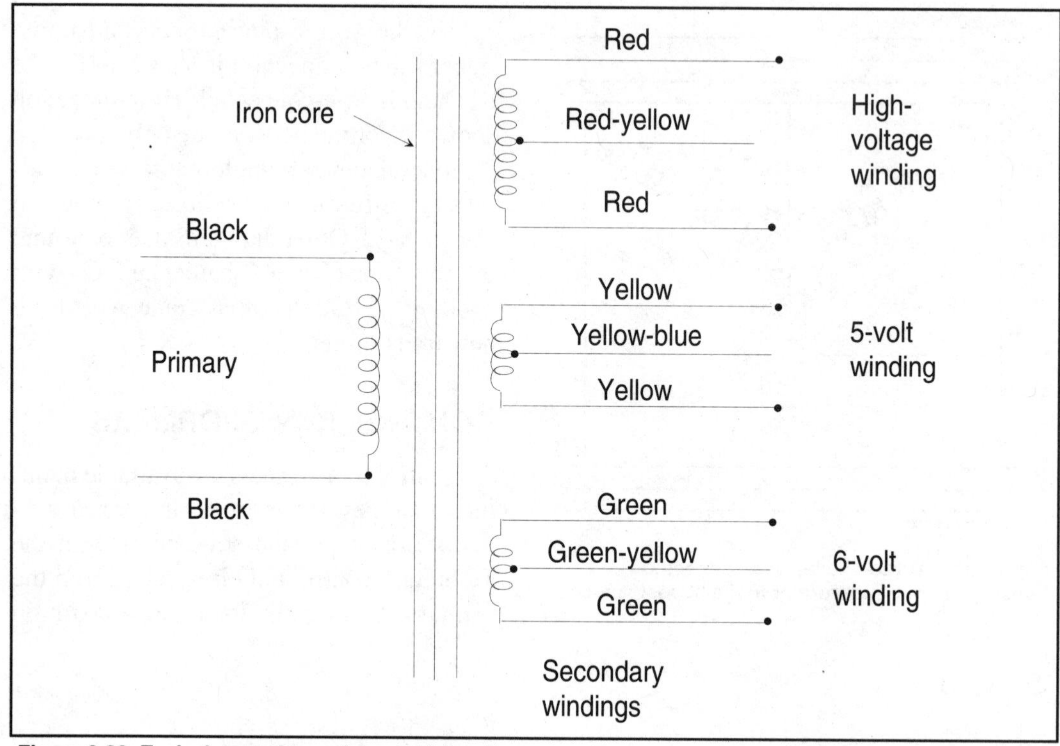

Figure 8-23: Typical control-transformer wiring diagram.

safety. *See* Figure 8-23. These transformers and other components are usually mounted within an enclosed control box or control panel, which has a pushbutton station or stations independently grounded as recommended by the *NEC*. Industrial control transformers are especially designed to accommodate the momentary current inrush caused when electromagnetic components are energized, without sacrificing secondary voltage stability beyond practical limits. *See NEC* Section 470-32.

Other types of control transformers, sometimes referred to as control and signal transformers, normally do not have the required industrial control transformer regulation characteristics. Rather, they are constant-potential, self-air-cooled transformers used for the purpose of supplying the proper reduced voltage for control circuits of electrically operated switches or other equipment and, of course, for signal circuits. Some are of the open type with no protective casing over the winding, while others are enclosed with a metal casing over the winding.

In seeking control transformers for any application, the loads must be calculated and completely analyzed before the proper transformer selection can be made. This analysis involves every electrically energized component in the control circuit. To select an appropriate control transformer, first determine the voltage and frequency of the supply circuit. Then determine the total inrush volt-amperes (watts) of the control circuit. In doing so, do not neglect the current requirements of indicating lights and timing devices that do not have inrush volt-amperes, but are energized at the same time as the other components in the circuit. Their total volt-amperes should be added to the total inrush volt-amperes.

POTENTIAL AND CURRENT TRANSFORMERS

In general, a potential transformer is used to supply voltage to instruments such as voltmeters, frequency meters, power-factor meters, and watt-hour meters. The voltage is proportional to the primary voltage, but it is small enough to be safe for the test instrument. The secondary of a potential transformer may be designed for several different voltages, but most are designed for 120 V. The potential transformer is primarily a distribution transformer especially designed for good voltage regulation so that the secondary voltage under all conditions will be as nearly as possible a definite percentage of the primary voltage.

Current Transformers

A current transformer (Figure 8-24) is used to supply current to an instrument connected to its secondary, the current being proportional to the

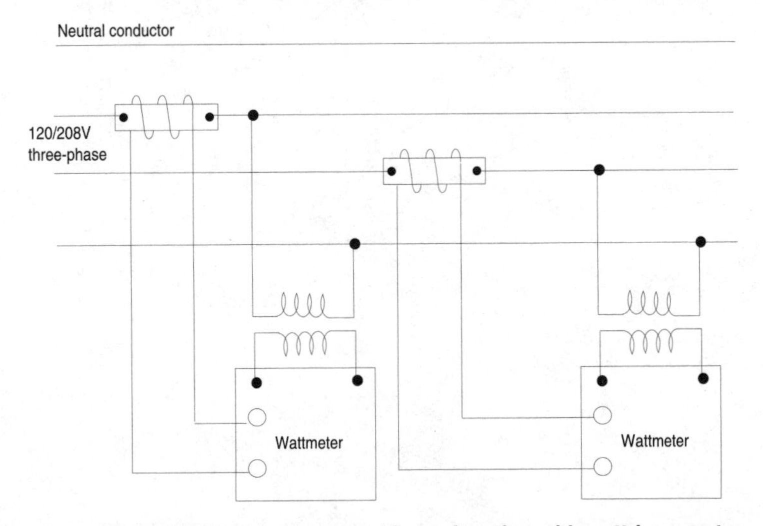

Figure 8-24: Current and potential transformers used in conjunction with watt-hour meter.

primary current, but small enough to be safe for the instrument. The secondary of a current transformer is usually designed for a rated current of 5 A.

A current transformer operates in the same way as any other transformer in that the same relation exists between the primary and the secondary current and voltage. A current transformer is connected in series with the power lines to which it is applied so that line current flows in its primary winding. The secondary of the current transformer is connected to current devices such as ammeters, wattmeters, watt-hour meters, power-factor meters, some forms of relays, and the trip coils of some types of circuit breakers.

When no instrument or other devices are connected to the secondary of the current transformer, a short-circuit device or connection is placed across the secondary to prevent the secondary circuit from being opened while the primary winding is carrying current, there will be no secondary ampere turns to balance the primary ampere turns, so the total primary current becomes exciting current and magnetizes the core to a high flux density. This produces a high voltage across both primary and secondary windings and endangers the life of anyone coming in contact with the meters or leads.

Chapter 9

Conductors and Wiring Methods

Several types of wiring methods are used for industrial electrical installations. The methods used on a given project are determined by several factors:

- The installation requirements set forth in the *NEC*
- Local codes and ordinances
- Type of building construction
- Location of the wiring in the building
- Importance of the wiring system's appearance
- Costs and budget

The following table lists the most common wiring methods in use today, along with the *NEC* reference for each.

WIRING METHODS		
Method	**Type**	**NEC Reference**
Armored cable	Cable	Article 333
Auxiliary gutters	Raceway	Article 374
Busways	Raceway	Article 364
BX cable	Cable	Article 333

WIRING METHODS *(Cont.)*		
Method	**Type**	**NEC Reference**
Cablebus	Assembly	Article 365
Cabletray, metallic	Assembled metallic structure	Article 318
Cabletray, nonmetallic	Assembled nonmetallic structure	Article 318
Cellular concrete floor raceways	Raceway	Article 358
Cellular metal floor raceways	Raceway	Article 356
Concealed knob-and-tube wiring	Conductors	Article 324
Electrical metallic tubing	Raceway	Article 348
Electrical nonmetallic tubing	Raceway	Article 331
EMT	Raceway	Article 348
Flat cable assembly	Cable assembly	Article 363
Flat conductor cable	Cable	Article 328
Flexible metal conduit	Raceway	Article 350
Flexible metallic tubing	Raceway	Article 349
IMC	Raceway	Article 345
Intermediate metal conduit	Raceway	Article 345
Liquidtight flexible metal conduit	Raceway	Article 351
Medium voltage cable	Cable	Article 326
Messenger supported wiring	Conductors and cable	Article 321
Metal-clad cable	Cable	Article 334
Mineral-insulated metal-sheathed cable	Cable	Article 330
Multioutlet assembly	Preassembled surface metal raceway	Article 353
Nonmetallic extensions	Cable assembly	Article 342
Nonmetallic-sheathed cable	Cable	Article 336
Nonmetallic underground conduit with conductors	Raceway assembly	Article 343

WIRING METHODS (*Cont.*)		
Method	**Type**	**NEC Reference**
Open wiring on insulators	Conductors	Article 320
Power and control tray cable	Conductors and cable	Article 340
Rigid metal conduit	Raceway	Article 346
Rigid nonmetallic conduit	Raceway	Article 347
Romex cable	Cable	Article 336
Service-entrance cable	Cable	Article 338
Surface metal raceway	Raceway	Article 352
UF cable	Cable	Article 339
Underfloor raceways	Raceway	Article 354
Underground feeder and branch-circuit cable	Cable	Article 339
Underground service-entrance cable	Cable	Article 338
USE cable	Cable	Article 338
Wireways, metal and nonmetallic	Raceway	Article 362

There are two basic wiring methods:

- Open wiring
- Concealed wiring

In open-wiring systems, the outlets and cable or raceway systems are installed on the surfaces of the walls, ceilings, columns, and the like where they are in view and readily accessible. Such wiring is often used in areas where appearance is not important and where it may be desirable to make changes in the electrical system at a later date. You will frequently find open-wiring systems in all mechanical rooms in commercial buildings; most wiring systems used in industrial applications will be exposed.

Concealed wiring systems have all cable and raceway runs concealed inside of walls, partitions, ceilings, columns, and behind baseboards or molding where they are out of view and not readily accessible. This type of wiring system is generally used in all new construction with finished interior walls, ceilings, floors and is the preferred type where good appearance is important.

CABLE SYSTEMS

Several types of cable systems are used to construct electrical systems in various types of occupancies, and include the following:

Type NM Cable: This cable is manufactured in two- or three-wire assemblies, and with varying sizes of conductors. In both two- and three-wire cables, conductors are color-coded: one conductor is black while the other is white in two-wire cable; in three-wire cable, the additional conductor is red. Both types will also have a grounding conductor which is usually bare, but is sometimes covered with a green plastic insulation — depending upon the manufacturer. The jacket or covering consists of rubber, plastic, or fiber. Most will also have markings on this jacket giving the manufacturer's name or trademark, the wire size, and the number of conductors. For example, "NM 12-2 W/GRD" indicates that the jacket contains two No. 12 AWG conductors along with a grounding wire; "NM 12-3 W/GRD" indicates three conductors plus a grounding wire. This type of cable may be concealed in the framework of buildings, or in some instances, may be run exposed on the building surfaces. It may not be used in any building exceeding three floors above grade; as a service-entrance cable; in commercial garages having hazardous locations; in theaters and similar locations; places of assembly; in motion picture studios; in storage battery rooms; in hoistways; embedded in poured concrete, or aggregate; or in any hazardous location except as otherwise permitted by the *NEC*. Nonmetallic-sheathed cable is frequently referred to as *Romex* on the job.

Romex is occasionally used in industiral applications for temporary wiring, small office areas, guard shacks, etc.

Type AC (Armored) Cable: Type AC cable — commonly called "BX" — is manufactured in two-, three-, and four-wire assemblies, with varying sizes of conductors, and is used in locations similar to those where Type NM cable is allowed. The metallic spiral covering on BX cable offers a greater degree of mechanical protection than with NM cable, and the metal jacket also provides a continuous grounding bond without the need for additional grounding conductors.

BX cable may be used for under-plaster extensions, as provided in the *NEC*, and embedded in plaster finish, brick, or other masonry, except in damp or wet locations. It may also be run or "fished" in the air voids of masonry block or tile walls, except where such walls are exposed or subject to excessive moisture or dampness or are below grade. This type of cable is a favorite for connecting 2 × 4 troffer-type lighting fixtures in industrial office-area installations.

Underground Feeder Cable: Type UF cable may be used underground, including direct burial in the earth, as a feeder or branch-circuit cable when provided with overcurrent protection at the rated ampacity as required by the *NEC*. When Type UF cable is used above grade where it will come in direct contact with the rays of the sun, its outer covering must be sun resistant. Furthermore, where Type UF cable emerges from the ground, some means of mechanical protection must be provided. This protection may be in the form of conduit or guard strips. Type UF cable resembles Type NM cable in appearance. The jacket, however, is constructed of weather-resistant material to provide the required protection for direct-burial wiring installations.

Service-Entrance Cable: Type SE cable, when used for electrical services, must be installed as specified in *NEC* Article 230. This cable is available with the grounded conductor bare for outside service conductors, and also with an insulated grounded conductor (Type SER) for interior wiring systems.

Type SE and SER cable are permitted for use on branch circuits or feeders provided all current-carrying conductors are insulated; this includes the grounded or neutral conductor. When Type SE cable is used for interior wiring, all *NEC* regulations governing the installation of Type NM cable also apply to Type SE cable.

Underground Service-Entrance Cable: Type USE cable is similar in appearance to Type SE cable except that it is approved for underground use and must be manufactured with a moisture-resistant covering. If a flame-retardant covering is not provided, it is not approved for indoor use.

Flat Conductor Cable: Type FCC cable consists of three or more flat copper conductors placed edge-to-edge and separated and enclosed within an insulating assembly. FCC systems consist of cable and associated shielding, connectors, terminators, adapters, boxes, and receptacles. These systems are designed for installation under carpet squares on hard, sound, smooth, continuous floor surfaces made of concrete, ceramic, composition floor, wood, and similar materials. If used on heated floors with temperatures in excess of 86°F, the cable must be identified as suitable for use at these temperatures.

FCC systems must not be used outdoors or in corrosive locations; where subject to corrosive vapors; in any hazardous location; or in residential, school, or hospital buildings.

Flat-Cable Assemblies: This is Type FC cable assembly and should not be confused with Type FCC cable; there is a big difference. A Type FC wiring system is an assembly of parallel, special-stranded copper conductors formed integrally with an insulating material web specifically designed for field installation in surface metal raceway. The assembly is made up of three- or four-conductor cable, cable supports, splicers, circuit taps, fixture hangers, insulating end caps, and other fittings. Guidelines for the use of this system are given in *NEC* Article 363. In general, the assembly is installed in an approved U-channel surface-metal raceway with one side open. Tap devices can be inserted anywhere along the channel. Connections from the tap devices to the flat-cable assembly are made by pin-type contacts when the tap devices are secured in place. The pin-type contacts penetrate the insulation of the cable assembly and contact the multistranded conductors in a matched phase sequence. These taps can then be connected to either lighting fixtures or power outlets.

Flat-cable assemblies must be installed for exposed work only and must not be installed in locations where they will be subjected to severe physical damage.

Mineral-Insulated Metal-Sheathed Cable: Type MI cable is a factory assembly of one or more conductors insulated with a highly compressed refractory mineral insulation and enclosed in a liquid-tight and gas-tight continuous copper sheath. It may be used for electric services, feeders, and branch circuits in dry, wet, or continuously moist locations. Furthermore, it may be used indoors or outdoors, embedded in plaster, concrete, fill, or other masonry, whether above or below grade. This type of cable may also be used in hazardous locations, where exposed to oil or gasoline, where exposed to corrosive conditions not deteriorating to the cable's sheath, and in underground runs where suitably protected against physical damage and corrosive conditions. In other words, MI cable may be used in practically any electrical installation. This type of cable would be used more frequently if it were not for the time-consuming terminations and splices required. *See* Figures 9-1 through 9-3.

Power and Control Tray Cable: Type TC power and control tray cable is a factory assembly of two or more insulated conductors, with or without associated bare or covered grounding conductors, under a nonmetallic sheath, approved for installation in cable trays, in raceways, or where supported by a messenger wire. The use of this cable is limited to commercial and industrial applications where the conditions of maintenance and supervision assure that only qualified persons will service the installation.

Metal-Clad Cable: Type MC cable is a factory assembly of one or more conductors, each individually insulated and enclosed in a metallic sheath of interlocking tape or a smooth or corrugated tube. This type of cable may be used for services, feeders, and branch circuits; power, lighting, control, and signal circuits; indoors or outdoors; where exposed or concealed; direct buried; in cable tray; in any approved raceway; as open runs of cable; as

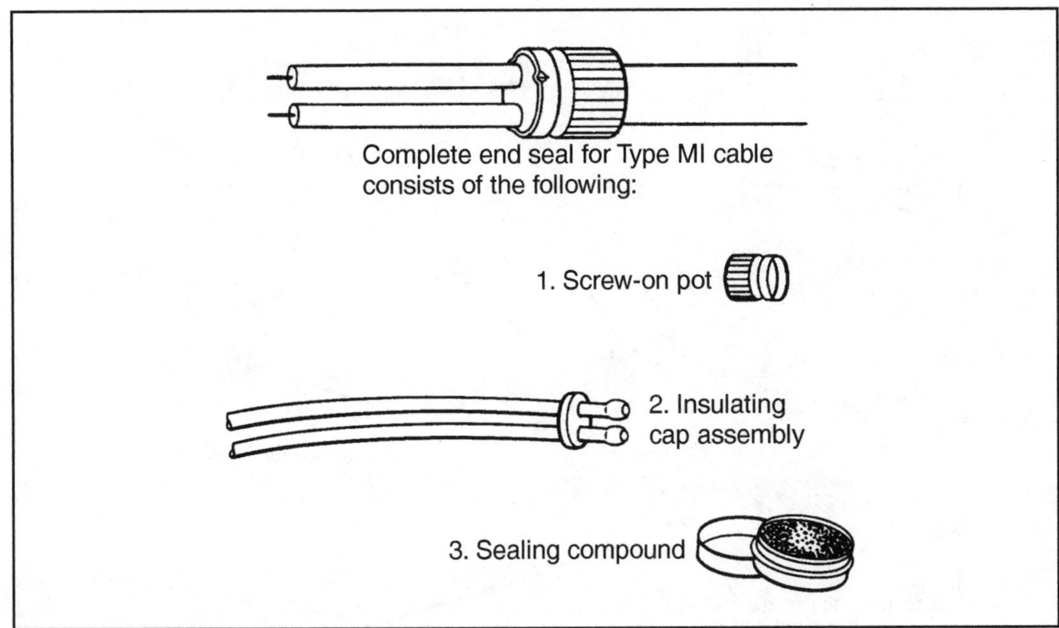

Figure 9-1: Components of MI cable end seals.

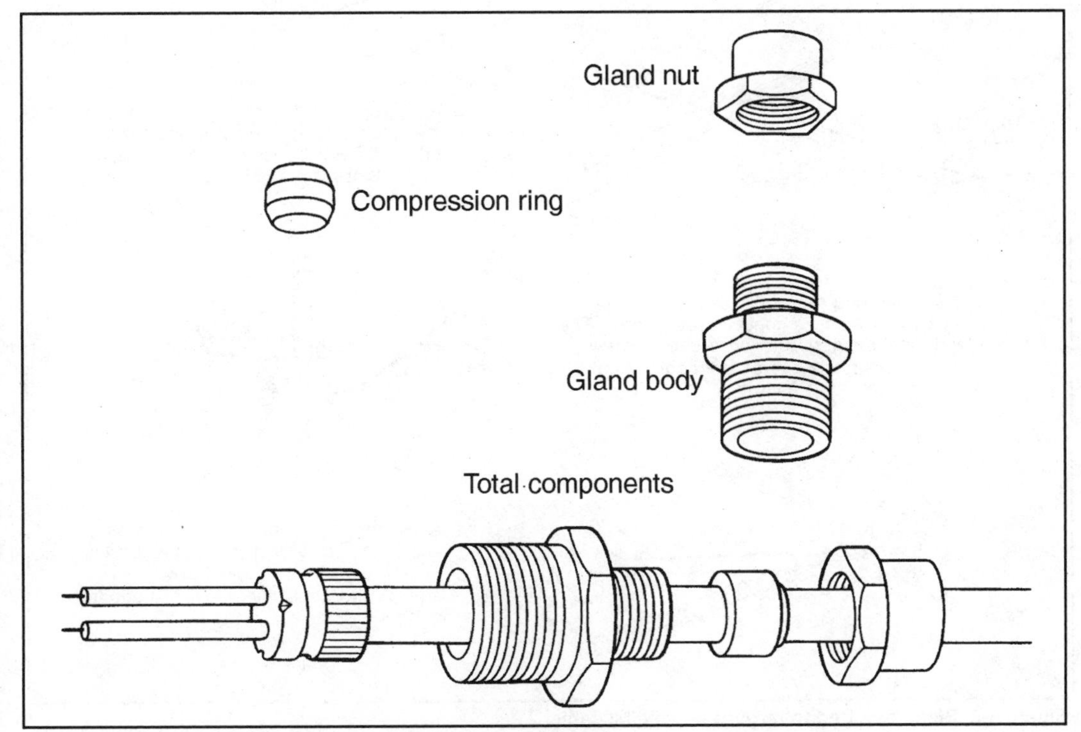

Figure 9-2: Components of threaded gland for Type MI cable.

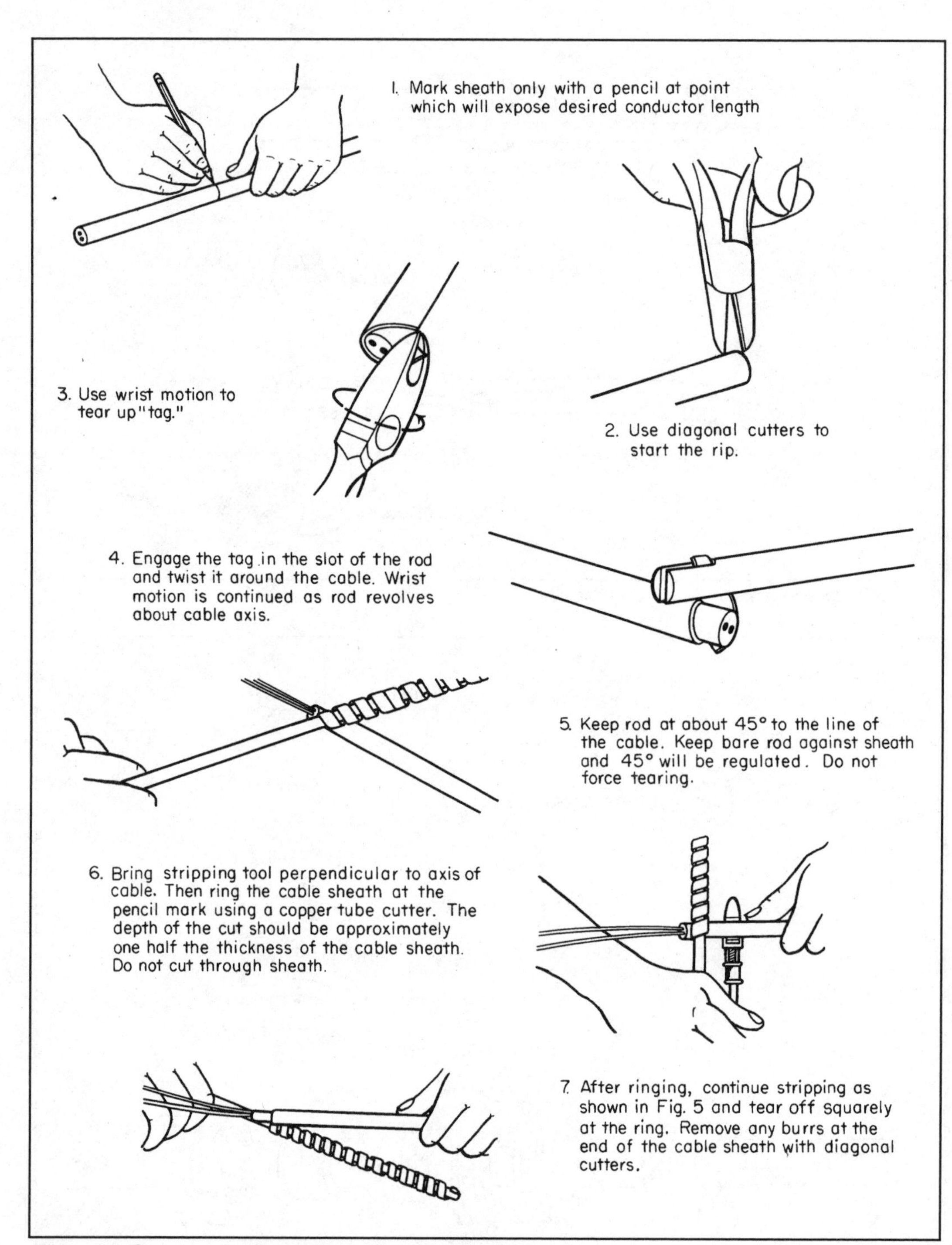

Figure 9-3: Steps required to terminate Type MI cable.

8. Slip the gland connector parts on to the cable in this order: gland nut, compression ring, gland body.
9. Engage the self-threading pot finger tight. See that pot is square. Then screw on with pipe wrench engaging all threads. Examine inside of pot for cleanliness and metallic slivers or dust. Test for alignment by bringing gland body up to enclose pot.

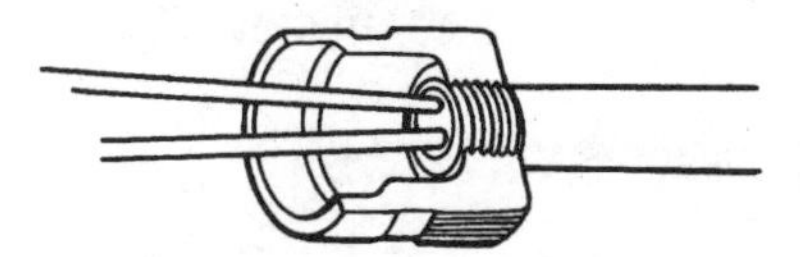

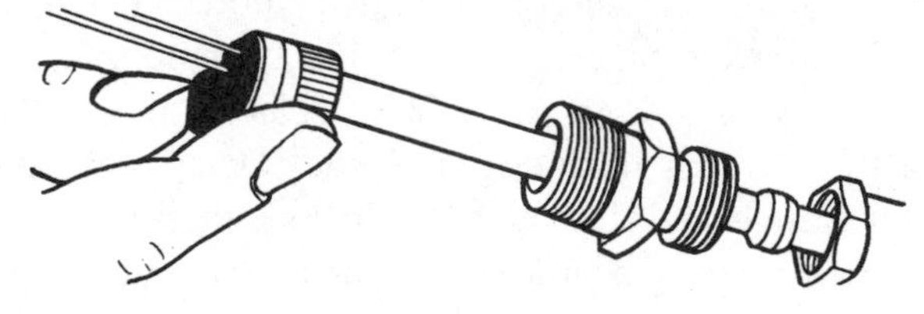

10. Press plastic compound into sealing pot until it is packed tightly. Important: Be sure the hands are free from metal dust or filings while feeding compound into pot.

11. Slip cap and sleeving sub assembly into position

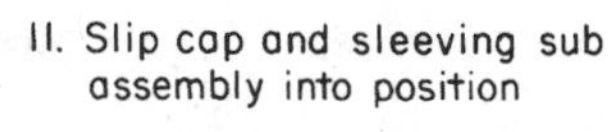

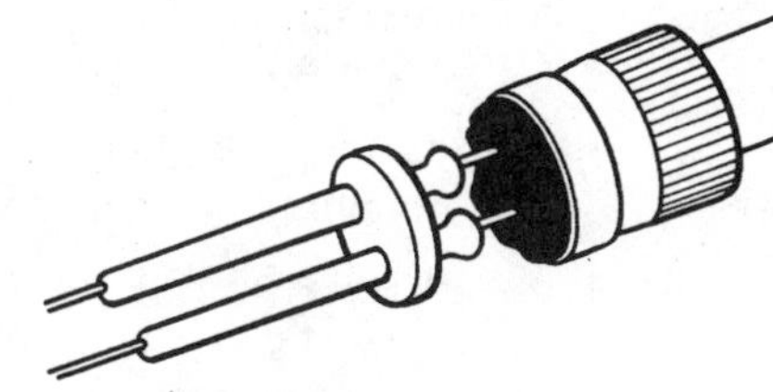

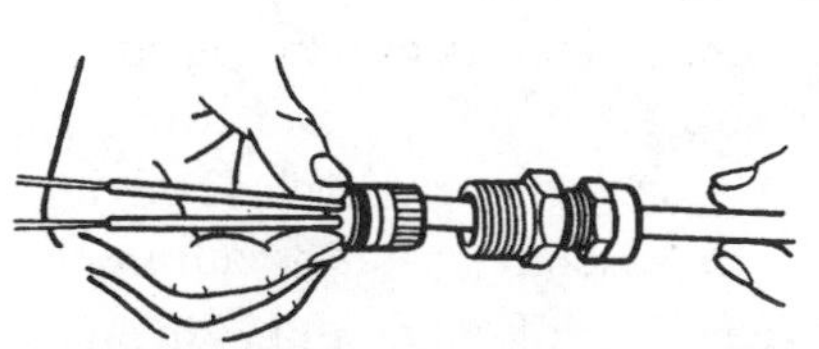

12. Force the insulating cap assembly into position on top of compound.

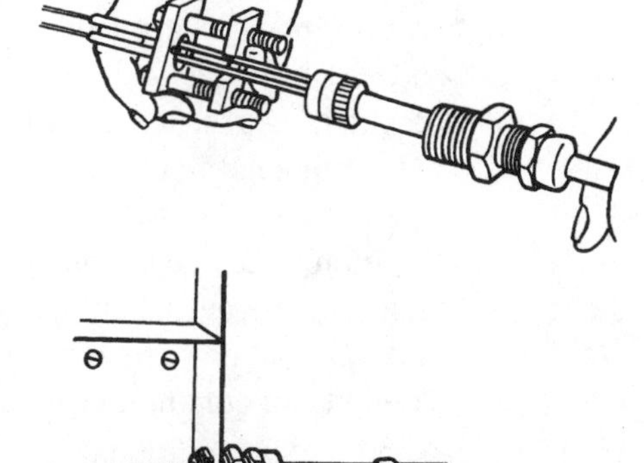

13. Slip the compression and crimping tool over the insulating leads

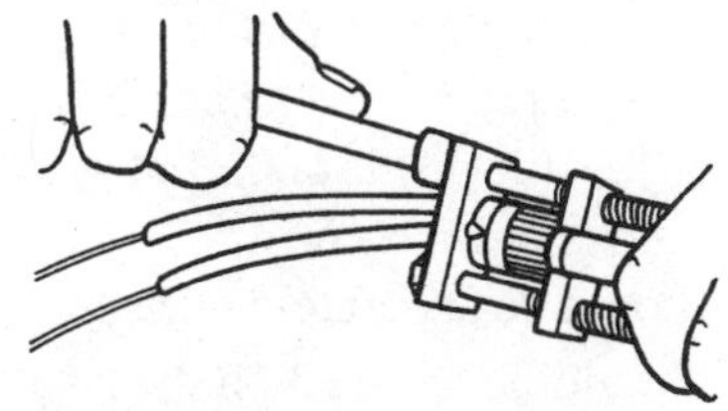

14. After positioning crimping tool on pot, compress the insulating washer into place flush with the top of the pot by tightening the two slotted cap screws. This operation compresses the compound in the pot and crimps the edge of the pot into the insulating cap, holding it in place. Remove compression tool.

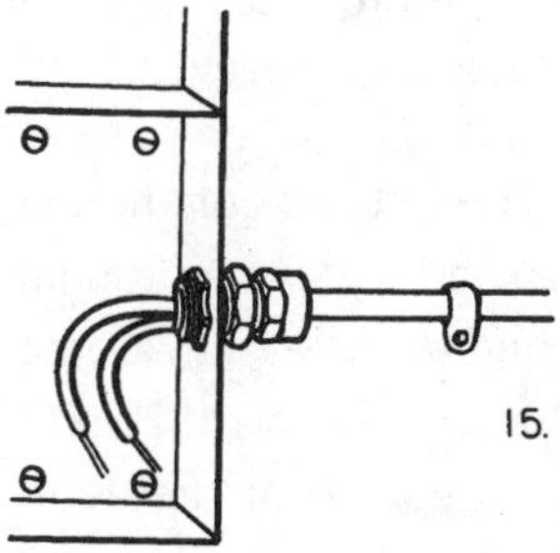

15. Completed termination in box utilizing standard lock nut.

Figure 9-3: Steps required to terminate Type MI cable. ***(Cont.)***

aerial cable on a messenger; in hazardous locations as permitted in *NEC* Articles 501, 502, and 503; in dry locations; and in wet locations under certain conditions as specified in the *NEC*.

RACEWAY SYSTEMS

A raceway wiring system consists of an electrical wiring system in which one or more individual conductors are pulled into a conduit or similar housing, usually after the raceway system has been completely installed. The basic raceways are rigid steel conduit, electrical metallic tubing (EMT), and PVC (polyvinyl chloride) plastic. Other raceways include surface metal moldings and flexible metallic conduit.

These raceways are available in standardized sizes and serve primarily to provide mechanical protection for the wires run inside and, in the case of metallic raceways, to provide a continuously grounded system. Metallic raceways, properly installed, provide the greatest degree of mechanical and grounding protection and provide maximum protection against fire hazards for the electrical system. However, they are more expensive to install.

Most electricians prefer to use a hacksaw with a blade having 18 teeth per in for cutting rigid conduit and 32 teeth per in for cutting the smaller sizes of conduit. For cutting larger sizes of conduit ($1\frac{1}{2}$ in and above), a special conduit cutter should be used to save time. While quicker to use, the conduit cutter almost always leaves a hump inside the conduit and the burr is somewhat larger than that made by a standard hacksaw. If a power band saw is available on the job, it is preferred for cutting the larger sizes of conduit. Abrasive cutters are also popular for the larger sizes of conduit.

Conduit cuts should be made square and the inside edge of the cut must be reamed to remove any burr or sharp edge that might damage wire insulation when the conductors are pulled inside the conduit. After reaming, most experienced electricians feel the inside of the cut with their finger to be sure that no burrs or sharp edges are present.

Lengths of conduit to be cut should be accurately measured for the size needed and an additional $\frac{3}{8}$ in should be allowed on the smaller sizes of conduit for terminations; the larger sizes of conduit will require approximately $\frac{1}{2}$ in for locknuts, bushings, and the like at terminations.

A good lubricant (cutting oil) is then used liberally during the thread-cutting process. If sufficient lubricant is used, cuts may be made cleaner and sharper, and the cutting dies will last much longer.

Full threads must be cut to allow the conduit ends to come close together in the coupling or to firmly seat in the shoulders of threaded hubs of conduit bodies. To obtain a full thread, run the die up on the conduit until the conduit barely comes through the die. This will give a good thread length adequate for all purposes. Anything longer will not fit into the coupling and will later corrode because threading removes the zinc or other protective coating from the conduit.

Clean, sharply cut threads also make a better continuous ground and save much trouble once the system is in operation.

Electrical Metallic Tubing

Electrical metallic tubing (EMT) may be used for both exposed and concealed work except where it will be subjected to severe damage during use, in cinder concrete, or in fill where subjected to permanent moisture unless some means to protect it is provided; the tubing may be installed a minimum of 18 in under the fill.

Threadless couplings and connectors are used for EMT installation and these should be installed so that the tubing will be made up tight. Both set-screw and compression types are commonly in use. Where buried in masonry or installed in wet locations, couplings and connectors, as well as supports, bolts, straps, and screws, should be of a type approved for the conditions.

Bends in the tubing should be made with a tubing bender so that no injury will occur and so the internal diameter of the tubing will not be effectively reduced. The bends between outlets or termination points should contain no more than the equivalent of four quarter-bends (360° total), including those bends located immediately at the outlet or fitting (offsets).

All cuts in EMT are made with either a hacksaw, power hacksaw, tubing cutter, or other approved device. Once cut, the tubing ends should be reamed with a screwdriver handle or pipe reamer to remove all burrs and sharp edges that might damage conductor insulation.

Flexible Metal Conduit

Flexible metal conduit generally is manufactured in two types, a standard metal-clad type and a liquid-tight type. The former type cannot be used in wet locations unless the conductors pulled in are of a type specially approved for such conditions. Neither type may be used where they will be subjected to physical damage or where any combination of ambient and/or conductor temperature will produce an operating temperature in excess of that for which the material is approved. Other uses are fully described in Articles 350 and 351 of the *NEC*.

When this type of conduit is installed, it should be secured by an approved means at intervals not exceeding $4\frac{1}{2}$ ft and within 12 in of every outlet box, fitting, or other termination points. In some cases, however, exceptions exist. For example, when flexible metal conduit must be finished in walls, ceilings, and the like, securing the conduit at these intervals would not be practical. Also, where more flexibility is required, lengths of not more than 3 ft may be utilized at termination points.

Flexible metal conduit may be used as a grounding means where both the conduit and the fittings are approved for the purpose. In lengths of more than 6 ft, it is best to install an extra grounding conductor within the conduit for added insurance.

Liquidtight flexible metal conduit is used in damp or wet locations and is covered in *NEC* Article 351. Please see the *NEC* book for further details.

Surface Metal Molding

When it is impractical to install the wiring in concealed areas, surface metal molding is a good compromise. Even though it is visible, proper painting to match the color of the ceiling and walls makes it very inconspicuous. Surface metal molding is made from sheet metal strips drawn into shape and comes in various shapes and sizes with factory fittings to meet nearly every application found in finished areas of commercial buildings. A complete list of fittings can be obtained at your local electrical equipment supplier.

The running of straight lines of surface molding is simple. A length of molding with the coupling is slipped in the end, out enough so that the screw hole is exposed, and then the coupling is screwed to the surface to which the molding is to be attached. Then another length of molding is slipped on the coupling.

Factory fittings are used for corners and turns or the molding may be bent (to a certain extent) with a special bender. Matching outlet boxes for surface mounting are also available, and bushings are necessary at such boxes to prevent the sharp edges of the molding from injuring the insulation on the wire.

Clips are used to fasten the molding in place. The clip is secured by a screw and then the molding is slipped into the clip, wherever extra support of the molding is needed, and fastened by screws. When parallel runs of molding are installed, they may be secured in place by means of a multiple strap. The joints in runs of molding are covered by slipping a connection cover over the joints. Such runs of molding should be grounded the same as any other metal raceway, and this is done by use of grounding clips. The current-carrying wires are normally pulled in after the molding is in place.

The installation of surface metal molding requires no special tools unless bending the molding is necessary. The molding is fastened in place with screws, toggle bolts, and the like, depending on the materials to which it is fastened. All molding should be run straight and parallel with the room or building lines, that is, baseboards, trims, and other room moldings. The decor of the room should be considered first and the molding made as inconspicuous as possible.

It is often desirable to install surface molding not used for wires in order to complete a pattern set by other surface molding containing current-carrying wires, or to continue a run to make it appear to be part of the room's decoration.

Wireways

Wireways are sheet-metal troughs with hinged or removable covers for housing and protecting wires and cables and in which conductors are held in place after the wireway has been installed as a complete system. They may be used only for exposed work and shouldn't be installed where they will be subject to severe physical damage or corrosive vapor nor in any hazardous location except Class II, Division 2 of the *NEC*.

The wireway structure must be designed to safely handle the sizes of conductors used in the system. Furthermore, the system should not contain more than 30 current-carrying conductors at any cross section. The sum of the cross-sectional areas of all contained conductors at any cross section of a wireway shall not exceed 20 percent of the interior cross-sectioned area of the wireway.

Splices and taps, made and insulated by approved methods, may be located within the wireway provided they are accessible. The conductors, including splices and taps, shall not fill the wireway to more than 75 percent of its area at that point.

Wireways must be securely supported at intervals not exceeding 5 ft, unless specially approved for supports at greater intervals, but in no case shall the distance between supports exceed 10 ft.

Busways

There are several types of busways or duct systems for electrical transmission and feeder purposes as shown in Figures 8-5 through 8-8. Lighting duct, trolley duct, and distribution bus duct are just a few. All are designed for a specific purpose, and electricians should become familiar with all types before an installation is laid out.

Lighting duct, for example, permits the installation of an unlimited amount of footage from a single working platform. As each section and the lighting fixtures are secured in place, the complete assembly is then simply transported to the area of installation and installed in one piece.

Trolley duct is widely used for industrial applications, and where the installation requires a continuous polarization to prevent accidental reversal, a polarizing bar is used. This system provides polarization for all trolley, permitting standard and detachable trolleys to be used on the same run.

Plug-in bus duct is also widely used for industrial applications, and the system consists of interconnected prefabricated sections of bus duct so formed that the complete assembly will be rigid in construction and neat and symmetrical in appearance.

Cable Trays

Cable trays are used to support electrical conductors used mainly in industrial applications, but are sometimes used for communication and data processing conductors in large commercial establishments. The trays themselves are usually made up into a system of assembled, interconnected sections and associated fittings, all of which are made of metal or other noncombustible material. The finished system forms into a rigid structural run to contain and support single, multiconductor, or other wiring cables. Several styles of cable trays are available, including ladder, trough, channel, solid-bottom trays, and similar structures.

Underfloor Raceway

Underfloor raceway consists of ducts laid below the surface of the floor and interconnected by means of fittings and outlet or junction boxes. Both metallic and nonmetallic ducts are used. Obviously, this system must be installed prior to the floor being finished.

Identifying Conductors

The *NEC* specifies certain methods of identifying conductors used in wiring systems of all types. For example, the high leg of a 120/240-volt grounded three-phase delta system must be marked with an orange color for identification; a grounded conductor must be identified either by the color of its insulation, by markings at the terminals, or by other suitable means. Unless allowed by *NEC* exceptions, a grounded conductor must have a white or natural gray finish. When this is not practical for conductors larger than No. 6 AWG, marking the terminals with white color is an acceptable method of identifying the conductors.

Color Coding

Conductors contained in cables are color-coded so that identification may be easily made at each access point. The following lists the color-coding for cables up through five-wire cable:

- Two-conductor cable: one white wire, one black wire, and a grounding conductor (usually bare)
- Three-conductor cable: one white, one black, one red, and a grounding conductor
- Four-conductor cable: fourth wire blue
- Five-conductor cable: fifth wire yellow
- The grounding conductor may be either green or green with yellow stripes

Although some control-wiring and communication cables contain 60, 80, or more pairs of conductors — using a combination of colors — the ones listed are the most common and will be encountered the most on electrical installations.

When conductors are installed in raceway systems, any color insulation is permitted for the ungrounded phase conductors except the following:

White or gray	Reserved for use as the grounded circuit conductor
Green	Reserved for use as a grounding conductor only

Changing Colors

Should it become necessary to change the actual color of a conductor to meet *NEC* requirements or to facilitate maintenance on circuits and equipment, the conductors may be reidentified with colored tape or paint.

For example, assume that a two-wire cable containing a black and white conductor is used to feed a 240-V, two-wire single-phase motor. Since the white colored conductor is supposed to be reserved for the grounded conductor, and none is required in this circuit, the white conductor may be marked with a piece of red tape at each end of the circuit so that everyone will know that this wire is not a grounded conductor.

For a complete listing of cables used in industrial wiring systems, see McGraw-Hill's *Handbook of Electrical Construction Tools and Materials* by Gene Whitson.

Chapter 10

Raceways, Boxes, and Fittings

A raceway is any channel used for holding wires, cables, or busbars, which is designed and used solely for this purpose. Types of raceways include rigid metal conduit, intermediate metal conduit (IMC), rigid nonmetallic conduit, flexible metal conduit, liquid-tight flexible metal conduit, electrical metallic tubing (EMT), underfloor raceways, cellular metal floor raceways, cellular concrete floor raceways, surface metal raceways, wireways, and auxiliary gutters. Raceways are constructed of either metal or insulating material.

Raceways provide mechanical protection for the conductors that run in them and also prevent accidental damage to insulation and the conducting metal. They also protect conductors from the harmful chemical attack of corrosive atmospheres and prevent fire hazards to life and property by confining arcs and flame due to faults in the wiring system.

One of the most important functions of metal raceways is to provide a path for the flow of fault current to ground, thereby preventing voltage build-up on conductor and equipment enclosures. This feature, of course, helps to minimize shock hazards to personnel and damage to electrical equipment. To maintain this feature, it is extremely important that all metal raceway systems be securely bonded together into a continuous conductive path and properly connected to a grounding electrode such as a water pipe or a ground rod.

A box or fitting must be installed at:

- Each conductor splice point

- Each outlet, switch point, or junction point
- Each pull point for the connection of conduit and other raceways

Furthermore, boxes or other fittings are required when a change is made from conduit to open wiring. Electrical workers also install pull boxes in raceway systems to facilitate the pulling of conductors.

In each case — raceways, outlet boxes, pull and junction boxes — the *NEC* specifies specific maximum fill requirements; that is, the area of conductors in relation to the box, fitting, or raceway system. This chapter is designed to cover these *NEC* requirements and apply these rules to practical applications.

CONDUIT FILL REQUIREMENTS

The *NEC* provides rules on the maximum number of conductors permitted in raceways. In conduits, for either new work or rewiring of existing raceways, the maximum fill must not exceed 40 percent of the conduit cross-sectional area. In all such cases, fill is based on using the actual cross-sectional areas of the particular types of conductors used. Other derating rules specified by the *NEC* may be found in Article 310. For example, if more than three conductors are used in a single conduit, a reduction in current-carrying capacity is required. Ambient temperature is another consideration that may call for derating of wires below the values given in *NEC* tables.

The allowable number of conductors in a raceway system is calculated as percentage of fill as specified in Table 1 of *NEC* Chapter 9 (Figure 10-1). When using this table, remember that equipment grounding or bonding conductors, where installed, must be included when calculating conduit or tubing fill. The actual dimensions of the equipment grounding or bonding conductor (insulated or bare) must be used in the calculation.

Number of conductors	1	2	Over 2
All conductor types	53	31	40

Figure 10-1: NEC Chapter 9, Table 1 — percent of cross section of conduit and tubing for conductors.

Conduit fill may be determined in one of two different ways:

- Calculating the fill as a percentage of the conduit's inside diameter (ID)
- Using tables in *NEC* Chapter 9

When determining the number of conductors (all the same size) for use in trade sizes of conduit or tubing ½ in through 6 in, refer to Table C1 in *NEC* Appendix C. For example, let's assume that four 500 kcmil THHN conductors must be installed in a rigid conduit run. What size of conduit is required for four 500 kcmil THHN conductors?

Turn to Table C1 in the *NEC* and scan down the left-hand column until the insulation type (THHN) is found. Once the insulation type has been located, move to the second column (Conductor Size AWG/kcmil) and scan down this column until 500 kcmil is found. Now, scan across this row until the desired number of conductors (four in this case) is found. When the number "4" has been located, move up this column to see the size of conduit required at the top of the page. In doing so, we can see that 3-in conduit is the size to use.

If compact conductors (all the same size) are used, refer to *NEC* Table C1A for trade sizes of conduit or tubing ½ in through 4 in. For example, let's assume that four 500 kcmil compact conductors are to be installed in a raceway system. What size of conduit is required?

Refer to *NEC* Table C1A. Scan down the left-hand column until the insulation type is found, then jump to the next column to find the wire size; move across this row until the number of conductors (4) is located. Note that this row jumps from 3 to 5 conductors. Therefore, the 5-conductor column must be used. Now scan upward until the required conduit size is located. Again, the conduit size is 3 in — the same as in our previous example. Therefore, no savings in conduit size will be realized in this case if compact conductors are used. Furthermore, the compact conductors will be more costly and it will be to everyone's advantage to stick with conventional conductors in this situation.

When working with conductors larger than 750 kcmil or combinations of conductors of different sizes, Tables in *NEC* Chapter 9 and the appendices should be used to obtain the dimensions of conductors, conduit, and tubing. These tables give the nominal size of conductors or tubing for use in computing the required size of conduit or tubing for various combinations of conductors. The dimensions represent average conditions only, and variations will be found in dimensions of conductors and conduits of different manufacturers.

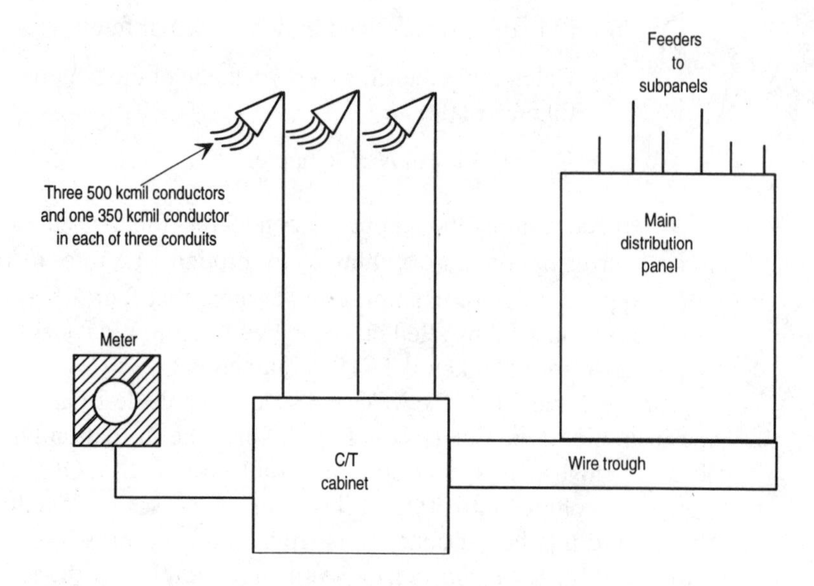

Figure 10-2: Power-riser diagram of a 1200-A service-entrance.

Note: *Where the calculated conductors, all of the same size (total cross-sectional area including insulation), include a decimal fraction, the next higher whole number must be used where the decimal is 0.8 or larger.*

Let's take a situation where a 1200-A service entrance is to be installed utilizing three parallel conduits, each containing three 500 kcmil THHN ungrounded conductors and one 350 kcmil grounded THHN conductor (neutral) as shown in Figure 10-2. What size of rigid conduit is required?

Step 1. Refer to *NEC* Chapter 9, Table 5 to determine the area (in^2) of 500 kcmil THHN conductor. The area is found to be .7073 in^2.

Step 2. Since there are three conductors of this area in each conduit, .7073 is multiplied by 3 to obtain the total in^2 for all three conductors.

$$.7073 \times 3 = 2.1219$$

Step 3. Refer again to *NEC* Chapter 9, Table 5 to obtain the area of the one 350 kcmil grounded conductor in each conduit. The area is found to be .5242 in.

Step 4. Add the total area of the three 500 kcmil conductors (2.1219) plus the area of the one 350 kcmil conductor (.5242) to obtain the total area occupied by all four conductors in the conduit.

$$2.1219 + .5242 = 2.646 \text{ in}$$

Step 5. Referring to *NEC* Table 4 of *NEC* Chapter 9, we look in the "Over 2 Wires, 40 percent" column and scan down until we come to a figure that is the closest to 2.646 in without going under this figure. In doing so, we find that the closest area is 3.538. Scan to the left from this figure (in the same column) and note that the conduit size is 3 in.

Therefore, each of the three conduits containing three 500 kcmil THHN conductors and one 350 kcmil THHN conductor must be 3 in ID to comply with the *NEC*.

Refer again to Figure 10-2 and note the conduit routed from the C/T cabinet to the meter location. If this conduit contains six #12 AWG THHN conductors, what size of conduit is required?

Since all conductors are the same size, refer to *NEC* Table C8, scan down the left-hand column until THHN insulation is found, and then move to the right one column and find #12 AWG conductor. Note that the next column to the right lists nine #12 conductors in ½-in conduit. Since six conductors are all that are in this conduit run, and since ½ in is the smallest trade size conduit, this will be the size to use.

CONDUIT BODIES, PULL BOXES, AND JUNCTION BOXES

The *NEC* specifically states that at each splice point, or pull point for the connection of conduit or other raceways, a box or fitting must be installed. The *NEC* further considers conduit bodies, pull boxes, and junction boxes and specifies the installation rules for these also.

Conduit bodies provide access to the wiring through removable covers. Typical conductors must have an area twice that of the largest conduit to which they are attached, but the number of conductors within the body must not exceed that allowed in the conduit. If a conduit body has entry for three or more conduits such as Type T or X, splices may be made within the conduit body. Splices may not be made in conduit bodies having one or two entries unless the volume is sufficient to qualify the conduit body as a junction box or device box.

When conduit bodies or boxes are used as junction boxes or as pull boxes, a minimum size box is required to allow conductors to be installed without undue bending. The calculated dimensions of the box depend on the type of conduit arrangement and on the size of the conduits involved.

Sizing Pull and Junction Boxes

Figure 10-3 shows a junction box with several conduits entering it. Since 4-in conduit is the largest size in the group, the minimum length required for the box can be determined by the following calculation:

Trade size of conduit x 8 (as per *NEC*)
= minimum length of box

$$4 \times 8 = 32 \text{ in}$$

Therefore, this particular pull box must have a length of at least 32 in. The width of the box, however, need be only of sufficient size to enable locknuts and bushings to be installed on all the conduits or connectors entering the enclosure.

Junction or pull boxes in which the conductors are pulled at an angle as shown in Figure 10-4 must have a distance of not less than six times the trade diameter of the largest conduit. The distance must be increased for

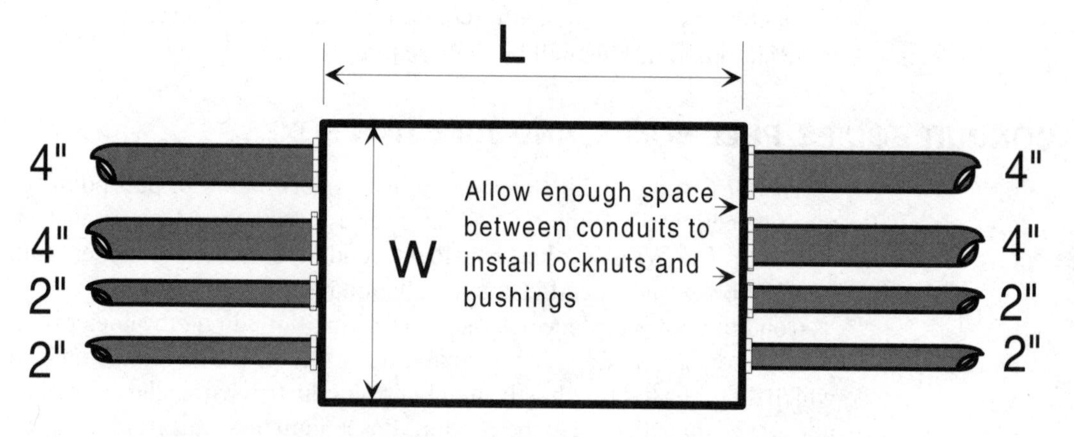

Figure 10-3: Pull box used on straight pulls.

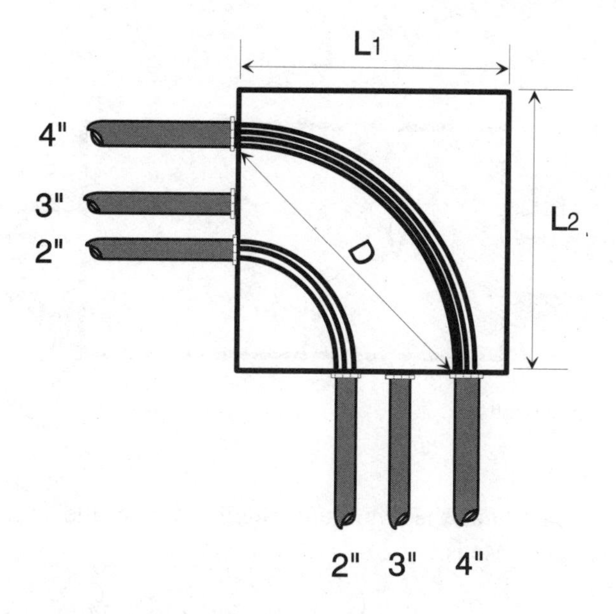

Figure 10-4: Junction box with conduit runs entering at right angles.

additional conduit entries by the amount of the sum of the diameter of all other conduits entering the box on the same side, that is, the wall of the box. The distance between raceway entries enclosing the same conductors must not be less than six times the trade diameter of the largest conduit.

The 4-in conduit is the largest of the lot in this case; consequently,

$$L1 = 6 \times 4 + (3 + 2) = 29 \text{ in}$$

Since the same number and sizes of conduit is located on the adjacent wall of the box, L2 is calculated in the same way; therefore, L2 = 29 in.

The distance (D) = 6 × 4 or 24 in, and this is the minimum distance permitted between conduit entries enclosing the same conductor.

The depth of the box need only be of sufficient size to permit locknuts and bushings to be properly installed. In this case, a 6-in deep box would suffice.

If the conductors are smaller than No. 4, the length restriction does not apply.

Figure 10-5 on the next page shows another straight-pull box. What is the minimum length if the box has one 3-in conduit and two 2-in conduits entering and leaving the box? Again, refer to *NEC* Section 370-28(a)(1)

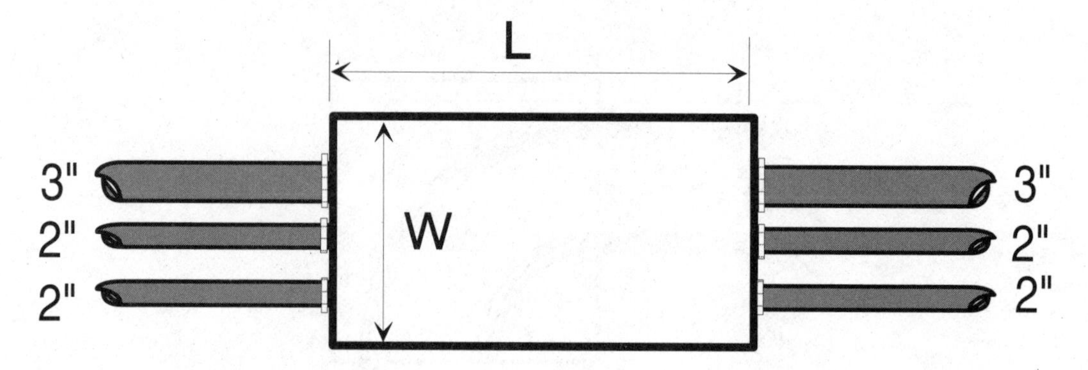

Figure 10-5: Typical straight-pull box.

and find that the minimum length is 8 times the largest conduit size which in this case is:

8 x 3 in = 24 in

Let's review the installation requirements for pull or junction boxes with angular or U-pulls. Two conditions must be met in order to determine the length and width of the required box.

The minimum distance to the opposite side of the box from any conduit entry must be at least six times the trade diameter of the largest raceway.

The sum of the diameters of the raceways on the same wall must be added to this figure.

Figure 10-6 shows the minimum length of a box with two 3-in conduits, two 2-in conduits, and two $1\frac{1}{2}$-in conduits in a right-angle pull. The minimum length based on this configuration is:

6×3 in	=	18 in
1×3 in	=	3 in
2×2 in	=	4 in
$2 \times 1\frac{1}{2}$ in	=	3 in
		28 in

Since the number and size of conduits on the two sides of the box are equai, the box is square and has a minimum size dimension of 28 in. However, the distance between conduit entries must now be checked to ensure that all *NEC* requirements are met; that is, the spacing (D) between conduits enclosing the same conductor must not be less than six times the

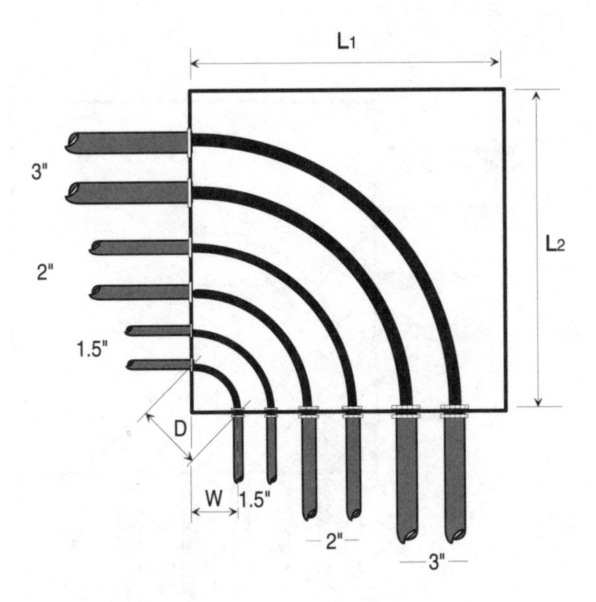

Figure 10-6: Minimum size pull box for angle conduit entries.

conduit diameter. Again refer to Figure 10-6 and note that the $1\frac{1}{2}$-in conduits are the closest to the left-hand corner of the box. Therefore, the distance (D) between conduit entries must be:

$$6 \times 1\frac{1}{2} = 9 \text{ in}$$

The next group is the two 2-in conduits which is calculated in a similar fashion; that is:

$$6 \times 2 = 12 \text{ in}$$

The remaining raceways in this example are the two 3-in conduits and the minimum distance between the 3-in conduit entries must be:

$$6 \times 3 = 18 \text{ in}$$

A summary of the conduit-entry distances is presented in Figure 10-7. However, some additional math is required to obtain the spacing (w) between the conduit entries. For example, the distance from the corner of

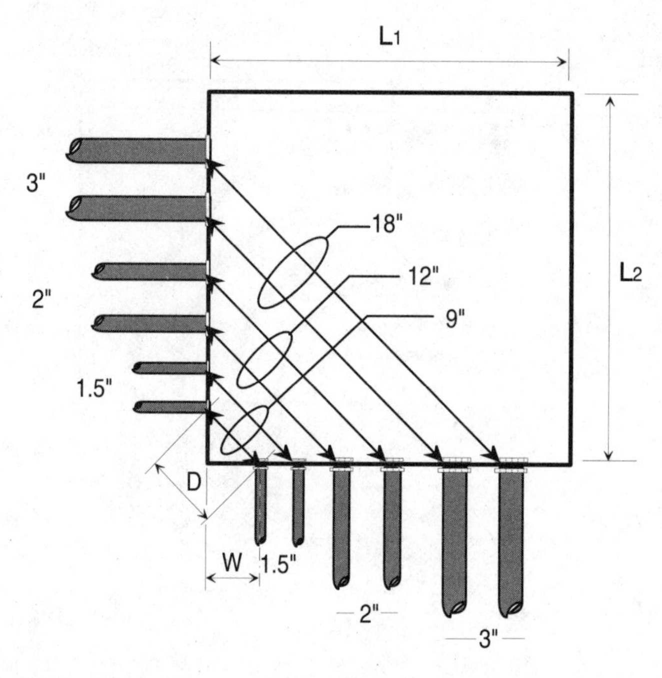

Figure 10-7: Required distances between conduit entries.

the pull box to the center of the conduits (w) may be found by the following equation:

$$\text{Spacing} = \frac{\text{Diagonal distance } (D)}{\sqrt{2}}$$

Consequently, the spacing (w) for the 1½-in conduit may be determined using the following equation:

$$\frac{9}{\sqrt{2}} = \frac{9}{1.414} = 6.4''$$

Therefore, the spacing (w) is 6.4 in. This distance is measured from the left lower corner of the box in each direction — both vertically and horizontally — to obtain the center of the first set of 1½-in conduits. This distance must be added to the spacing of the other conduits including locknuts or bushings.

Note: A rule-of-thumb is to allow ½ in clearance between locknuts.

Using all information calculated thus far, and using Figure 10-7 as reference, the required measurements of the pull box may be further calculated as follows:

Step 1. Calculate space (w):

$$D = 6 \times 1\tfrac{1}{2} = 9 \text{ in}$$

Step 2. Divide this number (9″) by the square root of 2 (1.414) and make the following calculation:

$$w = \frac{9''}{1.414} = 6.4''$$

Step 3. Measure from the left, lower corner of the pull box over 6.4 in to obtain the center of the knockout for the first 1½-in conduit. Measure up (from the lower, left corner) to obtain the center of the knockout for this same cable run on the left side of the pull box.

Step 4. Since there are two 3-in (inside diameter) conduits, each with an outside diameter of approximately 4.25 in, the space for these two conduits can be found by the following equation:

$$2 \times 4.25 = 8.5 \text{ in}$$

Step 5. The space required for the two 2-in (inside diameter) conduits, each with an outside diameter of approximately 3.12 in, may be determined in a similar manner; that is:

$$2 \times 3.12 = 6.24 \text{ in}$$

Step 6. The space required for the two 1.5-in (inside diameter) conduits, each with an outside diameter of approxiamtely 2.62 in, may be determined using the same equation:

$$2 \times 2.62 = 5.24 \text{ in}$$

Step 7. To find the required space for locknuts and bushings, multiply 0.5 in by the total number of conduit entries on one side of the box. Since there are a total of 6 conduit entries, use the following equation:

$$6 \times .5 = 3.0 \text{ in}$$

Step 8. Add all figures obtained in Steps 2 through 7 together to obtain the total required length of the pull box.

Clear space (w)	=	6.40 in
1.5-in conduits	=	5.24 in
2-in conduits	=	6.24 in
3-in conduits	=	8.50 in
Space between locknuts	=	3.00 in
Total length of box	=	29.38 in

Since the same number and size of conduits enter on the bottom side of the pull box and leave, at a right angle, on the left side of the pull box, the box will be square. Furthermore, although a box exactly 29.38 in will suffice for this application, the next larger standard size is 30 in; this should be the size pull box selected. Even if a "custom" pull box is made in a sheet-metal shop, the workers will still probably make it an even 30 in unless specifically ordered otherwise.

Cabinets and Cutout Boxes

NEC Article 373 deals with the installation requirements for cabinets, cutout boxes, and meter sockets. In general, where cables are used, each cable must be secured to the cabinet or cutout box by an approved method. Furthermore, the cabinets or cutout boxes must have sufficient space to accommodate all conductors installed in them without crowding.

NEC Table 373-6(a) gives the minimum wire-bending space at terminals along with the width of sizing gutter in inches. Figure 10-8 gives a summary of *NEC* requirements for the installation of cabinets and cutout boxes.

Other basic *NEC* requirements for cabinets and cutout boxes are as follows:

- Table 373-6(a) must apply where the conductor does not enter or leave the enclosure through the wall opposite its terminal.

- *Exception No. 1 states:* A conductor must be permitted to enter or leave an enclosure through the wall opposite its terminal provided the conductor enters or leaves the enclosure where the gutter joins an adjacent gutter that has a width that conforms to Table 373-6(b) for that conductor.
- *Exception No. 2 states:* A conductor not larger than 350 kcmil must be permitted to enter or leave an enclosure containing only a meter socket(s) through the wall opposite its terminal, provided the terminal is a lay-in type where either: (a) the terminal is directly facing the enclosure wall and offset is not greater than 50 percent of the bending space specified in Table 373-6(a), or (b) the terminal is directed toward the opening in the enclosure and is within a 45-degree angle of directly facing the enclosure wall.
- Table 373-6(b) must apply where the conductor enters or leaves the enclosure through the wall opposite its terminal.

NEC Article 374 covers the installation requirements for auxiliary gutters, which are permitted to supplement wiring spaces at meter centers, distribution centers, and similar points of wiring systems and may enclose conductors or busbars but must not be used to enclose switches, overcurrent devices, appliances, or other similar equipment.

In general, auxiliary gutters must not contain more than thirty (30) current-carrying conductors at any cross section. The sum of the cross-sectional areas of all contained conductors at any cross section of an auxiliary gutter must not exceed 20 percent of the interior cross-sectional area of the auxiliary gutter. We discussed earlier that conductors installed in conduits and tubing must not exceed 40 percent fill. Auxiliary gutters are limited to only 20 percent.

When dealing with auxiliary gutters, always remember the number "thirty." This is the maximum number of conductors allowed in any auxiliary gutter regardless of the cross-sectional area. This question will be found on almost every electrician's examination in the country. Consequently, this number should always be remembered.

The *NE*C specifies certain fill requirements for raceways, outlet boxes, pull and junction boxes, cabinets, cutout boxes, auxiliary gutters, and similar conductor-containing housings. In some cases, *NEC* tables may be used to determine the proper size of housing; in other cases, calculations are required in conjunction with tables and manufacturers' specifications.

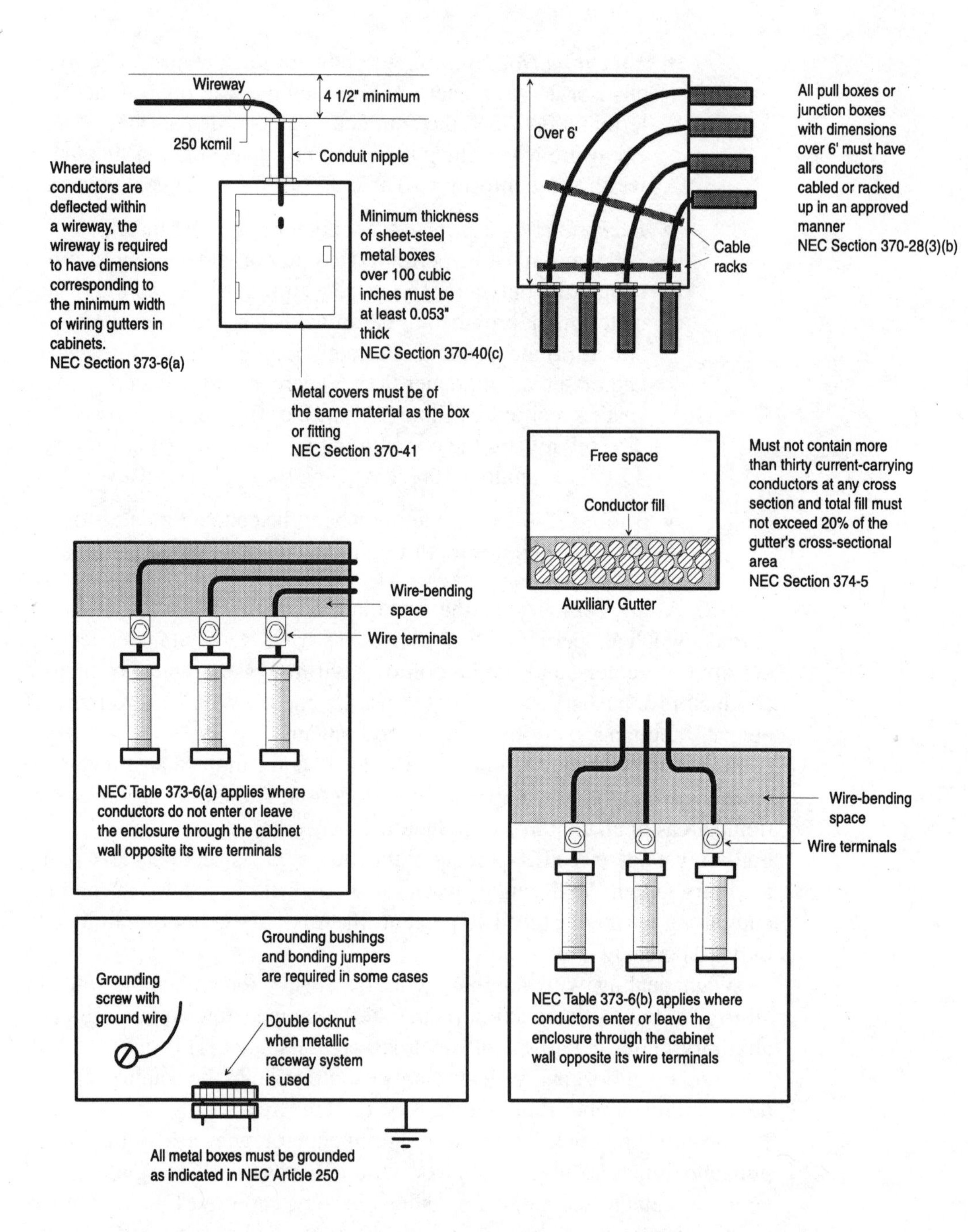

Figure 10-8: NEC installation requirements for cabinets, cutout boxes, and auxiliary gutters.

Chapter 11

Cable Tray

NEC Section 318-2 defines *cable tray system* as a unit or assembly of units or sections, and associated fittings, forming a rigid structural system used to support cables and raceways. *NEC* Article 318 — containing *NEC* Sections 318-1 through 318-13 — covers cable tray installations, along with the types of conductors to be used in such systems. Whenever a question arises concerning cable tray installations, this is the *NEC* Article to use.

Cable trays are the usual means of supporting cable systems in industrial applications. The trays themselves are usually made up into a system of assembled, interconnected sections and associated fittings, all of which are made of metal or noncombustible units. The finished system forms into a continuous rigid assembly for supporting and carrying single, multiconductor, or other electrical cables and raceways from their origin to their point of termination, frequently over considerable distances. Several styles of cable tray are available, including ladder, trough, channel, solid-bottom trays, and similar structures.

Cable tray is fabricated from both aluminum and steel. Some manufacturers provide an aluminum cable tray that is coated with PVC for installation in caustic environments. Relatively new all-nonmetallic trays are also available; this type of tray is ideally suited for use in corrosive areas and in areas requiring voltage isolation. Furthermore, cable tray is available in three basic forms:

- Ladder
- Trough
- Solid bottom

Ladder tray, as the name implies, consists of two parallel channels connected by rungs, similar in appearance to a conventional straight or extension ladder. Trough types consist of two parallel channels (side rails) having a corrugated, ventilated bottom. The solid-bottom cable tray is similar to the trough except that this type of trough has a corrugated, solid bottom. All of these types are shown in Figure 11-1. Ladder, trough, and solid-bottom trays are completely interchangeable; that is, all three types can be used in the same run when needed.

Cable tray is manufactured in 12- and 24-ft lengths. Common widths include 6, 9, 12, 24, 30 and 36 in. All sizes are provided in either 3-, 4-, 5-, or 6-in depths.

Cable tray sections are interconnected with various types of fittings. Fittings are also used to provide a means of changing the direction or dimension of the cable-tray system. Some of the more common fittings include:

- Horizontal and vertical tees
- Horizontal and vertical bends
- Horizontal crosses
- Reducers
- Barrier strips
- Covers
- Splice plates
- Box connectors

Again, all of these fittings are shown in Figure 11-1.

The area of a cable tray cross section which is usable for cables is defined by width (*W*) x depth (*D*), as shown in Figure 11-2. The overall dimensions of a cable tray, however, are greater than *W* and *D* because of the side flanges and seams. Therefore, overall dimensions vary according to the tray design. Cables rest upon the bottom of the tray and are held within the tray area by two longitudinal side rails as shown in Figure 11-3.

A *channel* is used to carry one or more cables from the main tray system to the vicinity of the cable termination (see Figure 11-4). Conduit is then used to finish the run from the channel to the actual termination.

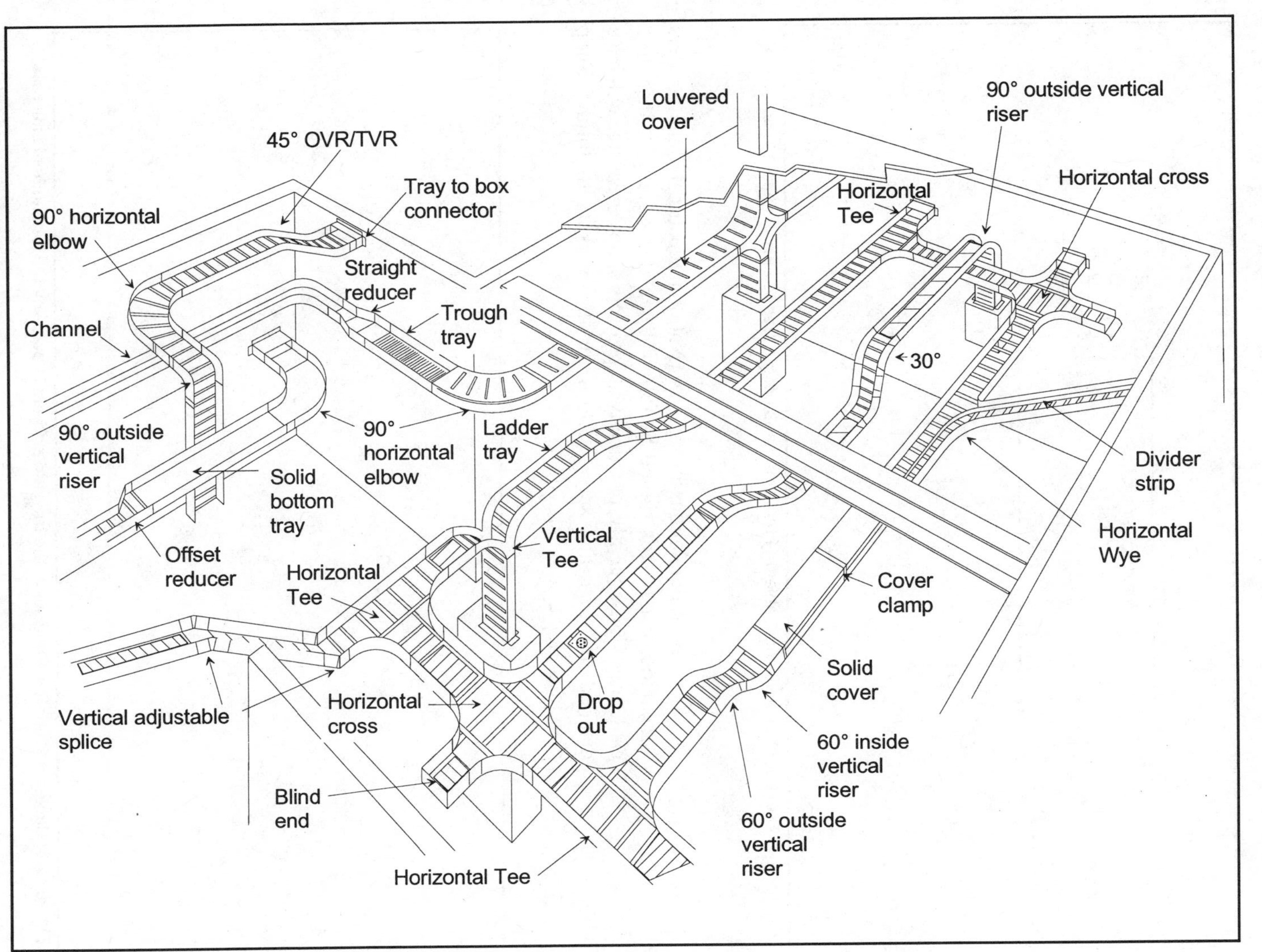

Figure 11-1: Typical cable tray system.

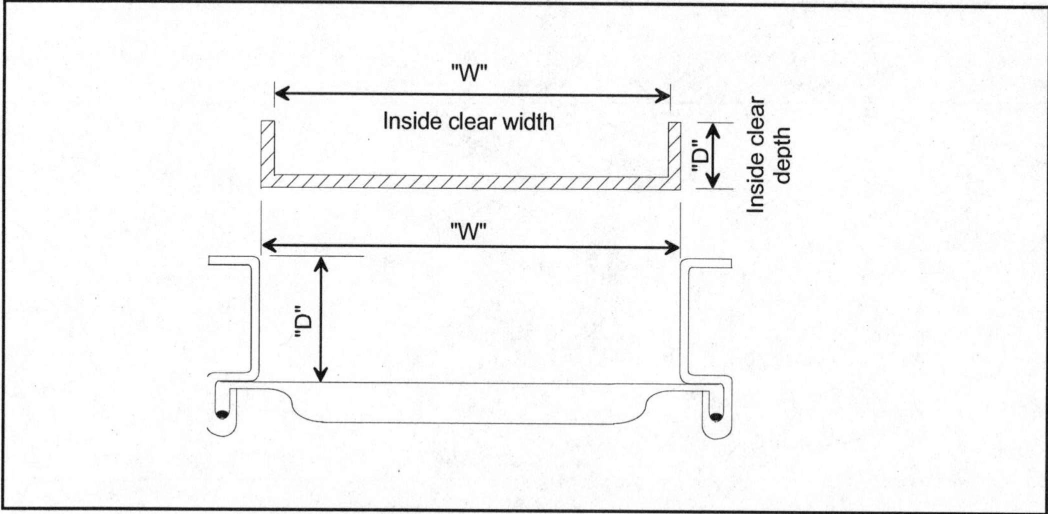

Figure 11-2: Cross-sectional view of cable tray that compares usable area to overall dimensions.

Certain *NEC* regulations and NEMA (National Electrical Manufacturers Association) standards should be followed when designing or installing cable tray. Consequently, practically all projects of any great size will have detailed drawings and specifications for the workers to follow. Shop drawings may also be provided.

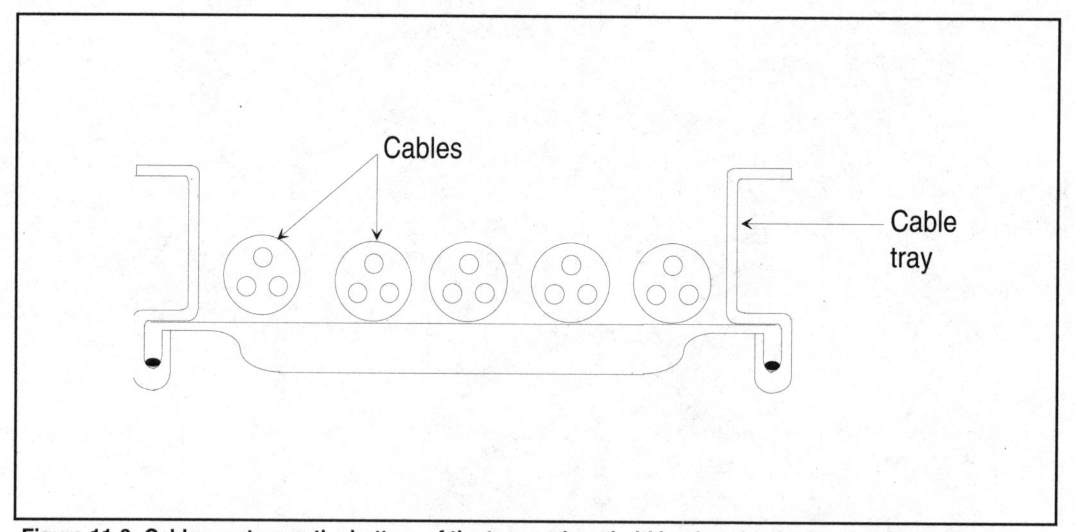

Figure 11-3: Cables rest upon the bottom of the tray and are held in place by the longitudinal side rails.

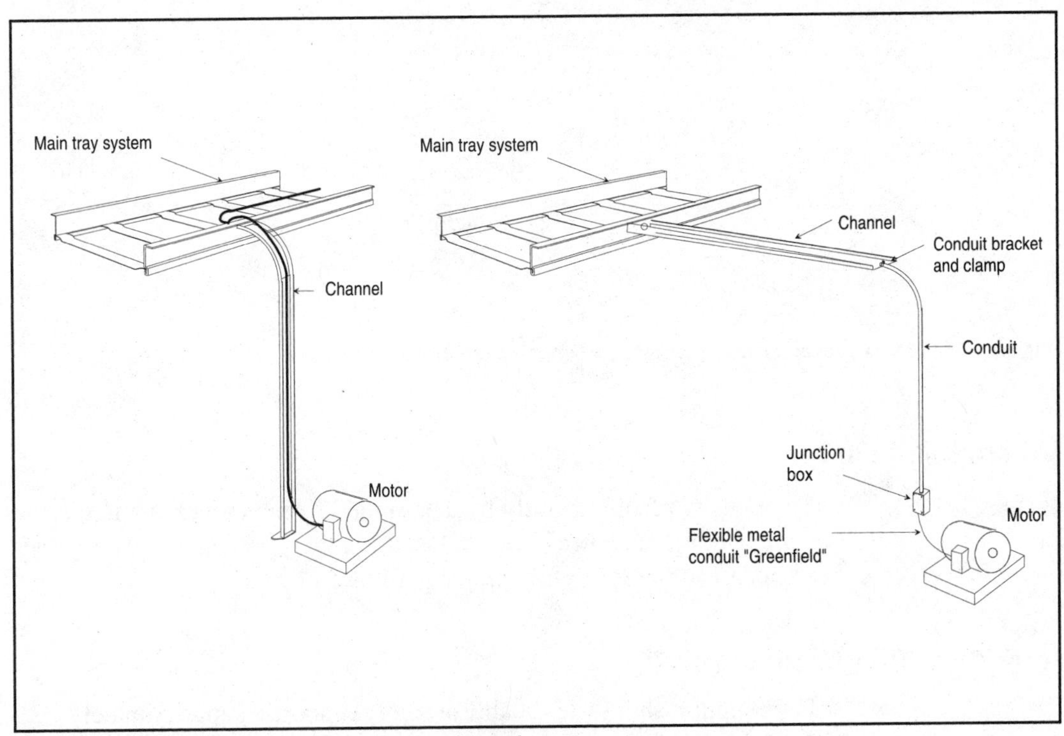

Figure 11-4: Two applications of cable tray channel.

NEMA STANDARDS

The National Electrical Manufacturer's Association (NEMA) is a non-profit organization supported by the manufacturers of electrical equipment and supplies. NEMA develops standards that are used when purchasing and installing electrical equipment. These standards help to insure that the product will match the application.

Load Factors

Load capacity of cable tray varies with each manufacturer and depends upon shape and thickness of the side rails, shape and thickness of the bottom members, spacing of rungs (if any), material used, safety factor used, method used to determine the allowable load, method of supporting tray, and the volume capacity. Consequently, each manufacturer publishes load data.

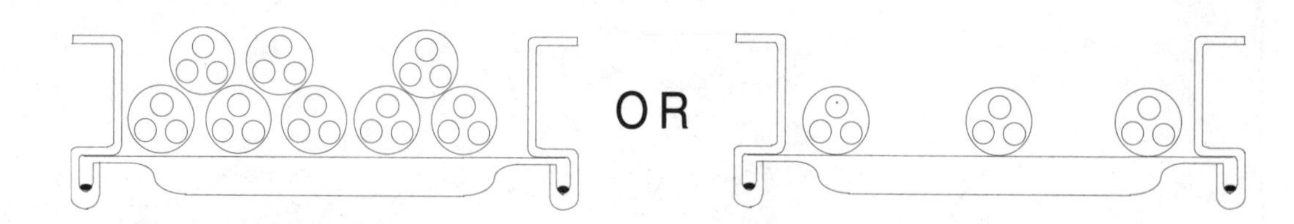

Figure 11-5: Cables packed closely together can impair each other's efficiency.

Determining Fill

The density of fill can only be determined by personnel laying out the system. In doing so, however, it is a fact that cables packed closely together can impair each other's efficiency. *See* Figure 11-5.

Determining Load on Support

Each support should be capable of safely supporting approximately 1.25 times the full weight of the cable and tray on a typical span as shown in Figure 11-6.

Uniform load of tray and cable

$$= \frac{\text{Full weight of tray and cable}}{\text{Total linear feet}}$$

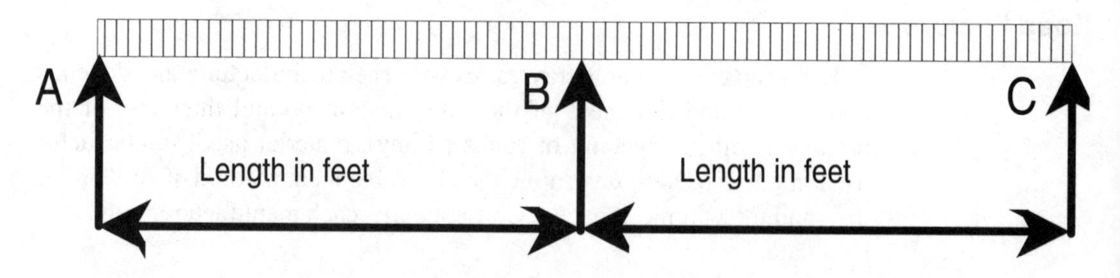

Figure 11-6: Each tray support should be capable of safely supporting the entire tray and cable assembly.

Deflection Under Load

Consider the case of a cable tray spanning only two supports (simple span). As the tray is loaded, the side rails take the deflected form. Simultaneously, in cross section, the side rails rotate inward or outward and the tray bottom deflects. The amount of inward or outward movement of the side rails is a critical factor in the ability of the tray to carry a load.

Failure Under Load

There are two types of failures that can occur with a loaded cable tray. These are longitudinal (side rail) failure and transverse (rung) failure. Transverse (rung) failure occurs when the load applied causes the fibers on the tension side (bottom edge) of the rung to stretch and permanently deform. Simultaneously, fibers on the compression side (top edge) are permanently crushed. Longitudinal (side rail) failure occurs either as a bending failure when the tray is supported on a larger span, or, on longer spans, buckling failure may occur because the side rails of the tray have little resistance to inward or outward movement. As the tray deflects, the side rails rotate and the top (compression) flanges of the tray buckle. Bending failure occurs on short spans because the side rails of the tray have greater resistance to rotation and remain reasonably upright. The tray does not fail until the load is such that it causes the fibers on the tension side (bottom edge) of the side rail to stretch and permanently deform. Simultaneously, fibers on the compression side (top edge) are permanently crushed. For any tray on an intermediate span it is difficult to anticipate whether bending or buckling failure would occur (*see* Figure 11-7).

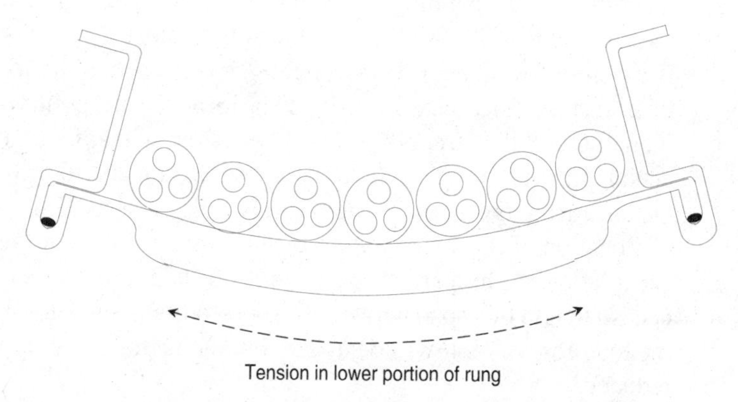

Figure 11-7: Two types of loaded cable tray failures include side rail and run failure.

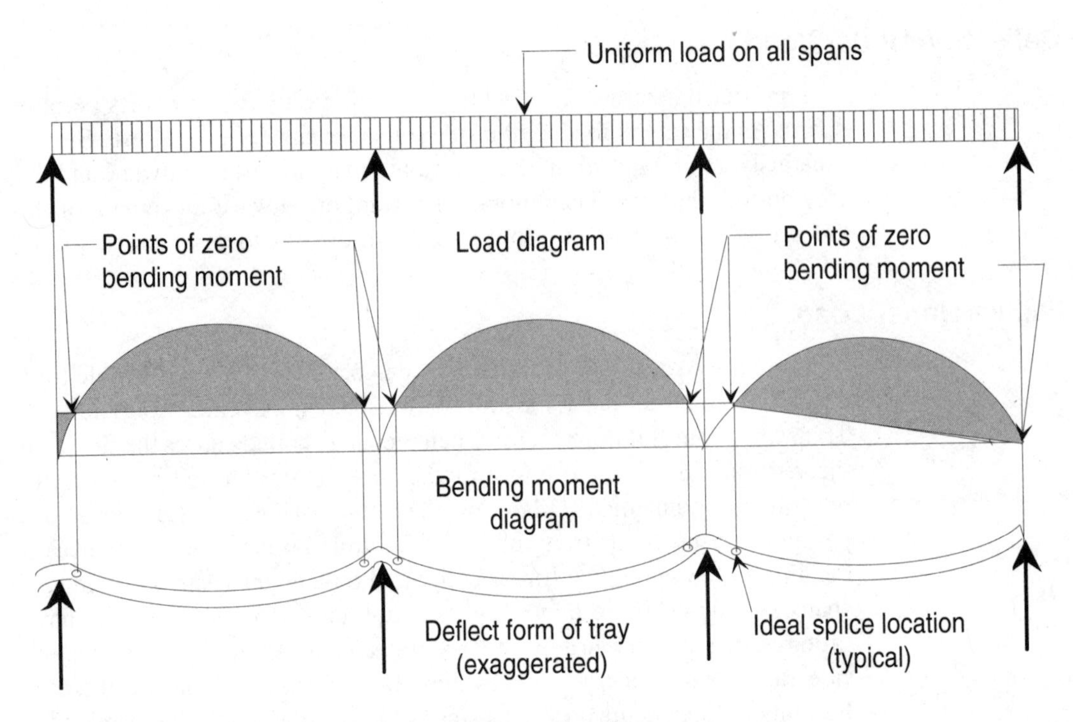

Figure 11-8: Load of cable creates bending moments along the span.

Splicing Straight Sections

In the diagram in Figure 11-8, the load on the cable tray creates bending moments along the spans. Stress in the side wall or rails of the tray is directly related to the bending moments at all points along the tray. The magnitude of the bending moment at any point is determined by measuring the vertical height of the shaded portion. In any cable tray system a splice is a point of weakness. Consequently, splices should be located at points of least stress. Ideally, splices would be located at the points of zero bending moment, and the strength of the tray system would be at a maximum. In actual practice, if the splice is located within one-fourth of the span's distance from the support, the result will be close to ideal.

Location of the splice within the one-fourth points of the span requires extra labor on the part of the installer. When a splice occurs within the central length of a span between the one-fourth points, the tray will support the load for which it was designed, but the factor of safety will be greatly reduced.

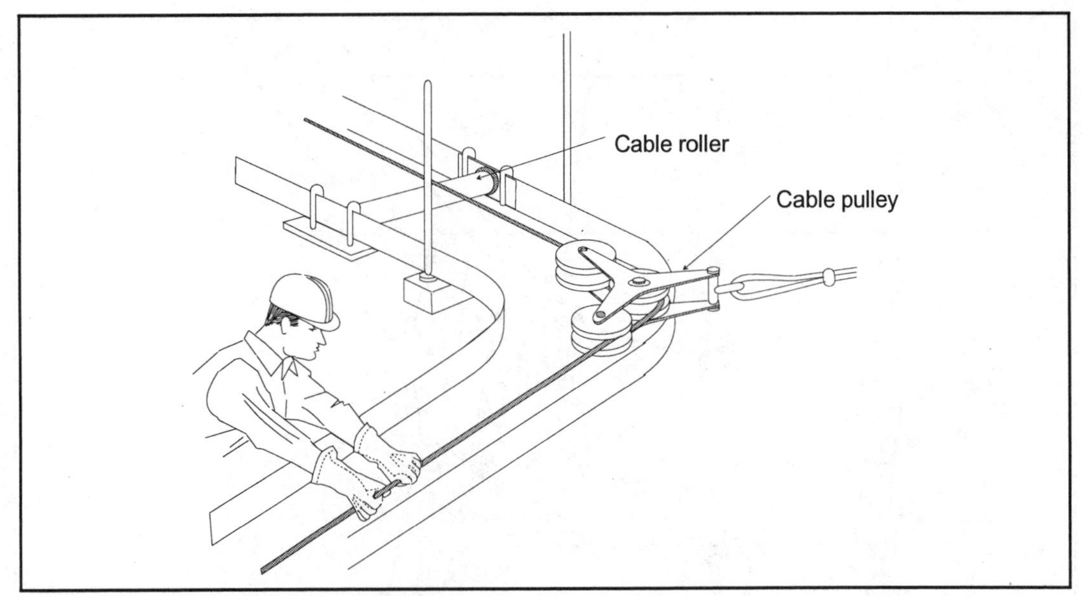

Figure 11-9: Cable pulley used to facilitate a cable pull in cable tray.

Cable Placement In Tray

Cables are placed in the tray either by being pulled along the tray or by being laid in over the side. Figure 11-9 shows a cable pulley being used to facilitate a cable pull in a tray supported by trapeze hangers.

When long lengths of cable are to be installed in raceways or cable trays, problems are frequently encountered, particularly when the cable has to be pulled directly into the tray and changes in direction of the tray sections are involved. An entire cable-pulling system has to be planned and set up so that the cable may be pulled into the trays without scuffing or cutting the sheathing and insulation, and also to avoid damaging the cable trays or the tray hangers. To accomplish a successful cable tray pull, a complete line of installation tools is available, developed through field experience, for pulling long lengths of cable up to 1000 ft or longer. These "tools" consist mainly of conveyor sheaves and cable rollers. Figure 11-10 shows a partial cable tray system with sheaves and rollers in place.

Short lengths of cable can be laid in place without tools or pulled with a basket grip. Long lengths of small cable, 2 in or less in diameter, can also be pulled with a basket grip. Larger cables, however, should be pulled by the conductor and the braid, sheath, or armor. This is done with a pulling eye applied at the cable factory or by tying the conductor to the eye of a basket grip and taping the tail end of the grip to the outside of the cable.

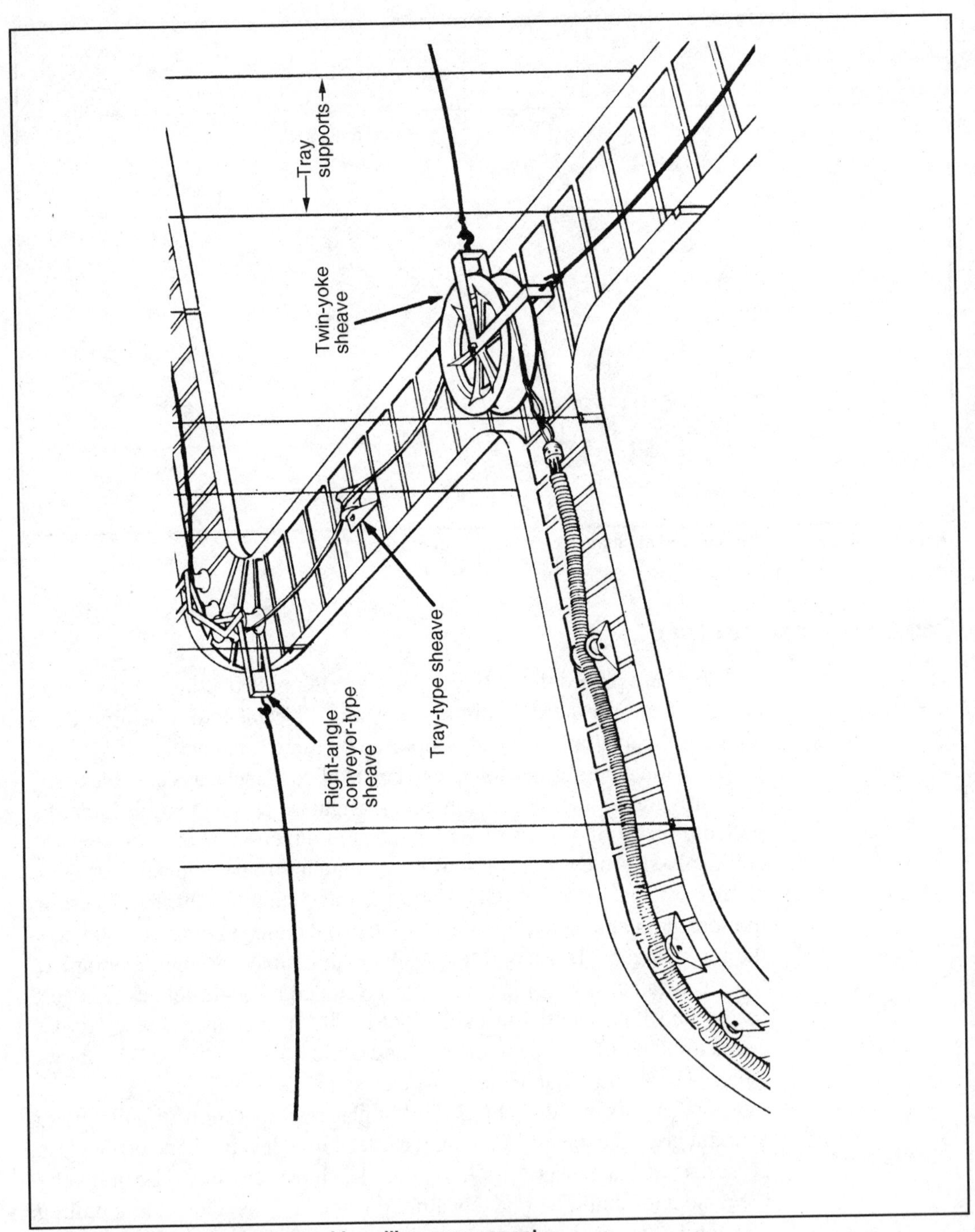

Figure 11-10: Typical cable tray cable pulling arrangement.

In general, the pull exerted on the cables pulled with a basket grip, not attached to the conductor, should not exceed 1000 pounds. For heavier pulls, care should be taken not to stretch the insulation, jacket, or armor beyond the end of the conductor nor bend the ladder, trough, or channel out of shape.

The bending radius of the cable should not be less than the values recommended by the cable manufacturer, which range from four times the diameter for a rubber-insulated cable 1-in maximum outside diameter without lead, shield, or armor, to eight times the diameter for interlocked armor cable. Cables or special construction such as wire armor and high-voltage cables require a larger radius bend.

Best results are obtained in installing long lengths of cable up to 1000 ft with as many as a dozen bends by pulling the cable in one continuous operation at a speed of 20 to 25 ft per minute. It may be necessary to brake the reel to reduce sagging of the cables between rollers and sheaves.

The pulling line diameter and length will, of course, depend on the pull to be made and the tools and equipment available. Winch and power units must be of adequate size for the job and capable of developing the high pulling speed required for best and most economical results.

In general, single or multicable rollers are placed in the bottom of trays to protect the cable as it is pulled along. Sheaves are placed at each change of direction—either horizontally or vertically. The bottom rollers may be secured to the tray bottoms except at vertical changes in direction. Extra support is necessary at these locations to prevent damage, or otherwise moving the tray system. *See* Figure 11-11 on the next page.

Sheaves must be supported in the opposite direction of the pull. For example, all right-angle conveyor sheaves should be supported at two locations as shown in Figure 11-12 to compensate for the pull of the cable.

Note: *Power cable pulls should not be stopped unless absolutely necessary. However, anyone associated with the pull — upon evidence of danger to either the cable or workers — may stop a cable-pulling operation. Communication is the most important factor in these cases.*

CAUTION!

Workers feeding a cable pull must carefully inspect the cable as it is paid off the cable reel. Any visible defects in the cable at the feeding end warrants stopping the pull.

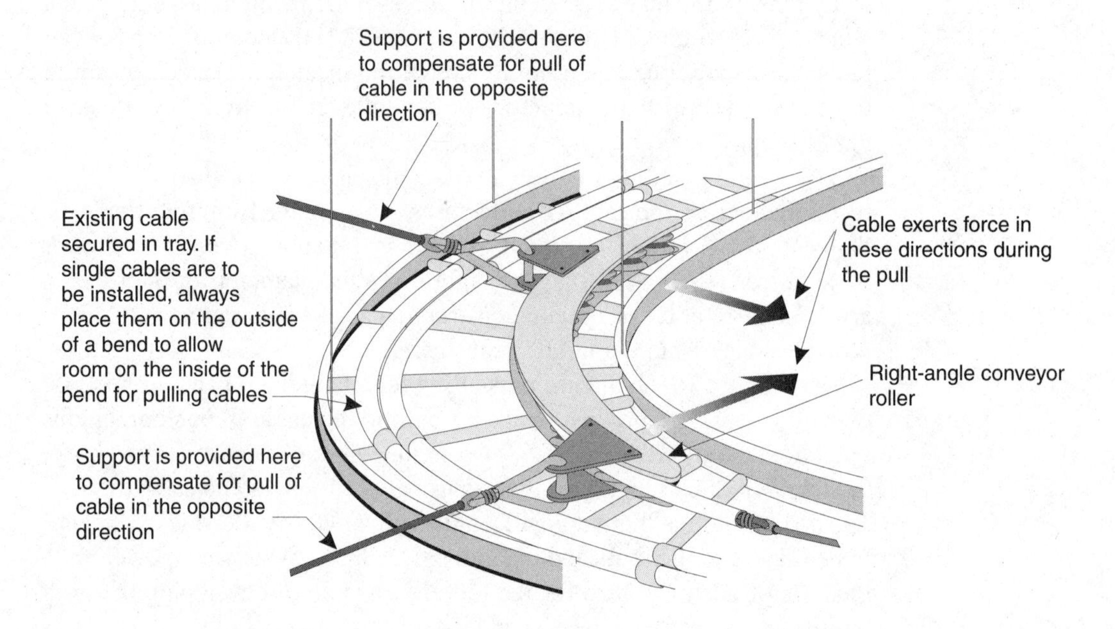

Figure 11-11: Support points for right-angle turns.

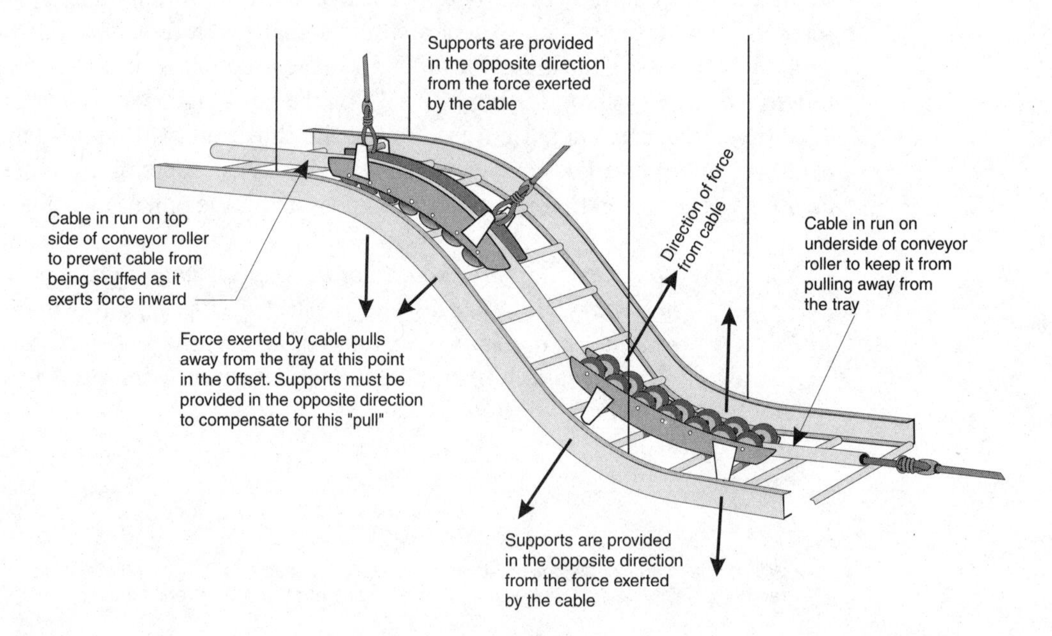

Figure 11-12: Typical support points for cable tray offset.

WARNING!

At the first sign of any type of malfunctioning equipment, broken sheaves, or other happenings that could cause danger to either the workers or the cable, the pull should be stopped. Make certain that communication equipment is in proper order.

Cable Tray Covers

Covers are used primarily for two reasons:

- To protect the insulation of the cables against damage that might be caused if an object were to fall into the tray. Prime hazards are tools, discarded cigarettes, and weld splatter.
- To protect certain types of cable insulation against the damaging effect of direct sunlight.

Cover Selection

When maximum protection is desired, solid covers should be used. However, if accumulation of heat from the cables is expected, caution should be used. Ventilated covers should be used if some protection of the cable is desired and provisions must be made to allow the escape of heat developed by the cables.

Cable Exit from Tray

Several different ways in which cables may exit from a cable tray are shown in Figure 11-13. While all of these methods are *NEC* approved, as well as endorsed by most cable tray manufacturers, engineering specifications on some projects may prohibit the use of some of these methods. Most notably is the "drop-out between runs (no plate)" method and the "drop-out from end of tray (no drop-out plate)" method. The cable radius may be too short with either of these methods. Since no drop-out plates are used, the cable or conductors are not protected. Although *NEC* Section 318-4(b) specifically requires that cable trays have smooth edges to insure that cable will not be damaged, accidents do occur. A tool might be dropped on a cable-tray rung during the installation which may cause a burr or other sharp edge on the rung. Then, after the cable is installed and the system is in use, vibrating machinery may cause this burr to cut into the cable insulation, resulting in a ground fault and possible outage of productive equipment. Consequently, always review the project specifications care-

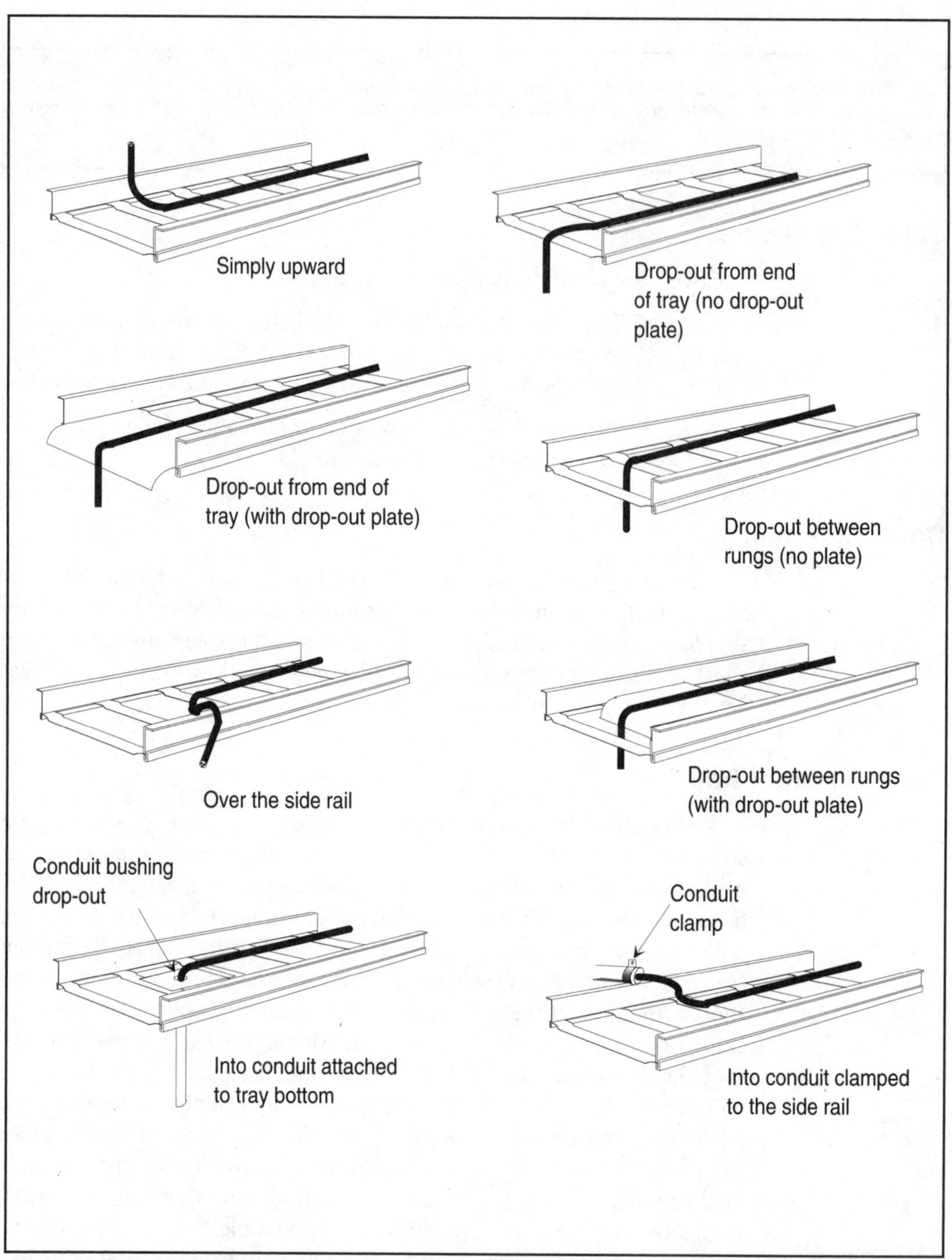

Figure 11-13: Several ways in which cables may exit from a cable tray.

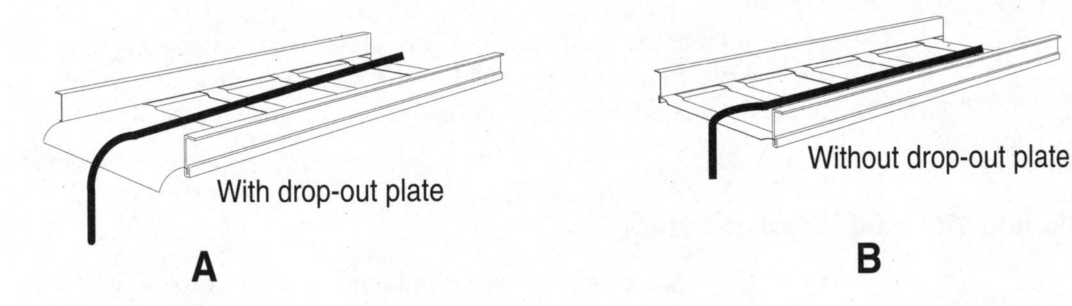

Figure 11-14: A drop-out plate provides a curved surface for the cable to follow as it leaves the tray.

fully and/or check with the project supervisor before using either of these two methods described above.

Drop-Out Plate

A drop-out plate provides a curved surface for the cable to follow as it passes from the tray (Figure 11-14A). Without a drop-out plate, cables can be bent sharply, causing damage to the insulation. See Figure 11-14B.

Cable Supports in Vertical Trays

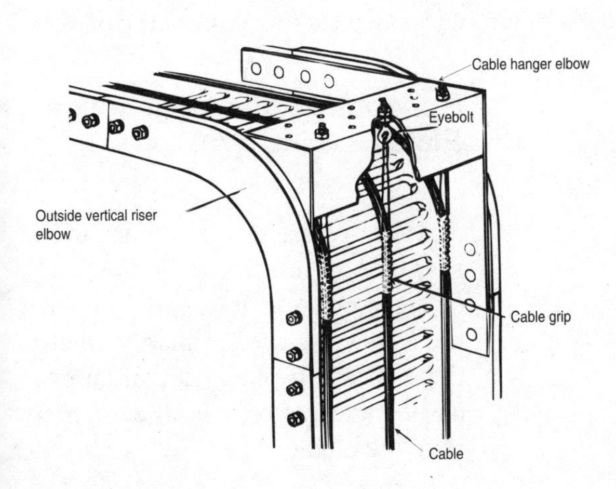

Figure 11-15: Typical application of supports in a vertical run.

A cable hanger elbow is used to suspend cables in long vertical runs. Care should be taken to ensure that the weight of the suspended length of cable does not exceed the cable manufacturer's recommendation for the maximum allowable tension in the cable.

In short vertical runs, the weight of cables can either be supported by the outside vertical riser elbow or by the vertical straight section when the cables are clamped to it. Figure 11-15 shows a typical application of supports in a vertical run.

Cable Edge Protection

The bottom of the solid bottom tray might be convex, concave, or flat. When two pieces of tray are butted together, the bottoms may be out of alignment. An alignment strip, also known as an H bar, is placed between the tray bottoms.

Vertical Adjustable Splice Plates

These plates are used to change elevation in a run of cable tray. They should not be used when it is important to maintain a cable bending radius.

These splice plates are particularly useful when the change in elevation is so slight or the angle so unusual that it would not be possible to install a standard outside vertical riser elbow and an inside vertical riser elbow in the space available.

In general, four pairs of swivel plates are used to build an offset in a cable tray system. Once the proper angles have been calculated, proceed as follows:

1. Bolt four pairs of swivel plates together at the proper angles, using the inner holes as the center or pivot hole. *See* Figure 11-16.

2. Using a flat surface such as a bench or concrete deck, space two pairs of swivel plates the proper center-to-center distance apart. Again, *see* Figure 11-16.

3. Measure and cut the amount of tray needed to complete the offset.

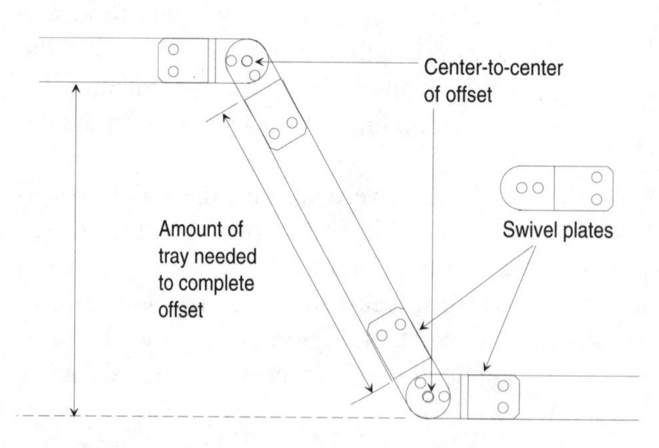

Figure 11-16: Fabricating a cable tray offset with swivel or pivot plates.

Horizontal Adjustable Splice Plates

These plates are sometimes used in place of horizontal elbows to change direction in a run of cable tray. They are used primarily where there is insufficient space or an unusual angle that prevents the use of a standard elbow.

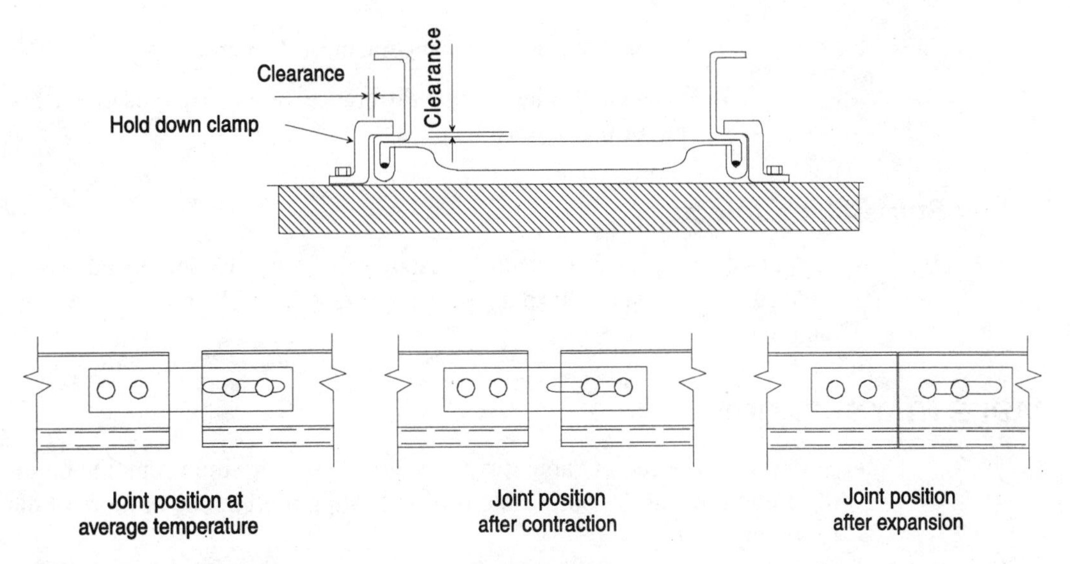

Figure 11-17: Examples of an expansion joint and splice plates.

Expansion Splice Plates

These plates are used at intervals along a straight run of cable tray to allow space for thermal expansion or contraction of the tray to occur, or where offsets or expansion joints occur in the supporting structure.

To enable the expansion joint to function properly, the cable tray must be allowed to slide freely on its supports. Any cable tray hold-down device used in an installation subject to expansion or contraction would give clearance to the tray. Examples of the expansion joint and splice plates are shown in Figure 11-17.

Divider Strip

Divider strips are used for various installations as a result of the nature of the installation, types of circuits used, type of equipment used, local codes, or the *NEC*. Some reasons for using divider strips are to:

- Separate or isolate electrical circuits.
- Separate or isolate cables of different voltages.
- Separate cable or wire runs from each other to prevent fire or ground-fault damage from spreading to other cables or wire in the same tray.

- Aid neatness in the arrangement of the cables.
- Warn electricians of the difference between cables on each side of the divider strip.

Divider Strip Cable Protector

A divider strip cable protector is used to bind any raw metallic edge over which a cable is to pass. Its purpose is to protect the cable insulation against damage.

CABLE TRAY SUPPORT

Proper supports for cable tray installations are very important in obtaining a good overall layout. Cable is usually supported in one or more of the following ways:

- Trapeze mounting
- Direct rod suspension
- Wall mounting
- Pipe rack mounting

Trapeze Mounting

When trapeze mounting is used, a structural member—usually a steel channel or Unistrut—is connected to the vertical supports to provide an appearance similar to a swing or trapeze. The cable tray is mounted to the structural member using anchor clips or J-clamps. Often, the underside of the channel or Unistrut is used to support conduit. See Figure 11-18.

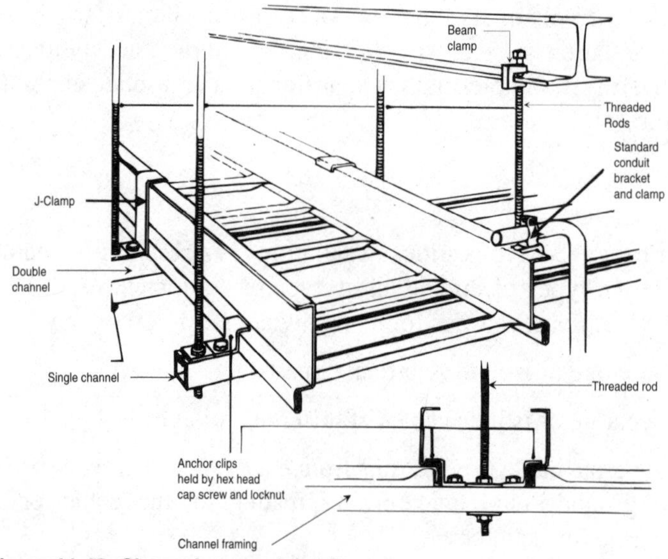

Figure 11-18: Channel support details.

Direct Rod Suspension

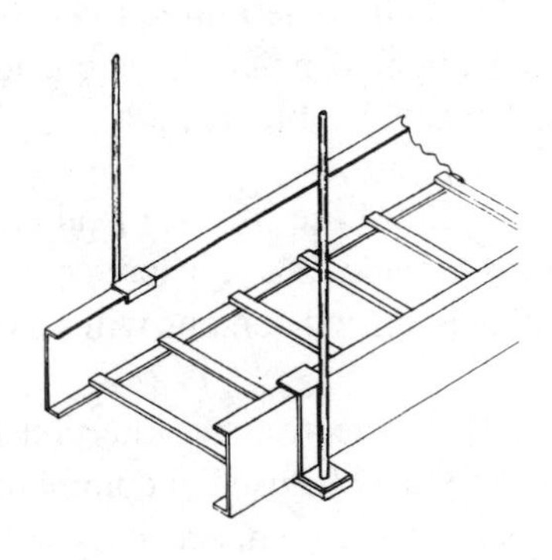

Figure 11-19: Direct rod suspension.

The direct rod suspension method of supporting cable tray uses threaded rods and hanger clamps. The threaded rod is connected to the ceiling or other overhead structure and is connected to the hanger clamps that are attached to the cable tray side rails as shown in Figure 11-19.

Wall Mounting

Wall mounting is accomplished by supporting the tray with structural members attached to the wall. This method of support is often used in tunnels (mining operations) and other underground or sheltered installations where large numbers of conductors interconnect equipment that is separated by long distances. When using this, and any other method of supporting cable tray, always examine the structure to which the hangers are attached, and make absolutely certain that the structure is of adequate strength to support the tray system.

Pipe Rack Mounting

Pipe racks are structural frames used to support the piping that interconnects equipment in outdoor industrial facilities. Usually, some space on the rack is reserved for conduit and cable tray. Pipe rack mounting of cable tray is often seen in petrol-chemical plants where power distribution and instrumentation wiring is routed over a large area, and for great distances.

NEC REQUIREMENTS

NEC Article 318 deals with cable tray systems along with the related wire, cable, and raceway installations therein. Article 340 covers *NEC* regulations governing power and control cable for use in cable tray. Everyone involved in industrial electrical systems should be thoroughly familiar with these Articles (and their related Sections) of the *NEC*.

Although the *NEC* allows the use of cable tray installations in other than industrial establishments, many of the allowable practices are based on certain conditions for use in industrial projects only. For example, *NEC* Section 318-3(b) states:

- In industrial establishments only, where conditions of maintenance and supervision assure that only qualified

> persons will service the installed cable tray system, any of the cables in (1) [single-conductor type TC cable] and (2) [multiconductor type TC cable] shall be permitted to be installed in ladder, ventilated trough, 4-in ventilated channel cable trays, or 6-in ventilated channel cable trays.

And then in Section 318-3(c), the *NEC* allows the use of metal cable tray as the equipment grounding conductor provided, ". . . where continuous maintenance and supervision assure that qualified persons will service the installed cable tray system."

Consequently, it is extremely important that electricians interpret the intent of the *NEC* in regard to where cable tray may be used in commercial buildings, many of which have no continuous maintenance program to ensure that the system will be serviced only by qualified personnel. On the other hand, most industrial establishments have electrical maintenance crews working 24 hours per day — ensuring that only qualified personnel will service the system.

Figure 11-20 summarizes *NEC* requirements for cable tray installations. For in-depth coverage, however, you will want to refer to the *National Electrical Code* itself . . . or better yet, use the *NEC Handbook* for reference. *NEC* Article 318 — containing *NEC* Sections 318-1 through 318-13 — covers cable-tray installations, along with the types of conductors to be used in such systems. Cable-tray manufacturers also have some excellent reference material available that details the installation of their products.

CABLE INSTALLATION

NEC Sections 318-8 through 318-12 cover the installation of cables in cable tray systems.

Section 318-8 covers the general requirements for all conductors used in cable tray systems; that is, splicing, securing, and running conductors in parallel. For example, cable splices are permitted in cable trays provided they are made and insulated by (*NEC*) approved methods. Furthermore, any splices must be readily accessible and must not project above the side rails of the tray.

In horizontal runs, and in most cases, the cables may be laid in the trays without further securing them in place. However, on vertical runs or any runs other than horizontal, the cables must be secured to transverse members of the cable trays.

Cables may enter and leave a cable tray system in a number of different ways, as discussed previously. In general, no junction box is required where

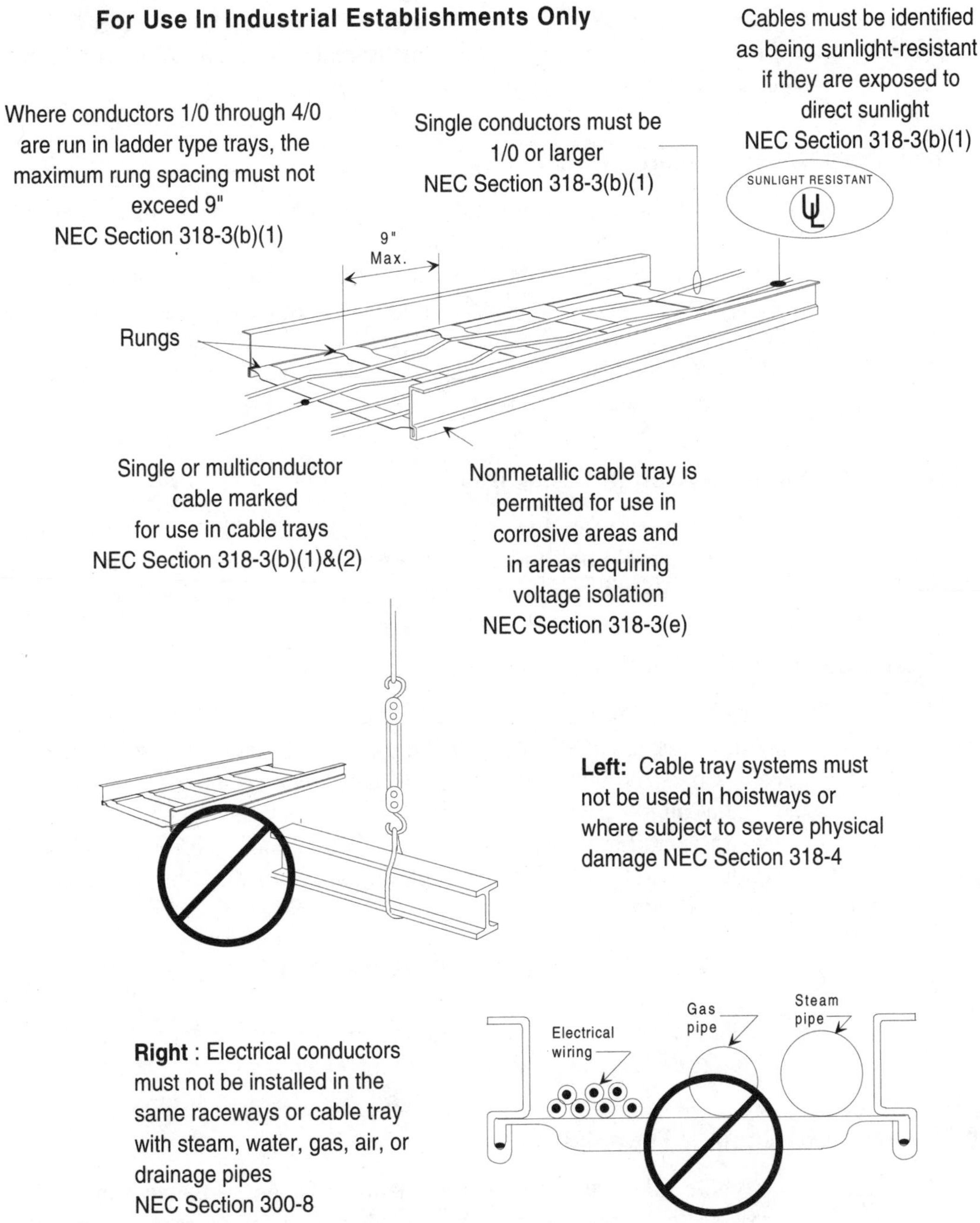

For Use In Industrial Establishments Only

Cables must be identified as being sunlight-resistant if they are exposed to direct sunlight
NEC Section 318-3(b)(1)

Where conductors 1/0 through 4/0 are run in ladder type trays, the maximum rung spacing must not exceed 9"
NEC Section 318-3(b)(1)

Single conductors must be 1/0 or larger
NEC Section 318-3(b)(1)

SUNLIGHT RESISTANT

9" Max.

Rungs

Single or multiconductor cable marked for use in cable trays
NEC Section 318-3(b)(1)&(2)

Nonmetallic cable tray is permitted for use in corrosive areas and in areas requiring voltage isolation
NEC Section 318-3(e)

Left: Cable tray systems must not be used in hoistways or where subject to severe physical damage NEC Section 318-4

Right : Electrical conductors must not be installed in the same raceways or cable tray with steam, water, gas, air, or drainage pipes
NEC Section 300-8

Electrical wiring

Gas pipe

Steam pipe

Figure 11-20: NEC regulations governing cable tray installations.

such cables are installed in bushed conduit or tubing. Where conduit or tubing is used, either must be secured to the tray with the proper fittings. Further precautions must be taken to insure that the cable is not bent sharply as it enters or leaves the conduit or tubing.

Conductors Connected in Parallel

Where single-conductor cables comprising each phase or neutral of a circuit are connected in parallel as permitted in *NEC* Section 310-4, the conductors must be installed in groups consisting of not more than one conductor per phase or neutral to prevent current unbalance in the paralleled conductors due to inductive reactance. Such conductors must be securely bound in circuit groups to prevent excessive movement due to fault-current magnetic forces.

Number of Cables Allowed in Cable Tray

The number of multiconductor cables, rated 2000 V or less, permitted in a single cable tray must not exceed the requirements of *NEC* Section 318-9. This section applies to both copper and aluminum conductors.

All Conductors Size 4/0 or Larger

Where all of the cables installed in the tray are No. 4/0 or larger, the sum of the diameters of all cables shall not exceed the cable tray width, and the cables must be installed in a single layer. For example, if a cable tray installation is to contain three 4/0 multiconductor cables (1.5 in diameter), two 250 kcmil multiconductor cables (1.85 in), and two 350 kcmil multiconductor cables (2.5 in), the minimum width of the cable tray is determined as follows:

$$3(1.5) + 2(1.85) + 2(2.5) = 13.2 \text{ in}$$

The closest standard cable tray size that meets or exceeds 13.2 in is 18 in. Therefore, this is the size to use.

All Conductors Smaller Than 4/0

Where all of the cables are smaller than No. 4/0, the sum of the cross-sectional area of all cables smaller than 4/0 must not exceed the maximum allowable cable fill area as specified in Column 1 of *NEC* Table 318-9; this gives the appropriate cable tray width. To use *NEC* Table 318-9, however, you must have the manufacturer's data for the cables being used. This will give the cross-sectional area of the cables.

The following shows the steps involved in determining the size of cable tray for multiconductors smaller than No. 4/0 AWG:

1. Calculate the total cross-sectional area of all cables used in the tray. Obtain the area of each from manufacturer's data.

2. Look in Column 1 of *NEC* Table 318-9 and find the smallest number that is at least as large as the calculated number.

3. Look at the number to the left of the row selected in Step 2. This is the minimum width of cable tray that may be used.

For example, let's determine the minimum cable tray width required for the following multiple conductor cables — all less than No. 4/0 AWG:

- Four @1.5 in diameter
- Five @1.75 in diameter
- Three @2.15 in diameter

Step 1. Determine the cross-sectional area of the cables from the equation:

$$A = \frac{\pi \times D^2}{4}$$

where:
A = area
D = diameter

The area of a 1.5" diameter cable is:

$$\frac{(3.14159)(1.5)^2}{4} = 1.7671 \text{ in}^2$$

The area of the four 1.5" cables is:

$$4 \times 1.7671 \text{ in}^2 = 7.0684 \text{ in}^2$$

The area of 1.75 in diameter cable is:

$$\frac{(3.14159)(1.75)^2}{4} = 2.4053 \text{ in}^2$$

The area of the five 1.75 in cables is:

$$5 \times 2.4053 \text{ in}^2 = 12.0264 \text{ in}^2$$

The area of the three 2.15 in cables is:

$$\frac{(3.14159)(2.15)^2}{4} = 3.6305 \text{ in}^2$$

Therefore, the total area of the three cables is:

$$3 \times 3.6305 \text{ in}^2 = 10.8915 \text{ in}^2$$

The total cross-section area is found by adding the above three totals to obtain:

$$7.0686 + 12.0264 + 10.8915 = 29.9865 \text{ in}^2$$

Step 2. Look in Column 1, *NEC* Table 318-9 and find the smallest number that is at least as large as 29.9865 in^2. The number is 35.

Step 3. Look to the left of "35" and see inside tray width of 30 in. Therefore, the minimum tray width that can be used for the given group of conductors is 30 in.

Combination Cables

Where No. 4/0 or larger cables are installed in the same cable tray with cables smaller than No. 4/0, the sum of the cross-sectional area of all cables smaller than No. 4/0 must not exceed the maximum allowable fill area resulting from computation in Column 2 of *NEC* Table 318-9, for the appropriate cable tray width. The No. 4/0 and larger cables must be installed in a single layer and no other cables can be placed on them.

To determine the size of tray for a combination of cables as discussed in the above paragraph, proceed as follows:

Step 1. Repeat the steps from the procedure used previously to determine the minimum width of tray required for the multiconductor cables having conductors sized No. 4/0 and larger. Call this number *Sd* for the "sum of diameters."

Step 2. Repeat the steps from the procedure used previously to determine the cross-sectional area of all multiconductor cables having conductors smaller than No. 4/0 AWG.

Step 3. Multiply the result of Step 1 by the constant 1.2 and add this product to the results of Step 2. Call this sum *A*. Search Column 2 of *NEC* Table 318-9 for the smallest number that is at least as large as A. Look to the left in that row to determine the minimum size cable tray required.

To illustrate these steps, let's assume that we desire to find the minimum cable tray width of two multiconductor cables each with a diameter of 2.54 in (conductors size 4/0 or larger); three cables each with a diameter of 3.30 in (conductors size 4/0 or larger), plus the following cable with conductors less than No. 4/0: eight cables each with a diameter of 1.92 in.

Step 1. The sum of all the diameters of cable having conductors 4/0 or larger is:

$$2(2.54) + 3(3.30) = \text{Sd} = 14.98 \text{ in}$$

Step 2. The sum of the cross-sectional areas of all cables having conductors smaller than 4/0 is:

$$\frac{(8)(3.14159)(1.92)^2}{4} = 23.1623 \text{ in}^2$$

Step 3. Multiply the results of Step 1 by 1.2 and add this product to the results of Step 2:

$$1.2 \times (14.98) + 23.1623 = 41.1383 \text{ in}^2$$

The smallest number in Column 2 of *NEC* Table 318-9 that is not larger than 41.1383 is 42. The tray width that corresponds to 42 is 36 in. Select a cable tray width of 36 in.

Solid Bottom Tray

Where solid bottom cable trays contain multiconductor power or lighting cables, or any mixture of multiconductor power, lighting, control, and signal cables, the maximum number of cables must conform to the following:

Where all of the cables are No. 4/0 or larger, the sum of the diameters of all cables must not exceed 90 percent of the cable tray width, and the cables must be installed in a single layer.

Where all of the cables are smaller than No. 4/0, the sum of the cross-sectional areas of all cables must not exceed the maximum allowable cable fill area in Column 3 of *NEC* Table 318-9 for the appropriate cable tray width.

Where No. 4/0 or larger cables are installed in the same cable tray with cables smaller than No. 4/0, the sum of the cross-sectional areas of all of the smaller cables must not exceed the maximum allowable fill area resulting from the computation in Column 4 of *NEC* Table 318-9 for the appropriate cable tray width. The No. 4/0 and larger cables must be installed in a single layer, and no other cables can be placed on them.

Where a solid bottom cable tray, having a usable inside depth of 6 in or less, contains multiconductor control and/or signal cables only, the sum of the cross-sectional areas of all cables at any cross section must not exceed 40 percent of the interior cross-sectional area of the cable tray. A depth of 6 in must be used to compute the allowable interior cross-sectional area of any cable tray that has a usable inside depth of more than 6 in.

In a previous example, we determined that the minimum tray size for multiple conductors cables with all conductors size 4/0 or larger was 18 in. The sum of all cable diameters for this example was 13.2 in. To see if an 18-in solid bottom tray can be used, multiply the tray width by .90 (90 percent):

$$18 \times .9 = 16.2 \text{ in}$$

Therefore, 16.2 in is the minimum width allowed for solid-bottom tray. Since we are using 18 in tray, this meets the requirements of *NEC* Section 318-9(c)(1).

When dealing with solid bottom trays, and using *NEC* Table 318-9, use Columns 3 and 4, instead of Columns 1 and 2 as used for ladder and trough type tray.

Single Conductor Cables

Calculating cable tray widths for single conductor cables (2000 V or under) is similar to the calculations used for multiconductor cables with the following exceptions:

Conductors 1000 kcmil and larger are treated the same as multiconductor cables, having conductors size 4/0 or larger.

Conductors that are smaller than 1000 kcmil are treated the same as multiconductor cables having conductors smaller than size 4/0.

NEC Section 318-10 covers the details of installing single conductor cables in cable tray systems.

Ampacity of Cable Tray Conductors

NEC Section 318-11 gives the requirements for cables used in tray systems with rated voltages of 2000 *V* or less. Section 318-12 covers cables with voltages of 2001 and over, while Section 318-13 deals with Type MV and Type MC cables rated over 2000 *V*.

CABLE TRAY DRAWINGS

For economical erection and satisfactory installation, working out the details of supports and hangers for a cable tray system is usually done beforehand by the engineering department or project engineer, and is seldom left to the judgment of a field force not acquainted with the loads and forces to be encountered. As a result, drawings and specifications will usually be furnished to the workers — giving details about the cable tray system. Consequently, all workers involved with the installation should know how to interpret these drawings.

The exact method of showing cable tray systems on working drawings will vary, so always consult the symbol list or legend before beginning the installation. Also study any shop or detail drawings that might accompany the construction documents. The standard symbol for cable tray is shown below.

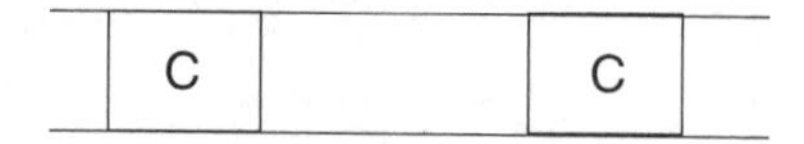

However, if space permits, many engineers prefer to draw the cable tray system as close to scale as possible, using modifications of this symbol to show the different types of cable tray to be installed.

Look at the floor-plan drawing in Figure 11-21. The cable tray system in this project originates at several power panels and motor control centers to feed and control motors in other parts of the building. The trays run from the motor control centers, offset to miss beams and other runs of cable tray, and then branch off to various parts of the building.

Although experienced workers in the industrial electrical trade will have little trouble in reading the information in this drawing, electricians new to this type of work may have some difficulty in visualizing the system. However, if supplemental detail drawings are provided with the floor plan drawing in Figure 11-21, they would provide a clearer picture of the system and leave little doubt as to how the cable-tray system is to be installed. Even new workers in the trade should be able to see how the system should be installed; this would take some of the load off experienced workers and give them more time to accomplish other tasks.

In actual practice, however, consulting engineering firms seldom furnish extensive detail drawings of cable tray systems; they merely show the layout in plan view. Consequently, plan views involving the construction details of cable tray systems must be studied carefully during the planning stage — before the work is begun.

PULLING CABLE IN TRAY SYSTEMS

When installing cables in tray systems, proper precaution must be taken to avoid damaging the cables. A complete line of installation tools is available, developed through field experience, for pulling long lengths of cable up to 1000 ft or longer. These tools save considerable installation time.

Short lengths of cable can be laid in place without power pulling tools, or the cable can be pulled manually using a basket grip and pulling rope. Long lengths of small cable, 2 in or less in diameter, can also be pulled with a basket grip and pulling rope. Larger cables, however, should be pulled by the conductor and the braid, sheath, or armor. This is usually done

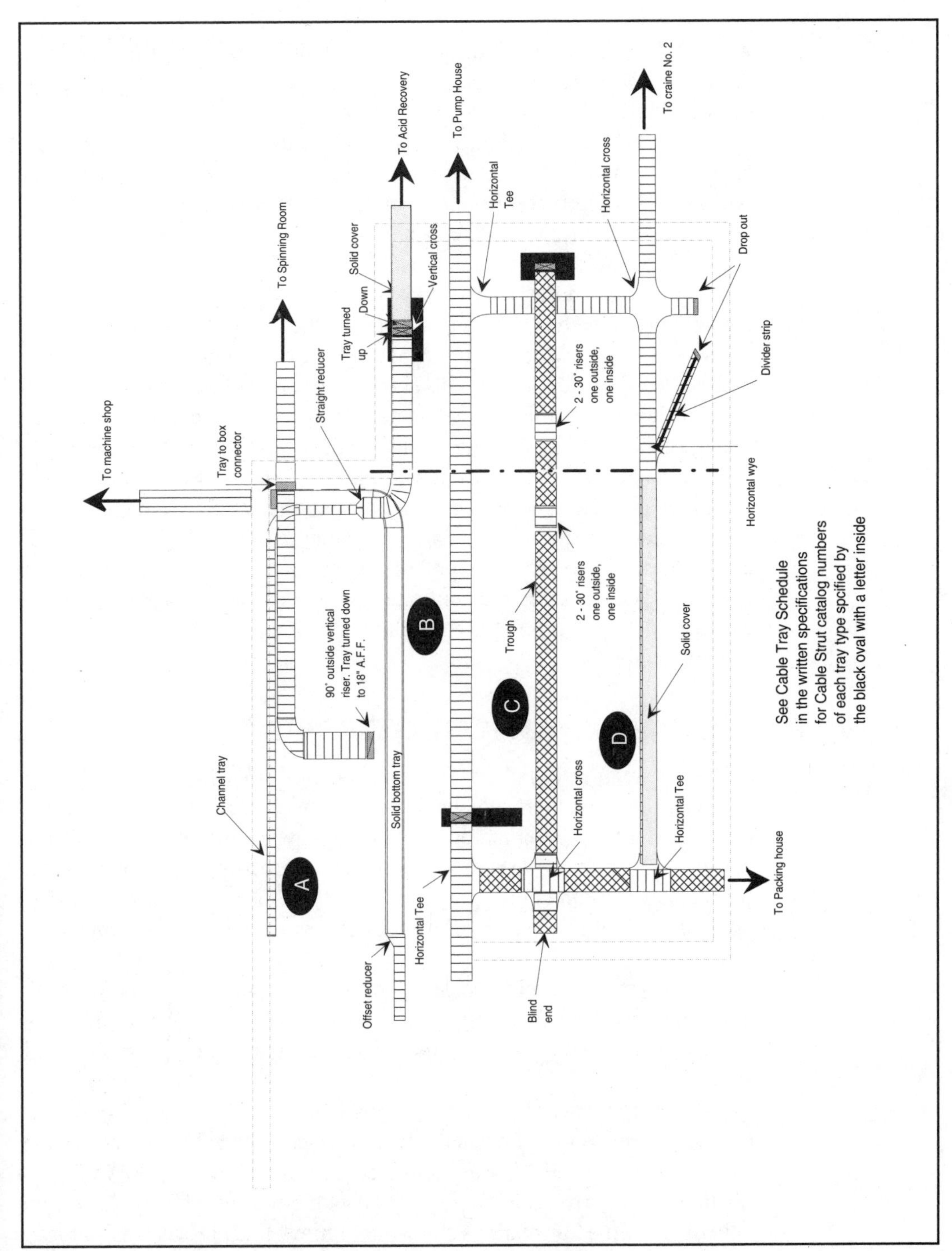

Figure 11-21: Sample floor plan showing layout of cable tray system.

with a pulling eye applied to cable at the factory, or by tying the conductor to the eye of a basket grip and taping the tail end of the grip to the outside of the cable.

In general, the pull exerted on the cables pulled with a basket grip, not attached to the conductor, should not exceed 1000 pounds. For heavier pulls, care should be taken not to stretch the insulation, jacket, or armor beyond the end of the conductor nor bend the ladder, trough, or channel out of shape.

The bending radius of the cable should not be less than the values recommended by the cable manufacturer, which range from four times the diameter for a rubber-insulated cable, 1-inch maximum outside diameter without lead, shield, or armor, to eight times the diameter for interlocked armor cable. Cables of special construction such as wire armor and high-voltage cables require a larger radius bend.

Best results are obtained in installing long lengths of cable up to 1000 ft with as many as a dozen bends by pulling the cable in one continuous operation at a speed of 20 to 25 ft per minute. The pulling line diameter and length will depend on the pull to be made and construction equipment available. Winch and power unit must be of adequate size for the job and capable of developing the high pulling speed required for best and most economical results.

CUT-AND-WELD METHOD OF FABRICATING OFFSETS

As a general rule, factory fittings should be used for most cable tray systems. However, there may be instances where an exact "factory fitting" is not available, or an emergency situation may arise where time does not permit waiting for a particular part to arrive. In such cases, an offset or other change-in-direction fitting may have to be fabricated on the job. One popular method of making on-the-job offsets is called the *cut-and-weld method*.

The cut-and-weld method of offset fabricating may be used for both horizontal and vertical bends in all sizes of cable tray. About the only additional tools required consist of a protractor and a conventional square.

The square is used to lay out a perpendicular line which is referred to as the "90° mean factor." This is the line to which is added or subtracted the offset degrees to be cut. For example, let's assume that a 30° offset is required in any size cable tray. A 30° offset requires two 15° angle cuts on either side of the 90° mark as shown in Figure 11-22. Proceed as follows:

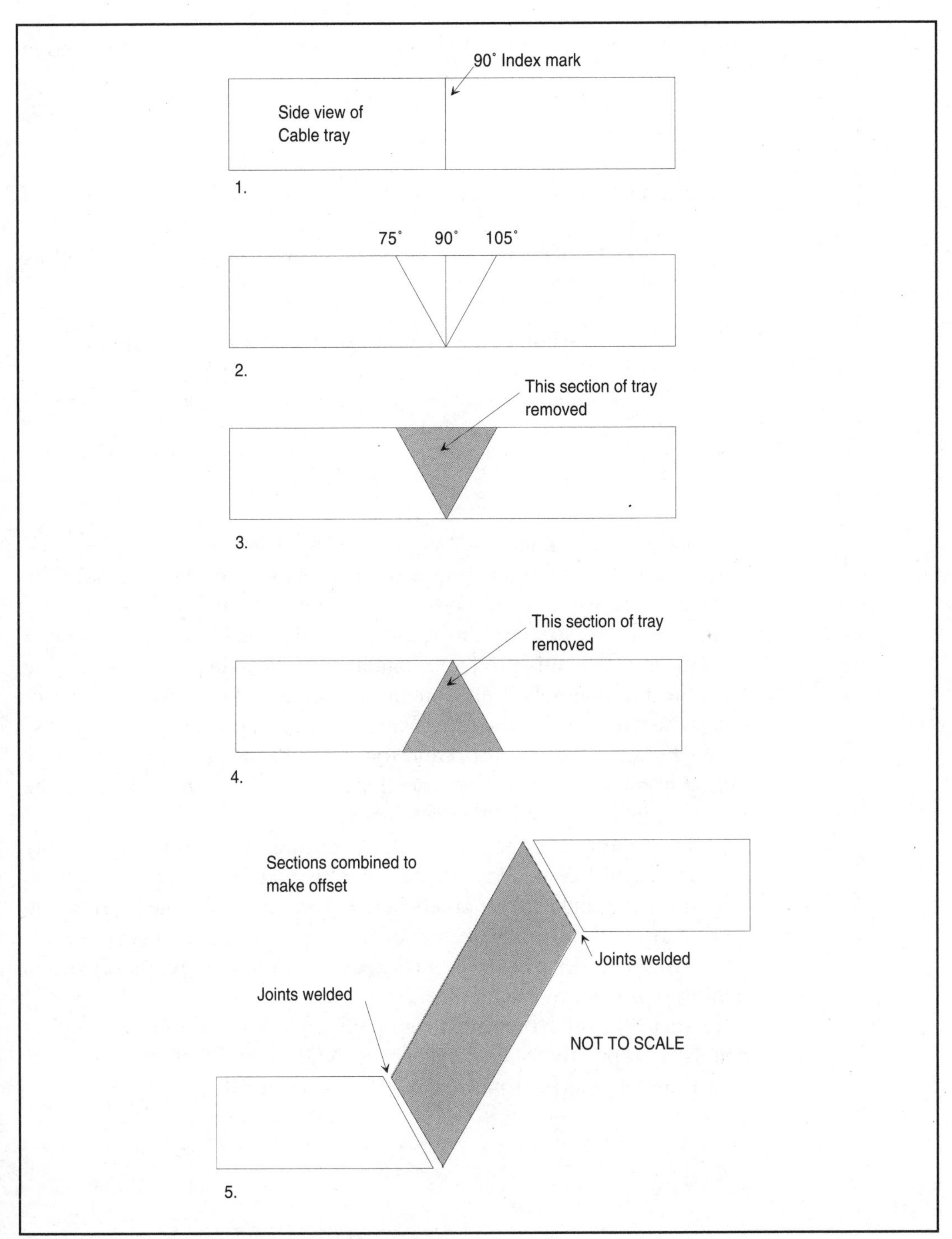

Figure 11-22: Steps in making a cut-and-weld offset.

1. Determine exactly where the 30° cutout needs to be made, and in which direction — up or down or across.

2. Use a square to make a 90° mark where the offset is needed.

3. To this 90° mark add and subtract 15° on either side.

4. Make the marks on both sides of the tray in exactly the same place as the other side.

5. Cut a "V" section out of the side of tray at the proper angles.

6. Prepare cuts for welding and make sure of a good fit prior to welding.

SAFETY

Working with cable tray means working at heights above the floor. Consequently, workers must take the necessary precaution to make the area, and their work habits, correspond to this condition.

In general, workers installing cable tray will use ladders, scaffolds, work from the tray assembly itself, or a combination of all three. When working from the tray assembly itself, care must be taken not to bend tray rungs or otherwise damage the structure. Walking boards, also called "pegboards" should be laid in the bottom of cable tray when the tray is used as a walkway during installation. Also make use of your life line when working on the tray assembly or when pulling conductors.

Keep the tray assembly uncluttered during installation. Tools, tray fittings, and the like are ideal obstacles for tripping workers. They may also fall off the tray and injure workers below. To help prevent the latter, set up work barriers beneath the section of tray assembly being worked on . . . and, if you're working on the ground, never penetrate or move these barriers until the work above is complete.

Learn to "tie off" your life line properly, and also make sure your safety belt is a proper fit. A belt that is too large — one that will slip — is sometimes more dangerous than not having any at all.

Chapter 12

Wiring Devices

A *wiring device* — by *NEC* definition — is a unit of an electrical system that is intended to carry, but not utilize, electric energy. This covers a wide assortment of system components that include, but are not limited to, the following:

- Switches
- Relays
- Contactors
- Receptacles
- Conductors

However, for our purpose, we will deal with switching devices and also those items used to connect utilization equipment to electrical circuits; namely, *receptacles*. Both are commonly known as *wiring devices*.

RECEPTACLES

A receptacle is a contact device installed at the outlet for the connection of a single attachment plug. Several types and configurations are available for use with many different attachment plug caps — each designed for a specific application. For example, receptacles are available for two-wire,120-V, 15- and 20-A circuits; others are designed for use on two- and three-wire, 240-V, 20-, 30-, 40-, and 50-A circuits. There are also many other types.

Receptacles are rated according to their voltage and amperage capacity. This rating, in turn, determines the number and configuration of the contacts — both on the receptacle and the receptacle's mating plug. Figure 12-1 shows the most common configurations, along with their applications. This chart was developed by the Wiring Device Section of NEMA and illustrates 75 various configurations, which cover 38 voltage and current ratings. The configurations represent existing devices as well as suggested standards (shown with an asterisk in the chart) for future design. Note that all configurations in Figure 12-1 are for general-purpose, nonlocking devices.

As indicated in the chart, unsafe interchangeability has been eliminated by assigning a unique configuration to each voltage and current rating. All dual ratings have been eliminated, and interchangeability exists only where it does not present an unsafe condition.

Each configuration is designated by a number composed of the chart line number, the amperage, and either "R" for receptacle or "P" for plug cap. For example, a 5-15R is found in line 5 and represents a 15-A receptacle.

A clear distinction is made in the configurations between "system grounds" and "equipment grounds." System grounds, referred to as grounded conductors, normally carry current at ground potential, and terminals for such conductors are marked "W" for "white" in the chart. Equipment grounds, referred to as grounding conductors, carry current only during ground-fault conditions, and terminals for such conductors are marked "G" for "grounding" in the chart.

Receptacle Characteristics

Receptacles have various symbols and information inscribed on them that help to determine their proper use and ratings. For example, Figure 12-2 on page 263 shows a standard duplex receptacle and contains the following printed inscriptions:

- The testing laboratory label
- The CSA (Canadian Standards Association) label
- Type of conductor for which the terminals are designed
- Current and voltage ratings, listed by maximum amperage, maximum voltage, and current restrictions

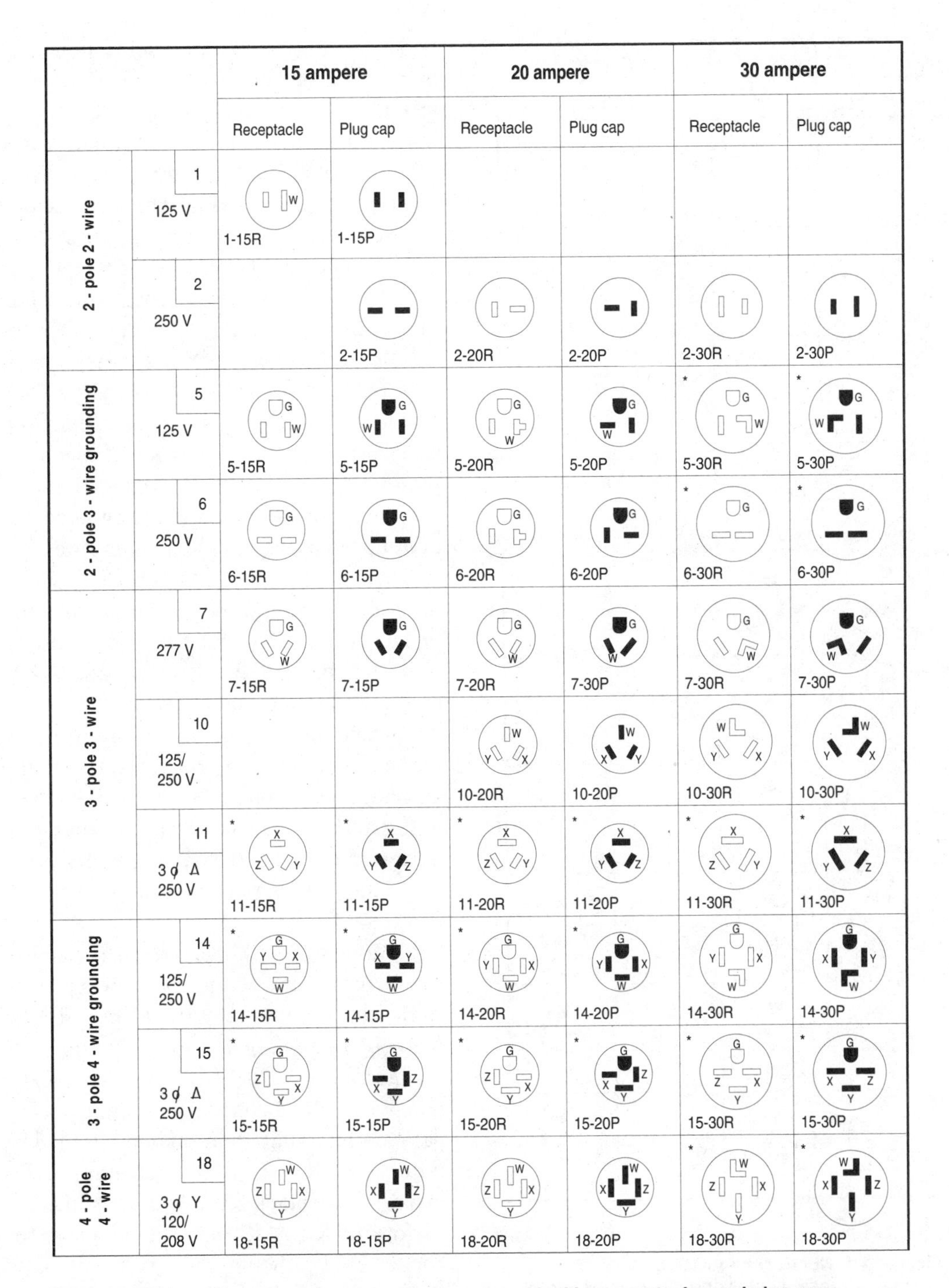

Figure 12-1: NEMA configurations for general-purpose, nonlocking receptacles and plug caps.

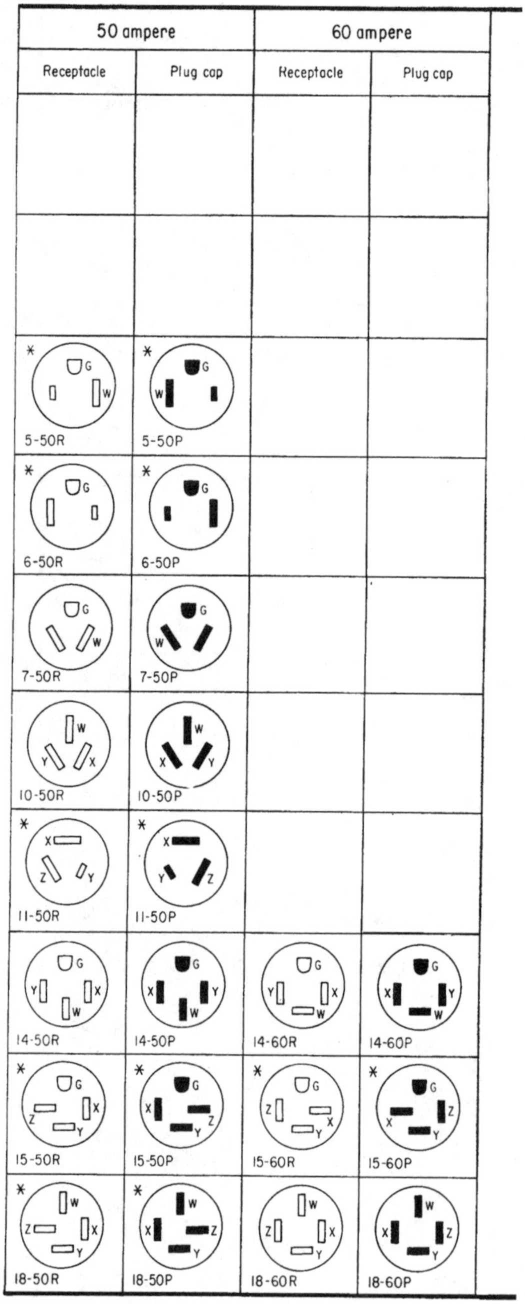

50 ampere		60 ampere	
Receptacle	Plug cap	Receptacle	Plug cap
* 5-50R	* 5-50P		
* 6-50R	* 6-50P		
7-50R	7-50P		
10-50R	10-50P		
* 11-50R	* 11-50P		
14-50R	14-50P	14-60R	14-60P
* 15-50R	* 15-50P	* 15-60R	* 15-60P
* 18-50R	* 18-50P	18-60R	18-60P

Figure 12-1: NEMA configurations for general-purpose, nonlocking receptacles and plug caps. ***(Cont.)***

The testing laboratory label is an indication that the device has undergone extensive testing by a nationally recognized testing lab and has met with the minimum safety requirements. The label does not indicate any type of quality rating. The receptacle in Figure 12-2 is marked with the "U.L." label which indicates that the device type was tested by Underwriters' Laboratories, Inc. of Northbrook, IL. ETL Testing Laboratories, Inc. of Cortland, NY, is another nationally recognized testing laboratory. They provide a labeling, listing and follow-up service for the safety testing of electrical products to nationally recognized safety standards or specifically designated requirements of jurisdictional authorities.

The CSA (Canadian Standards Association) label is an indication that the material or device has undergone a similar testing procedure by the Canadian Standards Association and is acceptable for use in Canada.

Current and voltage ratings are listed by maximum amperage, maximum voltage, and current restriction. On the device shown in Figure 12-2, the maximum current rating is 15 A at 125 V — the latter of which is the maximum voltage allowed on a device so marked.

Conductor markings are also usually found on duplex receptacles. Receptacles with quick-connect wire clips will be marked "Use #12 or #14 solid wire only." If the inscription "CO/ALR" is marked on the receptacle, either copper, aluminum, or copper-clad aluminum wire may be used. The letters "ALR" stand for "aluminum revised." Receptacles marked with the inscription "CU/AL" should be used for copper only, although they were originally intended for use with aluminum also. How-

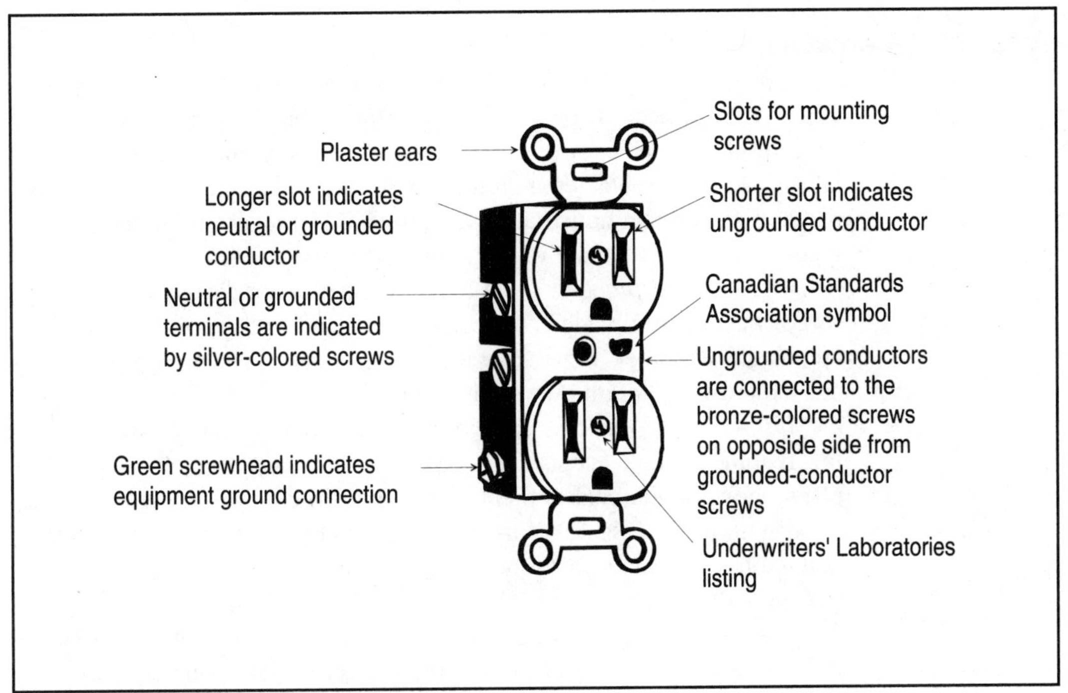

Figure 12-2: Characteristics of typical duplex receptacles.

ever, such devices frequently failed when connected to 15- or 20-A circuits. Consequently, devices marked with "CU/AL" are no longer acceptable for use with aluminum conductors.

The remaining markings on duplex receptacles may include the manufacturer's name or logo, "Wire Release" inscribed under the wire-release slots, and the letters "GR" beneath or beside the green grounding screw.

The screw terminals on receptacles are color-coded. For example, the terminal with the green screwhead is the equipment ground connection and is connected to the U-shaped slots on the receptacle. The silver-colored terminal screws are for connecting the grounded or neutral conductors and are associated with the longer of the two vertical slots on the receptacle. The brass-colored terminal screws are for connecting the ungrounded or "hot" conductors and are associated with the shorter vertical slots on the receptacle.

Note: *The long vertical slot accepts the grounded or neutral conductor while the shorter vertical slot accepts the ungrounded or hot conductor.*

Types of Receptacles

There are many types of receptacles. For example, the duplex receptacles that have been discussed are the straight-blade type which accept a straight blade connector or plug. This is the most common type of receptacle and such receptacles are found on virtually all electrical projects from residential to large industrial installations. Refer again to Figure 12-1 for types of receptacles that fall under this category.

Twist lock receptacles: Twist lock receptacles are designed to accept a somewhat "curved blade" connector or plug. The plug/connector and the receptacle will lock together with a slight twist. The locking prevents accidentally unplugging the equipment.

Pin-and-sleeve receptacles: Pin-and-sleeve devices have a unique locking feature. These receptacles are made with an extremely heavy-duty plastic housing that makes them highly indestructible. They are manufactured with long brass pins for long life and are color coded according to voltage for easy identification.

Low-voltage receptacles: These receptacles are designed for both ac and dc systems where the maximum potential is 50 V. Receptacles used for low-voltage systems must have a minimum current-carrying rating of 15 A.

440-V receptacles: Portable electrical equipment operating at 440 to 460 V is common on many industrial installations. Such equipment includes welders, battery chargers, and other types of portable equipment. Special " 440-V " plugs and receptacles are used to connect and disconnect such equipment from a power source. 440-V receptacles are available in 2-wire, single-phase; 3-wire, three-phase, and 4-wire, three-phase. Equipment grounding is required in all cases, and provisions are provided in each receptacle for such grounding.

WARNING!

Make certain that the plug-and-cord assembly is compatible with both the equipment and receptacle before connecting to a 440-V receptacle. Polarity and equipment-grounding checks on the plug-and-cord assembly should be made on a monthly basis and sooner if subject to hard usage.

SWITCHES

The purpose of a switch is to make and break an electrical circuit, safely and conveniently. In doing so, a switch may be used to manually control lighting, motors, fans, and other various items connected to an electrical circuit. Switches may also be activated by light, heat, chemicals, motion, and electrical energy for automatic operation. *NEC* Article 380 covers the installation and use of switches.

Although there is some disagreement concerning the actual definitions of the various switches that might fall under the category of *wiring devices*, the most generally accepted ones are as follows:

Bypass isolation switch: This is a manually operated device used in conjunction with a transfer switch to provide a means of directly connecting load conductors to a power source, and of disconnecting the transfer switch.

General-use switch: A switch intended for use in general distribution and branch circuits. It is rated in amperes, and it is capable of interrupting its rated current at its rated voltage.

General-use snap switch: A form of general-use switch so constructed that it can be installed in flush device boxes or on outlet box covers, or otherwise used in conjunction with wiring systems recognized by the *NEC*.

Isolating switch: A switch intended for isolating an electric circuit from the source of power. It has no interrupting rating, and is intended to be operated only after the circuit has been opened by some other means.

Motor circuit switch: A switch, rated in horsepower, capable of interrupting the maximum operating overload current of a motor of the same horsepower rating as the switch at its rated voltage.

Transfer switch: A transfer switch is a device for transferring one or more load conductor connections from one power source to another. This type of switch may be either automatic or nonautomatic.

Common Switch Terms

In general, the major terms used to identify the characteristics of switches are:

- Pole or poles
- Throw

The term *pole* refers to the number of conductors that the switch will control in the circuit. For example, a single-pole switch breaks the connection on only one conductor in the circuit. A double-pole switch breaks the connection to two conductors, and so forth.

The term *throw* refers to the number of internal operations that a switch can perform. For example, a single-pole, single-throw switch will "make" one conductor when thrown in one direction — the "ON" direction — and "break" the circuit when thrown in the opposite direction; that is, the "OFF"

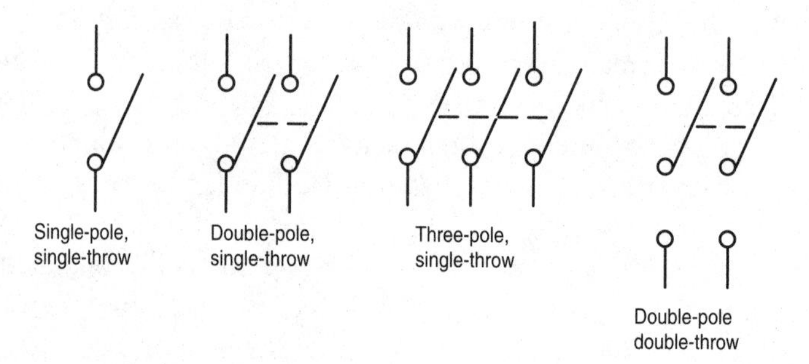

Figure 12-3: Common switch configurations.

position. The commonly used ON/OFF toggle switch is an SPST switch (single-pole, single-throw). A two-pole, single-throw switch opens or closes two conductors at the same time. Both conductors are either open or closed; that is, in the ON or OFF position. A two-pole, double-throw switch is used to direct a two-wire circuit through one of two different paths. One application of a two-pole, double-throw switch is in an electrical transfer switch where certain circuits may be energized from either the main electric service, or from an emergency standby generator. The double-throw switch "makes" the circuit from one or the other and prevents the circuits from being energized from both sources at once. Figure 12-3 shows common switch configurations.

Switch Identification

Wall switches vary in grade, capacity, and purpose. It is very important that proper types of switches are selected for the given application. For example, most single-pole toggle switches used for the control of lighting are restricted to ac use only. This same switch is not suitable for use on, say, a 32-V dc emergency lighting circuit. A switch rated for ac only will not extinguish a dc arc quickly enough. Not only is this a dangerous practice (causing arcing and heating of the device), the switch contacts would probably burn up after only a few operations of the handle, if not the first time.

Figure 12-4 shows a typical single-pole toggle switch — the type most often used to control ac lighting in all installations. Note the identifying marks. They are similar to those on the duplex receptacle discussed previously. The main difference is the " T " rating which means that the switch is rated for switching lamps with tungsten filaments (incandescent lamps).

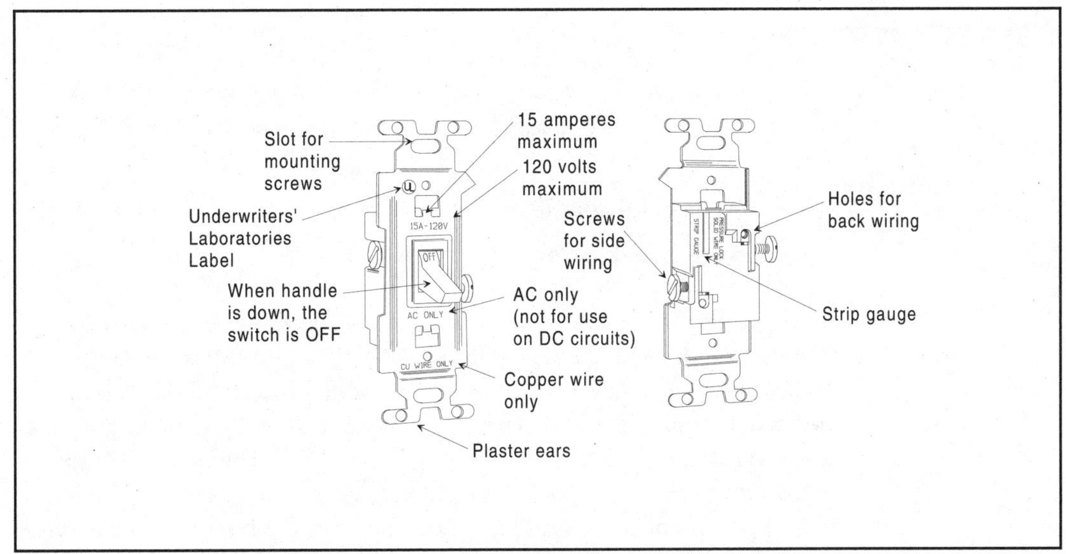

Figure 12-4: Typical identifying marks on a single-pole switch.

Screw terminals are also color coded on conventional toggle switches. Switches are typically constructed with a ground screw attached to the metallic strap of the switch. The ground screw is usually a green-colored hex-head screw. This screw is for connecting the equipment-grounding conductor to the switch. On three-way switches, the common or pivot terminal usually has a black or bronze screwhead.

The switch shown is the type normally used for residential construction. Heavier-duty switches are usually the type used on commercial wiring — some of which are rated for use on 277-V circuits with current-carrying ratings up to 30 A. Therefore, it is important to check the rating of each switch before it is installed.

The exact type and grade of switch to be used on a specific installation is often dictated by the project drawings or written specifications. Sometimes wall switches are specified by manufacturer and catalog number; other times they are specified by type, grade, voltage, current rating, and the like, leaving the contractor or electrician to select the manufacturer. The naming of a certain brand of switch for a particular project does not necessarily mean that this brand must be used.

SAFETY SWITCHES

Enclosed single-throw safety switches are manufactured to meet industrial, commercial, and residential requirements. The two basic types of safety switches are:

- General duty
- Heavy duty

Double-throw switches are also manufactured with enclosures and features similar to the general and heavy-duty single-throw designs.

The majority of safety switches have visible blades and safety handles. The switch blades are in full view when the enclosure door is open and there is visually no doubt when the switch is OFF. The only exception is Type 7 and 9 enclosures; these do not have visible blades. Switch handles, on all types of enclosures, are an integral part of the box, not the cover, so that the handle is in control of the switch blades under normal conditions.

Heavy-Duty Switches

Heavy-duty switches are intended for applications where ease of maintenance, rugged construction, and continued performance are primary concerns. They can be used in atmospheres where general-duty switches would be unsuitable, and are therefore widely used in industrial applications. Heavy-duty switches are rated 30 through 1200 A and 240 to 600 V ac or dc. Switches with horsepower ratings are capable of opening a circuit up to six times the rated current of the switch. When equipped with Class J or Class R fuses for 30 through 600 A switches, or Class L fuses in 800 and 1200 A switches, many heavy-duty safety switches are U.L. listed for use on systems with up to 200,000 RMS symmetrical amperes available fault current. This, however, is about the highest short-circuit rating available for any heavy-duty safety switch. Applications include use where the required enclosure is NEMA TYPE 1, 3R, 4, 4X, 5, 7, 9, 12, or 12K.

Switch Blade and Jaws

Two types of switch contacts are used by the industry in today's safety switches. One is the "butt" contact; the other is a knife-blade and jaw type. On switches with knife-blade construction, the jaws distribute a uniform clamping pressure on both sides of the blade contact surface. In the event of a high-current fault, the electromagnetic forces which develop tend to squeeze the jaws tightly against the blade. In the butt type contact, only one side of the blade's contact surface is held in tension against the conducting path. Electromagnetic forces due to high current faults tend to force the contacts apart, causing them to burn severely. Consequently, the knife blade and jaw type construction is the preferred type for use on all heavy-duty switches. The action of the blades moving in and out of the jaws aids in cleaning the contact surfaces. All current-carrying parts of these switches are plated to reduce heating by keeping oxidation at a minimum. Switch blades and jaws are made of copper for high conductivity. Spring-clamped blade hinges are another feature that help assure good contact surfaces and cool operations. "Visible blades" are utilized to provide visual evidence that the circuit has been opened.

WARNING!

Before changing fuses or performing maintenance on any safety swtich, always visibly check the switch blades and jaws to ensure that they are in the OFF position.

Fuse Clips

Fuse clips are plated to control corrosion and to keep heating to a minimum. All fuse clips on heavy-duty switches have steel reinforcing springs for increased mechanical strength and firmer contact pressure. *See* Figure 12-5.

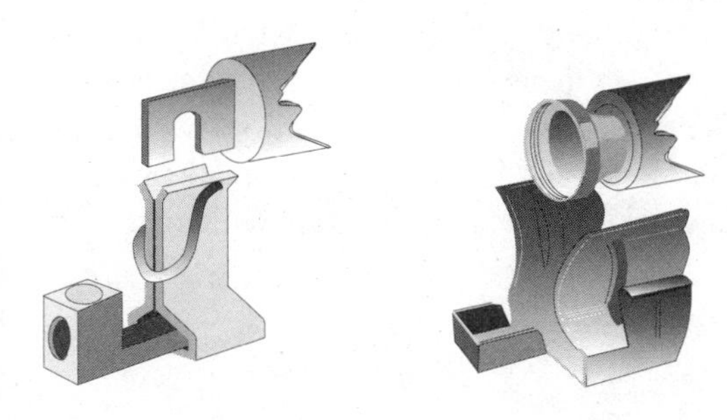

Figure 12-5: Typical fuse clips.

Terminal Lugs

Most heavy-duty switches have front removable, screw-type terminal lugs. Most switch lugs are suitable for copper or aluminum wire except NEMA TYPES 4, 4X, 5 stainless and TYPES 12 and 12K switches which have all copper current-carrying parts and lugs designated for use with copper wire only.

Insulating Material

As the voltage rating of switches is increased, arc suppression becomes more difficult and the choice of insulation material becomes more critical. Arc suppressors are usually made of insulation material and magnetic suppressor plates when required. All arc suppressor materials must provide proper control and extinguishing ability of arcs.

Operating Mechanism and Cover Latching

Most heavy-duty safety switches have a spring-driven, quick-make, quick-break mechanism. A quick-breaking action is necessary if the switch is to be safely switched OFF under a heavy load.

The spring action, in addition to making the operation quick-make, quick-break, firmly holds the switch blades in the ON or OFF position. The operating handle is an integral part of the switching mechanism and is in direct control of the switch blades under normal conditions.

A one-piece cross bar, connected to all switch blades, should be provided which adds to the overall stability and integrity of the switching assembly by promoting proper alignment and uniform switch blade operation.

Dual cover interlocks are standard on most heavy-duty switches where the NEMA enclosure permits. However, NEMA Types 7 and 9 have bolted covers and obviously cannot contain dual cover interlocks. The purpose of dual interlock is to prevent the enclosure door from being opened when the switch handle is in the ON position and prevents the switch from being turned ON while the door is open. A means of bypassing the interlock is provided to allow the switch to be inspected in the ON position by qualified personnel. However, this practice should be avoided if at all possible. Heavy-duty switches can be padlocked in the OFF position with up to three padlocks.

Enclosures

Heavy-duty switches are available in a variety of enclosures (*See* Figure 12-6) which have been designed to conform to specific industry requirements based upon the intended use. Sheet metal enclosures (that is, NEMA Type 1) are constructed from cold-rolled steel which is usually phosphatized and finished with an electrode deposited enamel paint. The Type 3R rainproof and Type 12 and 12K dusttight enclosures are manufactured from galvannealed sheet steel and painted to provide better weather protection. The Type 4, 4X, and 5 enclosures are made of corrosion resistant Type 304 stainless steel and requires no painting. Type 7 and 9 enclosures are cast

Enclosure	Explanation
NEMA Type 1 General Purpose	To prevent accidental contact with enclosed apparatus. Suitable for application indoors where not exposed to unusual service conditions
NEMA Type 3 Weatherproof (Weather Resistant)	Protection against specified weather hazards. Suitable for use outdoors
NEMA Type 3R Raintight	Protects against entrance of water from a rain. Suitable for general outdoor application not requiring sleetproof
NEMA Type 4 Watertight	Designed to exclude water applied in form of hose stream. To protect against stream of water during cleaning operations
NEMA Type 5 Dusttight	Constructed so that dust will not enter the enclosed area. Being replaced in some equipment by NEMA 12 Type
NEMA Type 7 Hazardous Locations A, B, C, or D Class I—letter or letters following type number indicates particular groups of hazardous locations per NEC	Designed to meet application requirements of NEC for Class I, hazardous locations (explosive atmospheres). Circuit interruption occurs in air
NEMA Type 9 Hazardous Locations E, F, or G Class II – letter or letters following type number indicates particular groups of hazardous locations per NEC	Designed to meet application requirements of NEC for Class II hazardous locations (combustible dusts, etc.)

Figure 12-6: NEMA classifications of enclosures.

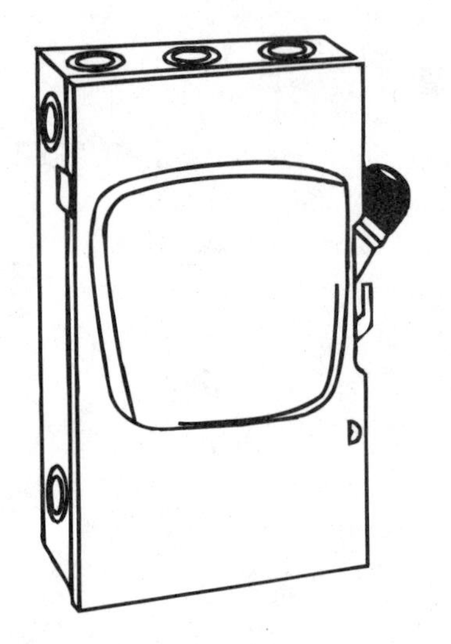

Figure 12-7: NEMA Type 1 general-duty safety switch.

from copper-free aluminum and finished with an enamel paint. Type 1 switches (Figures 12-7 and 12-8) are general purpose and designed for use indoors to protect the enclosed equipment from falling dirt and personnel from live parts. Switches rated through 200 A are provided with ample knockouts. 400 through 1200 A switches are provided without knockouts.

The following are the NEMA enclosure Types that will be encountered most often. Always make certain that the proper enclosure is chosen for the application.

Type 3R switches are designated "rainproof" and are designed for use outdoors. *See* Figure 12-9.

Type 3R enclosures for switches rated through 200 A have provisions for interchangeable bolt-on hubs at the top endwall. Type 3R switches rated higher than 200 A have blank top endwalls. Knockouts are provided (below live parts only) on enclosures for 200 A and smaller Type 3R switches. Type 3R switches are available in ratings through 1200 A.

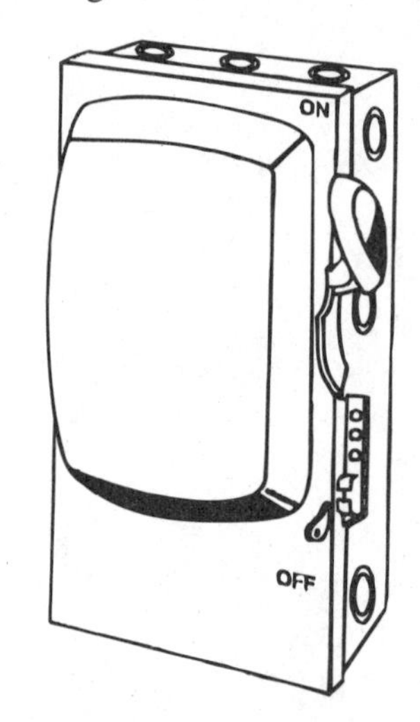

Figure 12-8: NEMA Type 1 heavy-duty safety switch.

Type 4, 4X, 5 stainless steel switches are designated dusttight, watertight and corrosion resistant and designed for indoor and outdoor use.

Common applications include commercial type kitchens, dairies, canneries, and other types of food processing facilities, as well as areas where mildly corrosive liquids are present. All Type 4, 4X, and 5 stainless steel enclosures are provided without knockouts. Use of watertight hubs is required. Available switch ratings are 30 through 600 A.

Figure 12-9: NEMA 3R raintight enclosure.

Figure 12-10: NEMA 12 switch enclosure.

Type 12 and Type 12K switches are designated dust-tight (except at knockout locations on Type 12K) and are designed for indoor use. In addition, NEMA Type 12 safety switches are designated as raintight for outdoor use when the supplied drain plug is removed. Common applications include heavy industries where the switch must be protected from such materials as dust, lint, flyings, oil seepage, etc. Type 12K switches have knock-outs in the bottom and top endwalls only. Available switch ratings are 30 through 600 A in Type 12 and 30 through 200 A in Type 12K.

NEMA Type 4, 4X, 5, Type 12, and 12K switch enclosures (Figure 12-10) have positive sealing to provide a dusttight and raintight (watertight with stainless steel) seal. Enclosure doors are supplied with oil resistant gaskets. Switches rated 30 through 200 A incorporate spring loaded, quick-release latches. 400 and 600 A switches feature single-stroke sealing by operation of a cover mounted handle. 30, 60, and 100 A switches in these enclosures are provided with factory installed fuse pullers.

Interlocked Receptacles

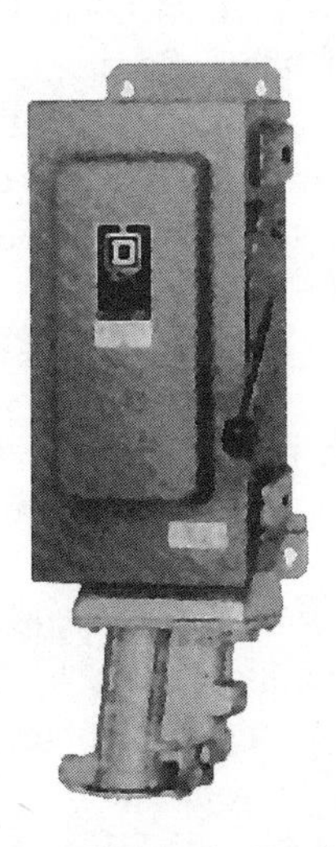

Figure 12-11: NEMA 1 heavy-duty switch enclosure with integral receptacle.

Heavy-duty, 60 A, Type 1 and Type 12 switches within interlocked receptacle, are also available. This receptacle provides a means for connecting and disconnecting loads directly to the switch. A nondefeating interlock prevents the insertion or removal of the receptacle plug while the switch is in the ON position. It also prevents operation of the switch if an incorrect plug is used. *See* Figure 12-11.

Accessories

Accessories available for field installation include Class R fuse kits, fuse pullers, insulated neutrals with grounding provisions, equipment grounding kits, watertight hubs for use with Type 4, 4X, 5 stainless or Type 12 switches, and interchangeable bolt-on hubs for Type 3R switches.

Electrical interlock consists of auxiliary contacts for use where control or monitoring circuits need to be switched in conjunction with the safety switch operation. Kits can be either factory or field installed, and they contain either one normally open and one normally closed contact or two normally open and two normally closed contacts. The electrical interlock is actuated by a pivot arm which operates directly from the switch mechanism. The electrical interlock is designed so that its contacts disengage before the blades of the safety switch open and engage after the safety switch blades close.

General-Duty Switches

General-duty switches for residential and light commercial applications are used where operation and handling are moderate and where the available fault current is 10,000 RMS symmetrical amperes or less. Some general-duty safety switches, however, exceed this specification in that they are U.L. listed for application on systems having up to 100,000 RMS symmetrical amperes of available fault current when Class R fuses and Class R fuse kits are used. Class T fusible switches are also available in 400, 600, and 800 A ratings. These switches accept 300 Vac Class T fuses only. Some examples of general-duty switch applications include residential, farm, and small business service entrances, and light-duty branch circuit disconnects.

General-duty switches are rated up to 600 A at 240 Vac in general purpose (Type 1) and rainproof (Type 3R) enclosures. Some general-duty switches are horsepower rated and capable of opening a circuit up to six times the rated current of the switch; others are not. Always check the switch's specifications before using under a horsepower-rated condition.

Switch Blades and Jaws

All current carrying parts of general-duty switches are plated to minimize oxidation and reduce heating. Switch jaws and blades are made of copper for high conductivity. Where required, a steel reinforcing spring increases the mechanical strength of the jaws and contact pressure between the blade and jaw. Good pressure contact maintains the blade-to-jaw resistance at a minimum, which in turn, promotes cool operation. All general-duty switch blades feature visible blade construction. With the door open, there is visually no doubt when the switch is OFF.

Fuse Clips

Fuse clips are normally plated to control corrosion and keep heating to a minimum. Where required, steel reinforcing springs are provided to increase the mechanical strength of the fuse clip. The result is a firmer, cooler connection to the fuses as well as superior fuse retention.

Terminal Lugs

Most general-duty safety switches are furnished with mechanical set screw lugs which are suitable for aluminum or copper conductors.

Insulating Material

Switch and fuse bases are made of a strong, noncombustible, moisture-resistant material which provides the required phase-to-phase and phase-to-ground insulation for applications on 240 Vac systems.

Operating Mechanism and Cover Latching

Although not required by either the UL or NEMA standards, some general-duty switches have spring-driven, quick-make, quick-break operating mechanisms. Operating handles are an integral part of the operating mechanism and are not mounted on the enclosure cover. The handle provides indication of the status of the switch. When the handle is up, the switch is ON. When the handle is down, the switch is OFF. A padlocking bracket is provided which allows the switch handle to be locked in the OFF position. Another bracket is provided which allows the enclosure to be padlocked closed.

Enclosures

General-duty safety switches are available in either Type 1 for general purpose, indoor applications, or Type 3R for rainproof, outdoor applications.

DOUBLE-THROW SAFETY SWITCHES

Double-throw switches are used as manual transfer switches and are not intended for use as motor circuit switches; thus, horsepower ratings are generally unavailable.

Double-throw switches are available as either fused or nonfusible devices and two general types of switch operation are available:

- Quick-make, quick-break
- Slow-make, slow-break

Figure 12-12 shows a practical application of a double-throw safety switch used as a transfer switch in conjunction with a stand-by emergency generator system.

NEC SAFETY-SWITCH REQUIREMENTS

Safety switches, in both fusible and nonfusible types, are used as a disconnecting means for services, feeders, and branch circuits. Installation requirements involving safety switches are found in several places throughout the *NEC*, but mainly in the following articles and sections:

- *NEC* Article 373
- *NEC* Article 380
- *NEC* Article 430-H
- *NEC* Article 440-B
- *NEC* Section 450-8(c)

When used as a service disconnecting means, the major installation requirements are listed in the following *NEC* Sections:

- *NEC* Section 110-16
- *NEC* Section 230-2, Ex. 4
- *NEC* Section 230-70
- *NEC* Section 230-71
- *NEC* Section 230-76
- *NEC* Section 230-81
- *NEC* Section 230-84

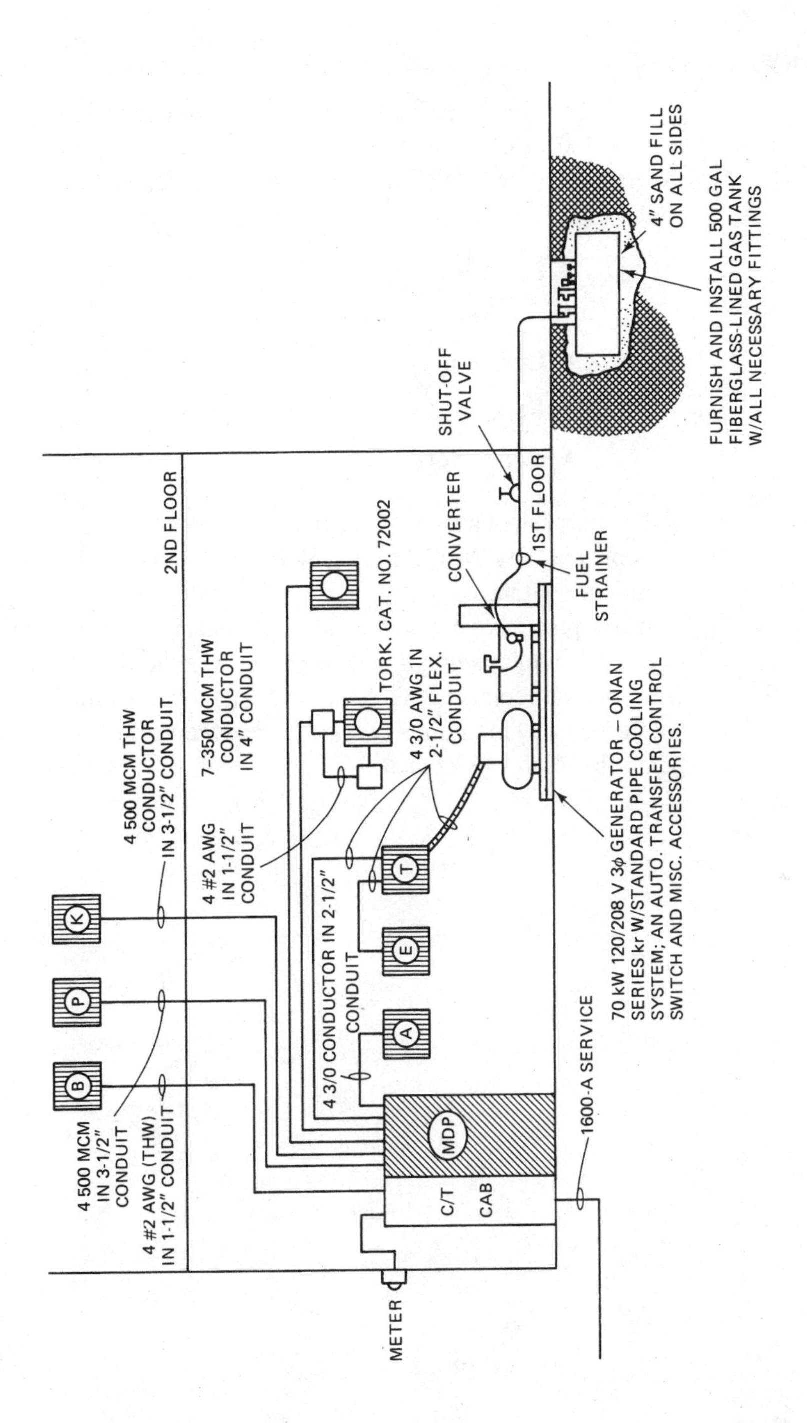

Figure 12-12: Automatic transfer switch used in conjunction with an emergency-standby generator.

Summary

A *device* — by *NEC* definition — is a unit of an electrical system that is intended to carry, but not utilize, electric energy. This covers a wide assortment of system components that include, but are not limited to, the following:

- Switches
- Relays
- Contactors
- Receptacles
- Conductors

The purpose of a switch is to make and break an electrical circuit, safely and conveniently. *NEC* Article 380 covers most of the installation requirements for switches.

A receptacle is a contact device installed at the outlet for the connection of a single attachment plug. A single receptacle is a single contact device with no other contact device on the same yoke. A multiple receptacle is a single device containing two or more receptacles — the most common being the *duplex receptacle*.

Chapter 13

Conductor Terminations and Splices

Anyone involved with electrical systems of any type should have a good knowledge of wire connectors and splicing, as it is necessary to make numerous electrical joints during the course of any electrical installation.

Splices and connections that are properly made will often last as long as the insulation on the wire itself, while poorly made connections will always be a source of trouble; that is, the joints will overheat under load and cause a higher resistance in the circuit than there should be.

The basic requirements for a good electrical connection include the following:

- It should be mechanically and electrically secure.
- It should be insulated as well as or better than the existing insulation on the conductors.
- These characteristics should last as long as the conductor is in service.

There are many different types of electrical joints for different purposes, and the selection of the proper type for a given application will depend to a great extent on how and where the splice or connection is used.

Electrical joints are normally made with solderless pressure connectors or lugs to save time, but electricians should also have a knowledge of the traditional splices. Most traditional splices appear in Figure 13-1 on the next page.

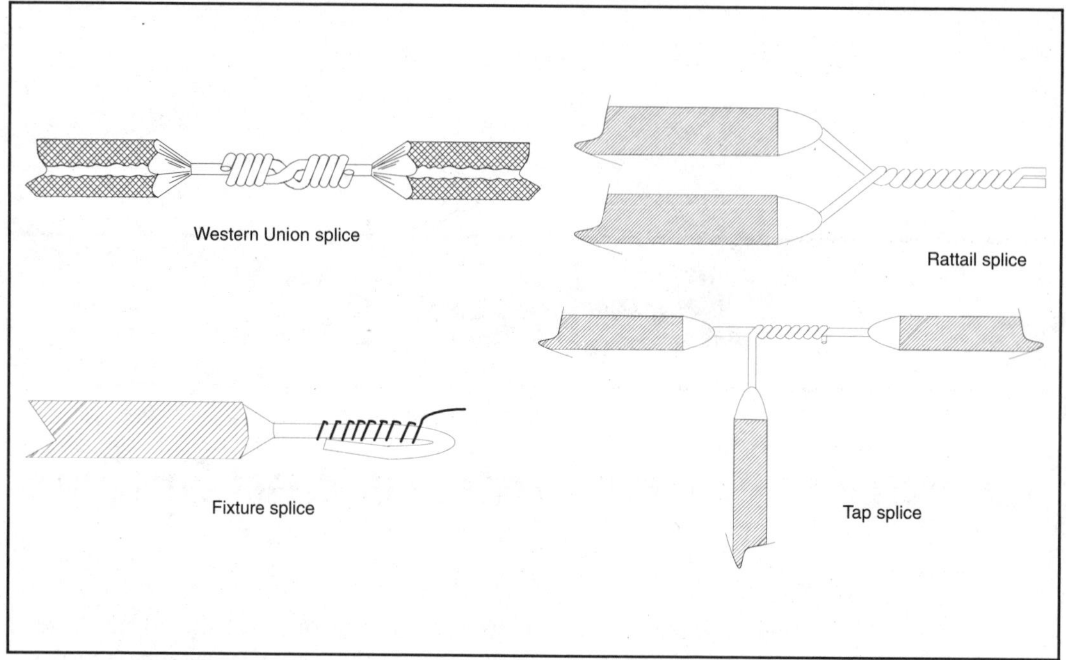

Figure 13-1: Traditional taps and splices.

STRIPPING AND CLEANING

Before any connection or splice can be made, the ends of the conductors must be properly stripped and cleaned. *Stripping* is the removal of insulation from the conductors at the end of the wire or at the location of the splice. Some electricians strip the smaller sizes of wire with a pocket knife or a pair of side-cutting pliers, but there are many handy tools on the market that will facilitate this operation. The use of such tools will also help to prevent cuts and nicks in the wire, which reduces the conductor area as well as weaken the conductor.

Poorly stripped wire can result in nicks, scrapes, or burnishes. Any of these can lead to a stress concentration at the damaged cross section. Heat, rapid temperature change, mechanical vibration, and oscillatory motion can aggravate the damage, causing faults in the circuitry or even total failure.

Lost strands are a problem in splices or crimp-type terminals, while exposed strands might be a safety hazard.

Slight burnishes on conductors, as long as they had no sharp edges, were acceptable at one time. Now, however, reliable experts feel that under certain conditions removing as little as 40 micro inches of conductor plating from some wires can cause failure.

Faulty stripping can pierce, scuff, or split the insulation. This can cause changes in dielectric strength and lower the wire's resistance to moisture and abrasion. Insulation particles often get trapped in solder and crimp joints. These form the basis for a defective termination. A variety of factors determine just how precisely a wire can be stripped: wire size, insulation concentricity, adherence, and others.

It is a common mistake to believe that a certain gauge of stranded conductor has the same diameter as a solid conductor. This is a very important consideration in selecting proper blades for wire strippers. The table in Figure 13-2 shows the nominal sizes referenced for the different wire gauges.

DIMENSIONS OF COMMON WIRE SIZES			
Size, AWG/kcmil	Area Circular Mils	Overall Diameter in Inches	
		Solid	Stranded
18	1620	0.040	0.046
16	2580	0.051	0.058
14	4130	0.064	0.073
12	6530	0.081	0.092
10	10380	0.102	0.116
8	16510	0.128	0.146
6	26240	-	0.184
4	41740	-	0.232
3	52620	-	0.260
2	66360	-	0.292
1	83690	-	0.332
1/0	105600	-	0.373
2/0	133100	-	0.419
3/0	167800	-	0.470

Figure 13-2: Dimensions of common wire sizes.

DIMENSIONS OF COMMON WIRE SIZES *(Cont.)*			
Size, AWG/kcmil	Area Circular Mils	Overall Diameter in Inches	
		Solid	Stranded
4/0	211,600	—	0.528
250	250,000	—	0.575
300	300,000	—	0.630
350	350,000	—	0.681
400	400,00	—	0.728
500	500,000	—	0.813
600	600,000	—	0.893
700	700,000	—	0.964
750	750,000	—	0.998
800	800,000	—	1.03
900	900,000	—	1.09
1000	1,000,000	—	1.15
1250	1,250,000	—	1.29
1500	1,500,000	—	1.41
1750	1,750,000	—	1.52
2000	2,000,000	—	1.63

Figure 13-2: Dimensions of common wire sizes. ***(Cont.)***

WIRE CONNECTIONS UNDER 600 V

Wire connections are used to connect a wire or cable to such electrically operated devices as fan-coil units, duct heaters, oil burners, motors, pumps, and control circuits of all types.

A variety of wire connectors are shown in Figure 13-3. These connectors are available in various sizes to accommodate wire from No. 22 AWG through 250 kcmil. They can be installed with crimping tools having a single indenter or double indenter. Wide range identification is normally stamped on the tongue of each terminal.

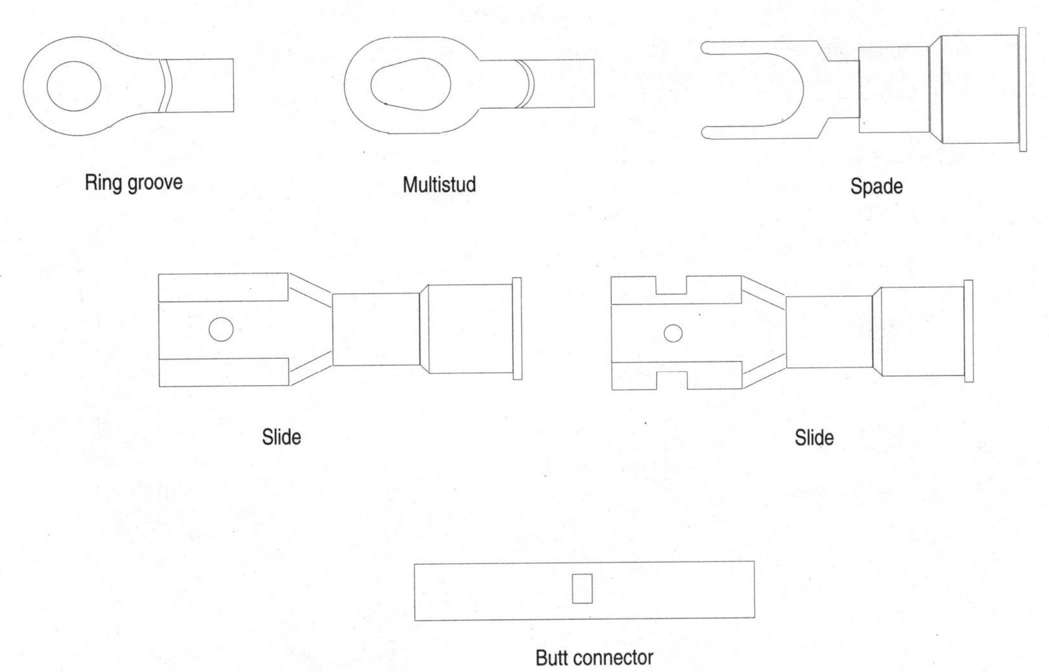

Figure 13-3: Several types of crimp connectors.

Compression type terminators are also available to accommodate wires from No. 8 AWG through 1000 kcmil. One-hole lugs, two-hole lugs, and split-bolt connectors are shown in Figure 13-4 on the next page.

Aluminum Connections

Aluminum has certain properties that are different from copper that must be understood if reliable connections are to be made. These properties are: cold flow, coefficient of thermal expansion, susceptibility to galvanic corrosion, and the formation of oxide film on the metal's surface.

Because of thermal expansion and cold flow of aluminum, standard copper connectors as found on the market today cannot be safely used on aluminum wire. Most manufacturers design their aluminum connectors with greater contact area to counteract this property of aluminum. Tongues and barrels of all aluminum connectors are larger or deeper than comparable copper connectors.

CAUTION!

Never use connectors designed for copper conductors only on aluminum conductors. Connectors designed for use on both metals will normally be marked "AL-CU."

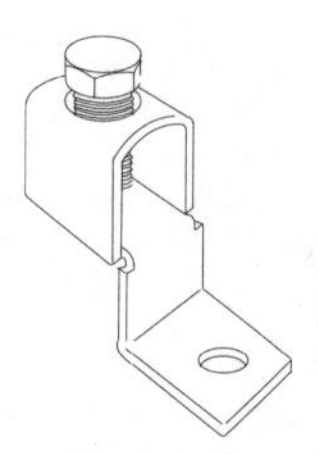

One barrel, offset tongue,
one hole, Type CB
No. 14 AWG through 1000 kcmil

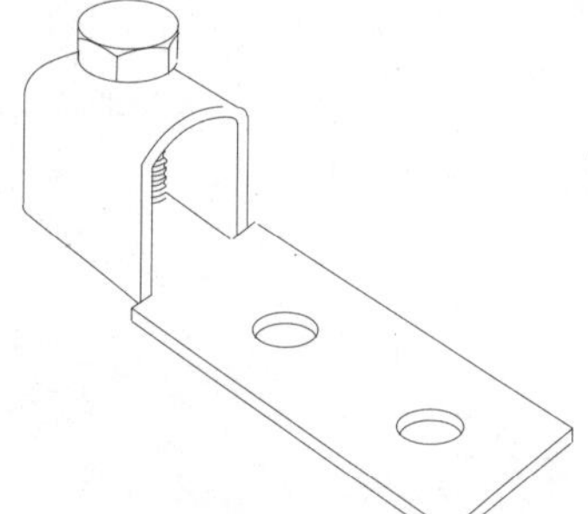

One barrel, offset tongue,
two hole, Type CO
No. 14 AWG through 1000 kcmil

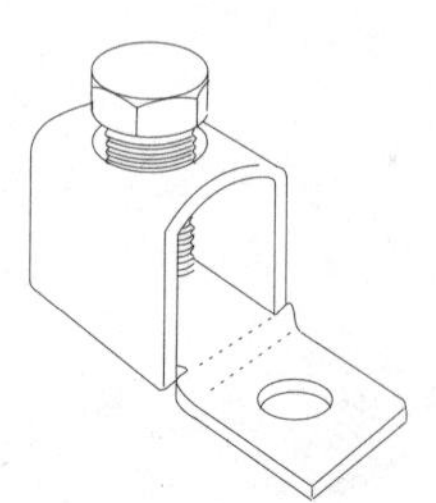

One barrel, fixed tongue,
one hole, Type CX
No. 14 AWG through 500 kcmil

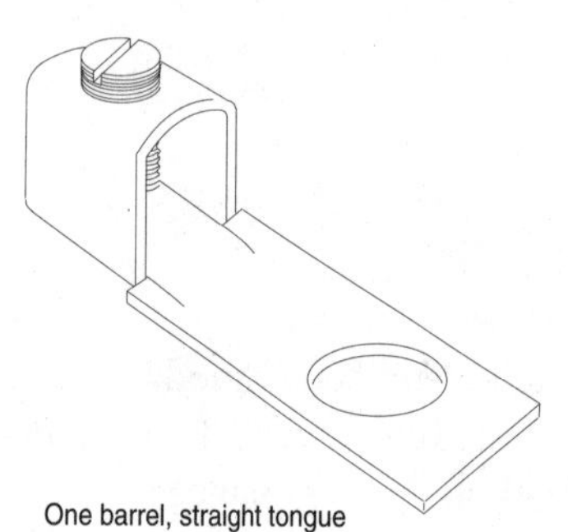

One barrel, straight tongue
one hole, Type CS
No. 14 AWG through 1000 kcmil

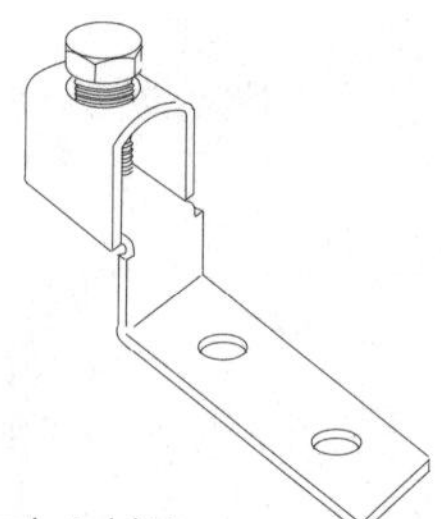

One barrel, straight tongue,
two hole, Type CD
No. 14 AWG through 1000 kcmil

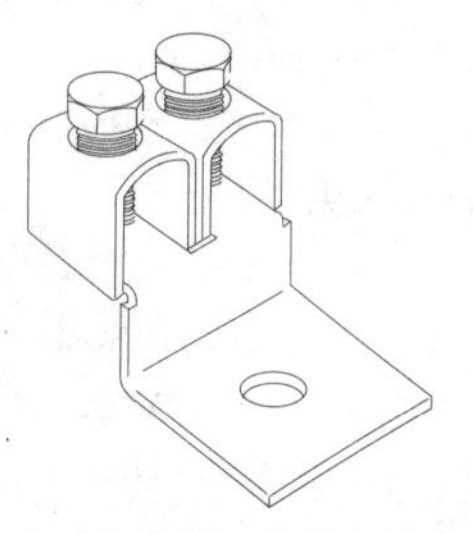

Two barrels, offset tongue,
one hole, Type DC
No. 6 AWG through 500 kcmil

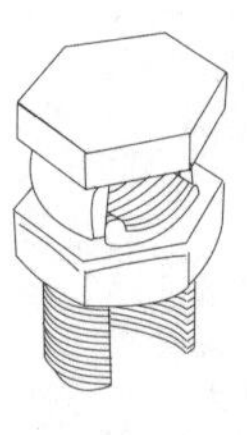

Split bolt connector, Type SBC
(2) No. 14 AWG through (2) 1000 kcmil
run and tap combinations

Figure 13-4: Types of compression connectors in current use.

The electrolytic action between aluminum and copper can be controlled by plating the aluminum with a neutral metal (usually tin). The plating prevents electrolysis from taking place and the joint remains tight. As an additional precaution, a joint sealing compound should be used. Connectors should be tin plated and prefilled with an oxide abrading sealing compound.

The insulating aluminum oxide film must be removed or penetrated before a reliable aluminum joint can be made. Aluminum connectors are designed to bite through this film as they are applied to conductors. It is further recommended that the conductor be wire brushed and preferably coated with a joint compound to guarantee a reliable joint.

- Connectors marked with just the wire size should only be used with copper conductors.
- Connectors marked with Al and the wire size should only be used with aluminum wire.
- Connectors marked with Al-Cu and the wire size may be safely used with either copper or aluminum.

Heat-Shrink Insulators

Heat-shrinkable insulators for small connectors provide skintight insulation protection and are fast and easy to use. They are designed to slip over wires, taper pins, connectors, terminals, and splices. When heat is applied, the insulation becomes semirigid and will provide positive strain relief at the flex point of the conductor. A vaporproof band will seal and protect the conductor from abrasion, chemicals, dust, gasoline, oil, and moisture. Extreme temperatures, both hot and cold, will not affect the performance of these insulators. The source of heat can be any number of types, but most manufacturers of these insulators also produce a heat gun especially designed for use on heat-shrink insulators. It closely resembles and operates the same as a conventional hair dryer with a fan blower as shown in Figure 13-5 on the next page.

In general, a heat-shrink insulator may be thought of as tubing with a memory. When initially manufactured, it is heated and expanded to a predetermined diameter and then cooled. Upon application of heat, through various methods, the tubing compound "remembers" its original size and shrinks to that smaller diameter. It is available in a range of sizes and is designed to shrink easily over any wire or device when heat is applied. This property enables it to conform to the contours of any object.

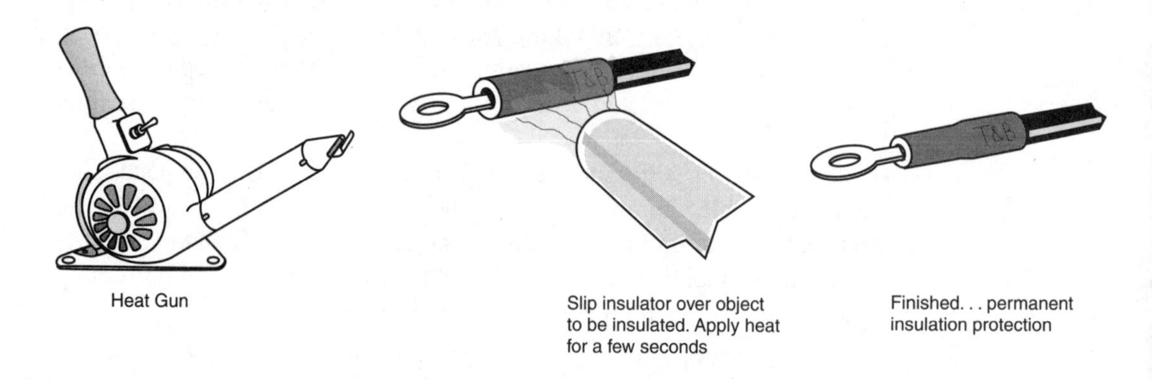

Figure 13-5: Method used to install heat-shrink insulators.

The following describes some of the types currently available. A tubing selector guide appears in Figure 13-6.

PVC: This type is a general-purpose, economical tubing that is widely used in the electronic industry. The PVC compound is irradiated by being bombarded with high-velocity electrons. This results in a denser, cross-linked material with superior electrical and mechanical properties. It also assures that the tubing will resist cracking and splitting.

Polyolefin: Polyolefin tubing has a wide range of uses for wire bundling, harnessing, strain relief, and other applications where cables and components require additional insulation. It is irradiated, flame retarding, flexible, and comes in a wide variety of colors.

Double wall: This type is available and designed for outstanding protective characteristics. It is semirigid tubing that has an inner wall that actually melts and an outer wall that shrinks to conform to the melted area.

Teflon®: This type is considered by many users to be the best overall heat-shrinkable tubing — physically, electrically, and chemically. Its high temperature rating of 250° C resists brittleness or loss of translucency from extended exposure to high heat and will not support combustion.

Neoprene: Components that warrant extra protection from abrasion require a highly durable, yet flexible, tubing. The irradiated neoprene tubing offers this optimal coverage.

Kynar®: Irradiated Kynar is a thin-wall, semirigid tubing with outstanding resistance to abrasion. This transparent tubing enables easy inspection

HEAT-SHRINK INSULATOR TUBING SELECTOR GUIDE								
Manhattan Number	Type	Material	Temperature Range, °C	Shrink Ratio	Max. Long. Shrinkage%	Tensile Strength, psi	Colors	Dielectric Strength, V/mil
MT-105	Nonshrinkable	PVC	+105	—	—	2700	White, red, clear, black	800
MT-150	Shrinkable	PVC	-35 to +105	2:1	10	2700	Clear, black	750
MT-200	Nonshrinkable	Teflon®	-65 to +260	—	—	2700	Clear	1400
MT-221	Shrinkable	Flexible polyolefin	-55 to +135	2:1	5	2500	Black, white, red, yellow, blue, clear	1300
MT-250	Nonshrinkable	Teflon®	-65 to +260	—	—	7500	Clear	1400
MT-300	Shrinkable	Polyolefin double wall	-55 to +110	6:1	5	2500	Black	1100
MT-350	Shrinkable	Kynar®	-55 to +175	2:1	10	8000	Clear	1500
MT-400	Shrinkable	Teflon®	+250	1.2:2	10	6000	Clear	1500
MT-500	Shrinkable	Teflon®	+250	11/2:1	10	6000	Clear	1500
MT-600	Shrinkable	Neoprene	+120	2:1	10	1500	Black	300

Figure 13-6: Tubing selector guide.

of components that are covered and retains its properties at its rated temperature.

Most tubing is available in a wide variety of colors and put-ups. Tubing selector guides are available from manufacturers that can help in the selection of the best tubing for any given application.

Wire Nuts

Ever since its invention in 1927, the wire nut (Figure 13-7) has been a favorite wire connector for use on branch-circuit applications where permitted by the *NEC*. Several varieties of wire nuts are available, but the following are the ones used the most:

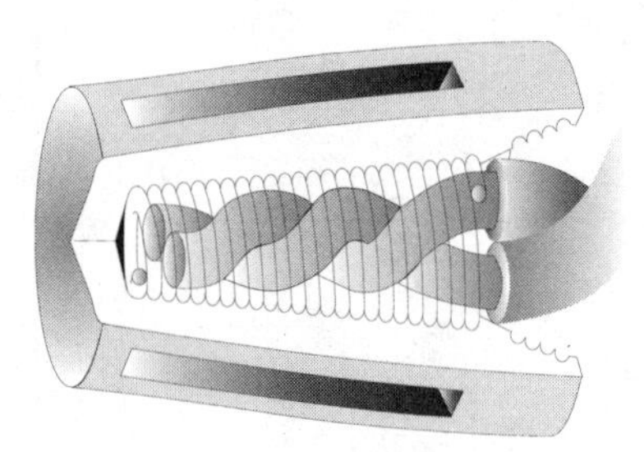

Figure 13-7: Typical wire nut showing the interior arrangement.

- For use on wiring systems 300 V and under.
- For use on wiring systems 600 V and under (1000 V in lighting fixtures and signs).

Most brands are U.L. listed for aluminum to copper in dry locations only; aluminum to aluminum only; or copper to copper only; 600 V maximum; 1000 V in lighting fixtures and neon signs. The maximum temperature rating is 105°C (221°F).

Wire nuts are frequently used for all types of splices in residential and commercial applications and are considered to be the fastest connectors on the market for this type of work.

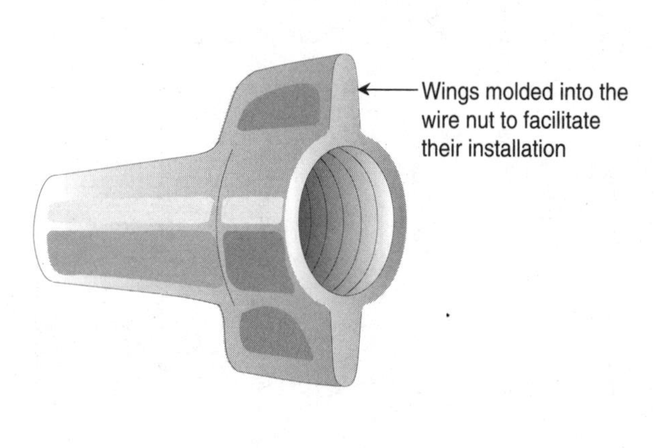

Figure 13-8: Some wire nuts have thin wings on each side to facilitate installation.

Traditionally, electricians form a pigtail splice on the ends of conductors with side-cutting pliers, trim the bare conductors with the pliers' cutters, and then screw on the wire nut. In doing so, the wire nut draws the conductors and insulation into the shirt of the connector which increases resistance to flashover. The internal spring is designed to tightly "thread" the conductors into the wire nut and then hold them with a positive grip. Some type of wire nuts have thin wings on each side of the connector to facilitate their installation. *See* Figure 13-8.

Wire nuts are made in sizes to accommodate conductors as small as No. 22 AWG up to as large as No. 10 AWG,

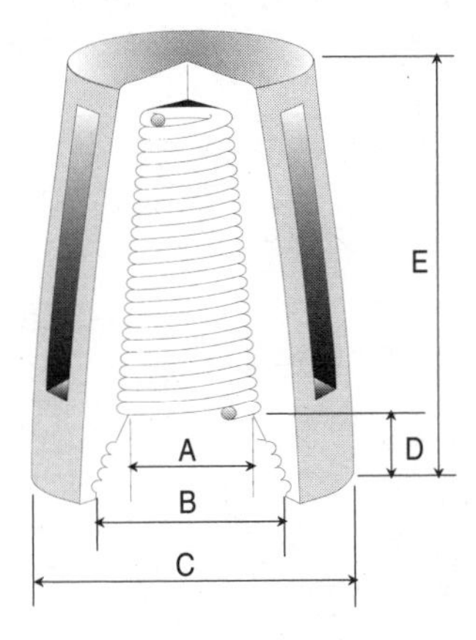

Model	A	B	C	D	E
71B	1/8	1/4	21/64	11/64	37/64
72B	9/64	9/32	25/64	11/64	45/64
73B	9/64	11/32	7/16	5/16	55/64
74B	11/64	7/16	35/64	17/64	61/64
76B	1/4	17/32	21/32	17/64	1 1/16

Figure 13-9: Dimensions of conventional wire nuts.

with practically any combination of those sizes in between.

The model numbers of wire nuts will vary with the manufacturer, but Ideal set a standard years ago with their 71B through 76B series. The table in Figure 13-9 gives the dimensions of this series.

The table in Figure 13-10 on the next pages gives the U.L. listings for various sizes of wire nuts, including the characteristics of each model.

TERMINATING CABLE

Type MI/mineral-insulated, metal-sheathed cable is a factory assembly of one or more conductors insulated with a highly compressed refractory mineral insulation and enclosed in a liquidtight and gastight continuous copper or alloy steel sheath. Although expensive, Type MI has a wide variety of use in the electrical industry, and may be used in practically any wiring situation except where exposed to destructive corrosive conditions.

Junction and outlet boxes designed especially for use with Type MI cable have one or more threaded openings for securing cable connectors. The covers for these boxes have gaskets which provide a moisture-tight and gas-tight seal for the boxes. Other types of boxes may also be used where their use complies with *NEC* regulations. Where knockouts only are present in the box or cabinet, special connectors must be used as described in the following paragraphs.

Model	300 Volt Max. Pressure-Type Wire Connector	600 Volt Max. Pressure-Type Wire Connector	Copper to Copper	Aluminum to Aluminum	Copper to Aluminum (Dry Location Only)	Maximum Temperature Rating
71B	●		●			105°C
72B	●		●			105°C
73B	●	I	I			105°C
74B	●	I	I	I	I	105°C
76B	●	I	I	I	I	105°C

Figure 13-10: U.L. listing of wire nuts.

Figures 13-11 and 13-12 show the component parts of a Type MI cable termination. The termination is designed to provide a good electrical contact between the metallic sheath of the cable and the junction box and to assure a moisture-tight and gas-tight seal at the termination. The purpose of the end seal is to seal the end of the mineral insulation, in which the conductors are encased, and to provide insulation extensions for the conductors so that the bare copper conductors will not be exposed inside the junction box. A sealing compound is applied to the end of the cable to complete the moisture-tight and gas-tight seal. The gland assembly provides a means to connect the cable to the junction box and seal the termination from moisture and gases. The brass gland body screws into the threaded junction-box opening. The brass gland nut tightens down on the compression ring and, thus, provides a moisture-tight and gas-tight seal for the termination.

WARNING!

Under no circumstances may the sheath of Type MI cable be used as a current-carrying conductor.

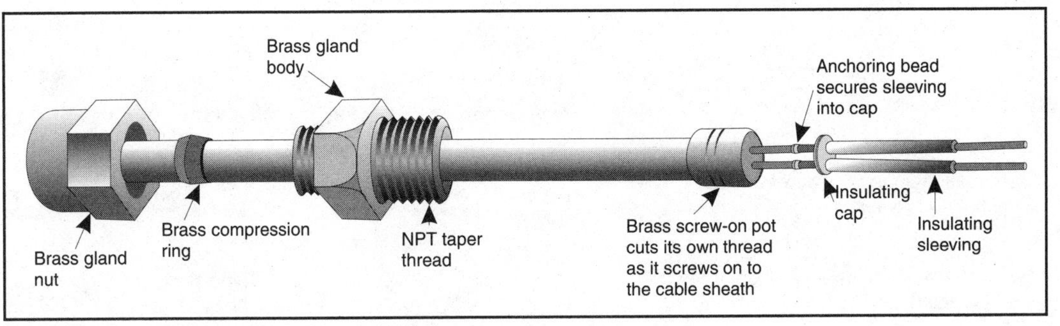

Figure 13-11: Details of MI cable termination.

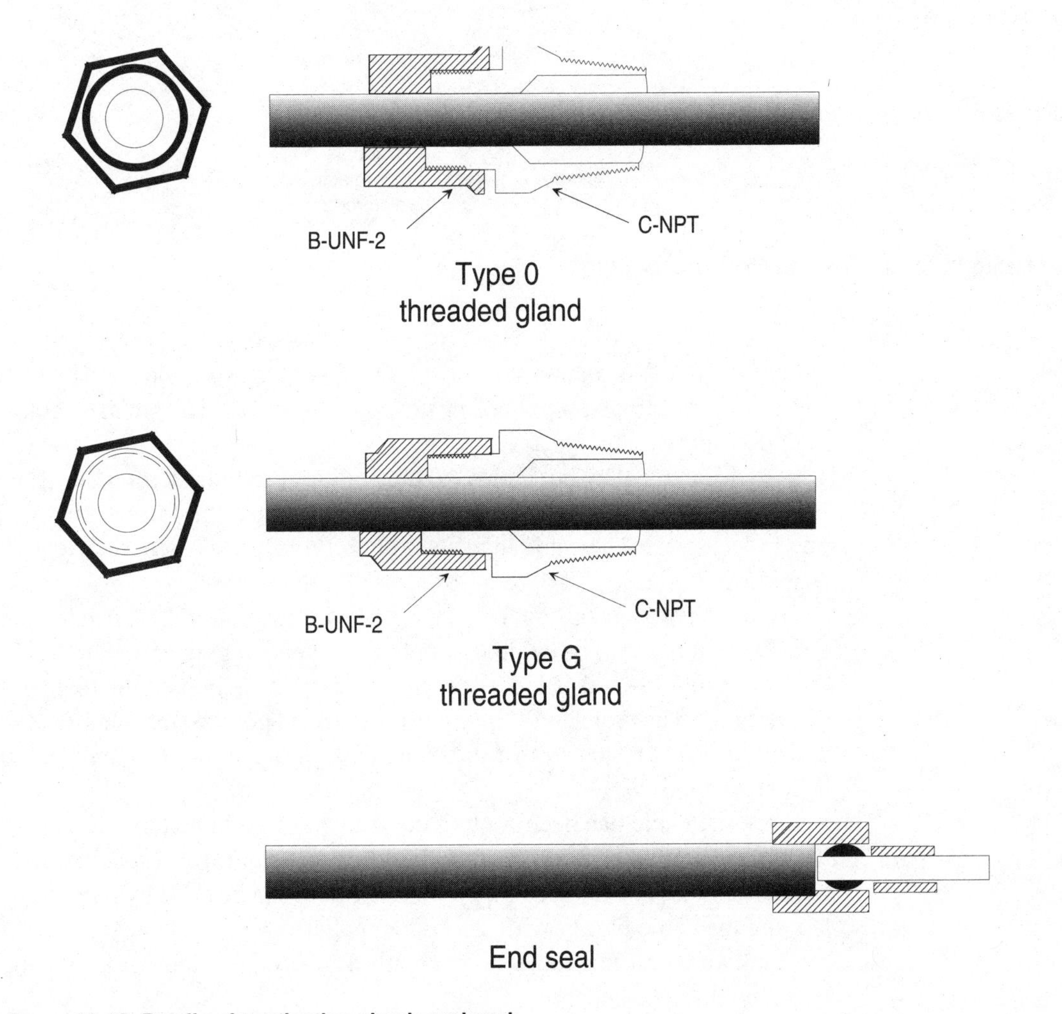

Figure 13-12: Details of termination glands and seal.

Applications	Seals			Gland
	Designation	Sleeving	Sealant	
General use up to 85° C, wet and dry location—U.L. listed	Type O	PVC	Type O plastic	Type O
Hazardous locations, wet and dry up to 85° C—U.L. listed	Type H	PVC	Type H epoxy	Type G
Low temperatures (below -10°C)	Low temperature	Silicone rubber (fiberglass reinforced)	Type O plastic	Type O
High temperatures (up to 125° C)	High temperatures	Silicone rubber (fiberglass reinforced)	Type O epoxy	Type O

Figure 13-13: MI cable termination summary.

Figure 13-13 is a summary of the terminations available for Type MI cable. The termination is chosen not only by cable size but by cable's intended application.

Type MI cable may be secured to the building structure with most of the common types of support hardware including straps, staples, hangers and one-hole clamps. The cable may also run in cable trays in accordance with *NEC* Article 318.

Guidelines for installing Type MI cable are given in *NEC* Article 330, Part B, Installation. The cable is supplied on reels ranging from 3 to 5 ft in diameter and in standard lengths up to 1900 ft, depending on the diameter of the cable. The reel should be positioned at one point of termination and placed on a reel frame so that the cable pays off, or rolls off, the reel as it is being routed.

Once the cable has been routed, the lead end should be terminated and the cable should be secured to the building structure in accordance with Section 330-12 of the *NEC*. The final alignment can be done by hand or by tapping the cable lightly with a wooden mallet.

Finally, the cable should be cut with a tube cutter and hacksaw, and terminated.

Bending Cable

Nearly all cable installations will require the cable to be bent at terminations and other points along the cable's route. These bends must comply with the *NEC*. Minimum bending radii are determined by the cable diameter and, in some instances, the construction of the metal sheath.

In general, bends in Type MI cable must be made so as not to damage the cable. The radius of the inner edge of any bend must not be less than the following:

- Five times the external diameter of the metallic sheath for cable not more than ¾ in in external diameter.
- Ten times the external diameter of the metallic sheath for cable greater than ¾ in but not more than 1 inch in external diameter.

TRAINING CONDUCTORS

Training is the positioning of cable so that it is not under tension. Bending is the positioning of cable which is under tension. When installing cable or any large-size conductors, the object is to limit these forces (tension) so that the cable's physical and electrical characteristics are maintained for the expected service life. Training conductors — rather than bending them—also reduces the tension on lugs and connectors—extending their service life considerably.

Two types of cable-bending tools are in common use:

- Ratchet
- Hydraulic

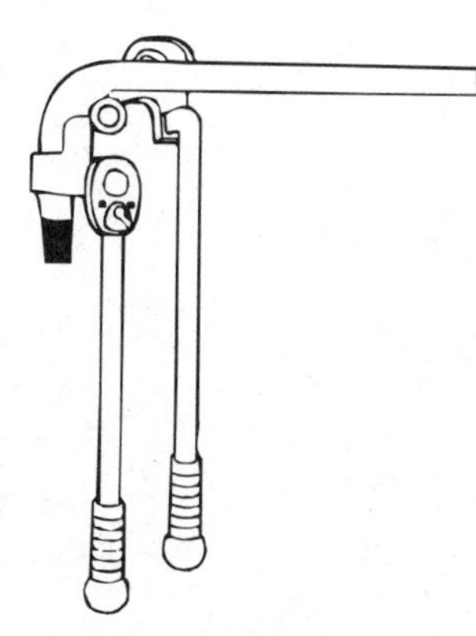

Figure 13-14: Ratchet cable bender.

The ratchet cable bender (Figure 13-14) bends 600-V copper or aluminum conductors up to 500 kcmil, while the hydraulic bender (Figure 13-15 is designed for cables from 350 kcmil through 1000 kcmil. In addition, the hydraulic bender is capable of one-shot bends up to 90° and automatically unloads the cable when the bend is finished. Either type simplifies and speeds cable installations.

Conductors at terminals or conductors entering or leaving cabinets or cutout boxes and the like must comply with certain *NEC* requirements — many of which are covered in *NEC* Article 373. The bending radii for various sizes of conductors that do not enter or leave an enclosure through the wall opposite its terminal are

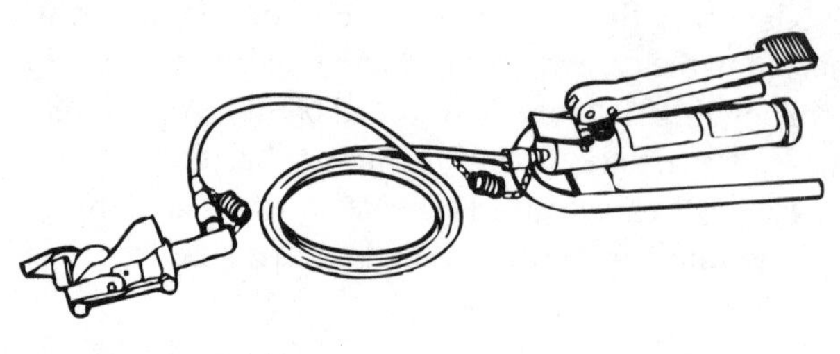

Figure 13-15: Hydraulic cable bender.

shown in Figure 13-16. When using this table, the bending space at terminals must be measured in a straight line from the end of the lug or wire connector (in the direction that the wire leaves the terminal) to the wall, barrier, or obstruction as shown in Figure 13-17.

A nonshielded cable can tolerate a sharper bend than a shielded cable. This is especially true for cables having helical metal tapes which, when bent too sharply, can separate, buckle, and cut into the insulation. The problem is compounded by the fact that most tapes are under jackets which conceal such damage. The shielding bedding tapes or extruded polymers have sufficient conductivity and coverage initially to pass acceptance testing, then fail prematurely due to corona at the shield/insulation interface.

AWG or Circular-Mil Size of Wire	Wires per Terminal				
	1	2	3	4	5
14 – 10	Not Specified	—	—	—	—
8 – 6	1½	—	—	—	—
4 – 3	2	—	—	—	—
2	2½	—	—	—	—
1	3	—	—	—	—
1/0 – 2/0	3½	5	7	—	—,

Figure 13-16: Minimum wire-bending space and gutter width. NEC Table 373-6(a).

AWG or Circular-Mil Size of Wire	Wires per Terminal				
	1	2	3	4	5
3/0 – 4/0	4	6	8	—	—
250 kcmil	4½	6	8	10	—
300 – 350 kcmil	5	8	10	12	—
400 – 500 kcmil	6	8	10	12	14
600 – 700 kcmil	8	10	12	14	16
750 – 900 kcmil	8	12	14	16	18
1000 – 1250 kcmil	10	—	—	—	—
1500 – 2000 kcmil	12	—	—	—	—

Figure 13-16: Minimum wire-bending space and gutter width. NEC Table 373-6(a). ***(Cont.)***

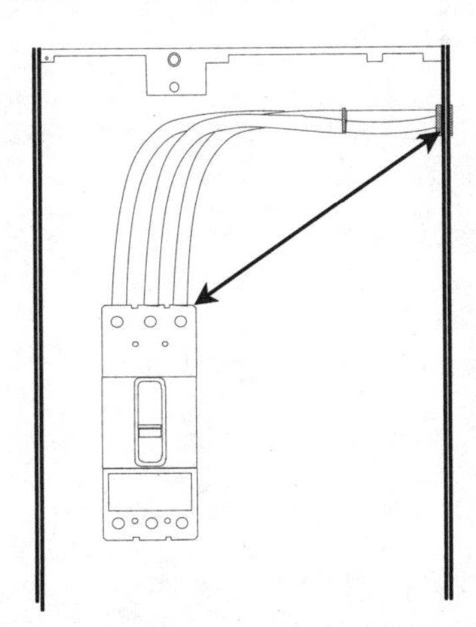

When using NEC Table 373-6(a) bending space at terminals must be measured in a straight line from the end of the lug or wire connector (in the direction that the wire leaves the terminals) to the wall, barrier, or obstruction

Figure 13-17: Bending space at terminals is measured in a straight line.

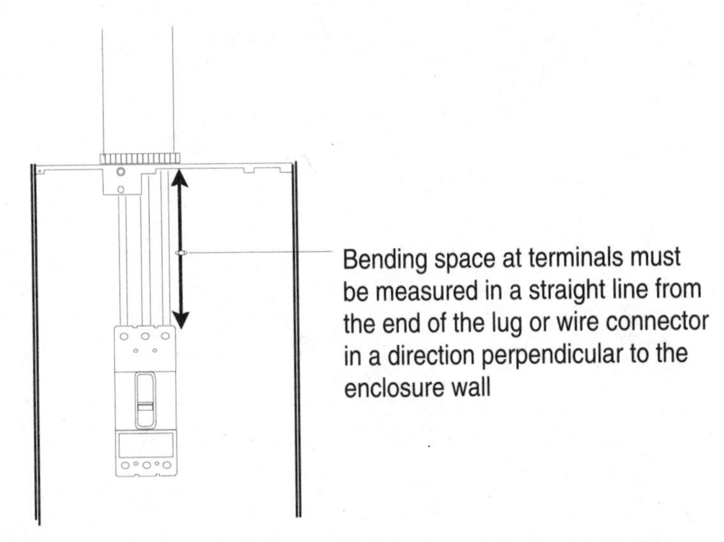

Figure 13-18: Conductors entering an enclosure opposite the conductor terminals.

When conductors enter or leave an enclosure through the wall opposite its terminals as shown in Figure 13-18, *NEC* Table 373-6(b) applies. *See* Figure 13-19. In using this table, the bending space at terminals must be measured in a straight line from the end of the lug or wire connector in a direction perpendicular to the enclosure wall. For removable and lay-in wire terminals intended for only one wire, the bending space in the table may be reduced by the number of inches shown in parentheses.

AWG or Circular-Mil Size of Wire	Wires per Terminal			
	1	2	3	4 or More
14 – 10	Not Specified	—	—	—
8	1½	—	—	—
6	2	—	—	—
4	3	—	—	—
3	3	—	—	—
2	31/2	—	—	—

Figure 13-19: Minimum wire-bending space at terminals to comply with NEC Section 373-6(b).

AWG or Circular-Mil Size of Wire	Wires per Terminal			
	1	2	3	4 or More
1	4½	—	—	—
1/0	5½	5½	7	—
2/0	6	6	7½	—
3/0	6½ (½)	6½ (½)	8	—
4/0	7 (1)	7½ (1½)	8½ (½)	—
250	8½ (2)	8½ (2)	9 (1)	10
300	10 (3)	10 (2)	11 (1)	12
350	12 (3)	12 (3)	13 (3)	14 (2)
400	13 (3)	13 (3)	14 (3)	15 (3)
500	14 (3)	14 (3)	15 (3)	16 (3)
600	15 (3)	16 (3)	18 (3)	19 (3)
700	16 (3)	18 (3)	20 (3)	22 (3)
750	17 (3)	19 (3)	22 (3)	24 (3)
800	18	20	22	24
900	19	22	24	24
1000	20	—	—	—
1250	22	—	—	—
1500	24	—	—	—
1750	24	—	—	—
2000	24	—	—	—

Figure 13-19: Minimum wire-bending space at terminals to comply with NEC Section 373-6(b). ***(Cont.)***

NEC TERMINATION REQUIREMENTS

There are several *NEC* requirements governing the termination of conductors as well as the installation of enclosures containing conductors. The main *NEC* coverage will be found in *NEC* Article 373. However, other Articles and Sections also apply.

Incoming Line Connections

In general, all ungrounded conductors in a motor control center (MCC) installation require some form of overcurrent protection to comply with *NEC* Section 240-20. Such overcurrent protection for the incoming lines to the MCC is usually in the form of fuses or a circuit breaker located at the transformer secondary that supplies the MCC. The conductors from the transformer secondary constitute the feeder to the MCC, and the "10-foot rule" and the "25-foot rule" of *NEC* Section 240-21 apply. The latter exceptions related to these rules allow the disconnect means and overcurrent protection to be located in the MCC, provided the feeder taps from the transformer are sufficiently short and other requirements are met.

Main Disconnect: A circuit breaker or a circuit interrupter combined with fuses controlling the power to the entire MCC may provide the overcurrent protection required as described above or may be a supplementary disconnect (isolation) means. Figure 13-20 shows a main disconnect with stab load connectors.

When the MCC has a main disconnect, the incoming lines (feeders) are brought to the line terminals of the circuit breaker or circuit interrupter. The load side of the circuit breaker or the load side of the fuses associated with the circuit interrupter is usually connected to the MCC bus bar distribution system. In cases where the main disconnect is rated 400 amperes or less, the load connection may be made with stab connections to vertical bus bars which connect to the horizontal bus distribution system.

Short Circuit Bracing

All incoming lines to either incoming line lugs or to main disconnects must be braced to withstand the mechanical force created by a high fault current. If the cables are not anchored sufficiently or the lugs tightened correctly, the connections become the weakest part of a panelboard or motor control center when a ground fault develops. In most cases, each incoming line compartment is equipped with a two-piece spreader bar located a distance from the conduit entry. This spreader bar should be used along with appropriate lacing material to tie cables together where they can

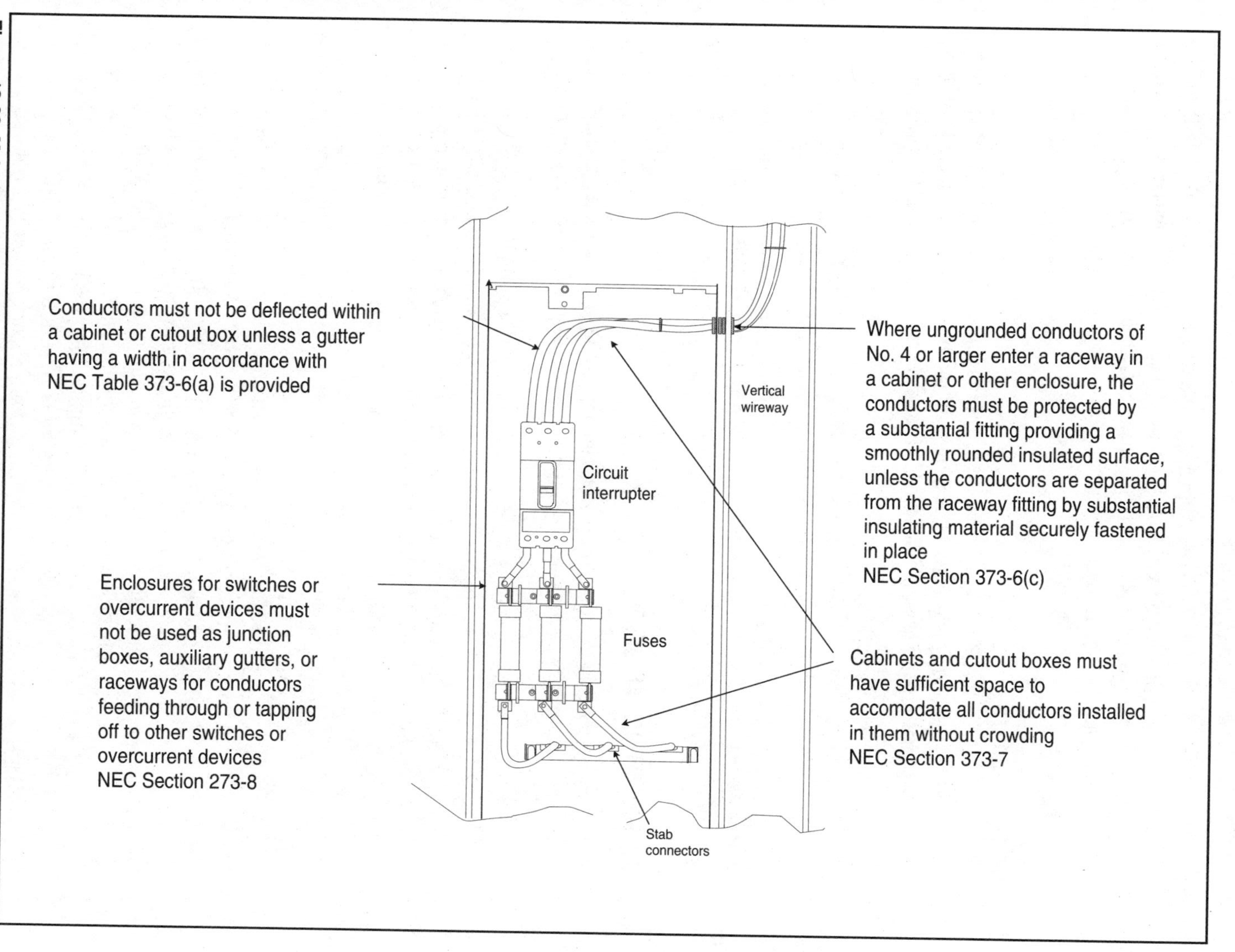

Figure 13-20: Main disconnect with stab load connectors.

be bundled and to hold them apart where they are separated. In other words, the incoming line cables should be first positioned, and then anchored in place.

Manufacturers of electrical panelboards and motor control centers normally furnish detailed information on recommended methods of short circuit bracing; follow these to a "T."

Making Connections

Before beginning work on incoming line connections, refer to all drawings and specifications dealing with the project at hand. Details of terminations are usually furnished to the workers on all of the larger installations.

CAUTION!

All incoming line compartments present an obvious hazard when the door is opened or covers are removed with power on. When working in this area, the incoming feeder should be de-energized.

TAPING ELECTRICAL JOINTS

All wire joints not protected by some other means should be taped carefully to provide the same quality of insulation over the splice as over the rest of the wires.

Spliced joints in rubber-insulated wires should be covered with rubber and friction tape while joints in thermoplastic-insulated wires should be covered with pressure-sensitive thermoplastic-adhesive tape such as Scotch No. 33 for indoor use or Scotch No. 88 for outside use.

For covering rubber-insulated wires, apply the rubber tape to the splice first to provide air-and moisture-tight insulation. The amount applied should be equal to the insulation that was removed. Then wrap the friction tape over the rubber tape to provide mechanical protection.

For joints in thermoplastic-insulated wires — 600 V or below — just the one tape is enough; that is, Scotch No. 33 or No. 88 or equivalent.

To start the taping of a splice or joint, start the end of the tape at one end of the splice (*see* Figure 13-21), slightly overlapping the insulation on the wires. Stretch it slightly while winding it on in a spiral. When the joint is completely covered with layers equal to the original insulation, press or pinch the end of the tape down tightly onto the last turn to make it stick. A properly wrapped electrical joint is shown in Figure 13-22.

Other splicing and taping techniques are shown in Figures 13-23 and 13-24.

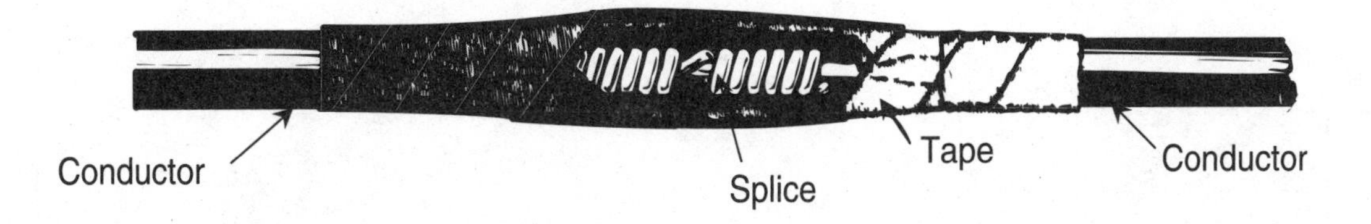

Figure 13-21: Method of taping a splice or joint.

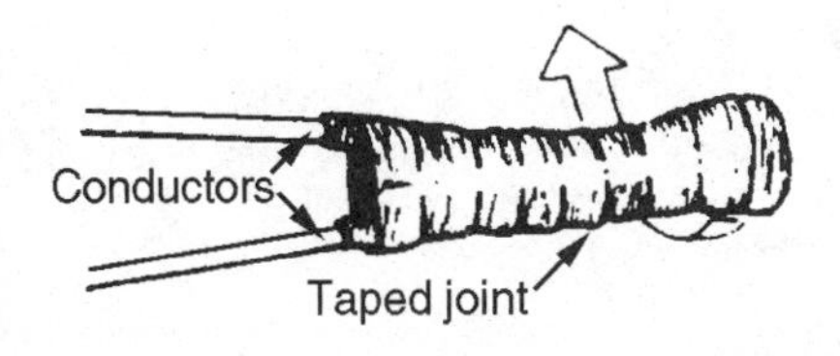

Figure 13-22: A properly wrapped electrical joint.

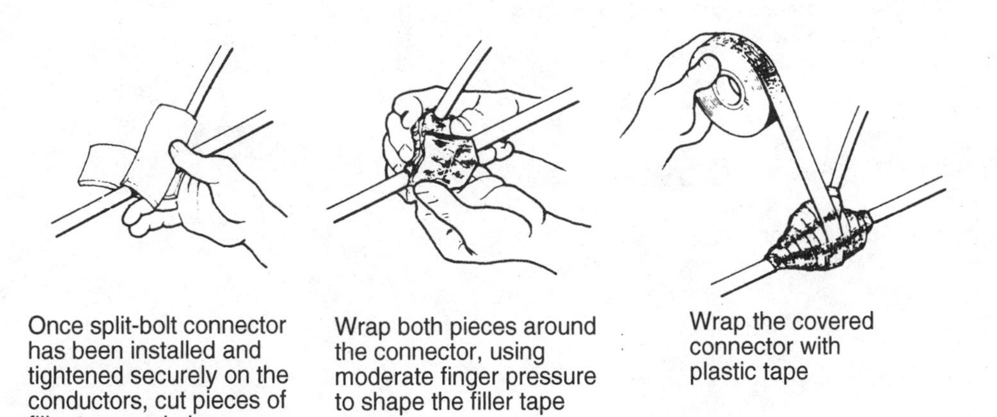

Figure 13-23: Method of insulating a split-bolt connector with filler tape.

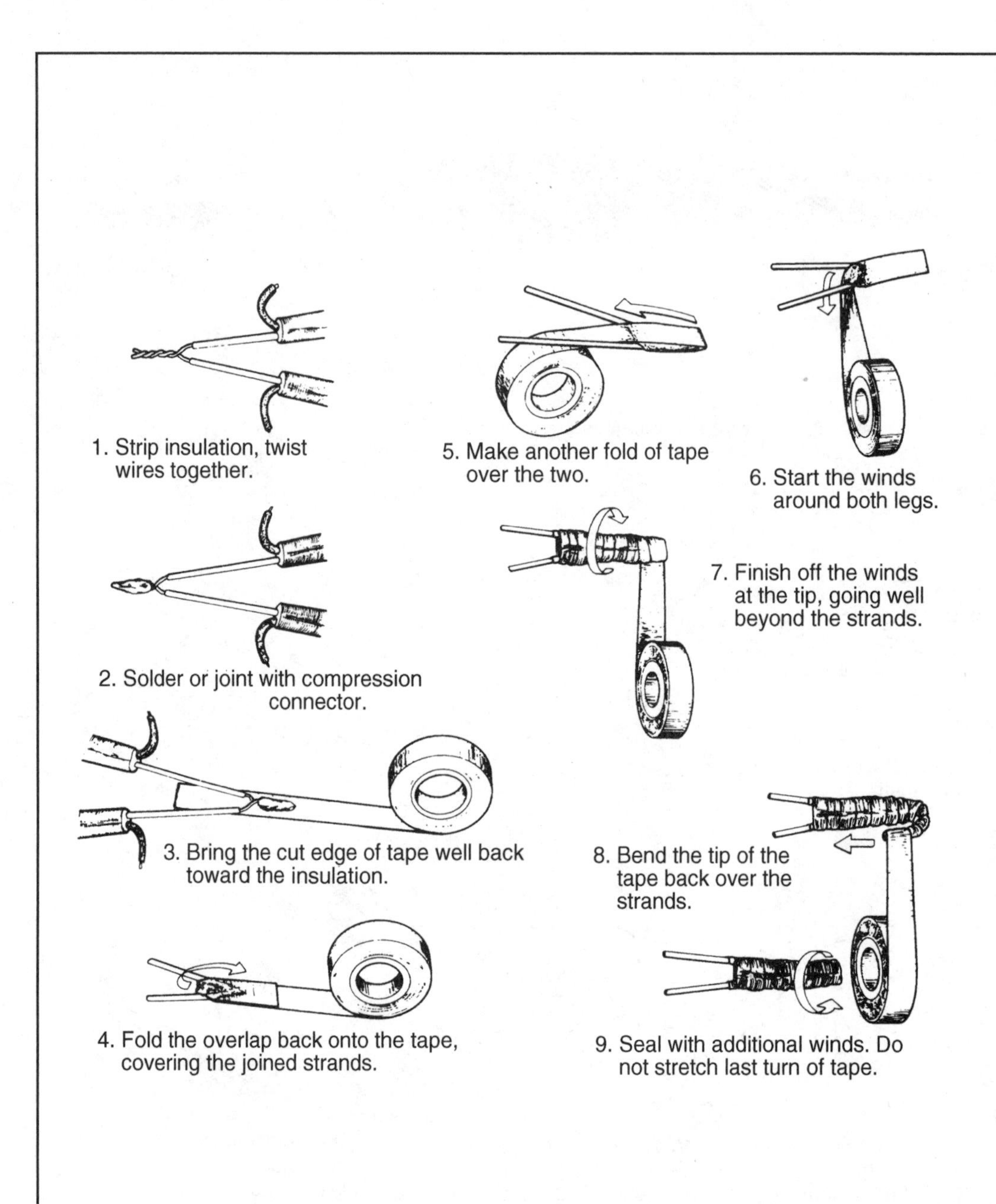

Figure 13-24: Typical low-voltage splicing methods.

Summary of Low-Voltage Splices and Joints

Solderless connectors are devices used to join wires firmly without the need of solder. Such connectors are convenient and save much time on the job. Consequently, many different types have been developed for any conceivable application.

Wire nuts are used to splice the smaller sizes of conductors (No. 10 AWG and smaller) on residential and some commercial installations. The wires should be twisted together before the wire nut is installed.

Split-bolt connectors provide a means of securing the larger sizes of conductors. The split bolt is slipped over the wires so the nut can be attached. Once attached, the nut is tightened to secure the wires.

Crimp connectors or terminals are used for wire sizes from, say, 250 kcmil or larger to as small as No. 22 AWG and smaller. They are very convenient for terminating conductors at terminal boards, control-wiring terminals, and the like.

Screw terminals are provided on many types of electrical devices. The wires may be attached directly to the screw or else first connected to eye wire terminals for a somewhat better, and more secure connection.

Lugs are provided for the larger wire sizes on panelboards and motor control centers. It is very important to tighten these lugs properly to provide a sound electrical connection as well as to provide short circuit bracing.

HIGH-VOLTAGE CONDUCTORS

High-voltage splicing is an art all to itself and only those specially trained should attempt such work. Most areas require a licensed high-voltage splicer to make terminations and splices on conductors carrying the higher voltages, or at least work under the direct supervision of a licensee. Years of experience and rigid tests are required to obtain such a license. However, an explanation of the techniques involved is in order at this time to round out your electrical training.

High voltage is normally considered to be any voltage over 600 V nominal. This can range from 601 V to 170,000 V or more. Although any conductor splice or termination must be done with care, extreme caution must be exercised with the higher voltages. In general, the higher the voltage, the more fine-tuning of the splice or termination is required. In general, every splice must provide mechanical and electrical integrity at least equal to the original conductor insulation. Furthermore, all high-voltage terminations and splices must be made under the best environmental conditions possible on the jobsite. The slightest amount of moisture, dirt, and similar foreign material can render a high-voltage termination or splice useless.

Straight Splices

A typical straight splice for a single, unshielded high-voltage conductor, ranging in voltages from 600 to 5000 V, is shown in Figure 13-25. In general, the conductor jacket and insulation should be removed one half of the connector length plus ¼ in. Then a protective wrap of friction tape is applied to the jacket ends to protect the jacket and insulation while cleaning the conductor and securing the connector (Figure 13-25B).

After performing the preceding operations, remove the friction tape and also remove additional jacket and insulation in an amount equal to 25 times the thickness of the overall insulation as shown in Figure 13-25C, extending in either direction from the first insulation cut. For example, if the thickness of the insulation is ¼ in, the outside edge of the splice will be:

$$.25 \times 25 = 6.25 \text{ or } 6\frac{1}{4} \text{ in}$$

Therefore, the outside edge of the splice will be 6¼ in beyond the insulation cut.

To continue, pencil the insulation with a sharp knife or insulation cutter, taking care not to nick the conductor.

Refer to Figure 13-25D. Starting from A or B, wrap the splice area with Plymozone or equivalent — an ozone-resistant splicing compound — stretching the rubber tape about 25 percent and unwinding the holland cloth interliner as your wrap. The finished splice should be built up 1½ times the factory insulation, if rubber (two times, if thermo-plastic), measured from the top of the connector.

For example, if the insulation is ¼ in, the finished splice would be ⅜ in thick when measured from the top of the connector, and should extend about 1½ in beyond the jacket edge.

To finish off the splice and to ensure complete protection, cover the entire splice with four layers of friction tape, or 6 layers of plastic tape, extending about 1½ in beyond the edge of the splicing compound.

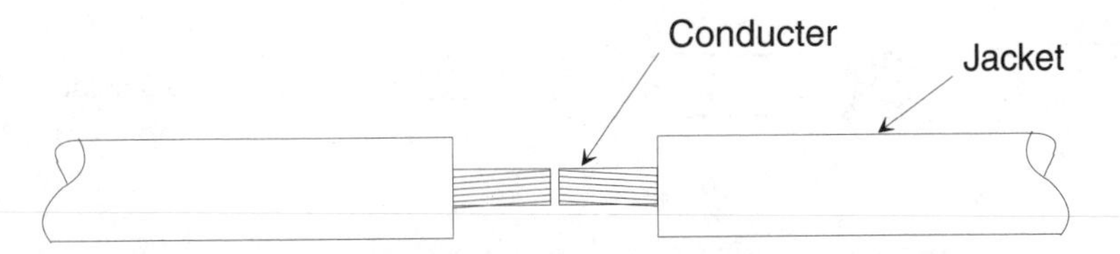

A. Typical straight splice for a single (unshielded) conductor.

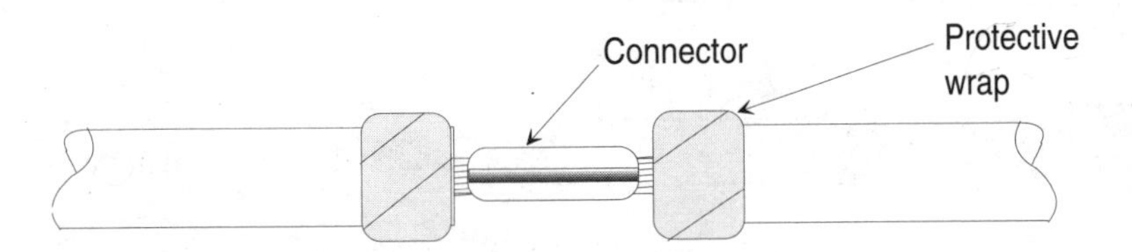

B. Apply a protective wrap of friction tape to the jacket ends to protect the jacket and insulation while cleaning and securing the connector.

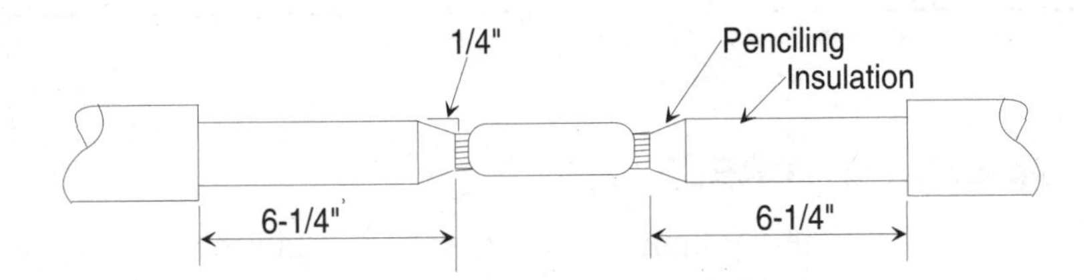

C. Remove additional jacket and insulation an amount equal to 25 times the thickness of the overall insulation.

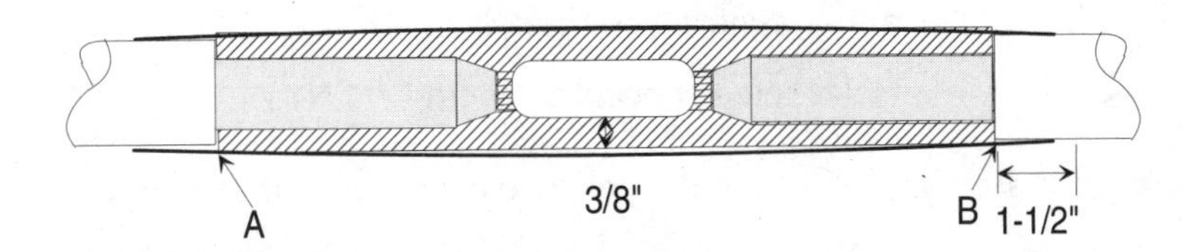

D. Method of insulating and taping the splice.

Figure 13-25: Steps in making a basic high-voltage splice.

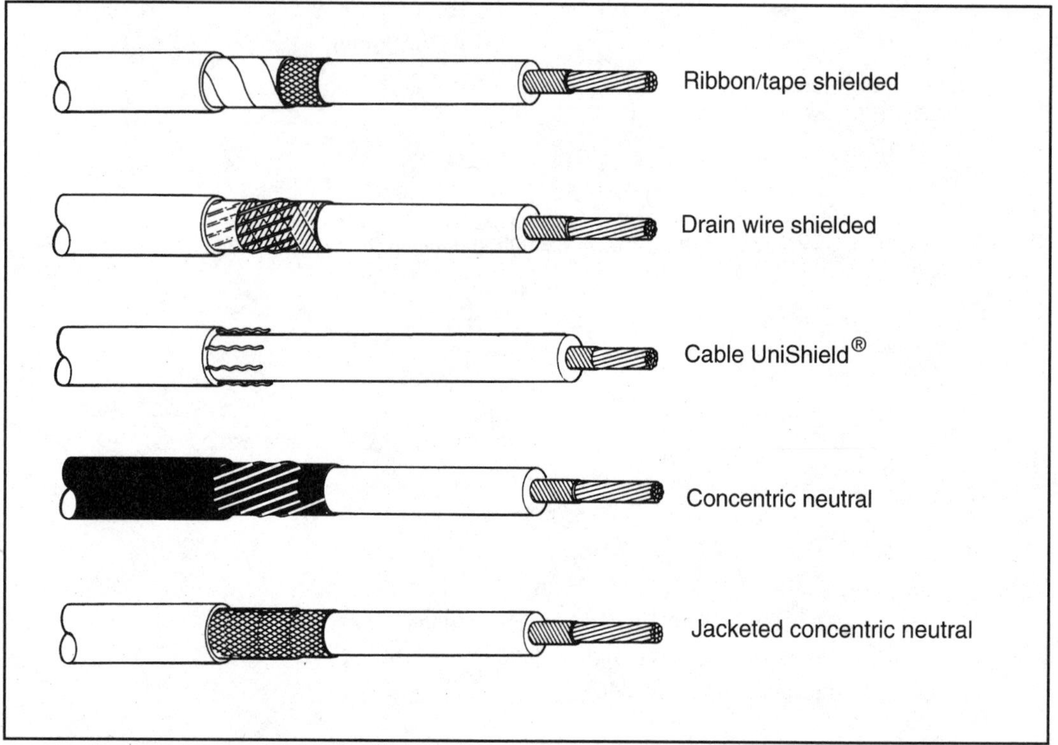

Figure 13-26: Basic types of high-voltage cable.

HIGH-VOLTAGE POWER CABLE

Of the nearly limitless variety of cables in use today, five of the most common are shown in Figure 13-26 and consist of the following:

- Ribbon or tape shielded
- Drain wire shielded
- Cable UniShield®
- Concentric neutral (CN)
- Jacketed concentric neutral (JCN)

Beyond cable shield types, two common configurations are used:

- Single conductor — consisting of one conductor per cable or three cables for a three-phase system.
- Three conductor — consisting of three cables sharing a common jacket.

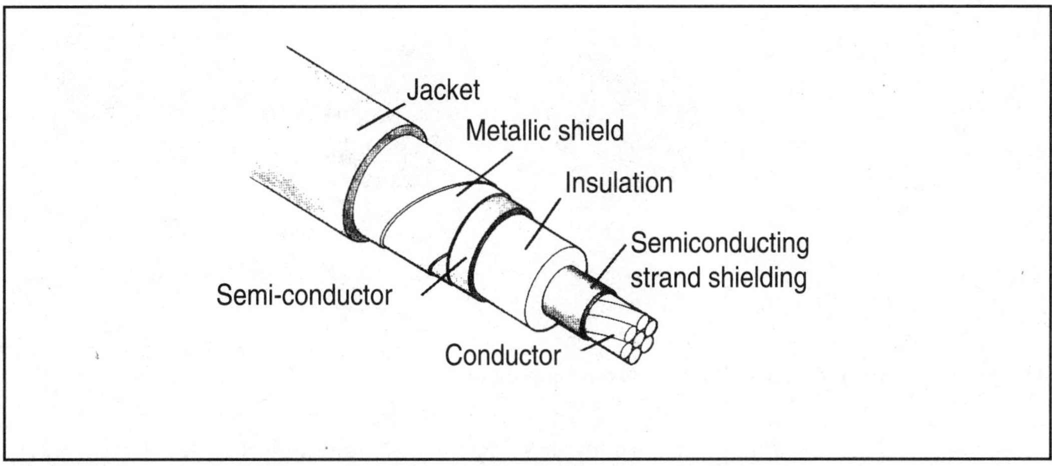

Figure 13-27: Basic components of all high-voltage cable.

Despite these visible differences, all power cables are essentially the same, consisting of the following:

- Conductor
- Strand shield
- Insulation
- Insulation shield system (semicon and metallic)
- Jacket

Each component is vital to an optimally performing power cable and must be understood in order to make a dependable splice or termination. *See* Figure 13-27.

High-Voltage Cable Components

Conductors used with modern solid dielectric cables are available in four basic configurations; all are shown in Figure 13-28 on the next page, and listed as follows:

- Concentric Stranding (Class B)
- Compressed Stranding
- Compact Stranding
- Solid Wire

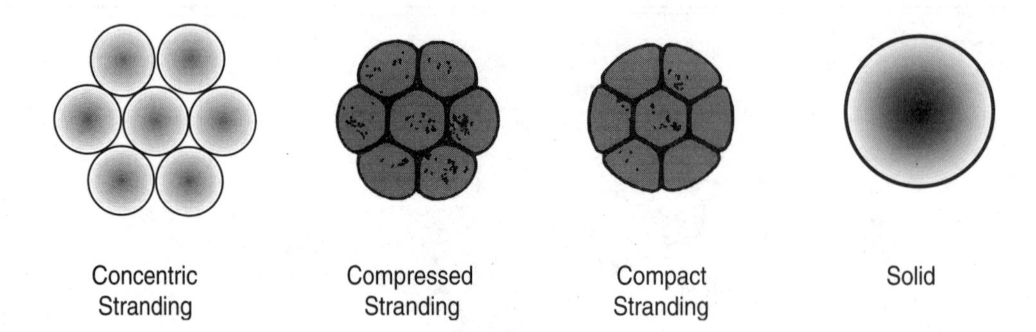

Figure 13-28: Four basic high-voltage cable configurations.

Concentric Stranding (Class B): Not commonly used in modern shielded power cables due to the penetration of the extruded strand shielding between the conductor strands, making the strand shield difficult to remove during field cable preparation.

Compressed Stranding: Compressed to 97 percent of concentric conductor diameters. This compression of the conductor strands blocks the penetration of an extruded strand shield, thereby making it easily removable in the field. For sizing lugs and connectors, sizes remain the same as with the concentric stranding.

Compact Stranding: Compacted to 90 percent of concentric conductor diameters. Although this conductor has full ampacity ratings, the general rule for sizing is to consider it one conductor size smaller than concentric or compressed. This reduced conductor size results in all of the cable's layers proportionally reduced in diameter, a consideration when sizing for molded rubber devices.

Solid Wire: This conductor is not commonly used in industrial shielded power cables.

Strand Shielding

The semiconductive layer between conductor and insulation compensates for air voids that exist between conductor and insulation.

Air is a poor insulator, having a nominal dielectric strength of only 76 volts/mil, while most cable insulations have dielectric strengths over 700 volts/mil. Without strand shielding an electrical potential exists that will over-stress these air voids.

As air breaks down or ionizes, it goes into corona (particle discharges). This forms ozone which chemically deteriorates cable insulations. The

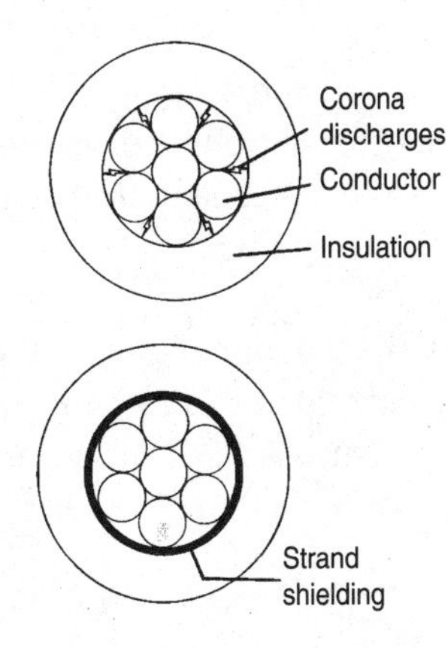

Figure 13-29: Strand shielding.

semiconductive strand shielding eliminates this potential by simply "shorting out" the air.

Modern cables are generally constructed with an extruded strand shield. *See* Figure 13-29.

Insulation

A third layer consisting of many different variations such as extruded solid dielectric, or laminar (oil paper or varnish cambric). Its function is to contain the voltage within the cable system. The most common solid dielectric insulations in industrial use today are:

- Polyethylene
- Cross-linked polyethylene (XLP)
- Ethylene propylene rubber (EPR)

Each is preferred for different properties such as superior strength, flexibility, temperature resistance, etc., depending upon the cable characteristics required.

The selection of the cable insulation level to be used in a particular installation must be made on the basis of the applicable phase-to-phase voltage and the general system category which follows:

1. *100 percent level:* Cables in this category may be applied where the system is provided with relay protection such that ground faults will be cleared as rapidly as possible, but in any case within one minute. While these cables are applicable to the great majority of cable installations which are on grounded systems, they may be used also on other systems for which the application of cable is acceptable provided the above clearing requirements are met in completely deenergizing the faulted section.

2. *133 percent level:* This insulation level corresponds to that formerly designated for ungrounded systems. Cables in this category may be applied in situations where the clearing time requirements of the 100 percent level category cannot be met, and yet there is adequate assurance that the faulted section will be deenergized in a time not exceeding one hour. Also, they may be used when additional insulation strength over the 100 percent level category is desirable.

3. *173 percent level*: Cables in this category should be applied on systems where the time required to deenergize a grounded section is indefinite. Their use is recommended also for resonant grounded systems. Consult the manufacturer for insulation thicknesses.

The percent insulation level is determined by the thickness of the insulation. For example, 15 kV cable at 100 percent has 175 mils of insulation while the same type cable at 133 percent has 220 mils. This variance holds true up to 1000 kcmil.

Insulation Shield System

The outer shielding is comprised of two conductive components: a semiconductive layer (semicon) under a metallic layer. The principal functions of the shield systems are to:

- Confine the dielectric field within the cable.
- Obtain a symmetrical radial distribution of voltage stress within the dielectric.
- Protect the cable from induced potentials.
- Limit radio interference.
- Reduce the hazard of shock.
- Provide a ground path for leakage and fault currents.

Note: *The shield must be grounded for the cable to perform these functions.*

The semi-conductive component is available either as a tape or as an extruded layer (some cables have an additional layer painted on between the semi-con and the cable insulation.) See Figure 13-30. Its function is similar to strand shielding: to eliminate the problem of air voids between the insulation and metallic component (in this case the metallic shielding). In effect, it "shorts out" the air that underlies metallic shield, preventing corona and its resultant ozone damage.

The metallic shield is the current carrying component that allows the insulation shield system to perform the functions mentioned earlier. This is the layer where various cable types differ most. Therefore, most cables are named after their metallic shield (e.g., tape shielded, drain wire shielded, UniShield®, etc.).

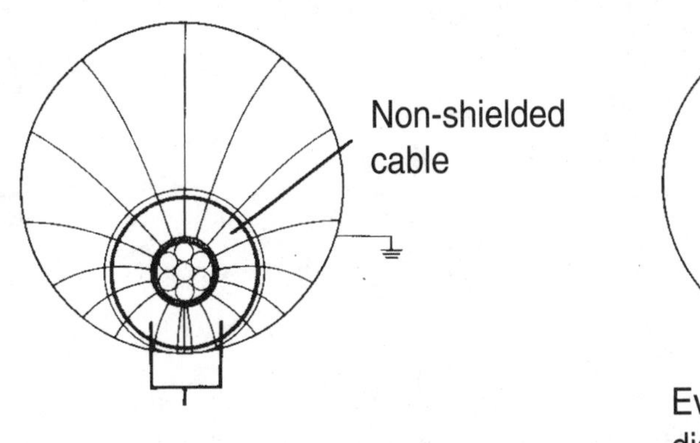

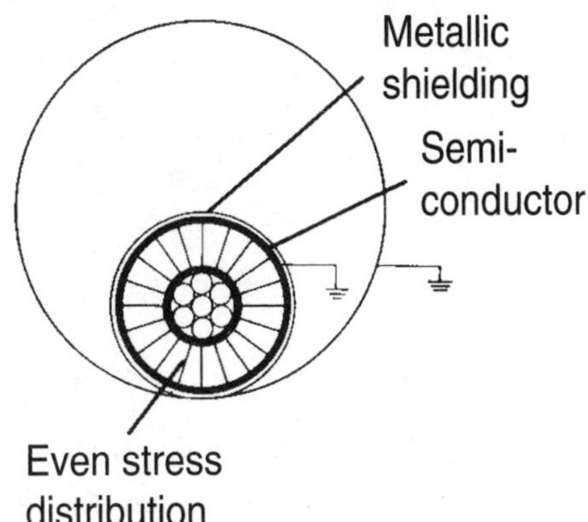

Figure 13-30: Insulation shield system.

Shield type (cable identification) thus becomes important information to know when selecting devices for splicing and terminating.

Jacket

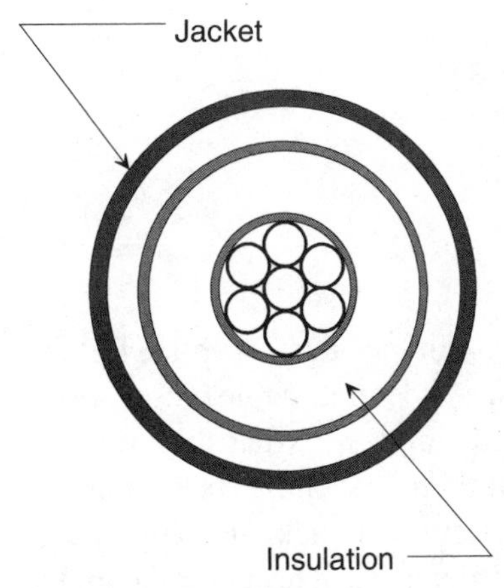

Figure 13-31: Jacket and insulation.

The tough outer covering for mechanical protection, as well as providing a moisture barrier, is called a *jacket*. Often the jacket serves as both an outer covering and the semiconductive component of the insulation shield system, combining two cable layers into one...the semi-conductive jacket. However, because the jacket is external, it cannot serve as the semi-conductive component of the system. Typical materials used for cable jackets are PVC, neoprene, lead, etc. Frequently, industrial three-conductor cables have additional protection in the form of an armor layer. *See* Figure 13-31.

SPLICING

Whenever possible, splicing is normally avoided. However, splicing is often an economic necessity. There can be many reasons for building splices such as:

- The supplied length of cable is not sufficient to perform the intended job when only so much cable can be wound on a reel (reel ends) or when only so much cable can be pulled through so much conduit, around so many bends, etc.

- Cable failures
- Cables damaged after installation
- There is not sufficient room for training radius of the cable
- Excessive twisting of cable would otherwise result
- A tap into an existing cable (tee or wye splices)

In all the above cases, the option is to either splice the cable or replace the entire length. The economy of modern splicing products in many cases makes splicing an optimal choice.

Whatever the reason to splice, good practice dictates that splices have the same rating as the cable. In this way the splice does not derate the cable and become the weak link in the system.

Splicing Steps

The previously quoted definition accurately develops five common steps in building a splice:

- Prepare surface
- Join conductors with connector(s)
- Reinsulate
- Reshield
- Rejacket

It should be recognized that the greatest assurance against splice failure remains with the person who makes the splice. Adequate cable preparation, proper installation of all components and good workmanship require trained skills performed by people adept at them. Much has been done in the last few years to develop products and systems that make splicing easier. Yet the expertise, skills, and care of the installer are still necessary to make a dependable splice.

Prepare the Surface

High quality products usually include detailed installation instructions. These instructions should be specifically followed. A suggested technique is to check off steps as they are completed. Good instructions alone do not qualify a person as a "cable splicer." Certain manufacturers offer "hands-on" training programs designed to teach proper installation of their products. It is highly recommended that inexperienced splice and termination installers take advantage of such programs.

It is necessary to begin with good cable ends. For this reason, it is common practice to cut off a portion of the cable after pulling to assure an undamaged end. A key to good cable preparation is the use of sharp, high quality tools. When the various layers are removed, cuts should extend only partially through the layer. For example, when removing cable insulation, the installer must be careful not to cut completely through and damage the conductor strands. Specialized tools are available to aid in the removal of the various cable layers. Another good technique for removing polyethylene cable insulations is to use a string as the cutting tool.

When penciling is required (not normally necessary for molded rubber devices), a full, smooth taper is necessary to eliminate the possibility of air voids.

Note: *Some extruded semiconductors will "peel" easier if they are heated slightly. A heatshrink blow gun works well.*

It is necessary to completely remove the semi-conductive layer(s) and the resulting residue. Two methods are commonly used to remove the residue: abrasives and solvent.

Abrasives: Research has proven a 120-grit to be optimum . . . fine enough for the high voltage interface and yet coarse enough to remove semi-con residue without "loading up" the abrasive cloth.

This abrasive must have a nonconducting grit. DO NOT use emery cloth or any other abrasive that contains conductive particles since these could imbed themselves into the cable insulation. When using an insulation diameter dependent device (e.g., molded rubber devices), care must be taken not to abrade the insulation below the minimums specified for the device.

Solvents: It is recommended that a nonflammable cable cleaning solvent be used. Any solvent that leaves a residue should be avoided. DO NOT use excessive amounts of solvent as this can saturate the semi-con layers and render them nonconductive. Know your solvent. Avoid toxic solvents which are hazardous to your health.

CAUTION!

Avoid toxic solvents which are hazardous to your health. Always wipe cable from the conductor back toward the jacket.

Cable preparation materials are available which contain solvent saturated rags appropriately filled with the proper quantity of nonflammable, nontoxic cable cleaning solvent. These kits also contain a 120-grit nonconductive abrasive cloth.

After the cable surfaces have been cleaned, the recommended practice is to reverse wrap (adhesive side out) a layer of vinyl tape to maintain the cleanliness of the cable.

Join Conductors with Connectors

After the cables are completely prepared, the rebuilding process begins. If a premolded splice (or other slip-on type) is being installed, the appropriate splice components must be slid onto the cable(s) before the connection is made. The first step is reconstructing the conductor with a suitable connector. A suitable connector for high-voltage cable splices is a compression or crimp type.

Do not use mechanical type connectors (e.g., split-bolts). Connector selection is based on conductor material — copper or aluminum.

For an aluminum conductor, connect with aluminum-bodied connector (marked Cu/Al). These must come preloaded with contact aid (anti-oxide paste) to break down the insulating aluminum oxide coating on both the connector and conductor surfaces.

When the conductor is copper, connect with either copper or aluminum bodied connectors.

It is recommended that a U.L. listed connector be used that can be applied with any common crimping tool. This connector should be tested and approved for use at high voltage. In this way, the choice of the high-voltage connector is at the discretion of the user, and is not limited by the tools available.

Reinsulate

Perhaps the most commonly recognized method for reinsulating is the traditional tape method. The most versatile approach, tape, is not dependent upon cable types and dimensions. Tape has a history of dependable service and is generally available. However, wrapping tape on a high-voltage cable can be time consuming and error prone since the careful buildup of tape requires accurate half-lapping and constant tension in order to reduce built-in air voids.

Technology has made available linerless splicing tapes. These tapes reduce both application time and error. Studies have shown time savings of 30 to 50 percent are possible since there is no need to stop during taping to tear off liner. This also allows the splicer to maintain a constant tape tension, thus reducing the possibility of taped-in voids. Tape splice kits which contain all the necessary tapes along with proper instructions are available. These versatile kits assure that the proper materials are available at the job, and make an ideal emergency splice kit.

Another method for reinsulating utilizes molded rubber technology. These factory-made splices are designed to be human engineered for the convenience of the installer. In many cases these splices are also factory tested and designed to be installed without the use of special installation tools.

Most molded rubber splices use EPR as the reinsulation material. This EPR must be cured during the molding process. Either a peroxide cure or a sulphur cure can be used. Peroxide cures develop a rubber with maximum flexibility (for easy installation on a wide range of cable) and most importantly, provides an excellent long-term live memory for lasting, more reliable splices.

Reshield

The cable's two shielding systems (strand shield and insulation shield system) must be rebuilt when constructing a splice. The same two methods are used as outlined in the reinsulation process: tape and molded rubber.

For a tape splice, the cable strand shielding is replaced by a semiconductive tape. This tape is wrapped over the connector area to smoothe the crimp indents and connector edges.

The insulation shielding system is replaced by a combination of tapes. Semicon is replaced with the same semiconducting tape used to replace the strand shield.

The cable's metallic shield is generally replaced with a flexible woven mesh of tin-plated copper braid. This braid is for electrostatic shielding only, and not designed to carry shield currents. For conducting shield currents, a jumper braid is installed to connect the cables' metallic shields. This jumper must have an ampacity rating equal to that of the cables' shields.

For a rubber molded splice, conductive rubber is used to replace the cable's strand shielding and the semiconductive portion of the insulation shield system. Again, the metallic shield portion must be jumpered with a metallic component of equal ampacity.

A desirable design parameter of a molded rubber splice is that it be installable without special installation tools. To accomplish this, very short electrical interfaces are required. These interfaces are attained through proper design shapes of the conductive rubber electrodes.

Laboratory field plotting techniques show that the optimum design can be obtained using a combination of logarithmic and radial shapes.

Rejacket

Rejacketing is accomplished in a tape splice by using a combination of the rubber splicing tape overwrapped with a vinyl tape.

In a molded rubber splice, rejacketing is accomplished by proper design of the outer semiconductive rubber, effectively resulting in a semiconductive jacket.

When a molded rubber splice is used on internally shielded cable (such as tape shield, drain wire shield, or UniShield® cables, a shield adapter is used to seal the opening that results between the splice and cable jacket.

As a general summary, for the versatility to handle practically any splicing emergency, or for situations where only a few splices need to be made, or when little detail is known about the cable, the most effective splice is made with tape or a tape kit.

For those times when cable size, insulation diameter, and shielding type are known and when numerous splices will be made, use molded rubber splices for dependability and simplicity as well as quick application.

Inline Tape Splice

The following detailed inline tape splicing procedures were furnished by Electro-Products Division, 3M Company and have been tested with both metallic and wire shield cable utilizing voltage ratings from 5 to 25 kV with a normal maximum temperature rating of 90°C and 130°C emergency operation.

Details of the cable types are shown in Figure 13-32, including the various cable sections. Become thoroughly familiar with these sections before continuing, and refer to Figure 13-32 as often as needed for reference.

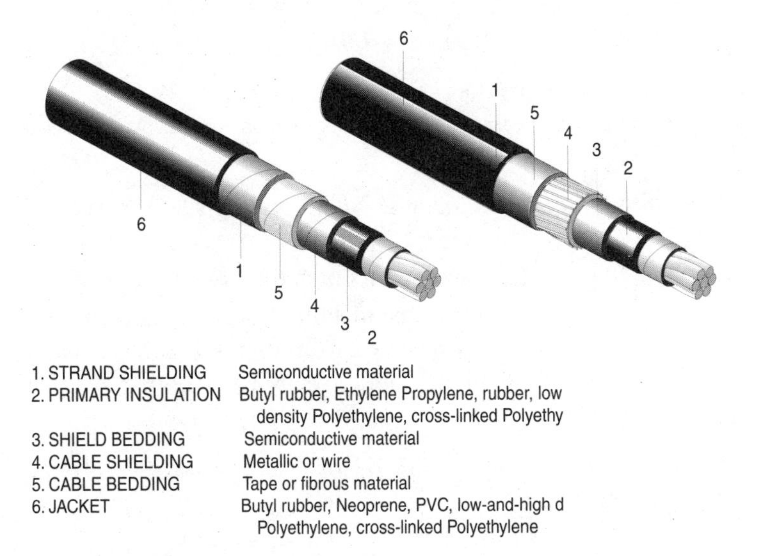

Figure 13-32: High-voltage cable specifications.

Prepare Cable

Step 1. Train cable in position and cut to proper length so cable ends will butt squarely.

Step 2. Thoroughly scrape jacket 3 in beyond dimension "A" in Figure 13-33 to remove all dirt, wax, and cable-pulling compound so tape will bond, making a moisture-tight seal.

Step 3. Remove cable jacket and nonmetallic filler tape from ends to be spliced for distance "A" (Figure 13-33) plus one-half connector length.

CAUTION!

Be careful not to ring cut into metallic shielding or insulation when removing jacket.

Step 4. Remove cable metallic shielding, leaving 1 in exposed beyond cable jacket. If wire shielded cable is used, see instructions in Figure 13-34.

CAUTION!

Do not nick semiconductive material and do not overheat thermosplastic insulation.

Step 5. Tack metallic shielding in place with solder.

Step 6. Remove semiconducting material, leaving ¼ in exposed beyond cable metallic shielding. Be sure to remove all traces of this material from exposed cable insulation by cleaning with nonconductive abrasive cloth.

CAUTION!

Do not use emery cloth, as it contains metal particles.

Step 7. Remove cable insulation from end of conductor for a distance of ½ in plus one-half length of connector. Be careful not to nick conductors.

Step 8. Pencil insulation at each end, using a penciling tool or a sharp knife, for a distance equal to "B" in Figure 13-32. Buff tapers with abrasive cloth so no voids will remain after joint is insulated.

Step 9. Taper cable jacket at ends smoothly for a distance equal to ¼ in. Buff tapers with abrasive cloth.

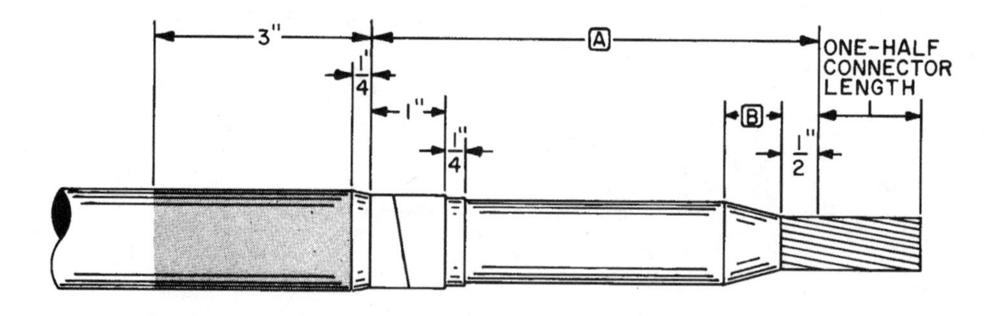

Figure 13-33: Recommended cable preparation procedures for inline splice.

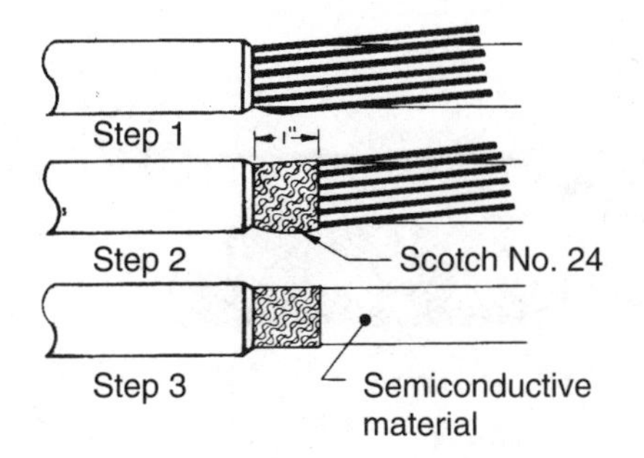

1. Remove cable jacket as per previous instructions. Be careful not to cut any of the wires.
2. Wrap two unstretched layers of Scotch No. 24 electrical shielding tape over shield wires for 1" beyond cable jacket. Tack in place with solder.
3. Cut shielding wires off flush with leading edge of tape.
4. Return to Step 6 in the installation instructions and continue cable preparation.

Figure 13-34: Wire shield procedures.

Step 10. Clean entire area of prepared splice by wiping with solvent-saturated cloth from cable preparation kit. Solvent may be used on thermosplastic insulation. If solvent is used, area must be absolutely dry and free of all solvent residue, especially in the conductor strands and underneath cable shield, before proceeding to the next step.

Connect Conductors

Step 1. Join conductors with connector. Use crimp connector to connect thermosplastic insulated cables. Follow connector manufacturer's directions. Use antioxidant paste for aluminum conductors. When using solder sweated connector, protect cable insulation with temporary wraps of cotton tape. Be sure to remove all antioxidant paste from connector area after crimping aluminum connector.

CAUTION!

Do not use acid-core solder or acid flux.

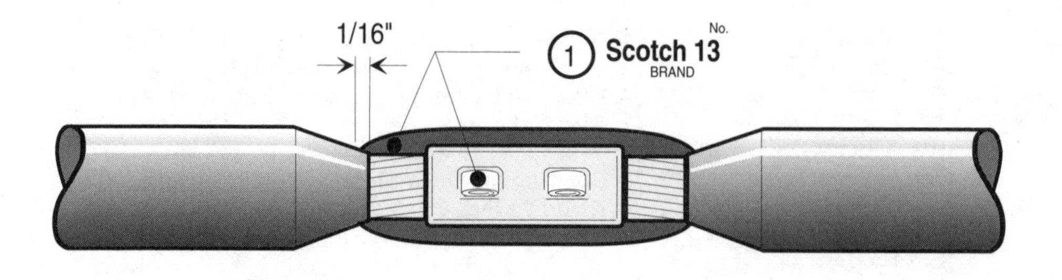

Figure 13-35: Connecting conductors.

Step 2. Fill connector indents with Scotch No. 13 Semiconducting tape as shown in "1" in Figure 13-35. Smoothly tape two half-lapped layers of No. 13 highly elogated over exposed conductor and connector area. Cover all threads of semiconducting strand shielding and overlap cable insulation with $\frac{1}{16}$ in semiconducting tape.

Note: *Stretching the Scotch No. 13 tape increases its conductance and will not harm it in any way.*

Apply Primary Insulation

Step 1. Apply Scotch No. 23 high voltage splicing tape (*see* "2" in Figure 13-36) half-lapped over hand-applied semiconducting tape and exposed insulation up to $\frac{1}{4}$ in from cable semi-conductive material. Continue to apply tape half-lapped in successive level wound layers until thickness over hand-applied semi-conducting tape is equal to dimension "D" in Figure 13-36. Outside surface should taper gradually along distance "C," reaching maximum diameter over penciled insulation.

Note *To eliminate voids in critical areas, highly elongate the splicing tape in connector area and at splice ends near cable semiconductive material. Stretch tape just short of its breaking point. Doing so will not alter its physical or electrical*

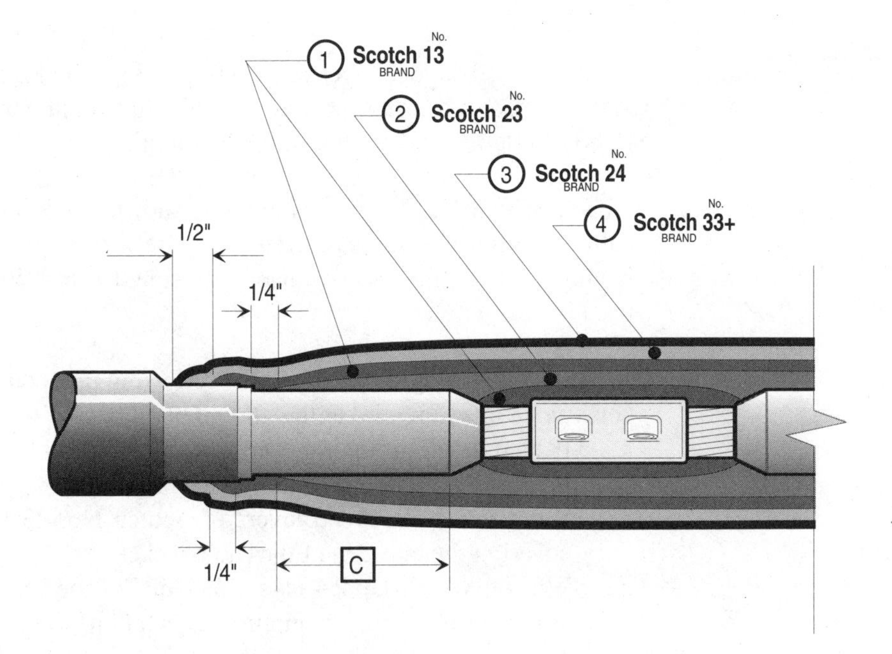

Figure 13-36: Applying primary insulation.

properties. Throughout the rest of the splice area, less elongation may be used. Normally, No. 23 tape is stretched to ¾ of its original width in these less critical areas. Always attempt to exactly half-lap to produce a uniform buildup.

Step 2. Wrap one half-lapped layer of Scotch No. 13 tape over splicing tape, extending over cable semiconductive material and onto cable metallic shielding ¼ in at each end. Highly elongate this tape when taping over the edge of cable metallic shield.

Step 3. Wrap one half-lapped layer of Scotch No. 24 electrical shielding tape over semiconducting tape, overlapping ¼ in onto cable metallic shielding at each end. Solder ends to metallic shielding.

Step 4. Wrap one half-lapped layer of Scotch No. 33+ vinyl plastic electrical tape, covering entire area of shielding braid. Stretch tightly to flatten and confine shielding braid.

Step 5. Attach Scotch No. 25 ground braid as shown, soldering to cable metallic shielding at each end. Use a wire or strap having at least the same ampacity as shield (#6 AWG is usually adequate).

Step 6. If a splice is to be grounded, use the following procedure to construct a moisture seal at the ground braid.

a. If a stranded ground is used, provide solder block to prevent moisture penetration into splice.
b. Wrap two half-lapped layers of Scotch No. 23 tape, covering the end 2 in of the cable jacket.
c. Wrap two half-lapped layers of No. 23 tape for 3 in along ground strap beginning at a point where the ground is soldered to the shield.
d. Lay the ground strap over the No. 23 tape which was applied on the jacket for 1 in, then bend the strap away from the cable. Press strap hard against the cable jacket.

Apply Outer Sheath

Step 1. Wrap four half-lapped layers of Scotch No. 23 tape or equivalent over entire splice 2 in. Highly stretch No. 23 tape to form moisture seal and eliminate voids. Make sure the application of No. 23 tape comes on either side of the ground braid completing the moisture seal.

Step 2. Wrap two half-lapped layers of Scotch 33+ over entire splice 1 in beyond the splicing tape and 1 along the grounding braid. A summary of the end preparation and taping is shown in Figure 13-37.

Tee Tape Splice

A Tee tape splice (tap) is performed basically the same way as the inline tape splice except a third conductor is attached as shown in Figure 13-38. Tee splices require a higher degree of skill to construct because the tape

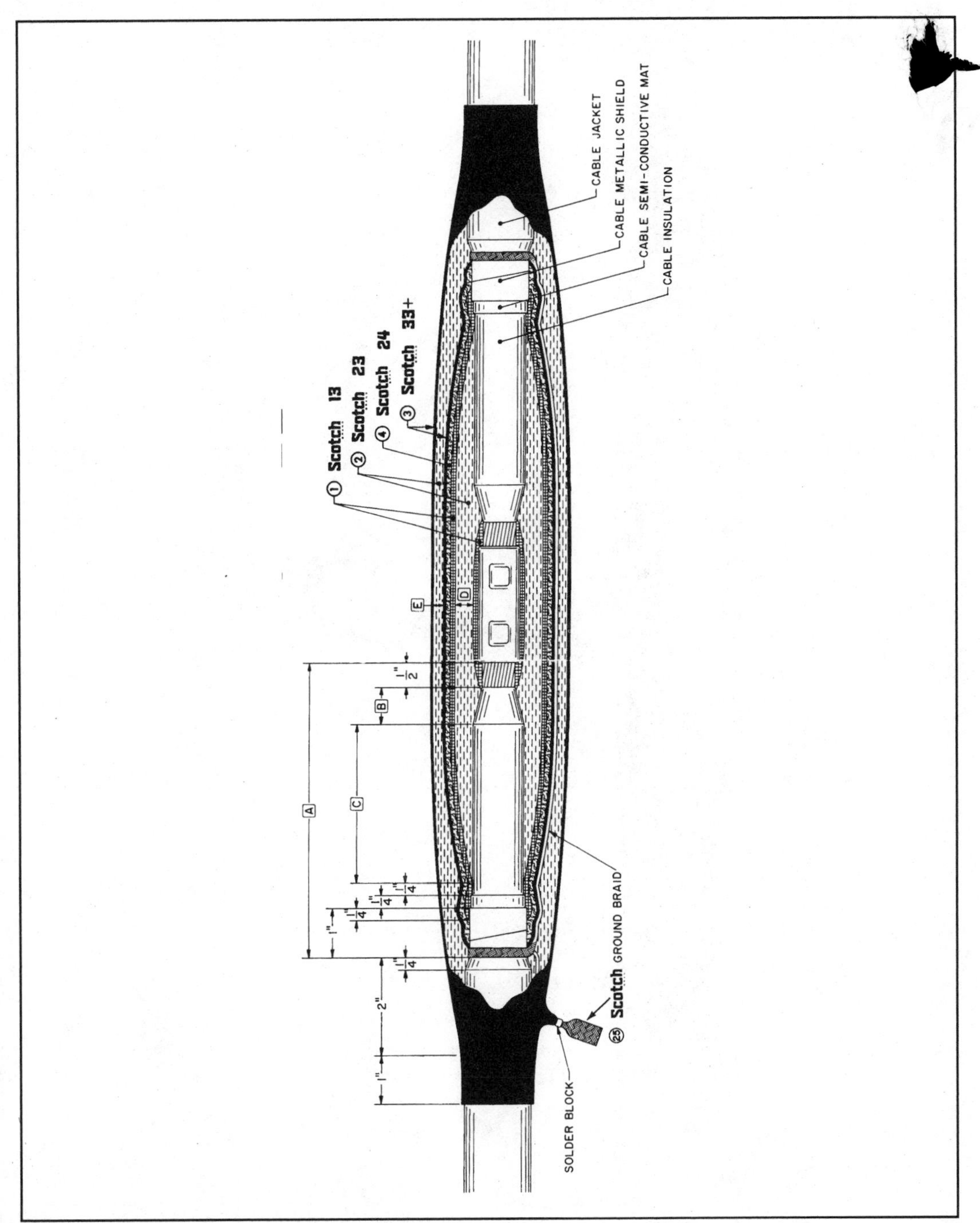

Figure 13-37: Summary of inline tape splice.

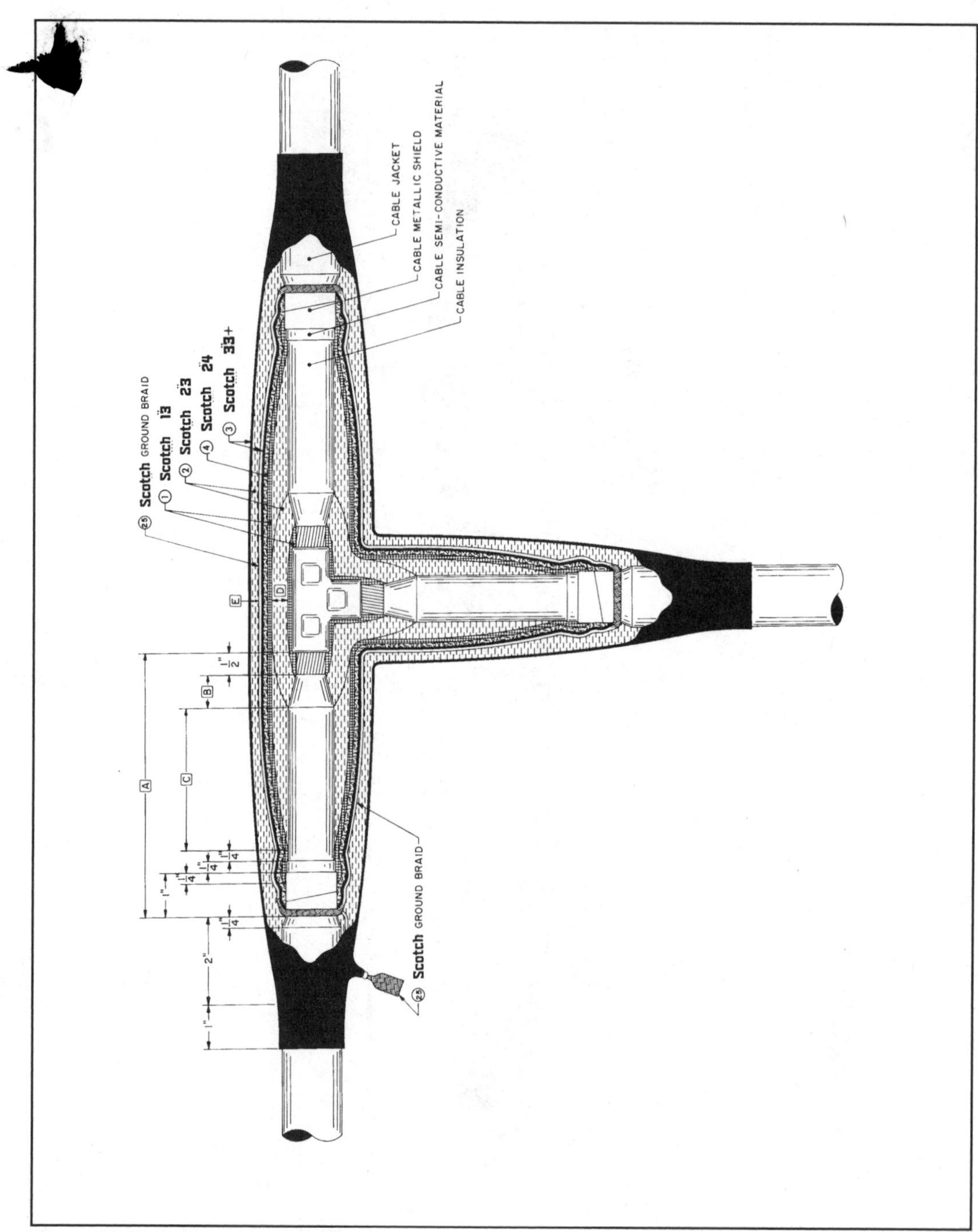

Figure 13-38: Summary of a Tee tape splice.

must be highly elongated when applied, so as to prevent voids in the crotch area.

QUICK INLINE SPLICING KITS

There are several relatively new types of high-voltage cable-splicing kits that have appeared on the market within the past few years. All greatly simplify cable splicing and should be used when the exact type and dimensions of the cable are known.

The following instructions are for a cold-shrink kit used on ribbon shielding cable. Kits are also available for other types; that is, wire shield and Unishield®. Complete instructions are included in each kit and once you understand the directions in this module for ribbon shielding cable, the directions for the other types should be readily understood also.

Prepare Cables According to Standard Procedures

Step 1. Clean cable jackets by wiping with a dry cloth up to approximately two ft from each end.

Step 2. Remove jacket to distance B and C for cables "X" and "Y," respectively. *See* Figure 13-39.

Note *The dimensions of A, B, C and D in Figure 13-39 will vary with the type of kit used. See the table in Figure 13-40 for details.*

Step 3. Remove metallic shielding for distance A in Figure 13-39 on the next page.

Step 4. Remove cable semiconductive material (semicon), allowing it to extend ¼ in beyond metallic shielding (dimension "A" in Figure 13-39).

Step 5. Remove insulation for dimension "D" in Figure 13-39 and taper edges ⅛ in at approximately 45°.

Step 6. Clean exposed insulation using enclosed cleaning pads. Do not use solvent or abrasive on cable semicon layer. If abrasive

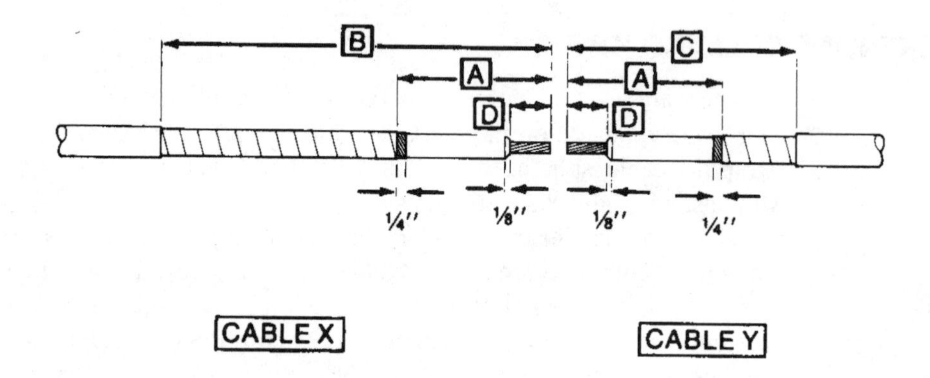

Figure 13-39: Cable preparation details.

must be used on insulation, do not reduce diameter below that specified for minimum splice application.

Step 7. Apply two highly elongted half-lapped layers of Scotch Brand 13 semiconductive tape starting 1 in on cable metallic shield and extending ½ in onto cable insulation. Leave a smooth leading edge and tape back to starting position. *See* Figure 13-41.

Step 8. On cable "C" apply one half-lapped layer (adhesive side out) orange vinyl tape (supplied in kit) over metallic shield begin-

Kit Number	A Insulation Level 100% .175 inch	A Insulation Level 133% .220 inch	B	C	D
5501	4⅜ inches	4⅛ inches	12 inches	7 inches	1¼ inches
5502	4½ inches	4½ inches	12 inches	7 inches	1⅝ inches

Figure 13-40: Dimensions for the letters in Figure 13-39.

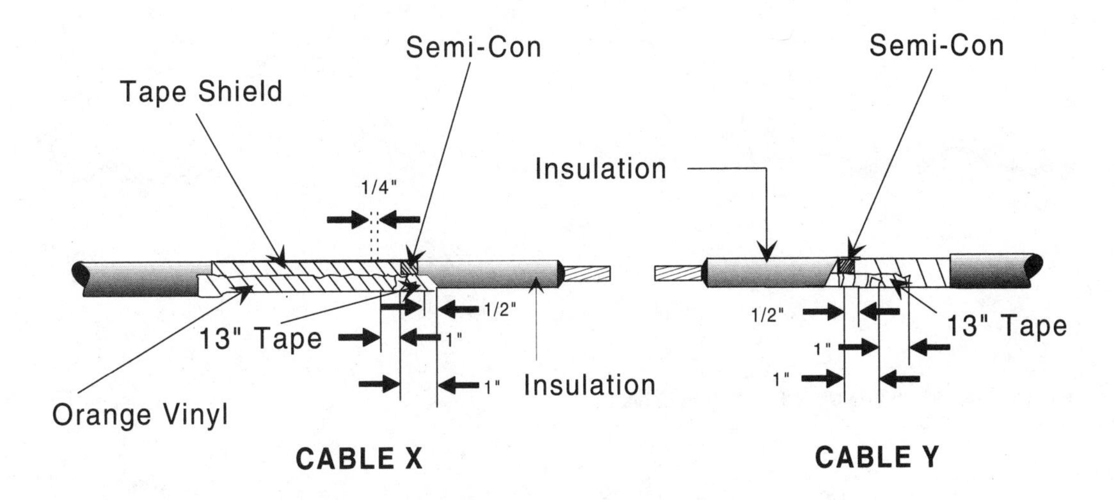

Figure 13-41: Components of a ribbon shielding cable splice.

ning approximately ¼ in on previously applied No. 13 tape and ending over cable jacket (Figure 13-41).

Splice Installation

Step 1. Slide longer PST cold shrink insulator onto cable "X" jacket and the shorter PST onto cable "Y" jacket directing pull tabs away from cable ends as shown in Figure 13-42.

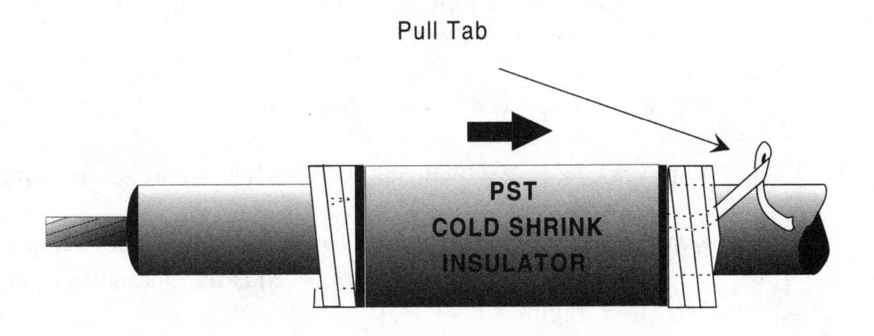

Figure 13-42: Installing PST cold-shrink insulator.

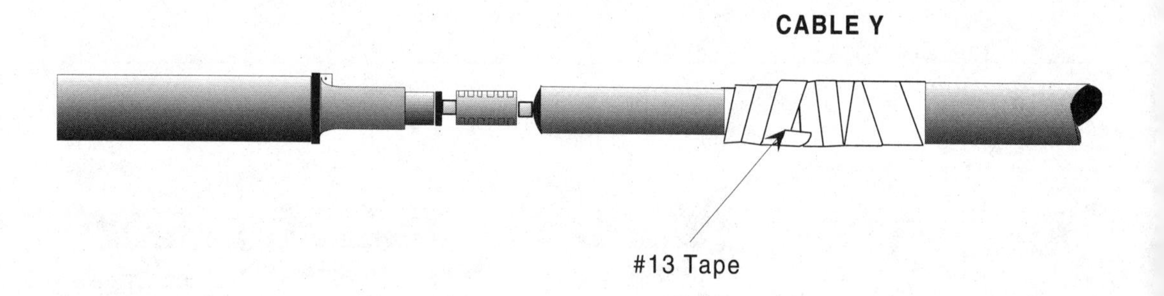

Figure 13-43: Wrapping No. 13 tape onto cable "Y."

Step 2. Apply a few wraps of vinyl tape over cable "X."

Step 3. Lubricate exposed insulation, No. 13 tape and orange vinyl tape of cables "X" and "Y" with silicone grease provided. Do not allow grease to extend onto metallic shield of cable "Y."

Step 4. Lubricate bore of splice with silicone grease and install onto cable "X" leaving conductor exposed for connector as shown in Figure 13-43.

Note: *Rotation of splice while pulling will ease installation. Make certain all splicing components (PSTs and splice body) are located properly on their respective cable end before installing connector.*

Step 5. Remove vinyl tape from cable "X" conductor.

Step 6. Install CI connector crimped per manufacturer's instructions.

Step 7. Slide splice body into final position over connector, using bumps formed on splice ends as guides for centering. *See* Figure 13-44.

Step 8. Remove previously applied orange vinyl tape and wipe off any remaining silicone grease.

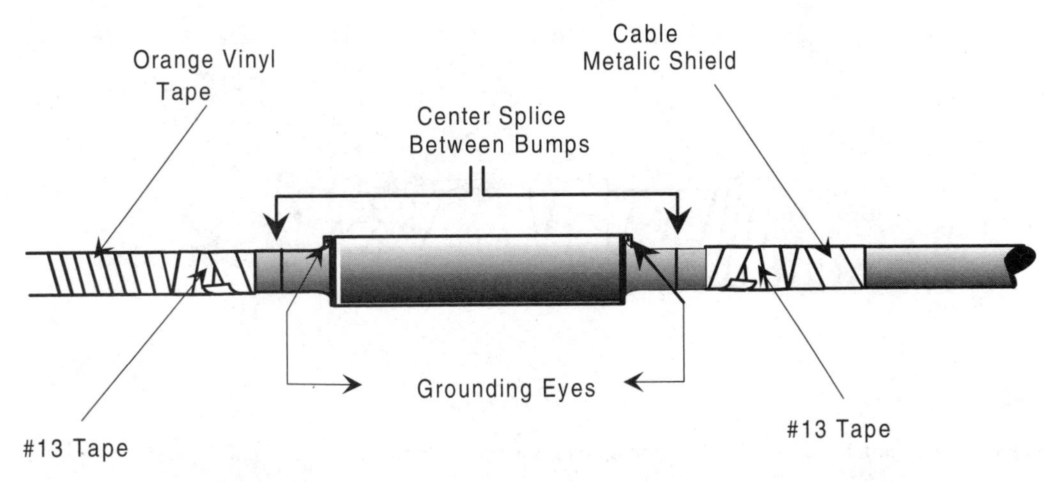

Figure 13-44: Use bumps formed on splice ends as guides for centering.

Install Shield Continuity Assembly

Step 1. Position shield continuity assembly over splice body and hold it in place with a strap or two of vinyl tape at each end of the splice body as shown in Figure 13-45. Form the shield continuity strap over the splice shoulder on each side as shown in Figure 13-46.

Note: *Coils of assembly must be facing cable and positioned so they will only make contact with metallic shielding when applied. Avoid positioning flat strap over splice grounding eyes.*

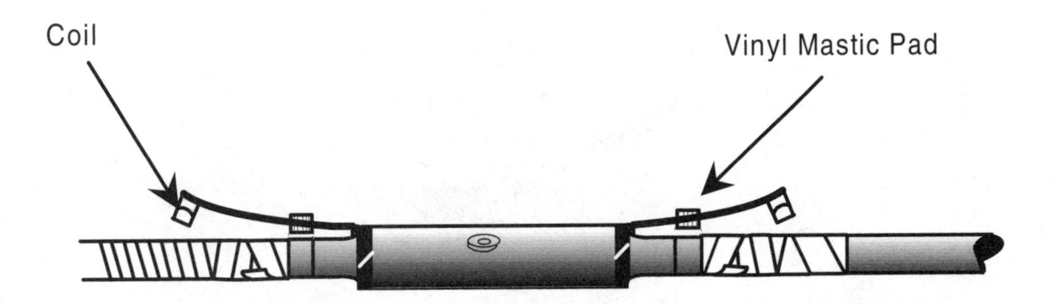

Figure 13-45: Wrap vinyl tape at each end of the splice body.

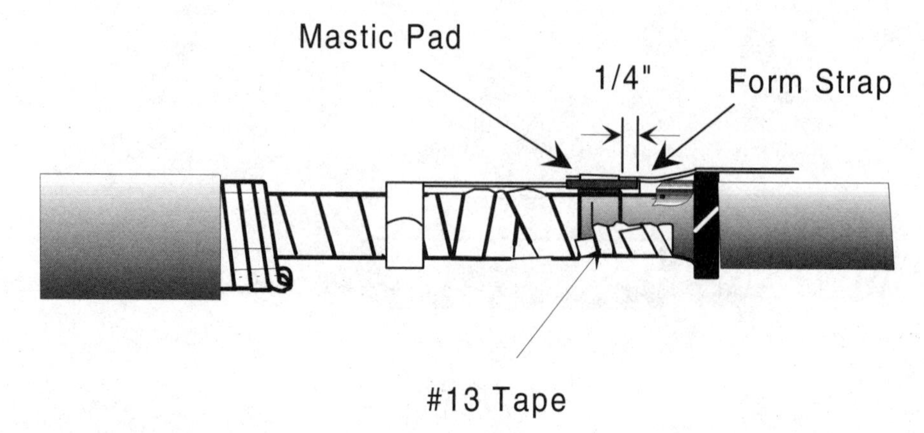

Figure 13-46: Form the shield continuity strap over the splice shoulder on each side.

Step 2. Coil application:

a. Unwrap coil and straighten for one to two in.

b. Hold the coil and shield strap in place with thumb (Figure 13-47). Pull (to unwrap) the coil around the cable and rewrap around cable metallic shield and itself.

Note: *Cinch (tighten) the applied coil after final wrap.*

c. Repeat "a" and "b" for other end of splice.

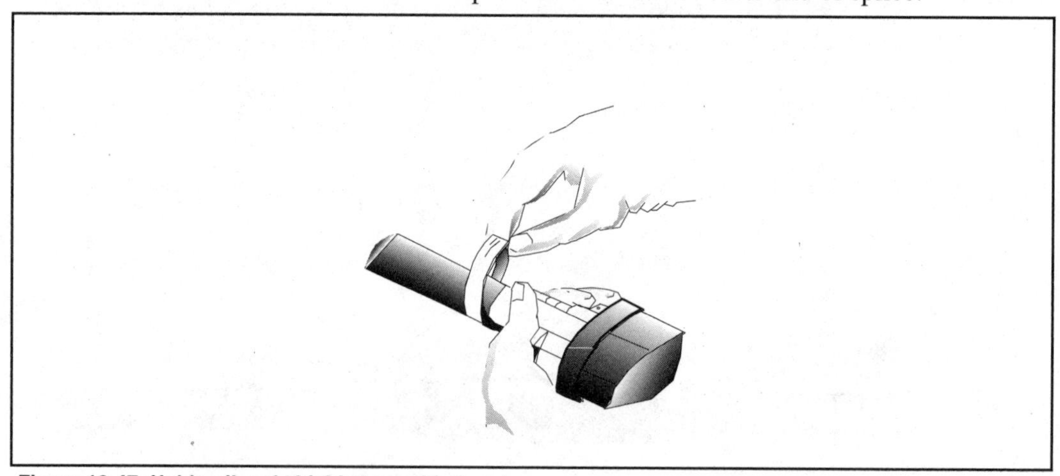

Figure 13-47: Hold coil and shield strap with thumb.

Step 3. Sealing shield strap:

a. Cut the supplied strip of vinyl mastic into four equal length pads.

b. After removing liner, place one pad under strap, mastic side toward strap and located at the end of the splice body (Figure 13-45). Place a second pad over the strap and first pad and press together, mastic to mastic, forming a sandwich. This latter operation will be wrapped with No. 13 tape.

c. Repeat step "b" for other end of splice body.

d. Beginning just beyond splice body, wrap No. 13 tape over vinyl mastic pads extending onto splice for approxiamtely 1/4 in and returning to starting point. Refer again to Figure 13-46.

Install PST Cold-Shrink Insulators

Step 1. Position each PST so its leading edge will butt against grounding eye of splice body. Remove core by unwinding counterclockwise and tugging occasionally as shown in Figure 13-48.

Grounding Splice

Step 1. If the practice calls for grounding the splice, fasten the alligator ground clamp (provided with most kits) to the

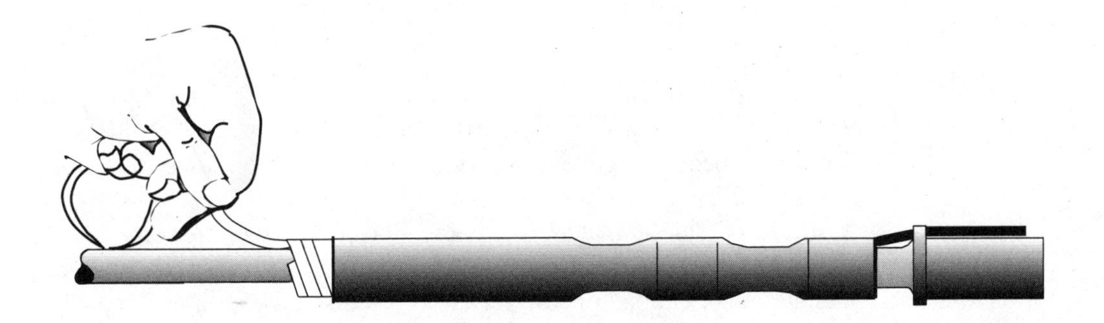

Figure 13-48: Remove core by unwinding counterclockwise.

exposed shield continuity strap approximately at the center of the splice. A ground strap conductor can then be fastened to the alligator clamp. Discard black plastic shoe packaged with ground clamp.

TERMINATIONS

A Class 1 High-Voltage Cable Termination (or more simply, a Class 1 termination) provides the following:

- Some form of electric stress control for the cable insulation shield terminus
- Complete external leakage insulation between the high-voltage conductor(s) and ground
- A seal to prevent the entrance of the external environment into the cable and to maintain the pressure, if any, within the cable system.

Note: This classification encompasses the conventional potheads for which the original IEEE Std 48-1962 was written. With this new classification or designation the term pothead is henceforth dropped from usage in favor of Class 1 termination.

A Class 2 termination is one that provides only some form of electric stress control for the cable insulation shield terminus and complete external leakage insulation, but no seal against external elements. Terminations falling into this classification would be, for example, stress cones with rain shields or special outdoor insulation added to give complete leakage insulation, and the more recently introduced skip-on terminations for cables having extruded insulation when not providing a seal as in Class 1.

A Class 3 termination is one that provides only some form of electric stress control for the cable insulation shield terminus. This class of termination would be for use primarily indoors. Typically, this would include hand-wrapped stress cones (tapes or pennants), and the slip-on stress cones.

Some Class 1 and Class 2 terminations have external leakage insulation made of polymeric material. It is recognized that there is some concern about the ability of such insulations to withstand weathering, ultraviolet radiation, contamination and leakage currents, and that a test capable of evaluating the various materials would be desirable. There are available today a number of test procedures for this purpose. However, none of them has been recognized and adopted by the industry as a standard. Consequently, this standard cannot and does not include such a test.

These IEEE Classes make no provision for "indoor" or "outdoor" environments. This is because contamination and moisture can be highly prevalent inside most industrial facilities (such as paper plants, steel mills, petro-chemical plants, etc.).

As a general recommendation, if there are airborn contaminates, or fail-safe power requirements are critical, use a Class 1 termination.

Stress Control

In a continuous shielded cable the electric field is uniform along the cable axis and there is variation in the field only in a radial section. This is illustrated in Figure 13-49 which shows the field distribution over such a radial section. The spacing of the electric flux lines and the corresponding equipotential lines is closer in the vicinity of the conductor than at the shield, indicating a higher electric stress on the insulation at the conductor. This stress increase, or concentration, is a direct result of the geometry of the conductor and shield in the cable section and is accommodated in practical cables by insulation thicknesses sufficient to keep the stress within acceptable values.

In terminating a shielded cable it is necessary to remove the shield to a point some distance from the exposed conductor as shown in Figure 13-50A. This is to secure a sufficient length of insulation surface to prevent breakdown along the interface between the cable insulation and the insulating material to be applied in the termination. The particular length required is determined by the operating voltage and the properties of the

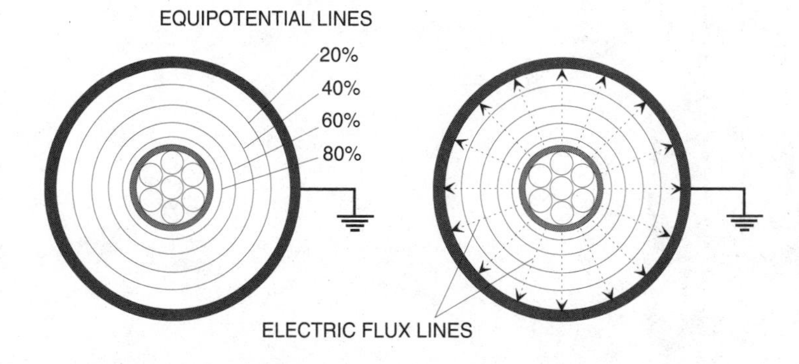

Figure 13-49: The electric stress is uniform along the cable axis; variations occur only in a radial section.

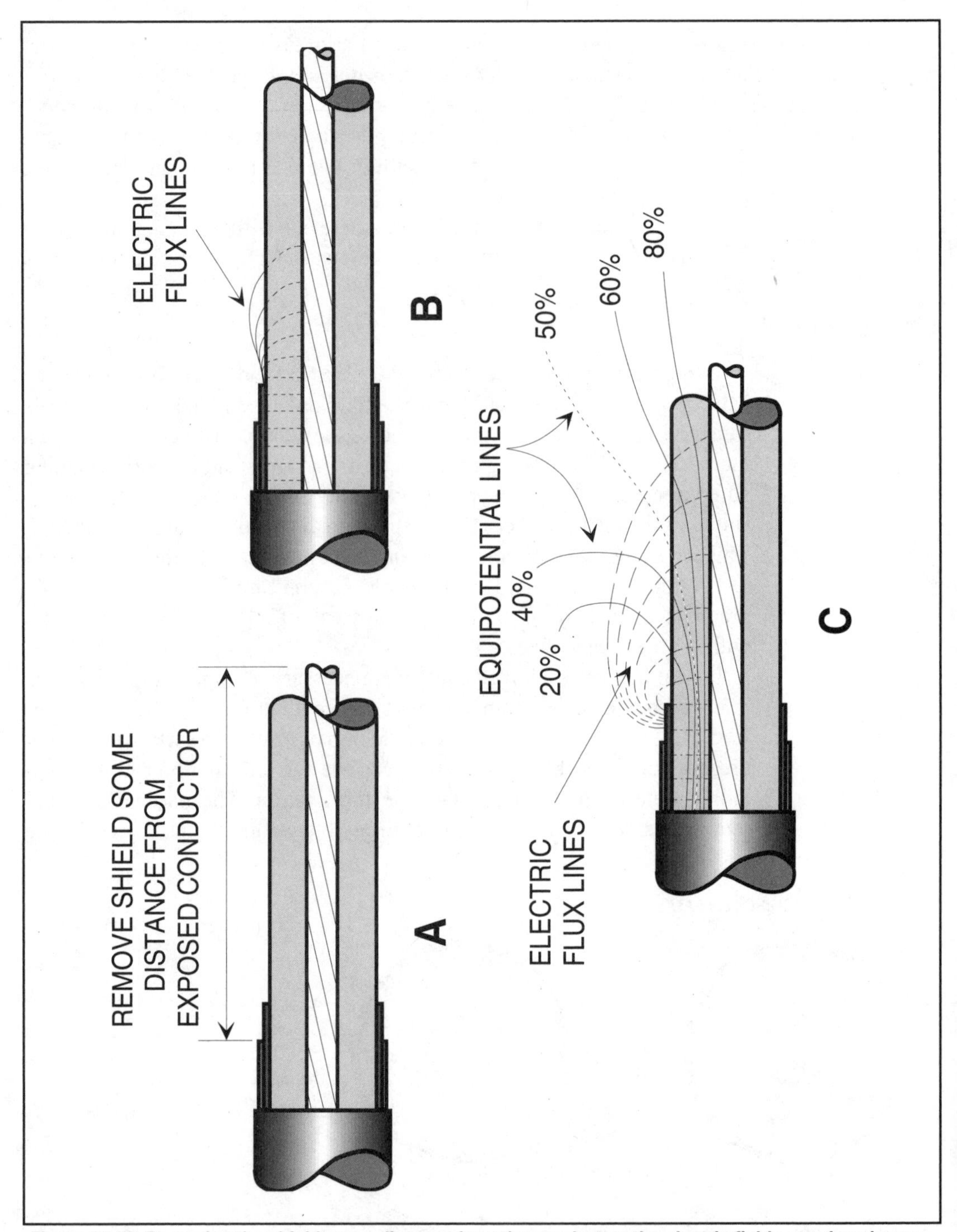

Figure 13-50: Removing the shield some distance from the conductor; the electric fields are also shown.

insulating materials. This removal of a portion of the shield results in a discontinuity in the axial geometry of the cable, with the result being that the electrical field is no longer uniform axially along the cable, but exhibits variations in three dimensions.

Figures 13-50B and 13-50C show the electric field in the vicinity of the shield discontinuity. The electric flux lines originating along the conductor are seen to converge on the end of the shield, with the attendant close spacing of the equipotential lines signifying the presence of high electric stresses in this area. This stress concentration is of much greater magnitude than that occurring near the conductor in the continuous cable, and as a result steps must be taken to reduce the stresses occurring near the end of the shield if cable insulation failure is to be avoided.

All terminations must at least provide stress control. This stress control may be accomplished by two commonly used methods:

- Geometric Stress Control
- Capacitive Stress Control

Geometric Stress Control

This method involves an extension of the shielding (Figure 13-51A) which expands the diameter at which the terminating discontinuity occurs and thereby reduces the stress at the discontinuity. It also reduces stresses by enlarging the radius of the shield end at the discontinuity. (Figures 13-51B and 13-51C).

Capacitive Stress Control

This method consists of a material possessing a high dielectric constant (K), generally in the range of K 30, and also a high dielectric strength.

Note: *Dielectric Constant = K and is a measurement of the ability of a material to store a charge.*

Material	K
Air	1
Cable insulation	3
130° C tape	3
High K material	30

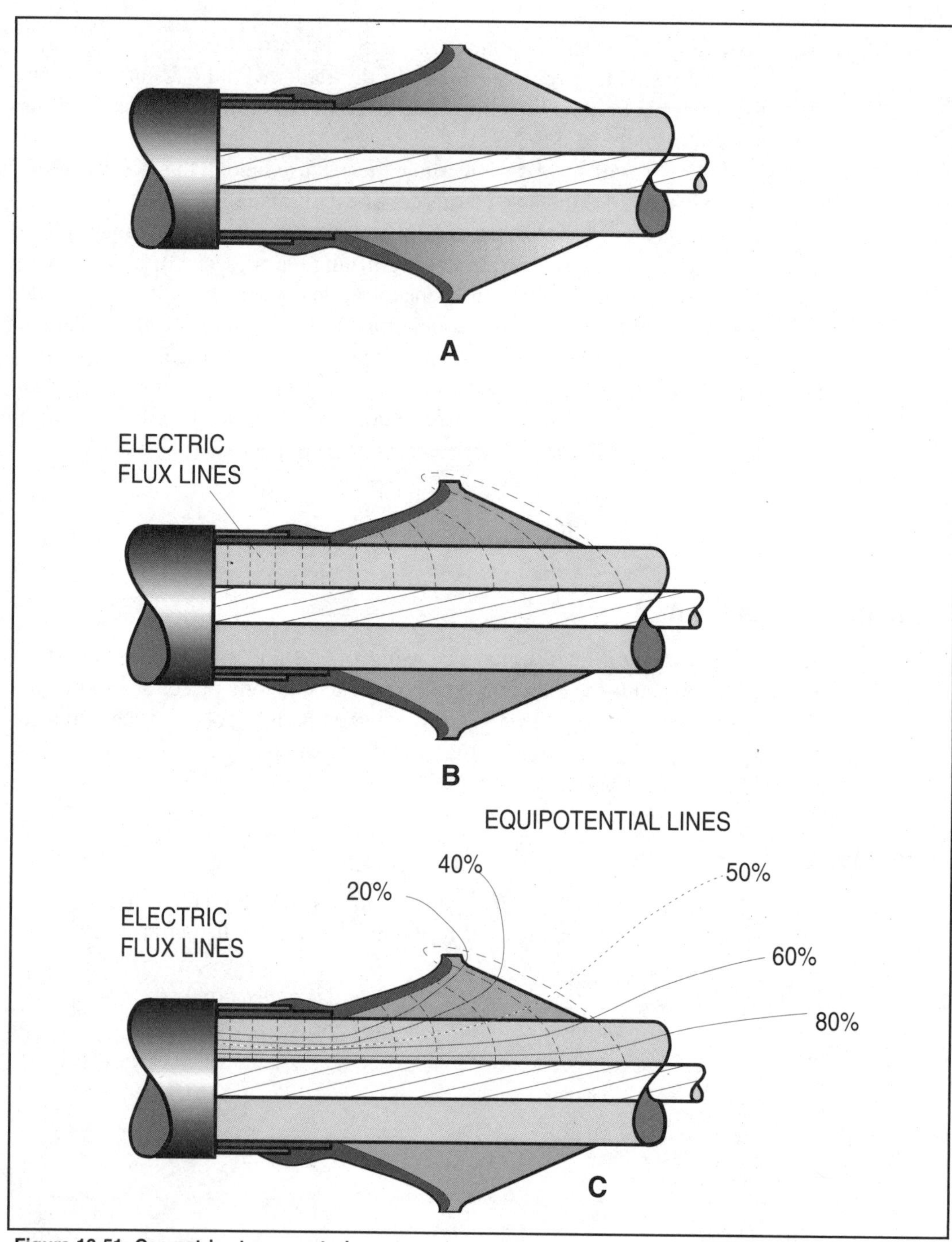

Figure 13-51: Geometric stress control.

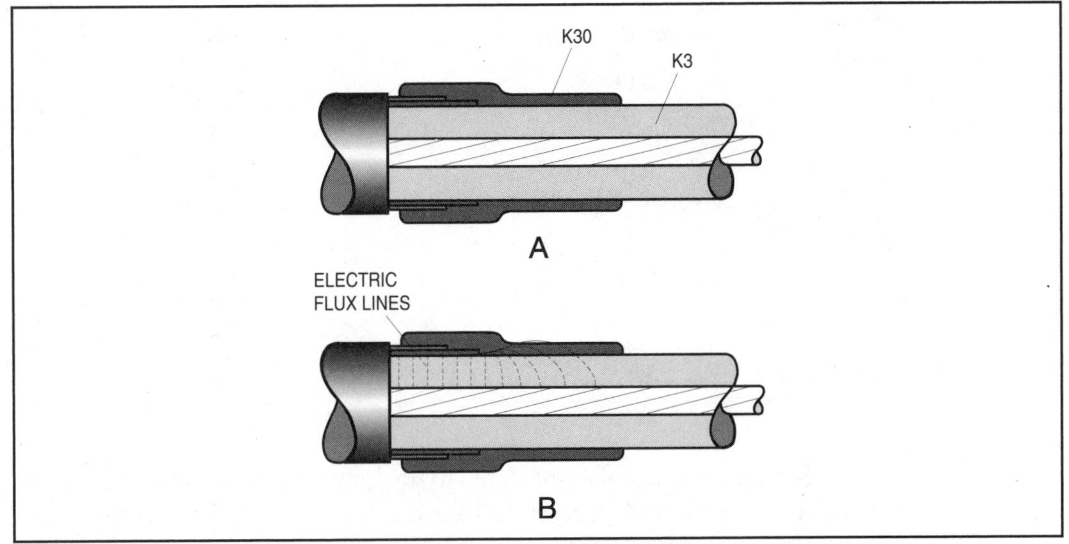

Figure 13-52: Lines of electrical flux are regulated to equalize the electrical stresses in a controlled manner.

This K is generally an order of magnitude higher than the cable insulation. Located at the end of the shield cut-back, the material capacitively changes the voltage distribution in the electrical field surrounding the shield terminus. Lines of electrical flux are regulated to equalize the electrical stresses in a controlled manner along the entire area where the shielding has been removed (Figures 13-52A and 13-52B).

By changing the electrical field surrounding the termination, the stress concentration is reduced from several hundred volts/mil to values found in continuous cable — usually less than 50 volts/mil at rated cable voltage.

External Leakage Insulation

This insulation must provide for two functions: protection from flashover damage and protection from tracking damage.

Terminations can be subjected to flashovers such as lightning induced surges or switching surges. A good termination must be designed to survive these surges. Terminations are assigned a BIL (Basic Lightning Impulse Insulation Level) according to their insulation class. As an example, a 15 kV class termination has a BIL rating of 110 kV-crest. See the following table.

Insulation Class (kV)	BIL (kV-crest)
5.0	75
6.7	95
15	110
25	150
34.5	200
46	250
69	350

Terminations are also subjected to tracking. Tracking can be defined as the process that produces localized deterioration on the surface of the insulator, resulting in the loss of insulating function by the formation of a conductive path on the surface.

A termination can be considered an insulator, having a voltage drop between the high-voltage conductor and shield. As such, there develops a leakage current between these points. The magnitude of this leakage is inversely proportional to the resistance on the insulation surface. *See* Figure 13-53.

Both contamination (dust, salt, airborn particles, etc.) and moisture (humidity, condensation, mist, etc.) will decrease this resistance. This results in surface discharges referred to as tracking. In an industrial environment it is difficult to prevent these conditions. Therefore, a track resistant insulator must be used to prevent failure.

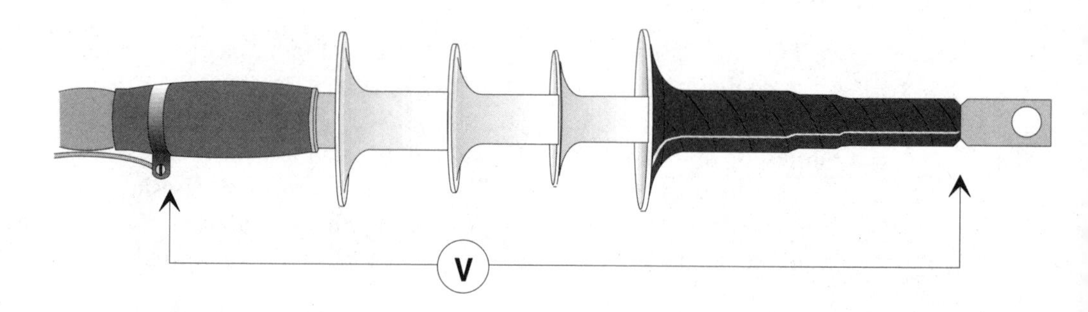

Figure 13-53: A terminator may be considered an insulator.

A high performance Class 1 termination is built with properly designed insulator skirts which are sized and shaped to form breaks in the moisture/contamination path, thus reducing the probability of tracking problems. Also, such skirt design can reduce the physical length of a termination by geometrically locating the creepage distance over the convolutions of the insulator, a factor when space is a consideration.

Good termination design dictates that the insulators be efficiently applied under normal field conditions. This means the product should be applied by hand with a minimum of steps, parts, and pieces. Preferably, no special installation tools or heat should be necessary.

Because they are inorganic materials and do not form carbon (conductive) residues, both porcelain and silicone are considered the best in track resistance.

SEAL TO THE EXTERNAL ENVIRONMENT

In order to qualify as a Class 1 termination, the termination must provide a seal to the external environment. Both the conductor/lug area and the shielding cut-back area must be sealed. These seals keep moisture out of the cable to prevent degradation of the cable components. Several methods are used to make these seals such as: factory-made seals, tape seals (silicone tape must be used for a top seal), and compound seals.

Grounding Shield Ends

All shield ends are normally grounded. However, individual circumstances can exist where it isn't desirable to ground both ends of the cable. The cable shield MUST BE grounded somewhere in the system. When using solid dielectric cables it is recommended that solderless ground connections be used (eliminating the danger of over-heating cable insulation when soldering).

As a general summary, for highly contaminated and exposed environments or on extremely critical circuits, porcelain or silicone rubber terminations are preferred. For potentially contaminated, moisture-prone areas — most likely the majority of cases — silicone rubber provides dependable Class 1 terminations.

HI-POT TESTING

Suitable dielectric tests are commonly applied to determining the condition of insulation in cables, transformers, rotating machinery, etc. Properly conducted tests will indicate such faults as cracks, discontinuity, thin spots or voids in the insulation, excessive moisture or dirt, faulty splices, faulty potheads, etc. An experienced operator can not only frequently predict the expected breakdown voltage of the item under test, but often can make a good estimate of the future operating life.

The use of direct current (dc) has several important advantages over alternating current (ac). The test equipment itself may be much smaller, lighter in weight, and lower in price. Properly used and interpreted, dc tests will give much more information than is obtainable with ac testing. There is far less chance of damage to equipment and less uncertainty in interpreting results. The high capacitance current frequently associated with ac testing is not present to mask the true leakage current, nor is it necessary to actually break down the material being tested to obtain information regarding its condition. Though dc testing may not simulate the operating conditions as closely as ac testing, the many other advantages of using dc make it well worthwhile.

The great majority of high-voltage cable testing is done with direct current — primarily due to the smaller size of machines and the ability to easily provide quantifiable values of charging, leakage, and absorption currents. While both ac and dc testers may damage a cable during tests, the damage in a dc test only occurs if increasing current is not interrupted soon enough and the capacitance stored is discharged across the insulation.

Hi-pot testers for high-voltage cable will also have a microammeter and a range dial to allow measurement of up to 5 milliamps, at which level the circuit breaker will trip or else the reactor will collapse.

WARNING!

Extreme caution must be exercised when conducting hi-pot testing of high-voltage cable. The cable will recharge itself as the absorption current "migrates" after the test unless grounds are constantly maintained.

Hi-pot testing may be divided into the following broad categories:

- *Design Tests* — These are the tests usually made in the laboratory to determine proper insulation levels prior to manufacturing.
- *Factory Tests* — These are the tests made by the manufacturer to determine compliance with the design or production requirements.

- *Acceptance Tests* — These are the tests made immediately after installation, but prior to putting the equipment or cables into service.
- *Proof Tests* — These are the HV tests made soon after the equipment has been put into service and during the guarantee period.
- *Maintenance Tests* — Maintenance tests are those performed during normal maintenance operations or after servicing or repair of equipment or cables.
- *Fault Locating* — Fault locating tests are made to determine the location of a specific fault in a cable installation.

The maximum test voltage, the testing techniques, and the interpretation of the test results will vary somewhat depending upon the particular type of test. Unfortunately, in many cases the specifications do not spell out the test voltage, nor do they outline the test procedure to be followed. Therefore, it is necessary to apply a considerable amount of common sense and draw strongly upon past experience in making these tests. There are certain generally accepted procedures, but the specific requirements of the organization requiring the tests are the ones that should govern. It is obvious that an acceptance test must be much more severe than a maintenance test, while a test made on equipment that is already faulted would be conducted in a manner different from a test being conducted on a piece of equipment in active service.

As a rough rule of thumb, acceptance tests are made at about 80 percent of the original factory voltage. Proof tests are usually made at about 60 percent of the factory test voltage. The maximum voltage used in maintenance testing depends on the age, previous history, and condition of the equipment, but an acceptable value would be approximately 50 to 60 percent of the factory test voltage.

Special testing instruments are available for all of the tests listed above. However, testing instruments will be used mostly by electricians to perform routine production tests and acceptance tests for newly installed systems.

The general criteria for acceptance of a cable system is a consistent leakage for each voltage step (linear) and a decrease in current over time at the final voltage. The level of charging current should also be consistent for each step.

Method of Application

DC hi-pot tests are performed primarily to determine the condition of the insulation. The extent of the highest breakdown voltage may or may not be of interest, but the relative condition of the insulation, in the neighborhood of and somewhat above operating voltage, is certainly required. While there are many ways of achieving these results, two of the most popular are the "Leakage Current vs. Voltage" test, and the "Current vs. Time" test. The "Current vs. Time" test is frequently made immediately after the "Current vs. Voltage" test and probably yields best results when performed in this manner.

Before performing a test, all accessory equipment, connected loads, potential transformers, etc., are disconnected from the cable or rotating machine being tested. The shield, frame, and unused phases are grounded and the hot side of the hypot tester is connected to the conductor under test.

The voltage is slowly raised in discrete steps allowing sufficient time at each step for the leakage current to stabilize. As the voltage is raised, the leakage current will first be relatively high, then will decrease with time until a steady minimum value is reached.

The initial high value of current is known as the charging current, and is dependent primarily on the capacitance of the item under test. The lower steady state current consists of the actual leakage current and the dielectric absorption current (which is relatively minor as far as these tests are concerned). The value of the leakage current is noted at each step of voltage, and as the voltage is raised a curve is plotted of "Leakage Current versus Voltage." As long as this curve is relatively flat (equal increments of voltage giving equal increments of current, Figure 13-54, point A to B), the item under test is considered to be in good condition. At some point the current will start rising at a more rapid rate (Figure 13-54, point C). This will show up on the plot as a knee in the curve. It is very important that the voltage increments be small enough so that the starting point of this knee can be noted. If the test is carried on beyond the start of the knee the current will increase at a much more rapid rate and breakdown will soon occur (Figure 13-54, point D). Unless breakdown is actually desired, it is usual practice to halt a test as soon as the beginning of the knee is observed.

With a little experience, the operator can extrapolate the curve and estimate the point of breakdown voltage. It is this procedure that enables the operator to anticipate the breakdown point without actually damaging the insulation, for if the test is stopped at the start of the knee, no harm is done.

While the important thing to watch for is the rate of change of current as the voltage is raised, much in formation may be obtained from the value

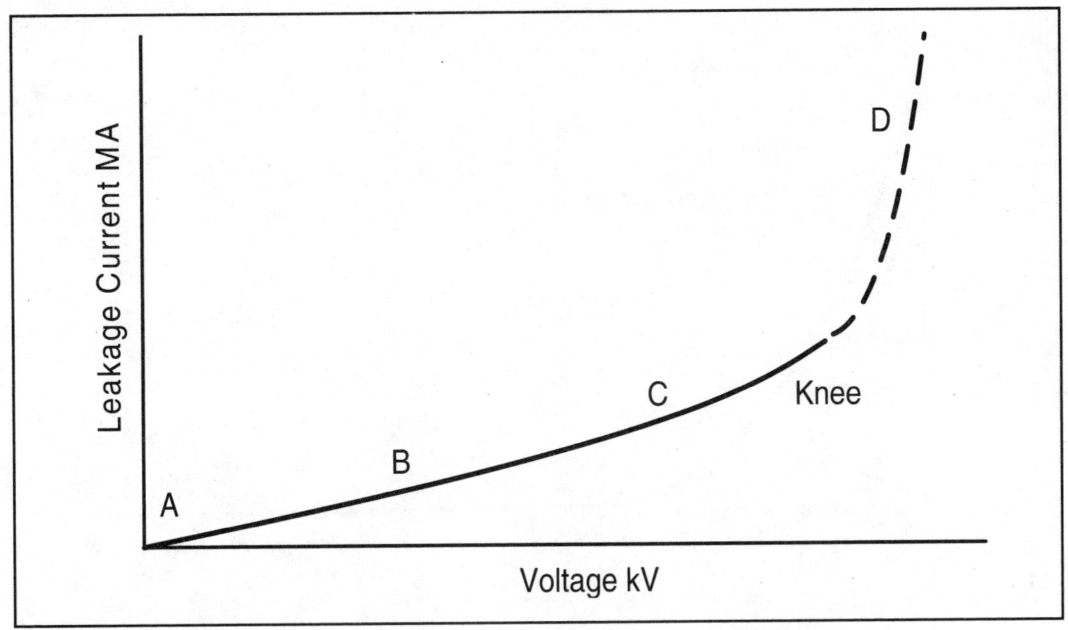

Figure 13-54: Leakage current versus voltage.

of the current. A comparison of the leakage current from phase to phase, or a comparison of leakage current with values measured in other equipment under similar conditions, or a comparison with the leakage current values obtained on tests made of the same equipment on prior occasions, will give an indication of the condition of the insulation. In general, higher leakage current is an indication of poorer insulation. However, this is a matter difficult to evaluate from a single test. But as long as the curve of "Leakage Current versus Voltage" is of the general shape shown in Figure 13-54, and as long as the knee occurs at a sufficiently high voltage (above the maximum test voltage required), the equipment being tested may be considered satisfactory.

On long cable runs or with rotating machinery having high capacitance between windings, or from winding to frame, the time for stabilization of the leakage current (decrease from E to F in Figure 13-55) may run as high as several minutes or several hours or more. To minimize the testing time under these conditions, it is common practice to time the decay of current and to take a reading at each voltage step after the same fixed interval. For example, voltage could be brought up to the first step, allowed to remain at that point for four minutes and then the current reading taken. The voltage is next raised to the second step, the same 4-minute wait is allowed and current reading again taken. This is continued throughout the test. While

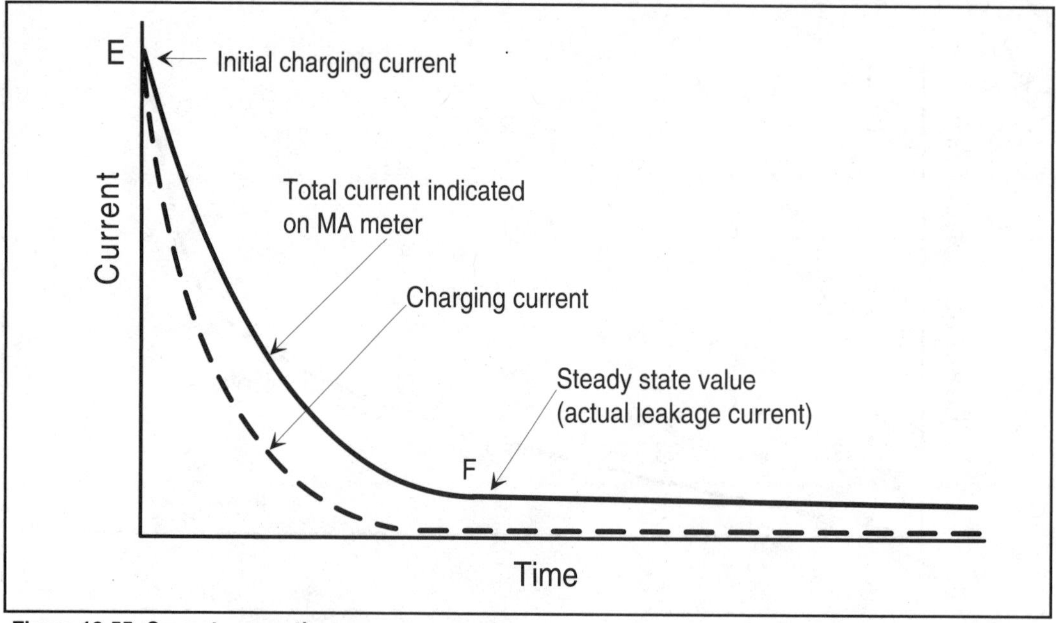

Figure 13-55: Current versus time curve.

this method will not give the true leakage current at any point, it will still give the properly shaped curve and will still permit much the same evaluation.

When the final test voltage is reached, the tester is left on and the "Current versus Time" curve (Figures 13-55 and 13-56) is plotted by recording the current at fixed intervals as it decays from the initial high charging value to the steady state leakage value. This curve for good cable should indicate a continuous decrease in leakage current with time or a stabilization of the current without any increase during the test.

Any increase of current during this test would be indicative of a bad cable or machine. The test should, of course, be stopped immediately if such an indication occurs (see Figure 13-57). After the current has stabilized and the last reading taken, shut off the high voltage. The kilovoltmeter on the tester will indicate the actual voltage present in the equipment under test as the tester's internal circuitry permits the charge to gradually leak off. After the voltage reaches zero (0), put an external ground on the cable or machine and disconnect it from the tester. It is important to wear rubber gloves at this point, since the effect of absorption currents may cause a buildup of voltage in the item under test until after it has been grounded for some time.

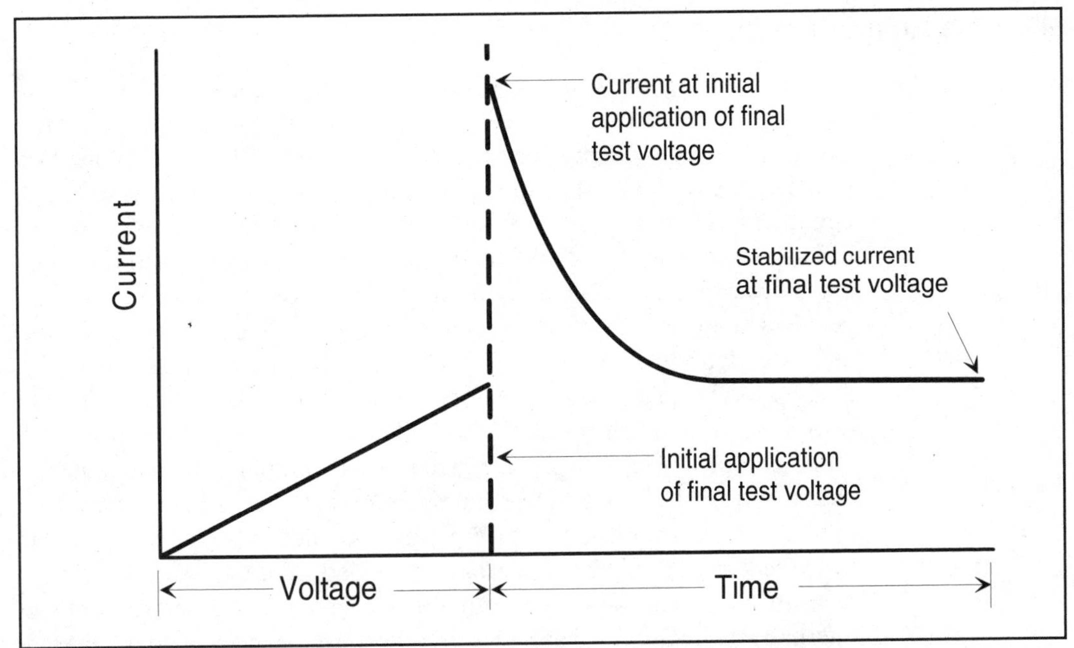

Figure 13-56: Current versus voltage and current versus time at maximum voltage.

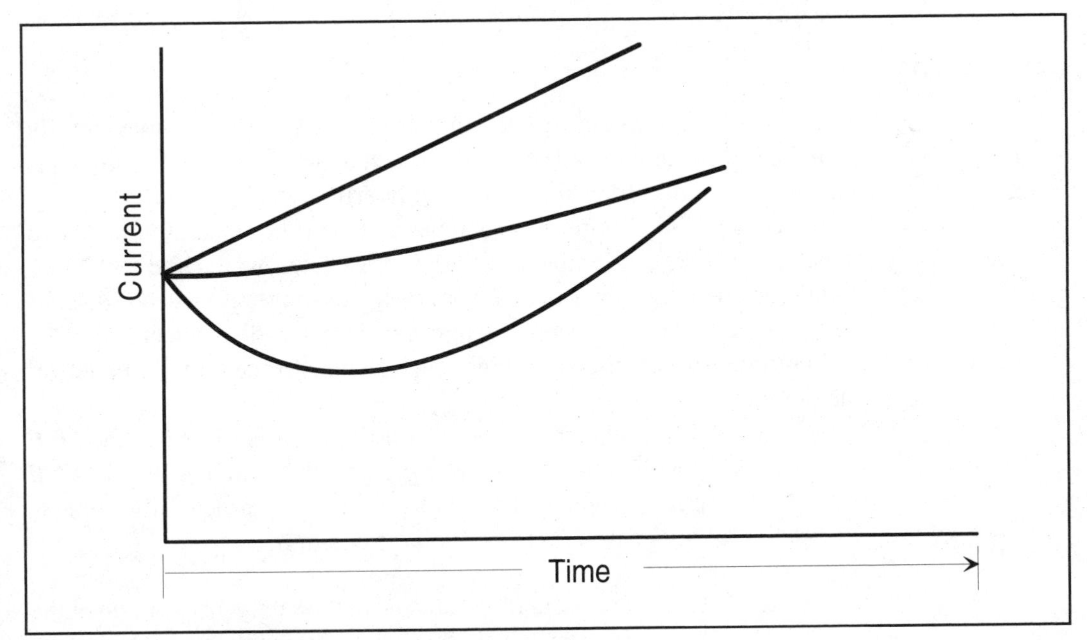

Figure 13-57: Faulty insulation.

Selective Guard Circuits

When making the "Current versus Voltage" test, a rapid climb in current may occur not due to defects in the equipment itself, but because of surface leakage or corona at the connections, cable ends, blades of connected switches, etc. All dc testers incorporate internal circuitry (selective guard circuit) to bypass this stray current so that it does not interfere with the leakage measurements. By proper use of the guard circuit and corona ring, when necessary, good leakage current readings even below one microampere can be made. This technique consists of providing a separate path for these stray currents, bypassing them around the metering circuit, and only feeding the leakage current to be measured through the metering circuit. In order to accomplish this, dc testers have two return paths: (1) the metered return binding post which provides a return path through the microammeter circuit and (2) the bypass return binding post which provides a return path bypassing the meter. Currents entering the first binding post will be measured, while currents entering the second binding post will not be measured. A panel switch is provided to connect the cabinet ground internally to either of these two binding posts (which enables ground currents to be measured or bypassed, depending on the test being performed).

Connections

Before performing any test, it is important that the test operator carefully plan the test procedure and connections so that the required measurements may be taken without misinterpretation or error.

All possible safety precautions must be taken. Equipment to be tested must be deenergized and grounded to eliminate any charge. Switches should be open, jumpers running from potheads to feeders must be disconnected, potential transformers and lightning arresters disconnected. If their insulation level is sufficiently high, current transformers might be left in the circuit.

Step 1. Make sure the main ON-OFF switch is turned to the OFF position; that the high-voltage ON switch is in the OFF position, and that the voltage control is turned fully counterclockwise to the zero voltage position.

Step 2. Connect a grounding cable from the safety ground stud of the tester to a good electrical ground, and make sure the connection is secure at both ends. It is essential that this connection be made. At no time should the tester be operated without it.

Step 3. Connect the return line from the item under test to the metered return or bypass return binding post of the tester, and operate the grounding switch on the tester panel to the appropriate position. The return lines connected to either the bypass return or metered return binding post should be insulated for about 100 V, so ordinary building grade wire may be used.

Step 4. Plug the shielded high-voltage cable furnished with the tester into the receptacle at the top of the unit. If the tester has an oil filled output receptacle, make sure that the oil level is at the mark indicated on the oil stick. If the oil level is low, add the proper amount of high grade transformer or other insulating oil. After the high voltage cable is plugged into the receptacle, tighten the clamping nut to hold it securely in place.

Step 5. Connect the shield grounding strap from the cable to the shield ground stud on the tester adjacent to the high-voltage receptacle.

Step 6. Connect the other end of the output cable to the item under test. This connection should be made mechanically secure avoiding any sharp edges. It is usually advisable to tape the connection using a high grade electrical tape to minimize corona at this point. If the voltages to be used are high, or if for any other reason corona is anticipated, its effect may be minimized by making use of a corona ring or corona shield.

Step 7. If a bench type tester or any other unit having an output bushing rather than a high-voltage receptacle is used, the same procedure is followed except that the output cable is fastened to the bushing and should be supported along its length, preferably using nylon cord so that it stays well away from any grounded or other conducting surface. The high voltage connection between the tester and the item under test should be as short and direct as possible.

Step 8. When shielded cable is used, it is important that the shield be trimmed back from each end for a distance of about an inch or more for every 10,000 V. The shield at the tester end should be fastened to the ground stud on the tester. The shield at the test end should be taped and no connection made to it.

Step 9. If the item under test is beyond the reach of the tester cable supplied, an extension running between the end of the tester cable and the item under test must be used. If this extension is made of shielded cable, it is necessary to splice not only the central conductor to the end of the tester cable, but a jumper should connect the shields of the two cables together at the splice, taking care to run the jumper well away from the splice itself to avoid leakage or even breakdown between the two.

The splice should be well taped. If no shielded high-voltage cable is available, the single wire extension used should be supported well away from all other objects, as described previously. When operating at voltages within the corona levels, it is preferable that the tester be brought close enough to the item under test so that an extension cable is not needed. It should be noted that the extension cable is a frequent cause of erratic readings, especially if the quality is not good enough at the high voltage used.

Step 10. Plug the line cord into the 115 V, 60 cycle, single-phase outlet and clip the ground wire coming from the line cord to the conduit receptacle mounting screw or other ground source.

Step 11. When line regulation is poor (because of location, heavy or intermittently loaded equipment in the vicinity, or due to the use of portable generating equipment) it may be necessary to connect a voltage regulator between the tester and the 115 V supply. The use of a regulator is particularly desirable when using a tester with sensitive vacuum tube current meter with ranges of less than 10 microamperes full scale, or when highly capacitive loads are being tested.

Since the dc output voltage of the tester is dependent on the ac line input voltage, an unstable line voltage will cause fluctuation of the tester output voltage with corresponding fluctuation of the current meter. This will be especially noticeable when measuring low current of the order of a few microamperes or when testing loads having considerable capacitance.

When the load is highly capacitive it charges up to the test potential and tends to maintain that potential because of the high leakage resistance. A sudden drop in line voltage causes the tester output voltage to also drop, so that the load capacitance is then charged to a higher value than the tester output. Since the load is at a higher voltage than the tester the direction of current flow will be reversed and the current meter will be driven down scale. A sudden increase in line voltage would cause the opposite effect, the tester voltage suddenly rising above the voltage to which the load has been charged, causing a rapid increase in the current meter reading. As a result the current meter will fluctuate so that it might be difficult to obtain a meaningful leakage current reading.

The use of a line voltage regulator under such conditions will greatly increase the usefulness and ease of operation of the equipment. In general the power rating of the regulator should be based on the maximum power output capability of the tester multiplied by the factor of about 1.3. For example, a tester rated at 75 kV @ 2.5 milliamps can deliver an output power of:

$$75 \times 2.5 \text{ or } 187.5 \text{ watts}$$

$$187.5 \times 1.3 = 243.75 \text{ watts}$$

Consequently, for this example, a 250 watt regulator should be used. The harmonic free type of regulator is preferable to the standard type unit though either may be used with satisfactory results. If considerable testing at very light loads is done, it might also be desirable to have a smaller regulator to use under those conditions, since a small regulator operating at half to three-quarters of its capacity will give better results than a lightly loaded large regulator. A method of connecting a voltage regulator is shown in Figure 13-58.

Step 12. Turn the microampere range switch to the highest range. On multiple voltage range units, turn the kilovolt range switch to the range that will allow testing to maximum voltage required. It should be noted in this connection that the microampere range may be changed while test is under way, but the kilovolt range may not. Safety interlocks within the unit will automatically shut off high voltage as the voltage range switch is turned while the test is in progress.

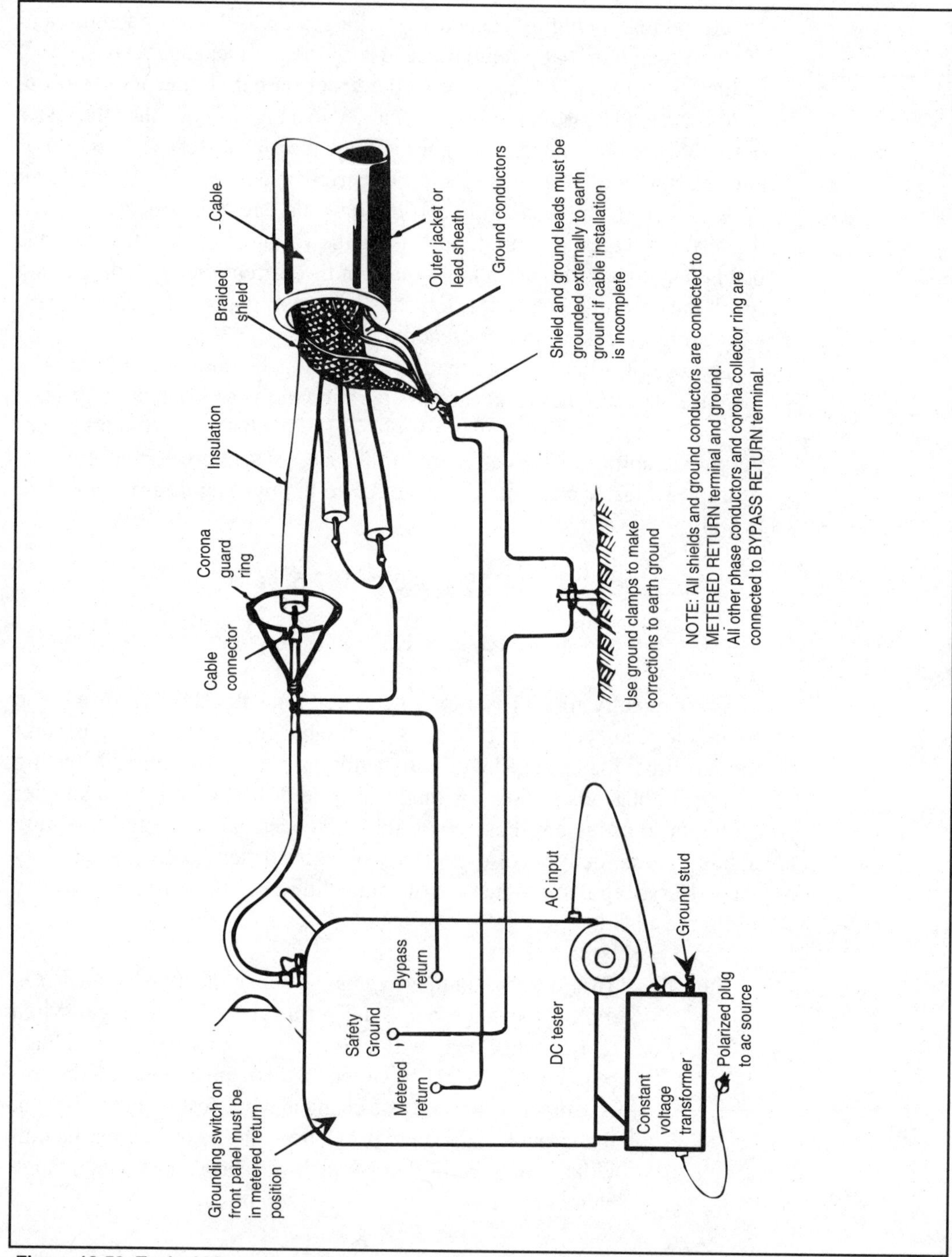

Figure 13-58: Typical hi-pot connections for testing multiple conductor shielded cable.

Step 13. The main power switch or circuit breaker may now be turned ON and the test begun.

The sketches given in Figures 13-59 through 13-65 illustrate typical connection diagrams. The experienced operator will soon develop the knack and technique of making connections most suitable to the test he is performing. The sketches are merely offered as a guide.

Note: *In the various connection diagrams given here, the guard circuit may be omitted by tying the connections shown going to the bypass return binding post to the metered return binding post instead. This is the difference shown by the connections in Figures 13-63 and 13-64.*

If the return path is grounded, set the panel grounding switch to the metered return position. Total load current (leakage, corona, etc.) will be read on the meter.

If the return path is not grounded, set the panel grounding switch to the bypass return position and the effect of corona currents on the meter reading will be minimized.

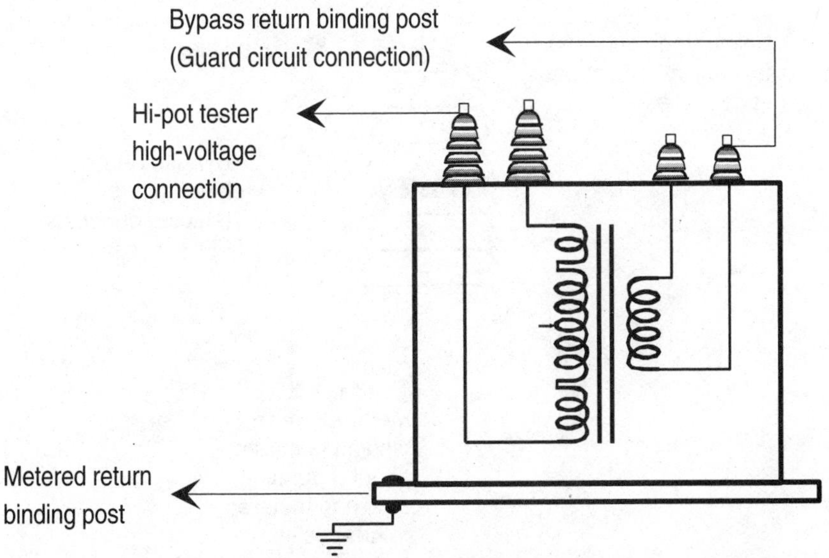

Figure 13-59: High-voltage wiring to grounded core or case.

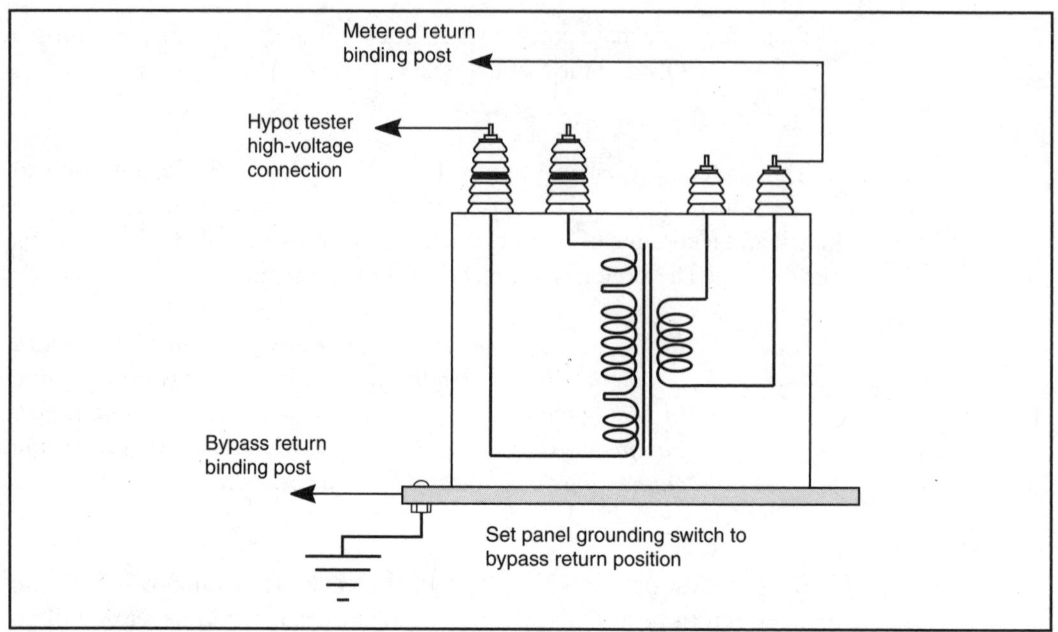

Figure 13-60: High-voltage winding to low-voltage winding.

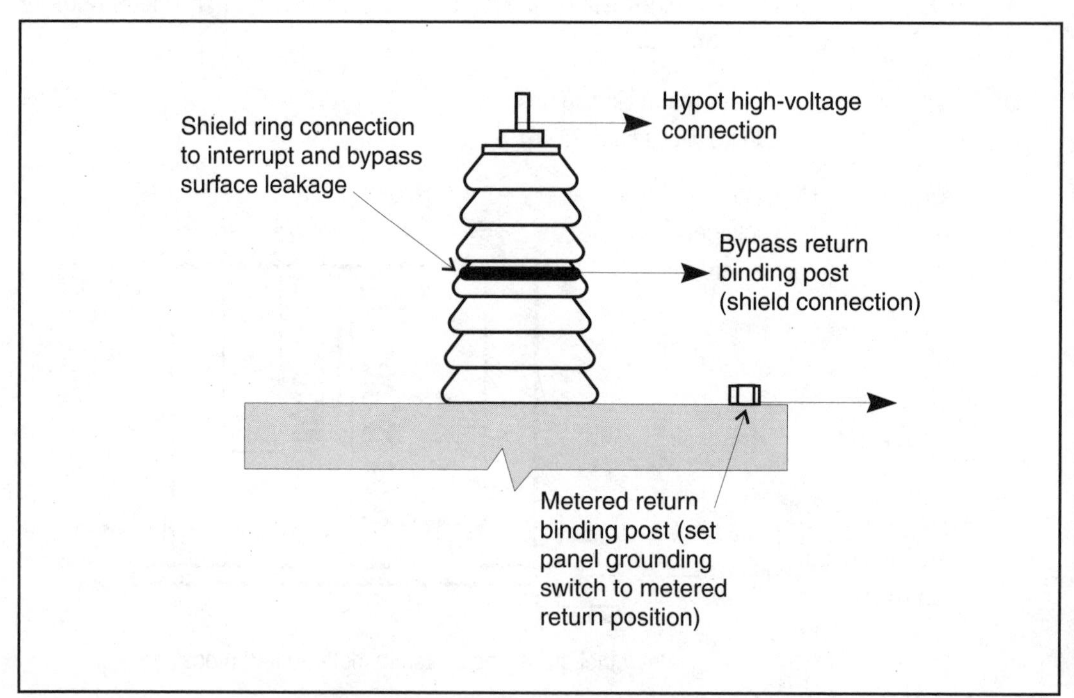

Figure 13-61: Measuring busing leakage.

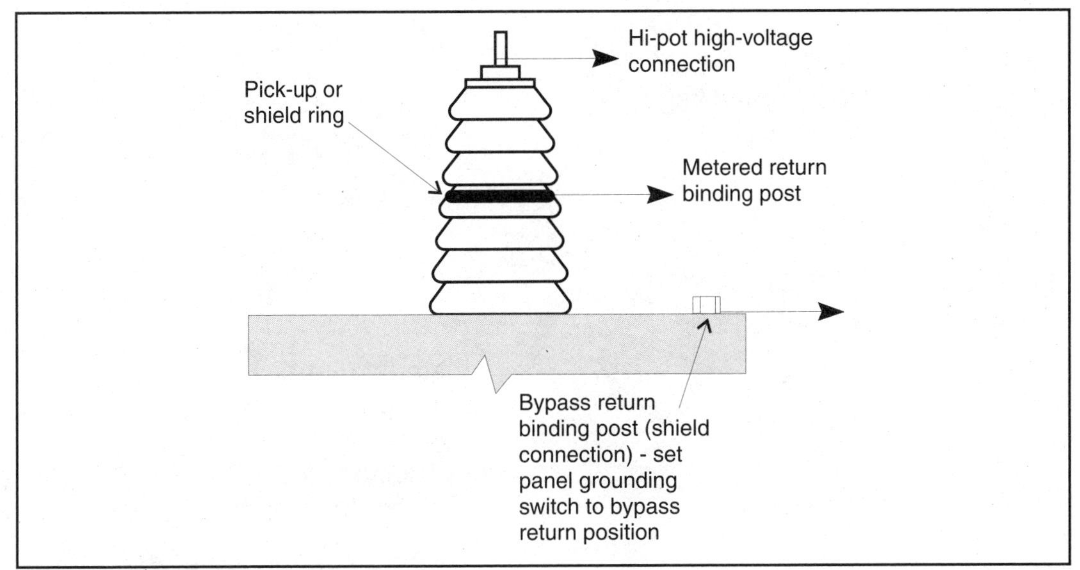

Figure 13-62: Measuring surface leakage.

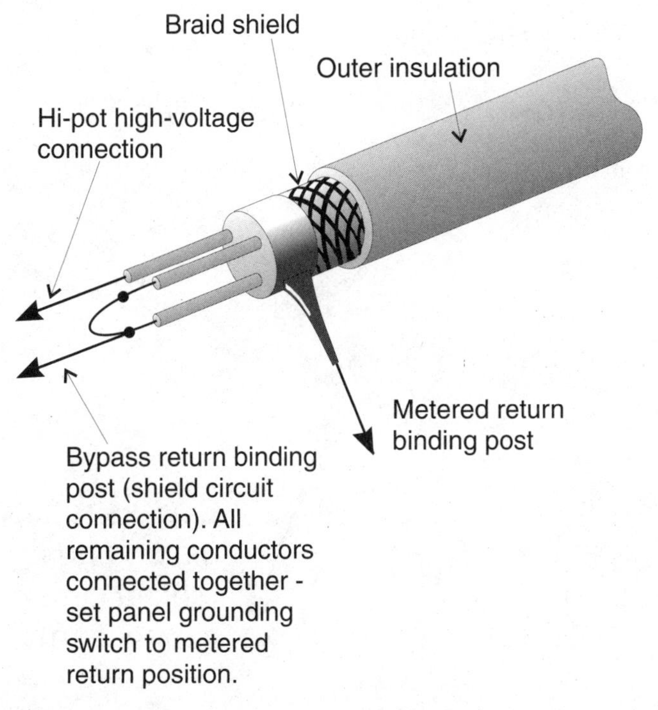

Figure 13-63: Test cable insulation between one conductor and shield.

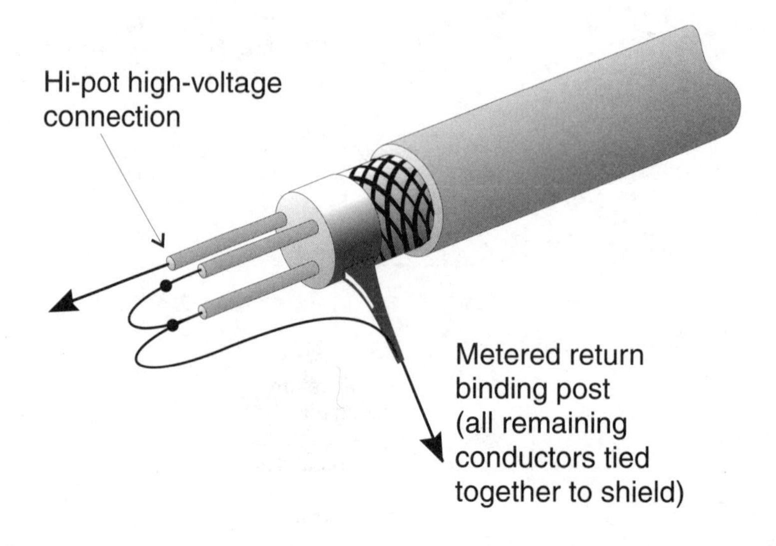

Figure 13-64: Testing cable insulation between one conductor and all others and shield.

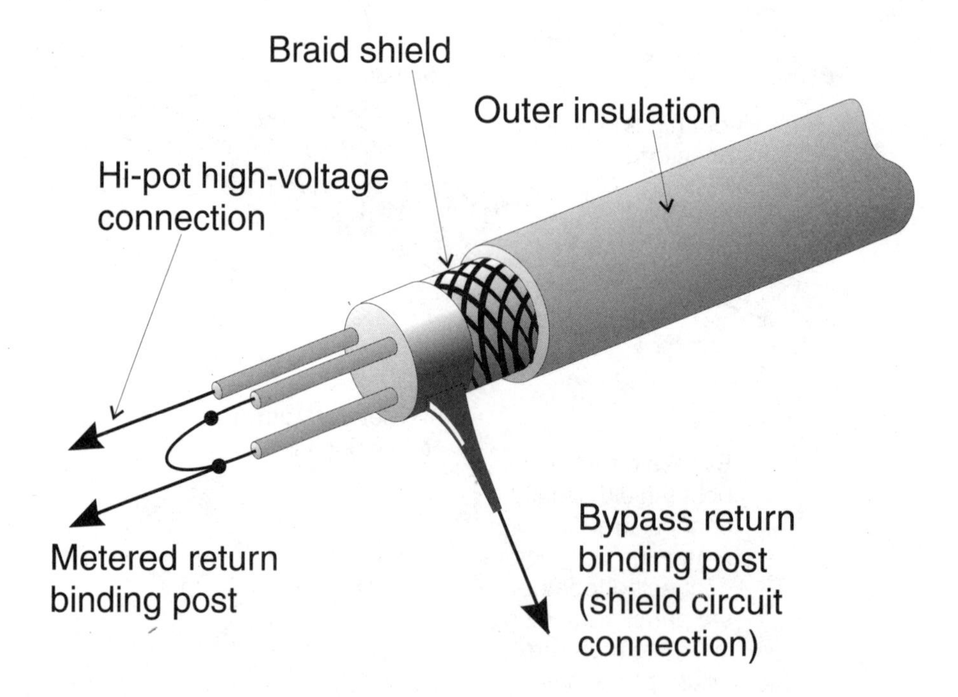

Figure 13-65: Testing cable insulation between conductors.

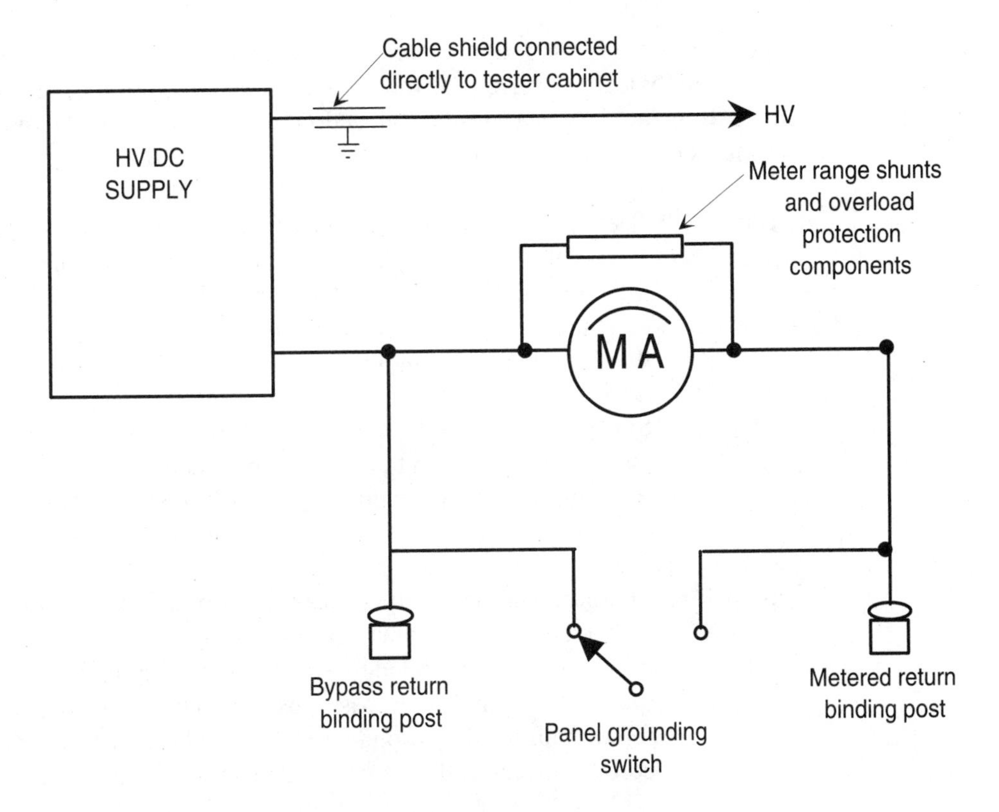

Figure 13-66: Simplified hi-pot tester output circuit diagram.

Selective Guard Circuit Connections

In the connection diagrams (Figures 13-59 through 13-65), frequent reference is made to the bypass return or metered return binding post. The simplified sketch of the tester output circuit shown in Figure 13-66 will aid in understanding the use of these terminals.

Examination of the sketch will show that any current to be measured should be returned to the tester through the metered return binding post and any current which is not to be measured, but is to be bypassed around the meter should be returned to the bypass return binding post.

As corona currents flow from the high-voltage connection to ground, they may be bypassed around the microammeter by grounding the bypass return binding post with the grounding switch on the control panel. This gives a direct return path back to the high-voltage supply for corona currents and bypasses them around the meter.

For purposes of illustration, assume a three-conductor shielded cable is to be tested for leakage current between one of the conductors and the outer shield. To make the proper tester connection, the test should be analyzed as follows (see Figure 13-66):

Step 1. The current which is to be measured must return to the low side of the tester through the metered return binding post; therefore, the HV connection should be made to the one conductor in question and the shield should be connected to the metered return binding post.

Step 2. Since the measurement of current flow between the HV connection and the other two conductors in the cable is not desirable, these two conductors should be connected together and returned to the bypass return binding post.

Step 3. In order to minimize the measurement of the corona current from the HV connection to ground, the bypass return binding post should be grounded by setting the grounding switch on the control panel to bypass position. This will provide a return path for the corona current around the meter and the corona current will not affect the meter readings.

A further analysis of the same cable will show the following:

Step 1. To measure the leakage current between the one conductor and the other two conductors (which are tied together), the return path for the two conductors should be through the metered return post and the shield should then be connected to the bypass return post. Again, to minimize the measurement of the corona current, the grounding switch should be in bypass return position.

Step 2. The corona current alone may be measured by connecting the shield and the two conductors together and connecting them to the bypass return binding post. If the grounding switch is then placed in the metered return position, the return path for the corona current will be through the microammeter and the corona current will be the only current being measured.

Generally speaking, the grounding switch on the control panel will either include or exclude the corona current in the meter reading, depending upon its position.

It was previously assumed that there were no restrictions placed on the cable with respect to grounding. This was done to illustrate the general case. In many cases, however, the shield is permanently grounded, and therefore certain modifications in the test setup must be made in order to achieve corona-free stability and accuracy in the leakage current readings.

If it is desired to measure the leakage current from one conductor to shield, the shield should be connected to the metered return binding post as before. But since the shield is now permanently grounded, the metered return binding post should also be grounded by placing the panel grounding switch in metered return position. If the panel grounding switch were placed in the bypass return position, then the metered return post and the bypass return post would both be grounded, shorting out the microammeter.

Therefore the grounding switch must be put in the metered return position when one side of the test item is permanently grounded.

As outlined above, all unwanted currents which flow between the HV point and any ungrounded elements in the cable may be returned through the bypass line and therefore be excluded from the leakage current reading. This does not, however, include corona current since corona current always flows between HV and ground. Since we cannot shunt the corona current around the microammeter by grounding the bypass return line as we did before, we must use some auxiliary method of intercepting the corona current before it reaches ground. This can easily be accomplished by the use of a "corona" or "guard" ring. A "corona" or "guard" ring is nothing more than a metal loop which encompasses the HV connection and intercepts the corona current. The ring is connected to the tester through the bypass return binding post and therefore keeps the corona current out of the meter reading.

Corona Guard Ring And Corona Shield

A simple but very effective guard ring may be constructed as illustrated in Figure 13-67. When using the corona ring, the diameter should be made as small as possible, without causing an arc to jump between the HV connection and the ring. This will insure as much corona current as possible being intercepted. A wire should be connected to the ring as shown in the illustration, and the other end should be connected to the bypass return binding post. This will bypass the intercepted corona current around the

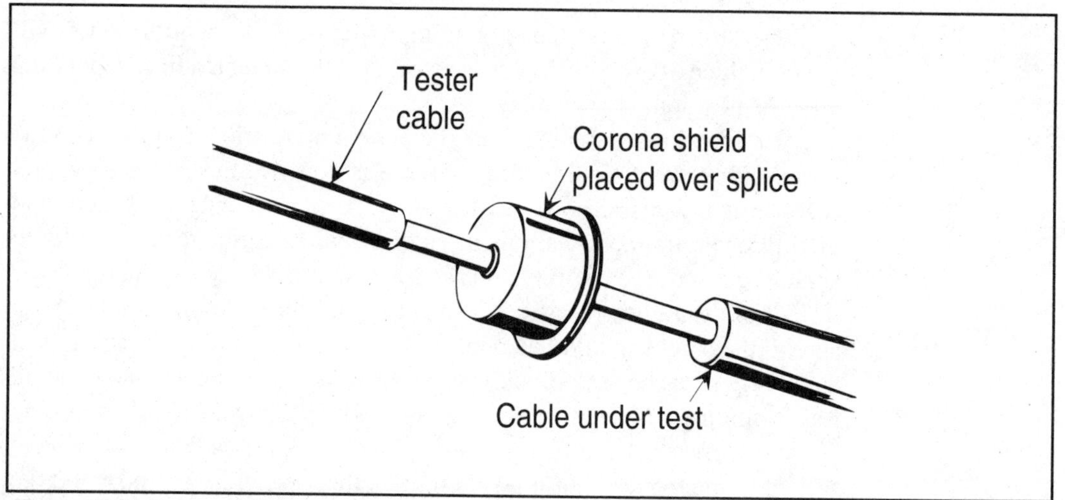

Figure 13-67: Corona shield.

microammeter and will allow essentially the same stability and accuracy which was obtainable previously.

Corona may also be minimized by electrically shielding or smoothing the connection. This may be done by smoothly wrapping it with metal foil or conductive tape or enclosing it in a round can or cover so that all jagged edges, wire points, or rough sur faces are within the metal enclosure. This corona shield must be electrically connected to the high-voltage and completely insulated from ground so that both the shield and high-voltage connection are at the same potential. A suitable corona shield is illustrated in Figure 13-67.

Detailed Operating Procedure

The step-by-step instructions to follow are meant to illustrate a typical operating procedure. In an actual test, account must be taken of special conditions peculiar to the installation, ranges, and operating controls of the tester available, and requirements of the operator.

Step 1. Assemble the necessary equipment and supplies at the test site. In addition to the tester, connecting cable, and so forth, a clip board or other writing surface, pencils, colored pencils, scratch paper, sheets of 8½ in by 11 in graph paper (which is very convenient for actually plotting the test results while the test is under way) and even a slide rule are desirable.

Step 2. Determine the type of test to be made; i.e., is the voltage going to be raised to breakdown, or is the voltage going to be raised to the specific test value only? If the item on the test is found to be defective, should the test stop before breakdown until the repairs can be made at a later date, or should the test be continued to the preselected value regardless of breakdown so that repairs could be made immediately?

Step 3. Determine the maximum test voltage to use. Since the value chosen depends not only on the type of test being made, but also on the material to be tested, it is necessary that all factors involved be considered. As a rough rule of thumb, proof testing is usually conducted at 60 percent, of the factory test voltage. Acceptance testing would call for a maximum value of about 80%, of the factory test voltage. For maintenance testing or on older work, the minimum test voltage chosen should be at least 1.7 times the operating voltage up to a maximum value of between 50 or 60 percent of the factory test voltage.

In many cases, the value of maximum test voltage to be used may be obtained from the IPCEA or AEIC specifications, or from manuals available from Rome Wire and Cable Company, Simplex Wire and Cable Company or others.

Step 4. Choose suitable increments so that the voltage may be raised to the final test value in about eight to ten steps (more or less to suit the individual conditions).

Step 5. Once the maximum test voltage and the increments have been chosen, the graph paper may be prepared so that the test results can be plotted as the readings are taken. In this way, the condition of the cable is under constant surveillance and if testing short of breakdown is decided upon, the operator has the necessary control to stop the test at the start of the knee of the current versus voltage curve. To prepare the graph paper, mark the current scale on the 8½ in side and the voltage scale along the horizontal 11 in side. Suitable voltage increments should have been chosen to give a convenient plot. The current scale should include values well above the expected maximum leakage current. If the magnitude of leakage current is not known, it may be advisable to either make a low voltage test run, or to hold off plotting the first few points until the magnitude is seen.

Step 6. Connect the tester to the equipment under test, as described previously.

Step 7. Make sure the adjustable voltage control is turned to the OFF position (fully counterclockwise); the high-voltage switch is in the center or OFF position, and the power switch is in the OFF position. Connect the input line cord to the 110-volt, 60 cycle receptacle and connect the alligator clip at the end of the line cord to ground. Turn the kilovoltmeter range switch and microammeter range switch to the appropriate positions.

Step 8. Turn the kilovoltmeter range switch and microammeter range switch to the appropriate positions.

Step 9. Turn the main ON-OFF switch or circuit breaker to the ON position.

Step 10. Operate the high-voltage ON switch. Note that the lever type high-voltage switch has a spring return from the DOWN position but will remain in the UP position. If it is desired to keep the high voltage under strict manual control, use the DOWN position of the high-voltage switch. As soon as it is released to the OFF position, high voltage will be turned off and it will become necessary to return the powerstat control to zero before the high voltage can be reapplied. If the high-voltage switch is placed in the UP position, the switch may be released and will remain in the UP position and high voltage will be available until the switch is manually returned to the center OFF position. On units with a key interlock, it is necessary to turn the key to the right to enable the UP position of the high-voltage lever switch to be operated.

Sep 11. Slowly rotate the variable voltage control from the extreme counterclockwise position (zero volts) to raise the voltage to the first increment as previously determined. The voltage should be raised at a slow enough rate to avoid having the microammeter pointer go off scale. The slower the voltage is raised, the lower will be the maximum charging current and microammeter reading. If the charging current is allowed to exceed the maximum current rating of the tester, the circuit breaker will trip. The circuit breaker will also trip if the unit under test fails. Since most testers are equipped with zero return interlocks for operator safety, it is necessary to return the voltage control to the zero position before high voltage can be reapplied once it is shut off, whether due to circuit breaker tripping, kilovolt range selector switch rotating or operation of the high-voltage ON lever switch to the OFF position.

Step 12. When the first increment of voltage is reached, watch the microammeter indicate gradually decreasing current as the cable or machine under test becomes charged. As the current drops to lower portions of the scale, switch the microammeter progressively to lower ranges until the current stabilizes. Note the length of time it takes for the current to stabilize, before accurate results can be obtained if the same stabilization period is allowed at each voltage step. After the current is stabilized, whether it takes 10 seconds, a minute, or 10 minutes, record the value of voltage and current.

Step 13. If the microampere scale for plotting the current versus voltage curve has been predetermined, the current reading at the first voltage step can now be plotted. If the scale has not yet been determined, it might be wise to just record this value and after a few more increments a scale could be determined and the plotting could take place.

Step 14. After the current is stabilized and a point plotted, turn the microammeter range switch back to the higher range and gradually raise the voltage to the next increment. Allow the current to stabilize the same length of time as previously determined. Take a second reading and continue the test, gradually raising the voltage step-by-step and plotting the current versus voltage at each step.

Step 15. Closely watch the rate of change of current at each voltage increment. The curve should indicate a linear (even) rate of change up to the final test voltage. If this is the case, the material being tested is all right. Upon any indication of a rapid rise in current, stop the test immediately, unless, of course, you wish to test to breakdown. Even a slight knee in the curve may be an indication of imminent breakdown.

Step 16. This step-by-step raising of voltage and plotting of current should be continued until the maximum test voltage previously selected is reached. If a current versus time plot is also desired, the time should be noted immediately upon reaching the final voltage step. Then, as the current slowly decays, readings of current and time should be taken at suitable increments and the current versus time curve may be plotted. For a safe insulation system, this curve should indicate a continuous decrease in leakage current with time until such time as the current stabilizes. At no time during this part of the test should there be any increase in current.

Step 17. At the end of the tester test, move the high voltage lever switch to the OFF position; allow the material being tested to discharge either through the internal discharge circuit of the tester or by using a hot stick and rubber gloves and grounding the output.

Step 18. The kilovoltmeter on the tester will give a direct indication of the voltage at the output terminal regardless of whether the tester is on OFF. The cable or equipment being tested should not be touched until the kilovoltmeter reads zero.

Step 19. After the kilovoltmeter reads zero, disconnect all test leads and immediately again connect all cable conductors or equipment terminals to ground and leave grounded. It should be noted that rubber gloves must be worn at all times during this por tion of the test! The dielectric absorption effect in cable in particular, and to some extent in rotating equipment, will tend to restore a dangerous voltage on the disconnected item until it has been adequately grounded for a considerable length of time.

Step 20. The other phases of a multiphase installation may next be tested in a similar manner.

Step 21. In many cases, once the fault occurs, its exact location may be determined without too much difficulty. However, in those cases where a fault occurs in a long run of cable, special fault finding techniques incorporating special testing equipment could be of great assistance.

Go, No-Go Testing

The tests described previously give good indications of the condition, and with proper interpretation may even allow prediction of the expected life of the equipment or cable. But to do the job properly takes time and a certain amount of experience and skill, both in conducting and evaluating the operation.

A much shorter and simpler test may be sufficient in those cases where the contractor, purchaser, or operator is only interested in knowing whether or not the installation meets a specific high-voltage breakdown requirement. That is, will the equipment or cable withstand "X" kilovolts without breakdown?

The procedure to be followed in conducting this Go, No-Go tester test is:

Step 1. Set the microammeter range switch to the highest range.

Step 2. Gradually raise the test voltage at a rate which will keep the charging current below the microammeter's full-scale point. The voltage is raised at this steady rate until the required test value is reached.

Step 3. Maintain the voltage at this value for the length of time required by the specification, or in the absence of any specification, until after the current has stabilized for a minute or more.

Step 4. If the leakage current does not become excessive (remains below the maximum specified value or is low enough to avoid tripping circuit breaker), the equipment passed the test.

Step 5. Failure is indicated by a gradual or abrupt increase of current, sufficient to trip the circuit breaker. It should be noted that this high leakage current should not be confused with a high charging current. High charging current may be minimized by raising the voltage at a slower rate.

Step 6. If the current seems to be rising too high, stop raising the voltage at that point and wait. If the current immediately starts decreasing, the indication is that of charging current, and the test may be continued.

Step 7. If the current holds steady or increases, it is either due to leakage current or to corona current. Methods of minimizing the effect of corona current were explained earlier.

Step 8. At the end of the test, reduce voltage to zero and follow procedures as recommended by the tester manufacturer.

This type of test is not by any means a thorough analysis of cable condition. However, it is a sufficient test and all that is necessary in a great number of cases. For example, the contractor installing the cable must immediately know if the cable and the installation is good. If it is defective, this test will reveal it in the shortest possible time. If the cable or the installation is not good, the test cannot harm it any further and the contractor may proceed to make the necessary repairs. The only information sought in this type of test is that the cable will not break down below a required test voltage level. When applicable, this type of test will save many hours of test time and its usefulness should not be overlooked.

Insulation Resistance Measurements

The dc tester is an invaluable tool for making insulation resistance measurements at higher voltages. Using the same connections and following the same precautions given for high potential testing, raise the voltage to the desired value, and read the microammeter after the current has stabilized (no further decay). Insulation resistance may then be calculated using Ohm's law:

$$R = \frac{E}{I}$$

Where: R is the insulation resistance in megohms

E is the voltage

(kV meter reading × 1000)

I is the current in microamperes

The insulation resistance may be calculated at each voltage step as the leakage current versus voltage test is made. Curves of insulation resistance versus voltage will yield much information to the experienced operator. It also helps to keep records of in sulation resistance at a specific voltage so that they can be compared from test to test or from year to year. This will give a good basis of comparison for proper insulation evaluation.

Insulation resistance may also be measured with standard VIBRO-TESTS. These instruments will give a direct reading of insulation resistance at a 500 volt (or higher, depending on the model) test potential. The use of this instrument should not be confused with a high potential test, since insulation resistance, in general, has little or no direct relationship to dielectric or breakdown strength. For example, an actual void in the insulation may show an exceedingly high value of insulation resistance even though it would permit breakdown at a relatively low voltage. Insulation resistance measurements using the VIBROTEST are of help in determining whether or not the items under test are even suitable for hypotting.

If the insulation resistance is abnormally low, the material is defective and there is no need to make the tester test. If the insulation resistance is high, then the tester test could be made.

Instead of waiting for the current to stabilize before calculating the insulation resistance, readings of voltage and current can be taken at fixed time intervals during the leakage current versus time test, and calculations made of insulation resistance as a function of time. A plot of the insulation resistance versus time is known as the absorption curve and is often used when testing rotating machinery. In a good insulation system, this curve will rise rapidly at first (because of the decaying charging current) and gradually level off (an indication of stabilization of leakage current). The more pronounced the rise, the better the insulation. A relatively flat curve will indicate a moist or dirty insulation.

The ratio of the insulation resistance at the end of ten minutes to the insulation resistance at the end of one minute is called the polarization index. If the value of polarization index is less than two, it is usually an indication of excessive moisture or contamination. Values as high as ten or more are commonly expected when testing large motors or generators.

Caution should be exercised in evaluating the importance of insulation resistance measurements. The absolute value is not critical, but the relative order of magnitude, or the change in value as the test progresses, is significant. Such factors as temperature, humidity, etc., can have a very large effect on the reading.

Summary

Of the nearly limitless variety of cables in use today, five of the most common are:

- Ribbon or tape shielded
- Drain wire shielded
- Cable UniShield®
- Concentric neutral (CN)
- Jacketed concentric neutral (JCN)

Beyond cable shield types, two common configurations are used:

- Single conductor—consisting of one conductor per cable or three cables for a three-phase system.
- Three conductor—consisting of three cables sharing a common jacket.

Despite these visible differences, all power cables are essentially the same, consisting of the following:

- Conductor
- Strand shield
- Insulation
- Insulation shield system (semicon and metallic)
- Jacket

Each component is vital to an optimally performing power cable and must be understood in order to make a dependable splice or termination.

Chapter 14

Anchors and Supports

Until the beginning of this century, an anchor was a piece of wood or a lead plug that was driven into a hole in masonry so that a nail or screw could then be driven into it. However, such anchors were unreliable and were soon replaced by lead and fiber anchors with hollow cores. In fact, *NEC* Section 110-13 specifically states:

> ". . . Electric equipment shall be firmly secured to the surface on which it is mounted. Wooden plugs driven into holes in masonry, concrete, or similar material shall not be used."

Many types of reliable anchors have been developed in recent years — for almost every conceivable job — but choosing just the right one may not always be an easy task. For this reason, selecting the type, size, and number of anchors to be used for a given application requires that you take the following factors into consideration:

- Base material
- Material composition of anchors
- Anchor spacing
- Anchor functioning
- Loading conditions
- Anchor installation torque

Base Material

The strength of the masonry material is a key factor in selecting an anchor. Maximum anchor performance requires that the material in which the anchor is installed can also sustain the load to which the anchor will be subjected. Anchors installed in stone and dense concrete can withstand far greater "pullout loads" than the same anchor installed in lightweight concrete, block, or brick. Medium-heavy to heavy loads cannot be safely installed in soft materials such as stucco, grout, plaster, or plasterboard. Materials should be fully cured prior to anchor installation.

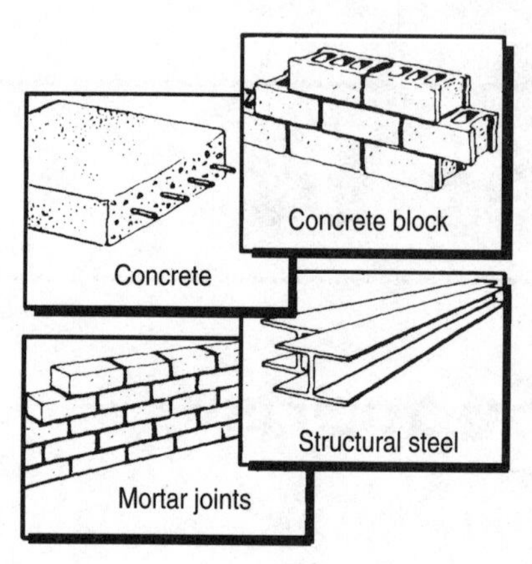

Figure 14-1: Base materials.

In general, base materials are metal, wood, and masonry of various types and hardnesses, such as those shown in Figure 14-1. It is very important that workers be able to determine the suitability of any material into which an anchor or fastener is installed. This fact is doubly important when using powder-actuated fastening systems.

Suitable base materials when pierced by a powder-actuated fastener will expand and/or compress and have sufficient hardness and thickness to produce holding power and not allow the fastener to pass completely through.

Unsuitable base materials can be put into three categories:

1. *Too hard:* Fastener will not be able to penetrate and could possibly deflect or break.

   Examples: Hardened steel, welds, cast steel, marble, spring steel, natural rock, etc.

2. *Too brittle:* Material will crack or shatter and fastener could deflect or pass completely through.

   Examples: Glass, glazed tile, brick, slate, etc.

3. *Too soft:* Material does not have the characteristics to produce holding power and fastener could pass completely through.

   Examples: Wood, plaster, drywall, composition board, plywood, etc.

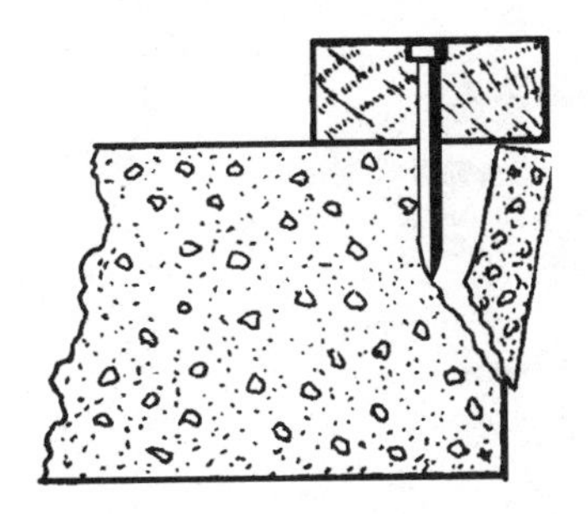

A rule-of-thumb for installing fasteners is do not fasten closer than 3" from edge of masonry.

Figure 14-2: Do not install fasteners too close to the edge of base materials.

The length, width, and thickness of the base material should also be considered. Normally, the width should be at least twice the recommended edge distance. The depth of anchor embedment should not exceed 80 percent of the base material thickness. See Figure 14-2. If masonry cracks, the fastener won't hold and there's a chance a chunk of masonry or the fastener could escape in an unsafe way.

MATERIAL COMPOSITION OF ANCHORS

When reviewing the basic considerations for selection of an anchor, the material from which the anchor is manufactured should be considered. The *Rawl Anchor Selection Guide* lists the basic materials from which each anchor is manufactured. Anchors selected should be manufactured from a material that is suitable for their intended use. Anchors manufactured from a material with a melting point of less than 1000°F are not recommended for overhead applications. Examples of these materials are zinc, lead, plastic, and adhesives. A steel anchor usually meets the requirements of most fire codes. Consideration should also be given to the strength of the anchor material in relation to the applied loads, including bending. The bolts used in conjunction with an anchor should be capable of sustaining the applied loads.

Spacing Recommendations

The load on a masonry anchor is transmitted to the material in which it is installed. Loading of anchors in closely spaced clusters of two or more can result in stress on the masonry material and lead to a reduction in anchor performance. The expansion anchor industry has established a minimum standard of 10 anchor diameters for spacing and 5 anchor diameters for edge distance to provide 100 percent anchor efficiency. These distances may be reduced by as much as 50 percent with a proportionate reduction in efficiency.

WARNING!

Setting fasteners too close together can cause masonry to crack.

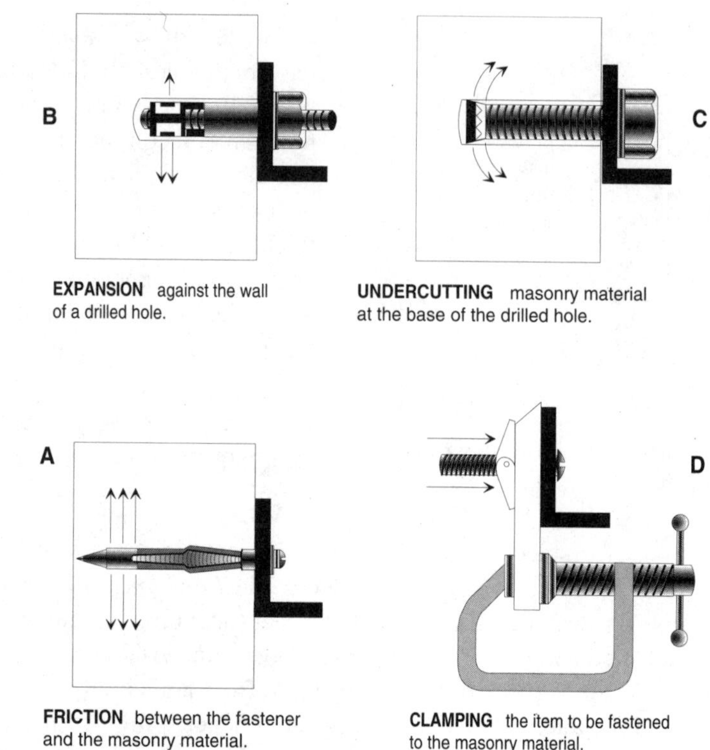

EXPANSION against the wall of a drilled hole.

UNDERCUTTING masonry material at the base of the drilled hole.

FRICTION between the fastener and the masonry material.

CLAMPING the item to be fastened to the masonry material.

Figure 14-3: How masonry anchors work.

Anchor Functioning

Anchor functioning is based on friction, compression, clamping, undercutting, and adhesion. In some cases, a combination of functions is used. Friction fasteners develop their load capacity by creating a friction force between the fastener and the base material. These types of fasteners may be driven into the base material without predrilling and are suitable for light-duty *static loads* (*see* Figure 14-3A).

Most anchors function by developing a compression force (expansion) against the wall of the drilled hole which resists the applied loads. This compression force is generated by a sleeve, ring, or wedge assembly which is actuated by a tapered cone, tapered plug, nail, or screw, depending upon the type (*see* Figure 14-3B).

Undercut anchors expand into the base material at the bottom of the drilled hole. As they undercut the base material, a large load surface is formed which can transfer greater loads to the base material. This type of anchor can generally sustain the greatest loads (*see* Figure 14-3C).

Clamping type anchors are used for hollow base materials. Tension loads are sustained by spreading the load over a large surface in the hollow walls, while shear is resisted by the friction developed between the fixture and the base material (*see* Figure 14-3D).

Loading Conditions

The holding power required should be calculated to include not only the load factor, but also the way the load is transmitted to the anchor. The conditions of use should be reviewed. Will the installed anchors be used to support a static load, or will they be subjected to vibration and/or shock loads? *See* Figure 14-4 for a description of each type of load condition. Trainees should review each item closely.

Anchor Installation Torque

Manufacturers publish recommended installation torque values based on standard product applications. The use of fixture coatings, lubrication of anchor components with sealants, strength of the base material, and other factors that can affect the torque or tension should be considered. The installation torque value used should be adjusted to fit the specific application.

As an anchor is expanded by applying a specified torque, a tension force is induced in the anchor. This tension force governs the load versus the deflection characteristics of the anchor. As load is applied, the induced tension force is relieved. When vibratory loads are applied to an anchor, it is standard practice to induce a tension force in the anchor that is equal to 1.5 to 2.0 times the working load.

To assist in the selection and specification of anchors, manufacturers provide load-test data for those products most frequently used in structural fastening and other critical applications. Such data is derived from tests performed in accordance with ANSI/ASTM Standard E488. A typical test setup device for direct axial loading of an embedded anchor as govered by the ASTM standard is shown in Figure 14-5. Detailed information about each specific anchor can be found in the product descriptions developed by the manufacturer.

Because of varying conditions of field installations, a 4:1 safety factor (25 percent of the ultimate value) is the minimum acceptable industry standard for static loads. Critical applications (vibratory loads, overhead installations, etc.) may require a safety factor of as much as 10:1 or more.

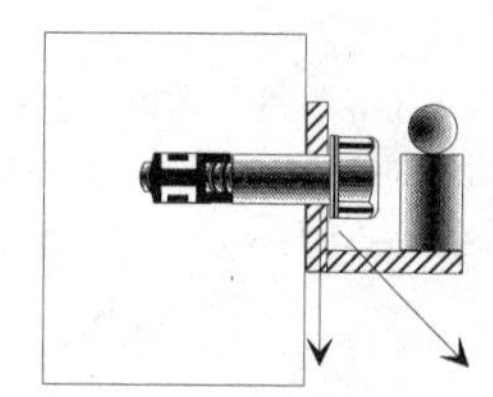

STATIC or Dead Load...constant and unchanging.

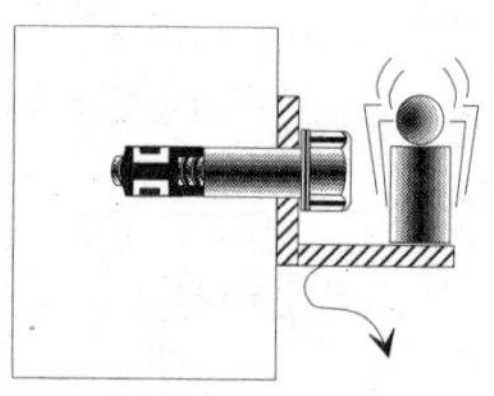

DYNAMIC or Vibration Load... intermittent and/or varying intensity.

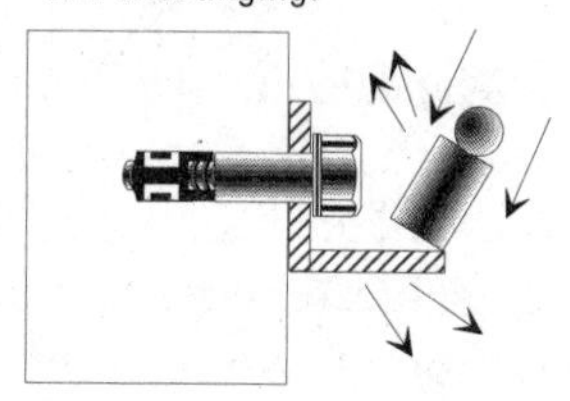

IMPACT or Shock Load...periodic load of substantial intensity.

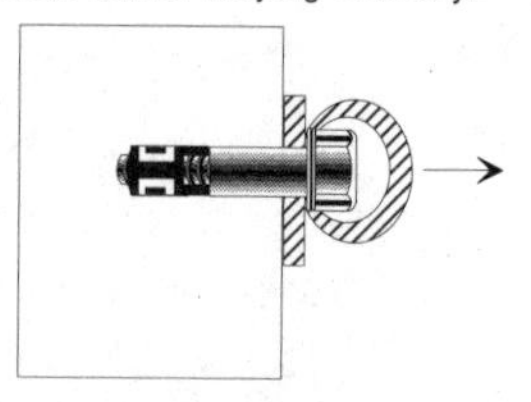

TENSION Load...direct axial load applied to installed anchor.

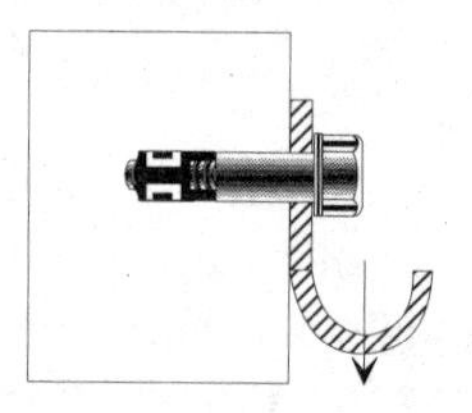

SHEAR Load...a load applied at a right angle to the installed anchor.

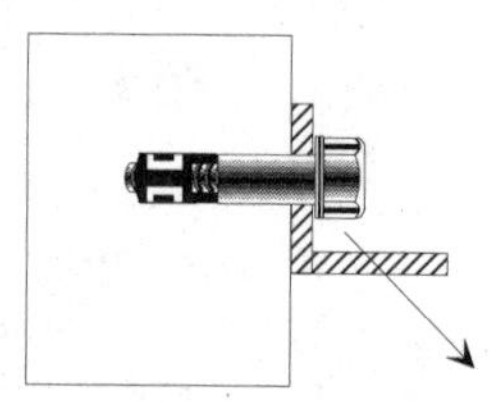

COMBINED Load...a load applied to the anchor at any angle between 0° and 90°.

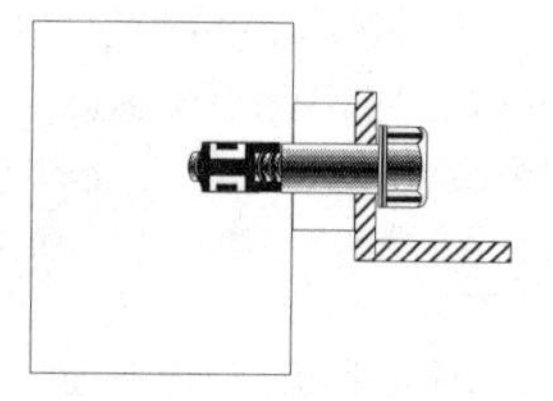

BENDING Load... a load applied at a right angle to the installed anchor at a distance above the base material.

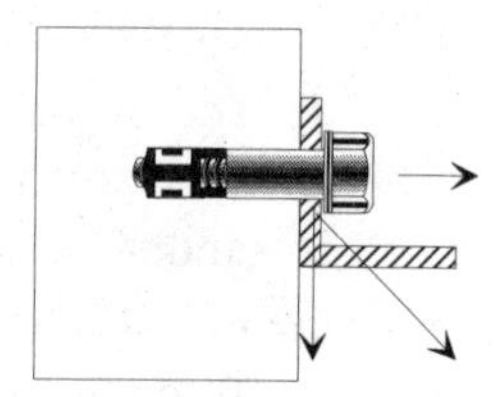

WALL...usually either combined or shear load.

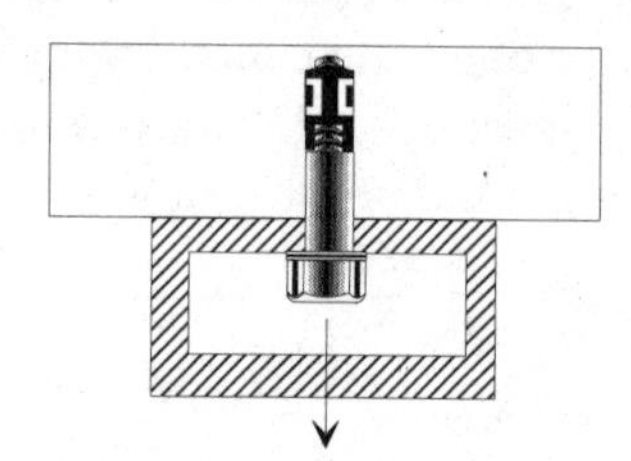

CEILING...usually diret axial (tension) load.

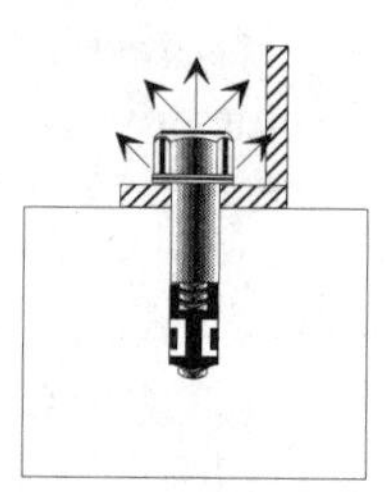

FLOOR...may be tension, shear, or combined load.

Figure 14-4: Load conditions.

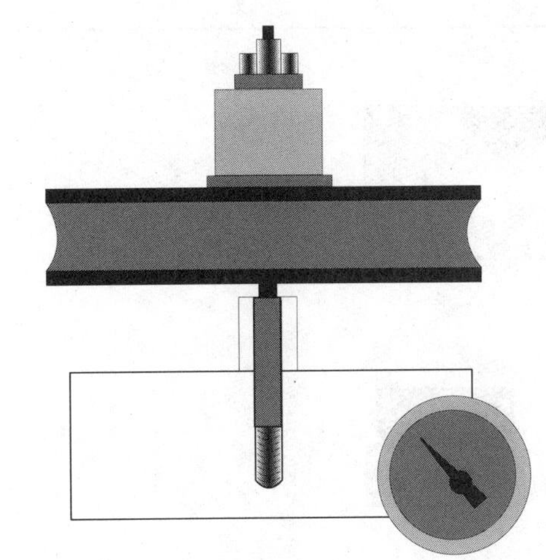

Figure 14-5: Load-capacity tester.

CONCRETE ANCHORING DEVICES

Lead anchors designed for use with machine screws are one of the oldest types of anchors for use in brick, concrete, and concrete block. Since lead is soft and when hit with a hammer, it would easily wedge and form itself to the hole. Unfortunately, this soft material could not withstand heavy loads. Consequently, steel anchors using tapered wedges were developed and, being harder, could handle heavier load requirements. They are known by several trade names, depending upon the manufacturer: calk-in, tap-in, tamp-in, etc. Some require a special setting tool, while others (noncalking types) are merely tapped in a predrilled hole with a hammer and then an inserted machine screw expands the lead against the hole's walls. Those requiring a setting tool are considered more reliable than those not requiring special setting. Rawl's calk-in (Figure 14-6), for example, utilizes a zinc alloy cone that is chamfered at the top for easy screw starting. A sharp collar prevents the cone from being drawn up through the sleeve when overloads are applied.

Figure 14-6: Rawl's calk-in anchor.

The calk-in anchors range in size from No. 6-32 to ¾-10 and require a predrilled hole. The anchor is inserted into the hole and then a special tool is used to expand the lead portion of the anchor to grip the side walls of the hole. A machine screw or bolt may then be installed into the threaded portion of the anchor. *See* Figure 14-7.

This type of lead anchor is capable of holding enormous loads and is normally the preferred type for securing panelboards, safety switches, and similar enclosures to masonry walls. No less than four anchors of the proper size should be used for each switch or panelboard installed.

Another type of calk-in anchor is the multi-calk anchor. This type of anchor is made in short and long styles and both are available either threaded or plain. The multi-calk type anchor is made of zinc alloy material and is used with machine screws as discussed previously. Sizes of the anchor range from ¼ in to ¾ in. The short style is used for hard masonry material while the long style is used for soft masonry material. *See* Figure 14-7.

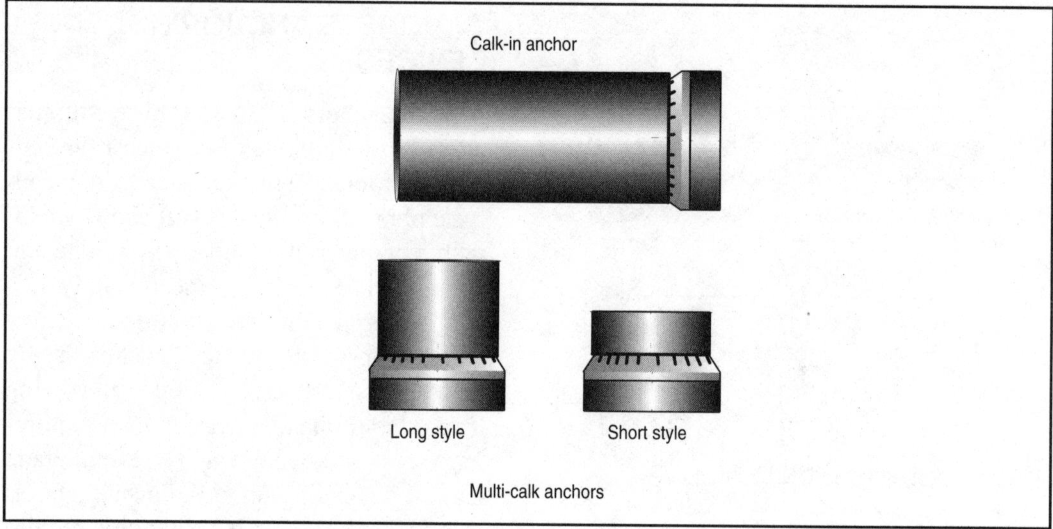

Figure 14-7: Types of Rawl multi-calk anchors.

Self-Drilling Anchors

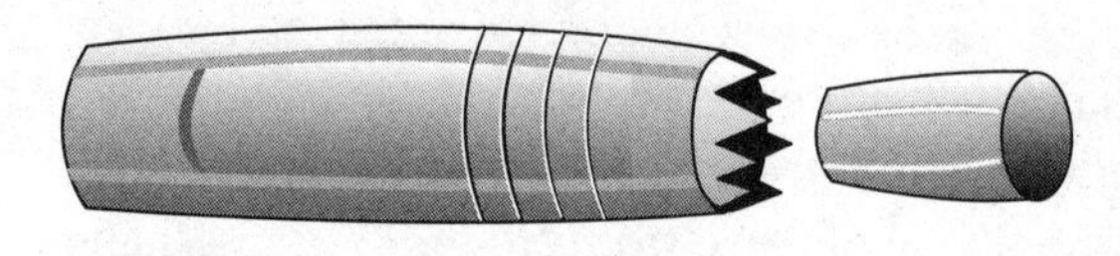

Figure 14-8: Self-drilling anchor.

Drilling with masonry bits or star drills was a slow process and quickly dulled the drilling tools, so the invention of a self-drilling anchor, each one with a fresh set of cutting teeth, was welcomed by trade workers.

The construction of a self-drilling anchor (see Figure 14-8) requires the following:

- A tapered " snap off " shoulder to insert into the drill chuck for positive holding, to be snapped off and discarded after the hole is drilled and the anchor is seated.
- Internal threading to accept a bolt and self-drilling teeth, hardened to resist wear.
- Taper wedge to spread anchor in the drilled hole.

To use self-drilling anchors with hand-rotating chucks and electric hammer, perform the following operations:

Step 1. Mark position of holes.

Step 2. Move equipment.

Step 3. Insert anchor into chuck.

Step 4. Drill hole while rotating hand-held chuck in 180° arc. Continue drilling to full depth of an anchor, and remove anchor.

Step 5. Blow out cuttings. Air blowing bulbs are available, or a simple tube can be inserted in the hole and then blow on the other end with the mouth.

Step 6. Insert taper plug into anchor.

Step 7. Using hammer, seat anchor firmly on taper plug.

Step 8. With firm, sharp action, pull hammer backwards, causing anchor head to snap off.

Step 9. Using drift pin and hammer, remove disposable snap-off end.

Step 10. Reposition equipment.

Step 11. Insert bolts.

Later model hammers with self-rotating features have eliminated manual rotation (*see* Figure 14-9), but all other steps remain the same.

Drop-In Anchors

Once fast-drilling rotary hammers were developed, holes were being drilled faster and anchor manufacturers developed drop-in anchors to fill those holes faster.

Since holes could be drilled faster with a rotary hammer than with self-drilling anchors, the self-drilling anchor took a different shape, eliminating the tapered snap-off end and the self-drilling teeth. The basic anchor and holding power remained the same. Since this requires less time to manufacture, it also costs less. It is much faster to drill a hole and hand-set this anchor than it was using a self-drilling anchor.

To install the drop-in anchor, proceed as follows:

Step 1. Drill hole to exact bolt size. Drill deeper than needed to insert anchor. (Overdrilling eliminates need to clean out remaining

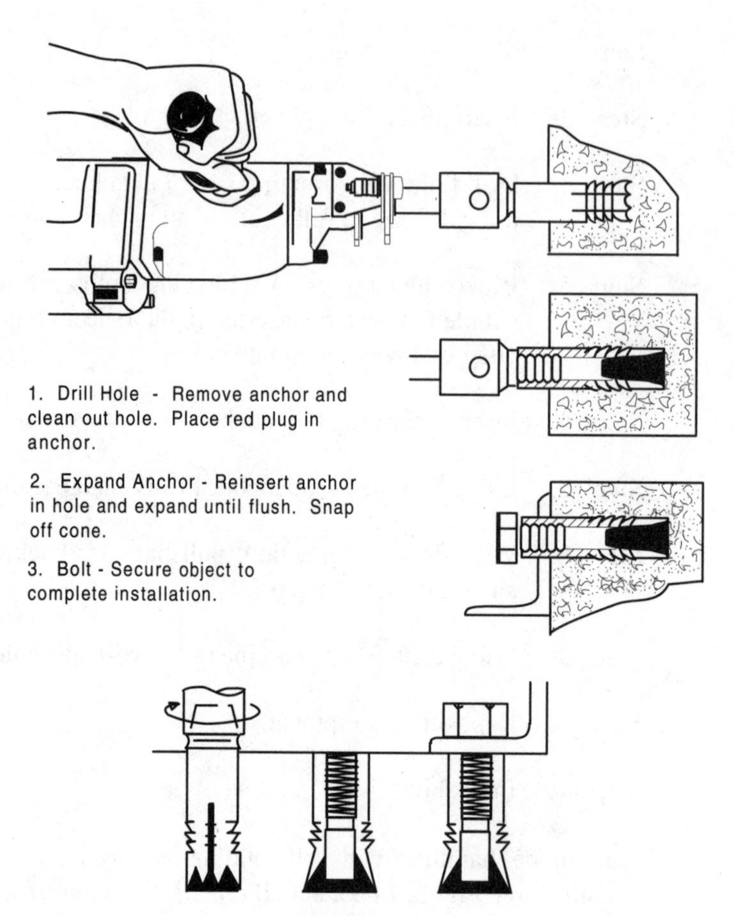

Figure 14-9: Procedures for using self-drilling anchors.

concrete dust. Also, if equipment is moved later, remove nut and drive anchor completely below surface of concrete.)

Step 2. Insert anchor. Tap with hammer if required.

Step 3. Tighten nut.

When drilling holes for drop-in anchors, try to drill the holes with the equipment to be secured in place. Doing so will ensure that the anchor hole will always line up with the equipment. Methods of installing various types of drop-in anchors are shown in Figures 14-10 and 14-11.

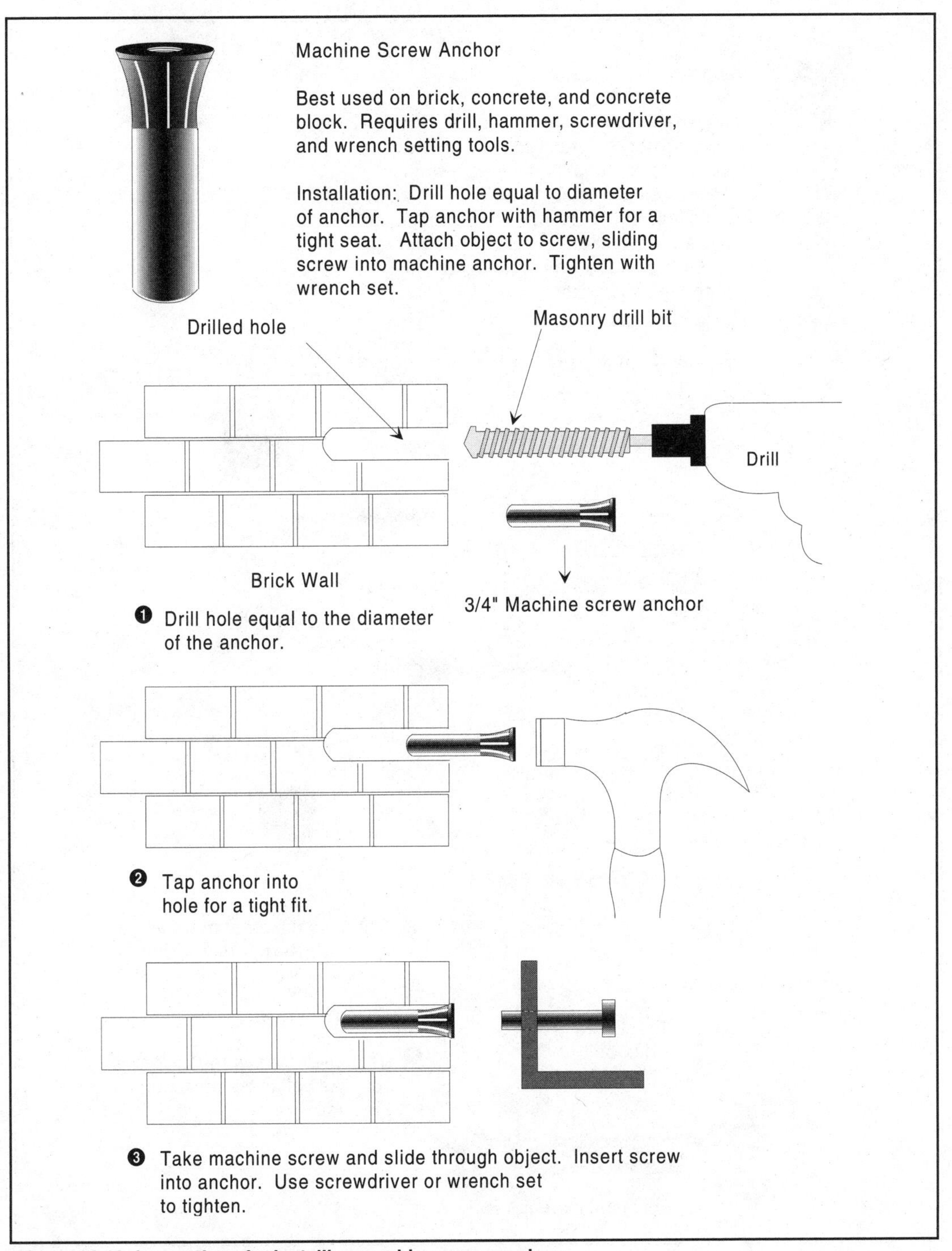

Figure 14-10: Instructions for installing machine screw anchors.

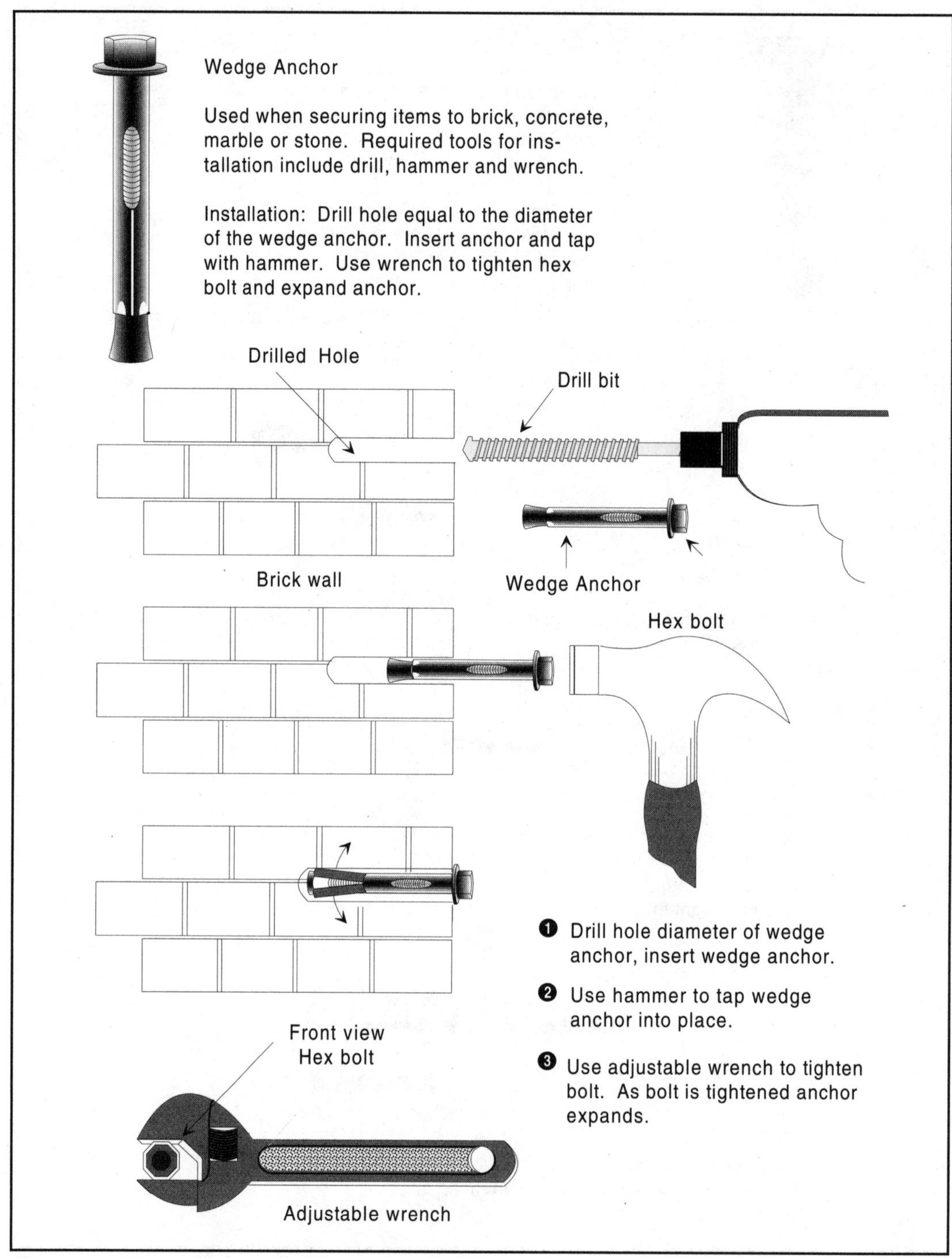

Figure 14-11: Instructions for installing wedge anchors.

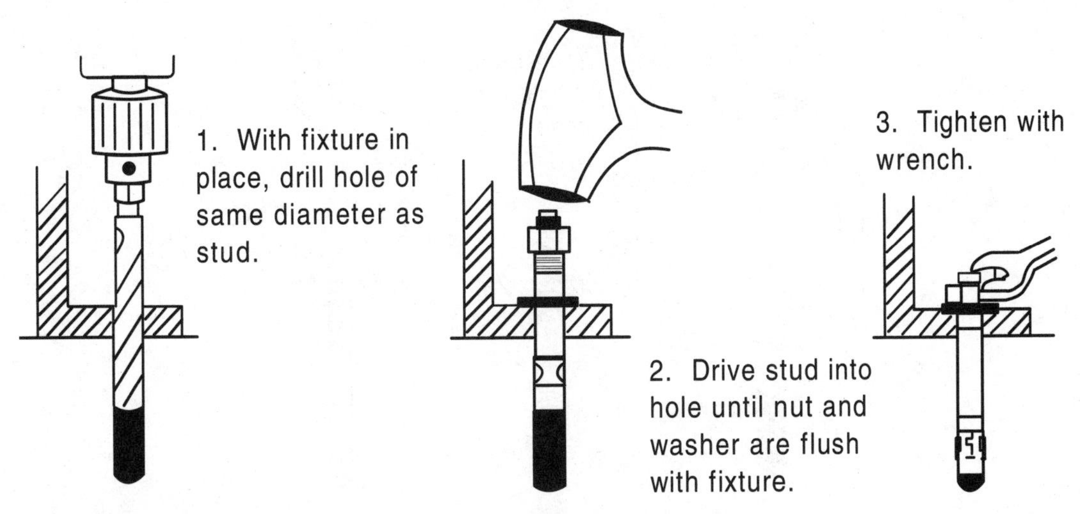

Figure 14-12: Installation of masonry expansion stud.

Masonry Expansion Stud

The masonry expansion stud is a one-piece anchor bolt available in standard low-carbon steel or stainless steel for installations in highly corrosive environments. The dual pressure expansion collar distributes the load equally in the lateral planes of the concrete — preventing bolt cocking or premature rupture of the concrete due to distortion or uneven distribution of the load.

Masonry expansion studs range in size from $\frac{1}{4} \times 1\frac{3}{4}$ to $1\frac{1}{4} \times 12$ in and are used in heavy-duty construction work. Stainless steel anchors are used in food processing plants, chemical plants, and other environments where the stud may be exposed to corrosive elements. *See* Figure 14-12 for typical installation details of masonry expansion studs.

Masonry Expansion Bolt

The masonry expansion bolt is a single-unit, vibration-resistant, removable anchor bolt assembly with a finished hex or flat head design. As this anchor is driven into the hole, the slotted, over-sized annular ring on the bottom of the cone is compressed until it mates with the hole. Expansion occurs as the cone is pulled into the large expansion sleeve, developing a mid-level, load-bearing capacity over a large surface area.

Further turning of the hand bolt causes the threaded bolt to advance into the threads at the compressed end of the cone, forcing the four sections of

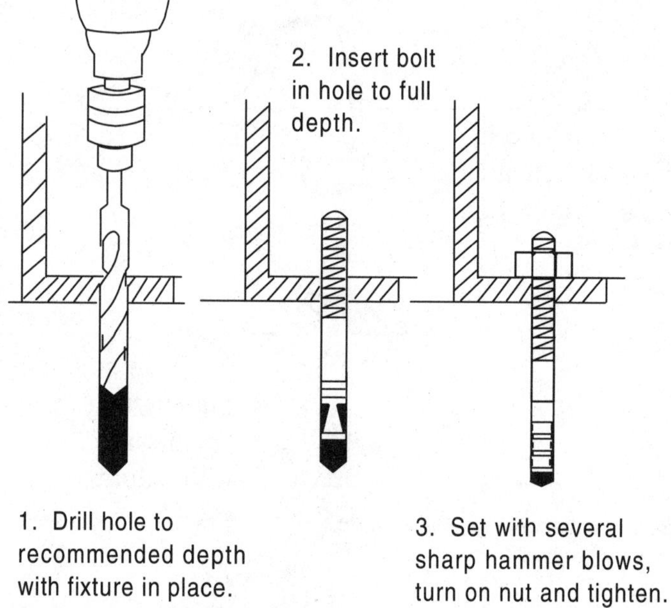

Figure 14-13: Installation instructions for a masonry expansion bolt.

the cone outward, driving them into the masonry. This action develops a lower-level undercut load-bearing capacity deep in the hole over a full 360° area, which greatly increases the holding power of the anchor. Anchors of this type range in size from 1/4 in × 1 3/4 in to 3/4 in × 8 1/4 in and are a good choice for heavy loading capacities, even under severe vibratory conditions. The hole drilled to accept the bolt must be as specified to ensure a proper fit. *See* Figure 14-13.

Single-Bolt Anchor

A Rawl Single is a standard non-calking machine bolt anchor made of rustproof zinc alloy. As the bolt is tightened, the threaded cone is drawn up into the expansion shield which develops a wedging action deep in the masonry where the masonry material is the strongest. The Rawl Single ranges in size from 1/4 in to 3/4 in and requires a predrilled hole before the anchor may be inserted. A machine screw may then be used to secure the fixture, cabinet, and the like. *See* Figure 14-14.

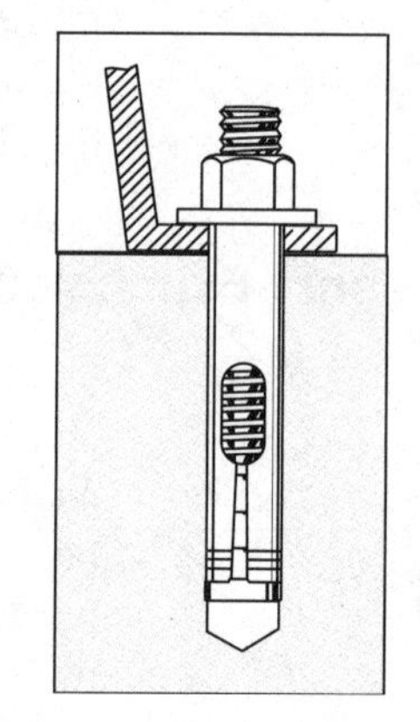

Figure 14-14: Single-bolt anchor.

Double-Bolt Anchor

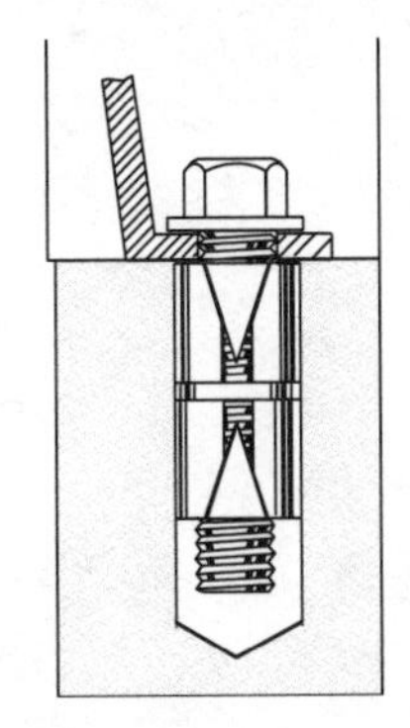

Figure 14-15: Double-bolt anchor.

The double-bolt anchor is designed for use in masonry materials of questionable strength. Setting the anchor causes opposing wedges at either end to be drawn tightly into the anchor, providing full-length expansion against the walls of the hole.

The top wedge-shaped cone acts as a bearing sleeve for the bolt when subjected to shear loads. The double is used in concrete, brick, and stone. Double-bolt anchors range in size from ¼ in to 1 in and require layout or hole-spotting. Double anchors are made of zinc alloy material, except for the ⅞ in and 1 in sizes which are made of malleable iron. Double-bolt anchors are used in conjunction with machine screws and bolts. *See* Figure 14-15.

EXPANSION ANCHORS

Expansion-screw type anchors are designed for use in either hollow or solid walls, depending upon the type. Most types are designed to be installed flush with the surface of the base material. Some are used in conjunction with sheet metal, wood, or lag screws, while others are self-contained devices.

Lag Shield

A lag shield is designed to be used with a lag bolt and is generally used in a mortar joint or concrete. The lag shield is made of zinc alloy material and ranges in size from ¼ in to ¾ in. The lag shield requires a predrilled hole before the anchor is inserted.

The short style lag-shield is used for hard masonry material and the long style lag shield is used for soft masonry material. When installing the lag shield, the anchor is inserted into a predrilled hole until it is flush with the surface of the base material. The fixture is positioned and a lag bolt is inserted and tightened. Bolt length should be selected to assure full thread engagement in the anchor. *See* Figure 14-16 for installation details.

Installation details for a similar type anchor — the lag-screw shield — is shown in Figure 14-17.

Rawlplug

The rawlplug is a popular masonry anchor that is designed for use in conjunction with wood, sheet metal, and lag screws. The plug is formed of

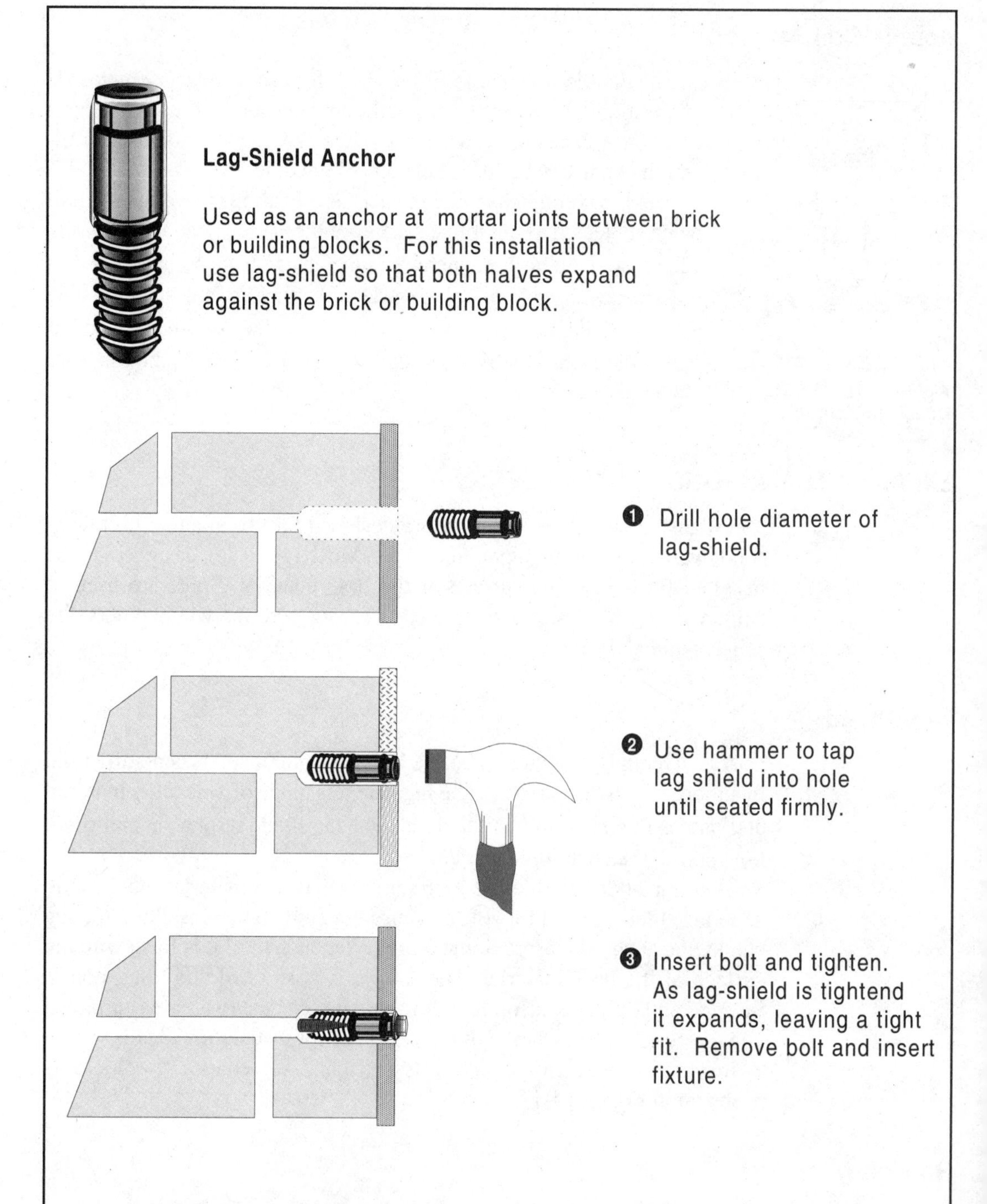

Figure 14-16: Lag shield installation details.

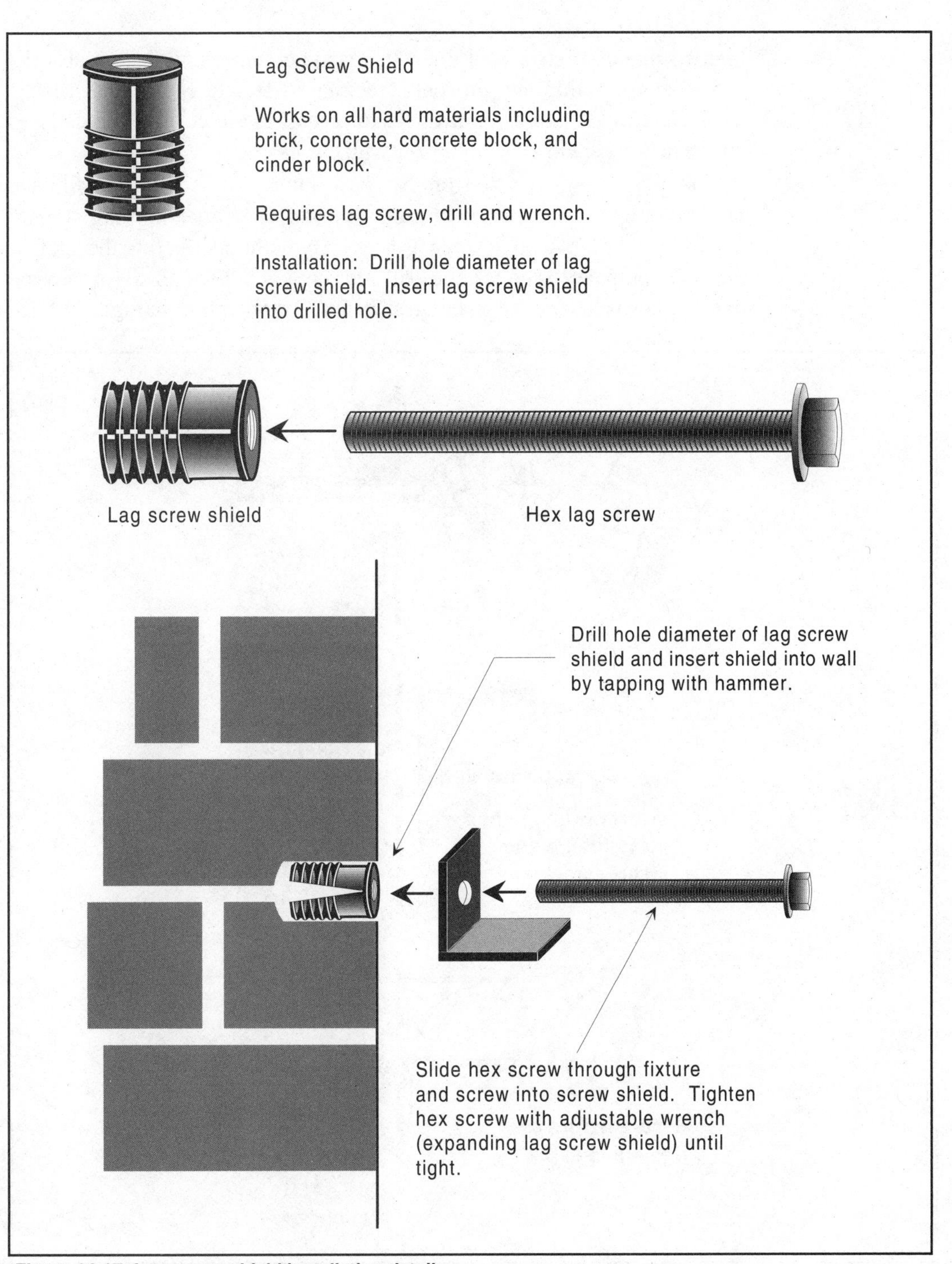

Figure 14-17: Lag-screw shield installation details.

braided jute and has a lead liner. Its elastic-compression design absorbs shock and vibration and prevents cracking of fragile or brittle masonry. Therefore, rawlplugs are a favorite for securing outlet boxes directly to concrete blocks and similar masonry materials.

Rawlplugs range in size from No. 6 × ¾ in to 3 in lengths and all have high holding power. The screw used — wood, sheet metal, or lag — may be removed or replaced as often as necessary without affecting the anchor. For maximum performance, the drill and screw should be sized for the type of rawlplug used. *See* the installation details for rawlplugs in Figure 14-18.

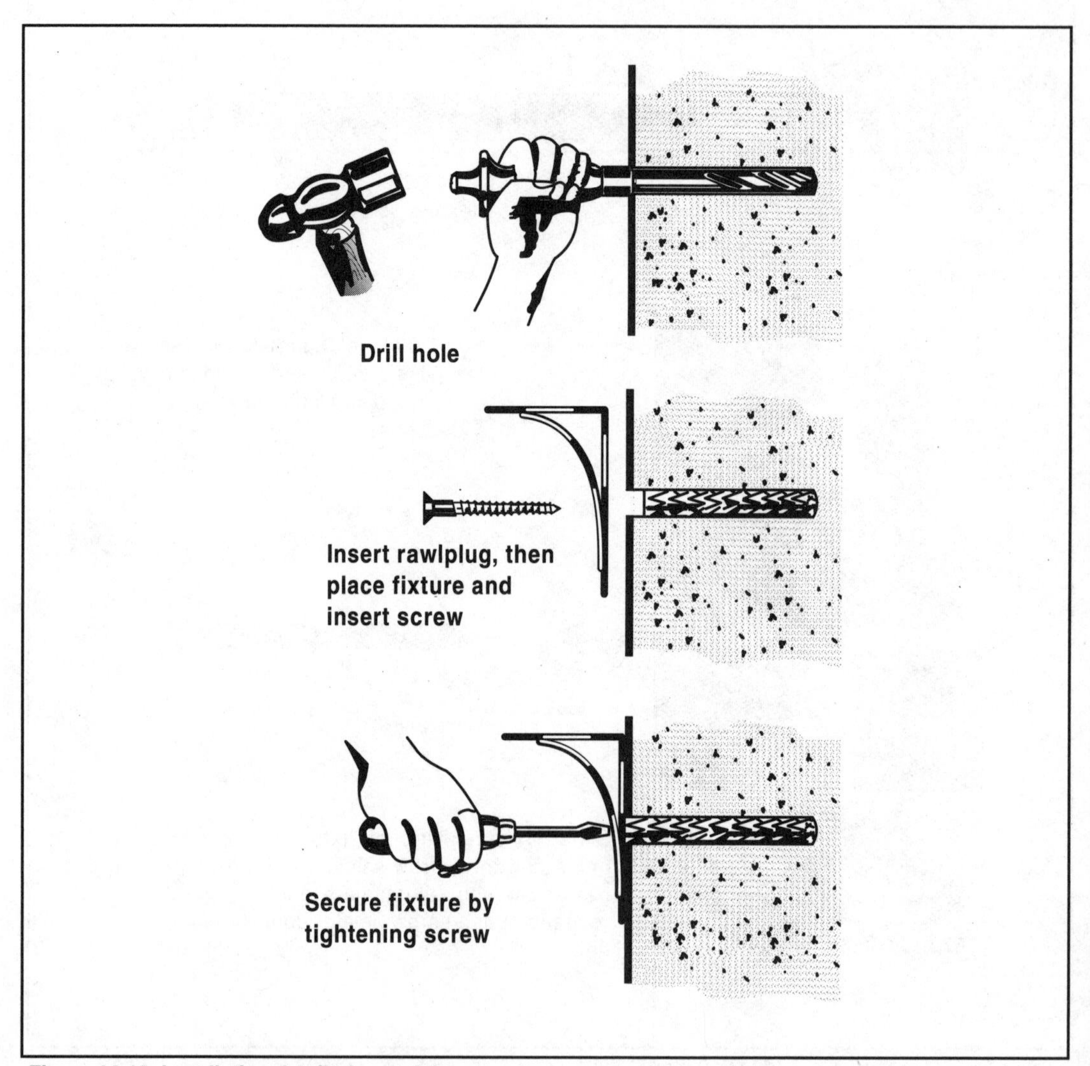

Figure 14-18: Installation details for rawlplugs.

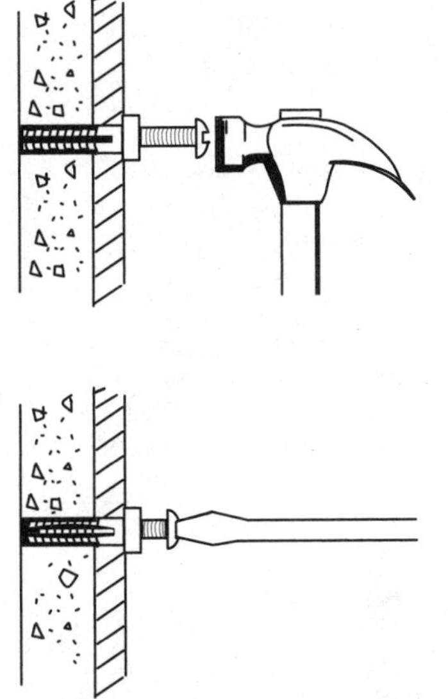

1. Drill hole same size as nylon anchor. Set anchor in hole, insert nail and drive flush with top of anchor.

2. If necessary to remove fixture unscrew the threaded nail and anchor body can be easily removed.

Note: For hollow wall installation follow the same steps as above.

Figure 14-19: Installation instructions for nylon anchors.

Nylon Anchors

The nylon anchor is a multisize lightweight anchor for use in conjunction with wood and sheet metal screws. This type of anchor is economical and can be used in a variety of light-duty applications from wallboard to concrete. The flange makes this type of anchor easy to install in hollow walls.

Nylon anchors range in size from No. 6-8 × $^{3}/_{4}$ in to 14-16 × $1^{1}/_{2}$ in and requires predrilling. This type of anchor is recommended for light-duty static applications where holding power is not a critical factor. *See* Figure 14-19.

Molly Bolt Anchor

The molly bolt anchor is a multisize lightweight plastic fastener for light-duty applications. It is an economical and easy to use hollow-wall anchor. In addition to the short, medium, and long anchors, a specially designed mini-molly has been introduced for use in $^{1}/_{8}$-in hollow doors, an extra long for use in $^{3}/_{4}$ in wallboard, and a super long for use in 1 in wallboard or similar surfaces.

Molly bolt anchors range in size from $^{1}/_{8}$ in to 1 in and accepts No. 6 through No. 12 machine screws. Holes must be drilled prior to installing the anchor. The molly bolt anchor is recommended for light-duty static applications where holding power is not a critical factor. *See* Figure 14-20.

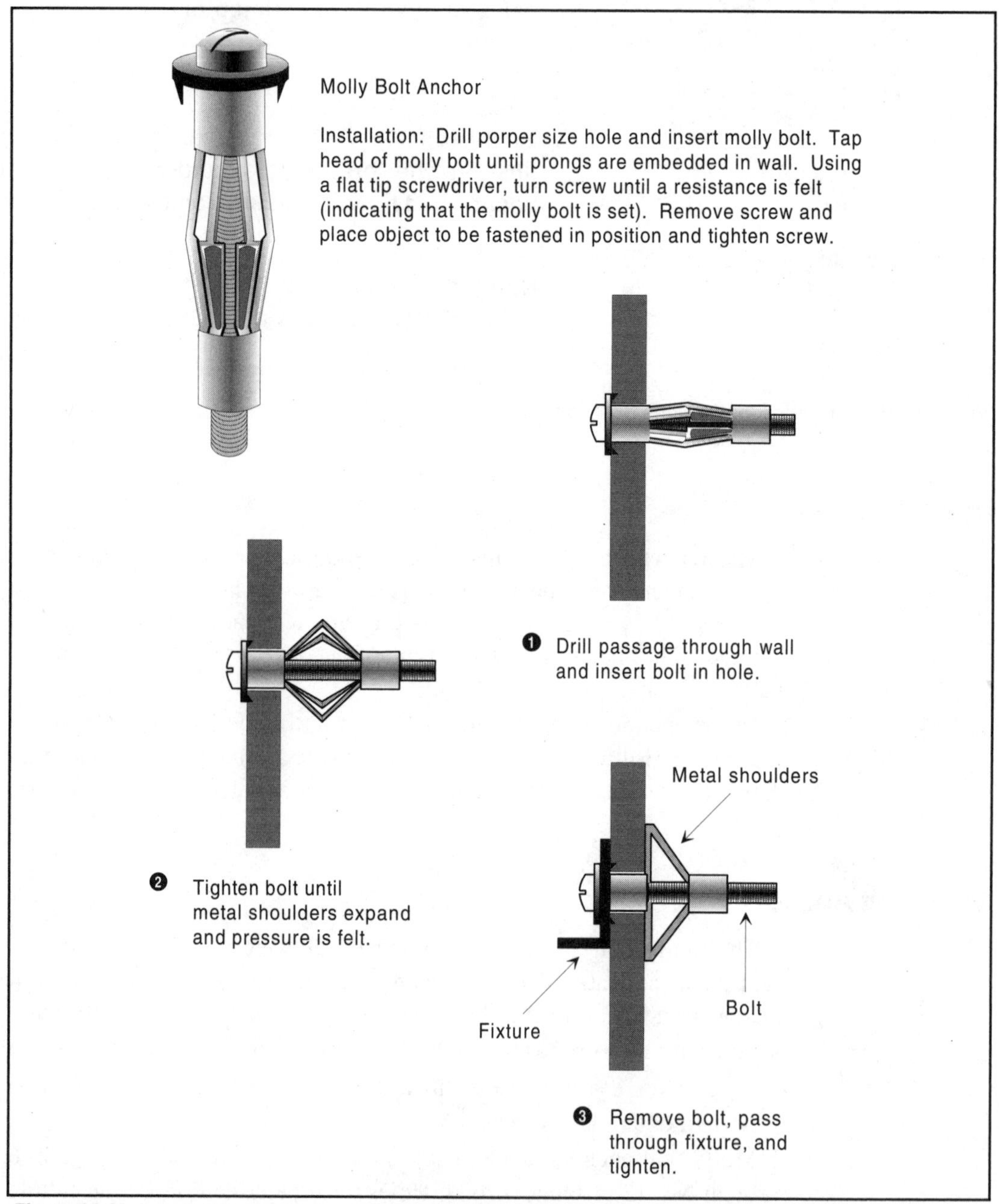

Figure 14-20: Installation details for molly bolt anchors.

Toggle Bolts

Toggle bolts are used to fasten a part or fixture to a hollow wall or panel. Toggle bolts have a hinged and spring-acting mechanism attached to a standard nut. A machine screw is threaded part way into the nut and then the hinged mechanism is squeezed together and inserted into the opening. The spring action then extends the mechanism, and when the machine screw is tightened, the mechanism holds tight against the back of the wall or panel. *See* Figure 14-21. Installation instructions for a hammer bolt anchor and a drive anchor are shown in Figures 14-22 and 14-23, respectively. Installation instructions for a gravity toggle bolt are shown in Figure 14-24.

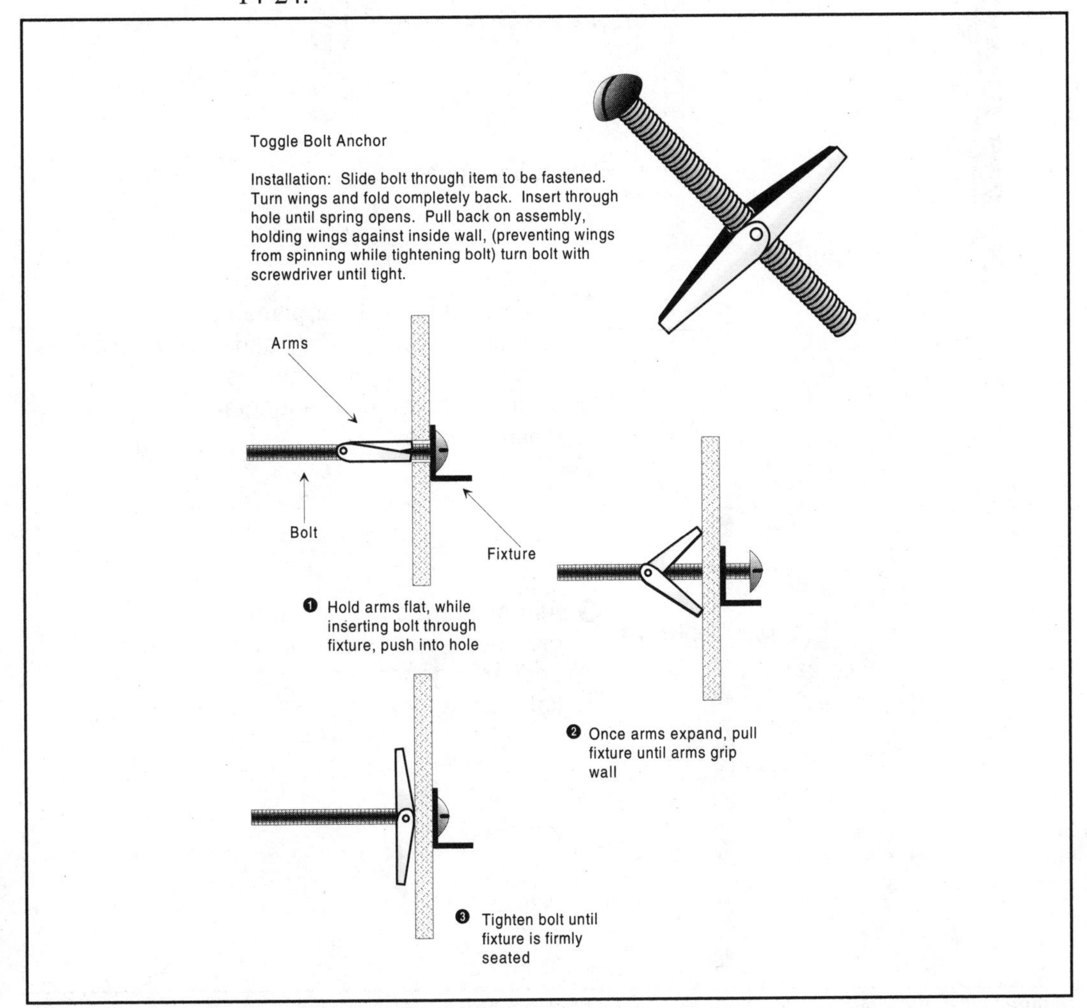

Figure 14-21: Application of toggle bolt anchor.

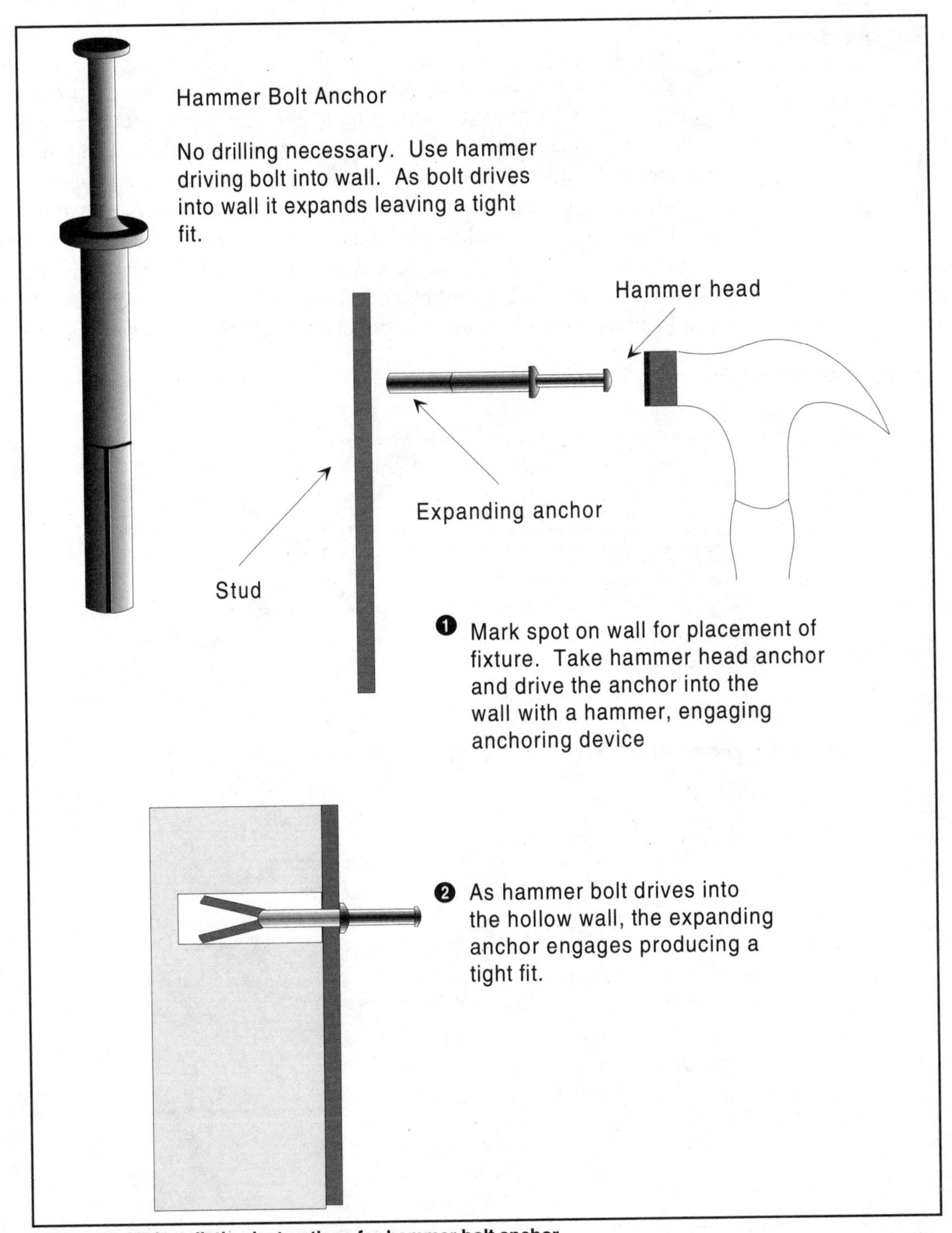

Figure 14-22: Installation instructions for hammer bolt anchor.

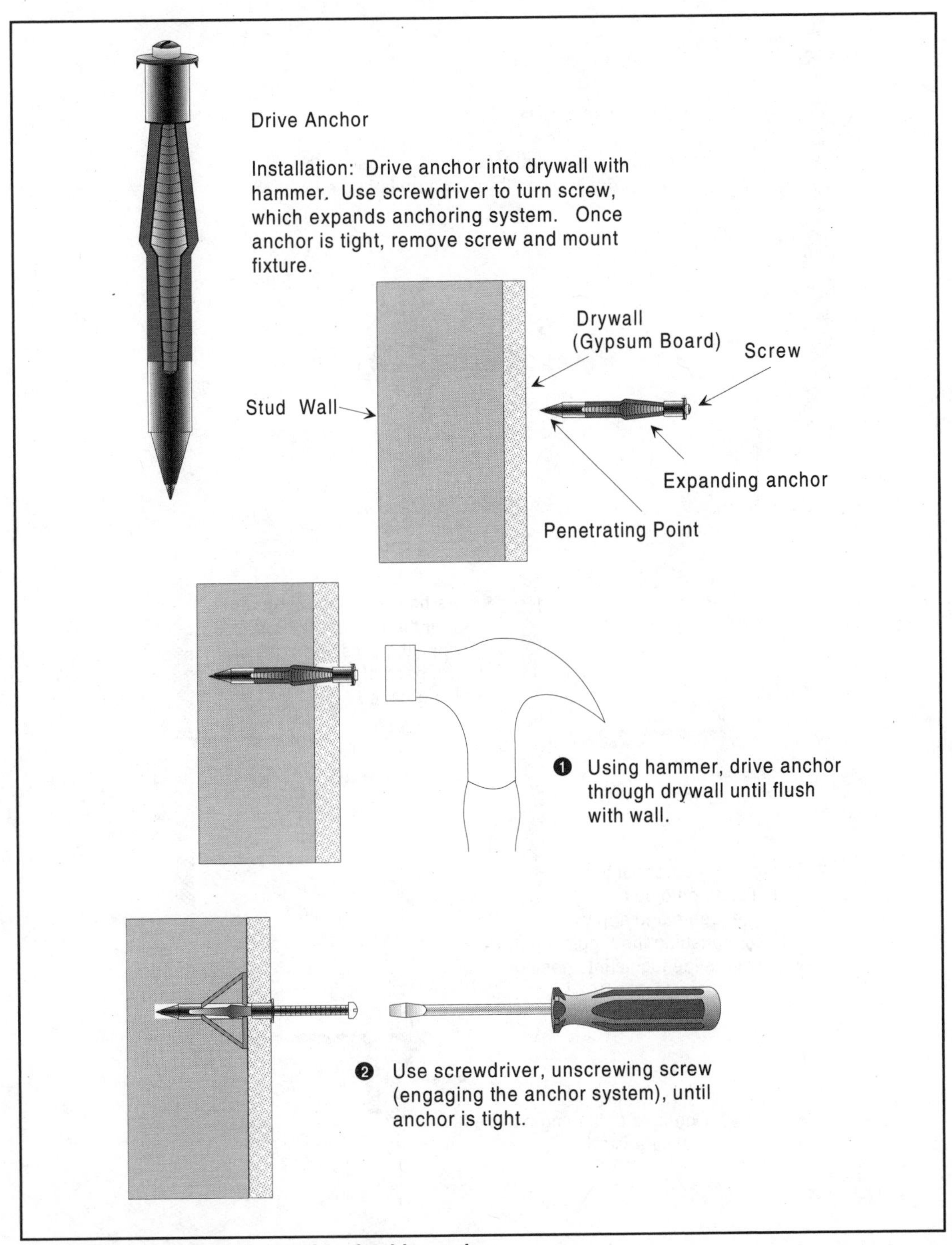

Figure 14-23: Installation instructions for drive anchor.

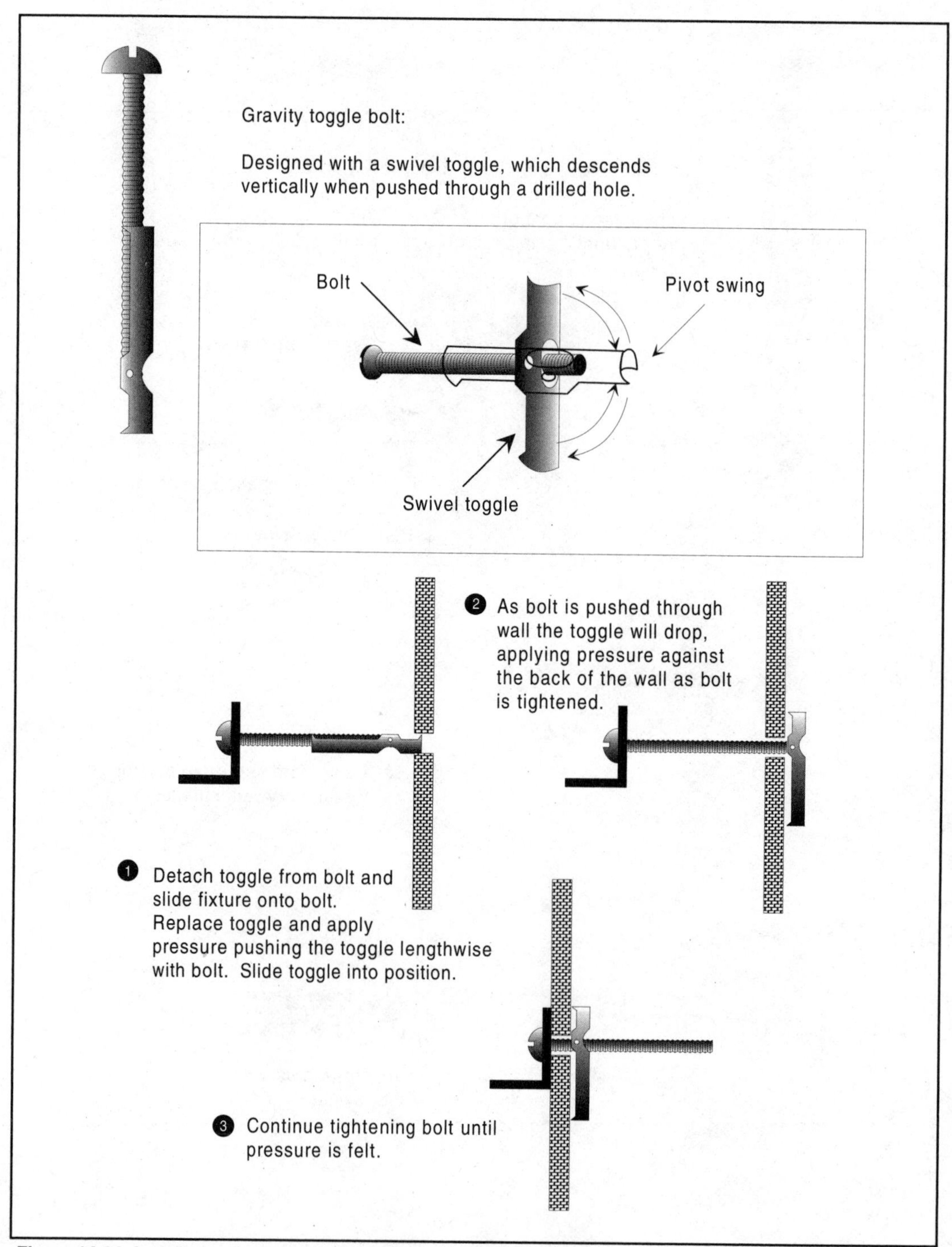

Figure 14-24: Installation instructions for a gravity toggle bolt.

CONCRETE SCREWS

One of the unique fastening systems to come along in recent years is a concrete screw called TAPCON®. The screw itself is the "complete anchor," although a "system" is recommended for ease of installation. The holding power is adequate for many applications that do not involve excessive loads.

The TAPCON System

The TAPCON concrete fastening system consists of the TAPCON anchor, designed to tap threads in concrete, and a CONDRIVE® installation tool that both drills the required pilot hole and drives the anchor. The system provides a combination of high-speed installation and exceptional holding strength. TAPCON anchors are available in a variety of diameters, head styles and lengths to accommodate the fastening of a wide range of components to such materials as:

- Poured concrete—concrete block—cinder block
- Brick—precast panels—stone
- Lightweight concrete—mortar

Typical applications for TAPCON anchors include outlet boxes, conduit straps, insulation, fluorescent fixtures, furring strips, joist hangers, and many other components. *See* Figure 14-25.

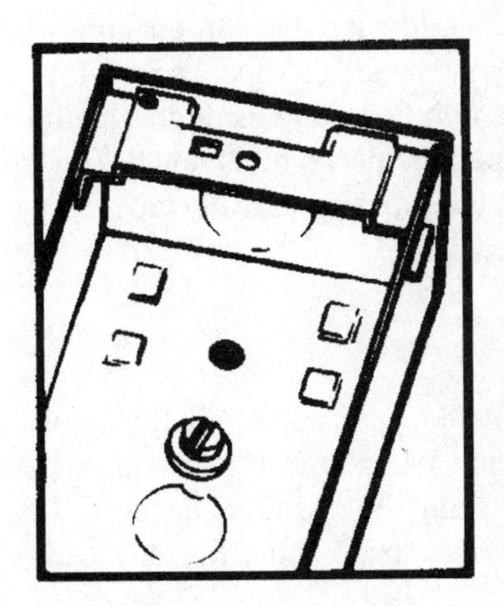

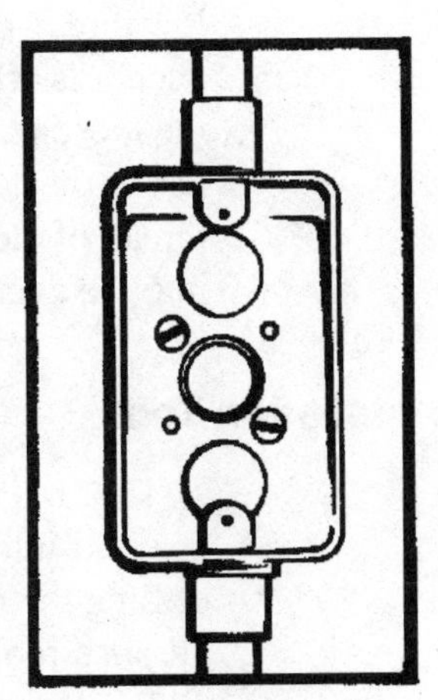

Figure 14-25: Outlet boxes, conduit, and fittings are easily installed with TAPCON anchors.

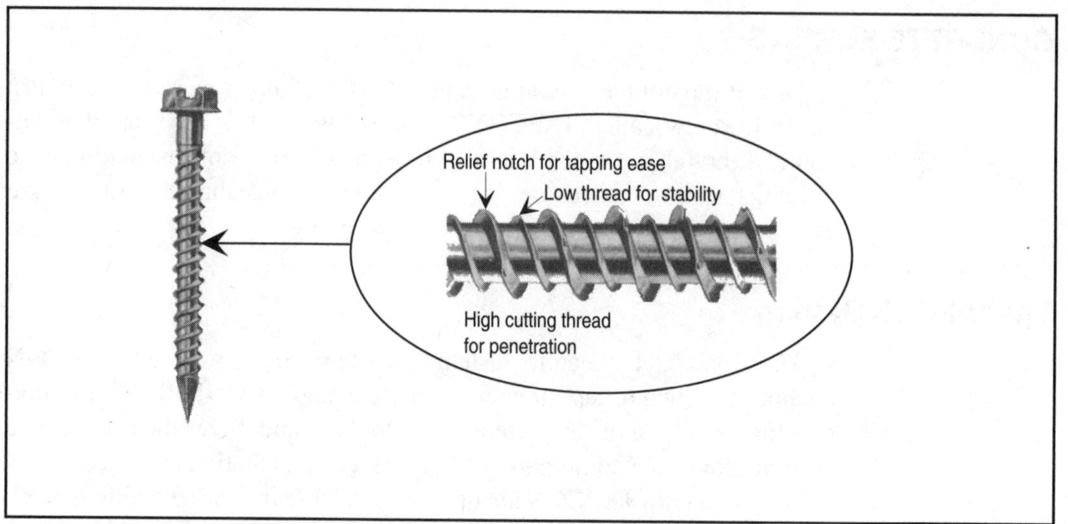

Figure 14-26: Thread characteristics of TAPCON anchors.

The specially heat treated TAPCON anchor has a Hi-Lo thread form, tapered and notched to create a "broaching" action, that actually cuts threads in masonry materials. The TAPCON anchor provides a lot more than a friction fit. It produces a positive mating of anchor and masonry material that generates consistently high pullout strength. No inserts of any kind are needed; therefore there is no need to prespot holes.

The high, sharp thread on a TAPCON anchor displaces a minimal amount of material while cutting into it deeply and easily. The alternating low thread on the anchor provides stability and prevents cocking when the anchor is driven into the hole. The low thread also applies pressure against the loose material, packing it more tightly around the high, flat threads to add to the holding strength of the anchor. Special relief notches on the high thread of the anchor act as cutting edges to progressively tap a deep thread into the concrete. *See* Figure 14-26.

The Installation Tool

The TAPCON system is completed with an installation tool that is inserted into the chuck of a standard hammer/drill and is used both for drilling the required hole and for driving the TAPCON anchors. This dual purpose tool contributes greatly to the speed and efficiency with which TAPCON anchors are installed.

A drill bit is included with each box of TAPCON anchors to insure that a properly sized hole is drilled every time.

There are two basic tools available in the system:

- The CONDRIVE 2000 provides a highly automated means for rapid, repetitive hole drilling and anchor installation. After drilling the hole, the outer sleeve is quickly and easily slid forward to the drive position for anchor installation. The entire operation is completed with a single tool. The CONDRIVE 2000 is designed specifically for installing TAPCON anchors having a slotted hex head.
- The CONDRIVE 1000 is a multipurpose tool with components that can be used to drive hex head or Phillips flat head TAPCON anchors. After drilling is completed, this tool requires the attachment of a separate outer drive sleeve over the carbide tipped drill bit for anchor installation.

To use either the 2000 or 1000 tool, proceed as follows:

Step 1. Rotate the tool sleeve and pull it back to expose the drill bit.

Step 2. Drill the hole deep enough so anchor will not bottom.

Step 3. Slide sleeve forward over drill bit to expose hex socket and rotate sleeve to lock position. Then place head of TAPCON anchor in socket, insert point in drilled hole and drive anchors.

Small anchoring devices have made hammer drills popular, since it is now possible to use small ¼ in and ⅜ in anchors in places that used to require much larger holes. High-speed hammer drills can drill small diameter holes in less time than it takes to read this sentence.

Anchor manufacturers will continue to develop new exotic anchors that work faster and have more holding abilities, and each one will require a hole made by a high-speed hammer.

SCREWS

Self-tapping screws are normally used for fastening sheet metal to sheet metal or sheet metal to steel structures.

Installation of self-tapping screws requires two operations:

- Predrill hole to recommended size.

- Insert and drive screw into the predrilled hole. The hardened screw threads will cut and form a thread in the metal.

Predrilling a pilot hole too small will cause the thread to bind and possibly snap the head off the screw. Too large a pilot hole would limit the holding power of the screw.

Self-Drilling — Self-Fastening Screws

The self-tapping screw represented a big breakthrough in many applications, since it eliminated the need to tap the predrilled hole. However, the ultimate screw would be one that could drill its own hole. That obviously would not only require that each screw have its own bit, but also that it be inexpensive to make it practical for general use. The "Tek" screw was developed to meet those requirements, and both construction and power screwdrivers underwent a great change. Selecting the right screw and right power screwdriver now permits the operator to:

- Drill the exact size pilot hole through material thicknesses in metals to ½ in thick.
- Tap the hole.
- Seat and disengage the screw at the proper depth (positive clutch driver), and also at desired (inch-pounds) torque setting (adjustable clutch driver).

The advantages are obvious. This would now eliminate:

- Drill motor — drill bit — time required for this operation.
- Tap gun (most often done with an impact wrench) — taps — time required for this operation.
- The possibility of material shifting and misaligning with predrilled holes.

While the Tek screw initially cost more than self-tapping screws, the cost saving in both equipment and time more than offset the difference.

Screws are case hardened, giving the drill point the ability to drill most materials. Special stainless steel and aluminum screws are available for special applications.

Speed is very important, since the primary requirement is to drill a hole fast and efficiently. These screws are available in ¼ in diameter or less;

therefore you should select the same high speed that is recommended for a ¼ in H.S. drill bit. A standard of 2500 RPM has been selected for most Tek applications. Strangely enough, Tek screws perform poorly at slow speeds, resulting in frequent drill point failure.

Think of this screw as a drill bit. Firm pressure is desirable, but excessive pressure will only result in drill point failure and operator fatigue.

The pitch on the drill point is designed to remove more material per revolution than a H.S. drill bit. The reasons are obvious.

- A high-speed ¼ in is expected to drill 100 or more holes.
- A ¼ in Tek screw is expected to drill one hole.

Result: The Tek will drill much faster (reducing labor costs) than the equivalent high-speed drill bit.

Since the Tek always drills the exact size pilot hole for its self-tapping thread, it creates a surprisingly tight fit, and requires some effort to remove.

Drilling Characteristics

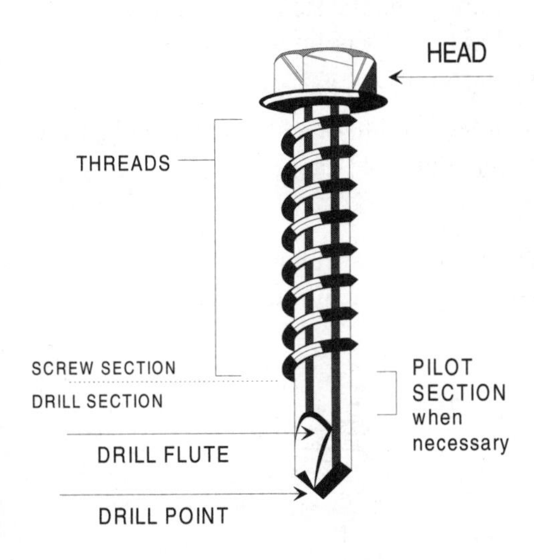

Figure 14-27: Parts of a Tek fastener.

The Tek self-drilling fasteners penetrate metal up to ½ thick. The drill point, the drill flutes (that provide for clearance and removal of the metal chips), and the pilot section are the key performance features of the drilling section. The length of the flutes and the unthreaded pilot section determine the drilling capability of ther fastener. *See* Figure 14-27.

When a Tek drills a ⅛ in hole into steel, for example, maximum efficiency is achieved when the point advances 0.004 in per revolution. And the thickness of the chip being removed is, therefore 0.004 in. When a thread is engaged, it advances one pitch per revolution. The rate of advancement for the No. 8-18 fastener, for example, is 0.055 in per revolution. Therefore, the screw advances 0.055 in per revolution after the thread starts to engage. If drilling hasn't been completed prior to thread engagement, the point will be required to advance 0.055 in and remove a chip 0.055 in thick; this is 14 times as thick as that removed during the drilling operation. Neither the Tek fastener nor the driver is designed for this.

The drilling action must be completed before the first thread engages in the metal and begins to form or cut a thread in the drilled hole. In selecting the proper Tek for the metal to be drilled, the length of the unthreaded Teks point must be equal to or greater than the thickness of the material to be drilled.

It must be understood that when specifying a Tek point for applications involving more than one thickness of material, calculations of the total thickness to be drilled must include voids or spaces between thicknesses.

Other Point Styles

Several self-drilling fasteners described have points designed to meet specialized fastening requirements. *See* Figure 14-28.

- Type W—For fastening drywall to conventional wood framing materials. The sharp point makes a small hole fast, minimizes damage to wood fibers, reduces splitting, and holds securely.
- Type S—Special forged point that fastens dry wall and composition to maximum 22 gauge steel stud with little effort. Extrudes metal at fastener point to provide maximum thread engagement and rigidity.
- Spoiler—Sharp point designed to quickly penetrate drywall and metal stud up to maximum 22 gauge.
- Stitch Teks—Special undersize drill point extrudes metal when fastening two thin sheets, creates higher resistance to the joint.

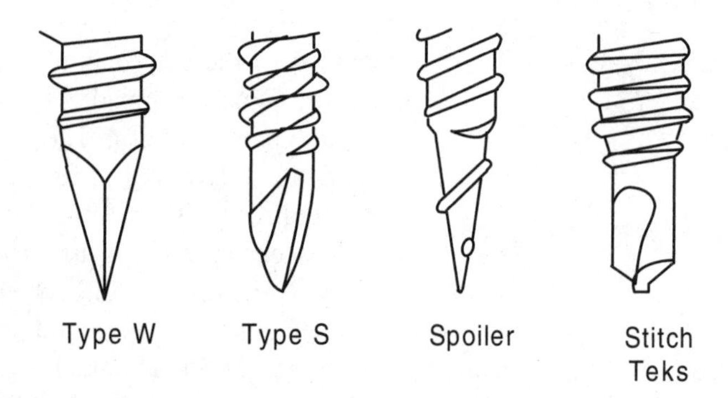

Figure 14-28: Samples of Tek point styles.

Head Styles

A number of fastener head styles are available to meet the varied range of surface and materials requirements found in the construction industry.

- Indented Hex Washer—Washer face provides a bearing surface for driving sockets, with less chance of marring surface due to driver camout.
- Phillips Wafer—Large head provides necessary bearing surface to fasten plywood and other soft materials, seats flush for clean finished appearance.
- Phillips Bugle—Used mainly in drywall applications. Seats flush without crushing materials or tearing paper to simplify filling and taping.
- Phillips Pan — Conventional head for general applications.
- Phillips Flat—Used primarily in wood to countersink and seat flush without causing splintering.
- Low Profile Torx Drive System Recess— Reduces camout by providing positive engagement between fastener and driver bit. Extends bit life.
- Panelmate Nylon Head—Super-tough nylon head molded in colors to match metal panels. Eliminates rust.
- Taper Resistant Torx—Torx drive system recess with center post that discourages tampering.
- Also available with Pozidriv recess, a special recess that reduces camout, end thrust, operator fatigue, and bit wear.
- Color-matched painted heads are available on request.

Thread Forms

A variety of fastener thread forms are available to meet a wide range of applications. *See* Figure 14-29.

- Fine—Used to reduce friction and driving torques; normally used with thicker materials.
- Spaced—Fast installation in thinner materials, high holding power.

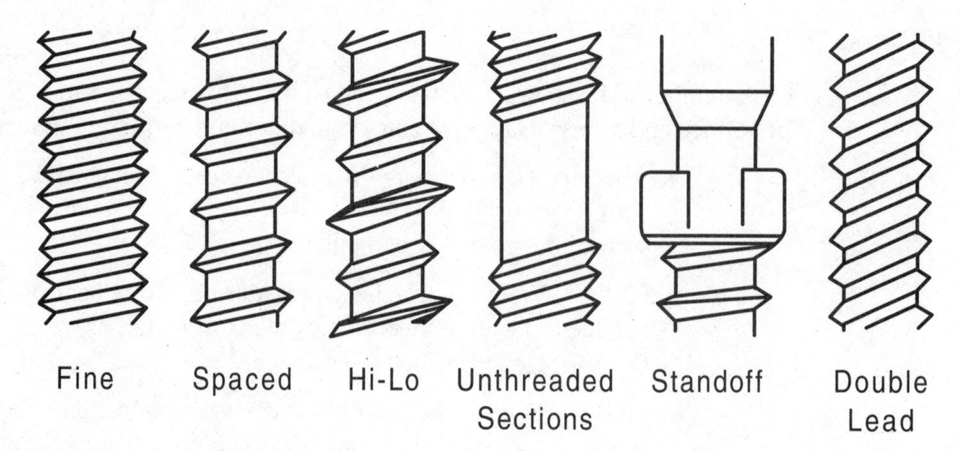

Figure 14-29: Basic thread forms.

- Hi-Lo—Advances rapidly, maintains maximum holding power with low driving torque. Provides driving stability in wood, gypsum, thin studs, and other soft materials.
- Unthreaded Sections—Can be provided to meet special needs.
- Standoff—For positive spacing between outer sheet and structural member when installing rigid insulation. Other designs to provide positive spacing can be produced.
- Double Lead—Two threads for very rapid fastener advancement.

POWDER-ACTUATED FASTENING SYSTEMS

This section provides only the basic knowledge of powder-actuated fastening systems. To become a qualified operator, additional training covering operation, maintenance, and recommended practices for each manufacturer's tool is necessary. Operators should also read and be familiar with any local and state regulations applicable to this type of fastening system.

The operating principle of the powder-actuated tool is to fire a fastener into material and anchor or make it secure to another material. Some applications are wood to concrete, steel to concrete, wood to steel, steel to steel and numerous applications of fastening fixtures, electrical outlet boxes and the like to concrete or steel.

This type of system is very competitive with other fastening systems as discussed previously in this module, and there are areas where it may or may not be preferred.

A powder-actuated tool in simple terms is a "pistol." A pistol fires a bullet composed of two elements:

- Cartridge with firing cap and powder
- Bullet

When you pull the trigger on a pistol, the firing pin detonates the firing cap and powder and sends the bullet in free flight to its destination.

The direct-acting type powder-actuated tool can be described in the same manner as a pistol with one small change. The "bullet" is loaded as two separate parts.

- Cartridge (firing cap and powder)
- Bullet (fastening device)

Figure 14-30 shows the basic operating procedure for using a powder-actuated fastening system; the fastening device inserted into the barrel, and then the correct cartridge is inserted. The tool is pressed against the work surface, the trigger is pulled and the fastener then travels in free flight to

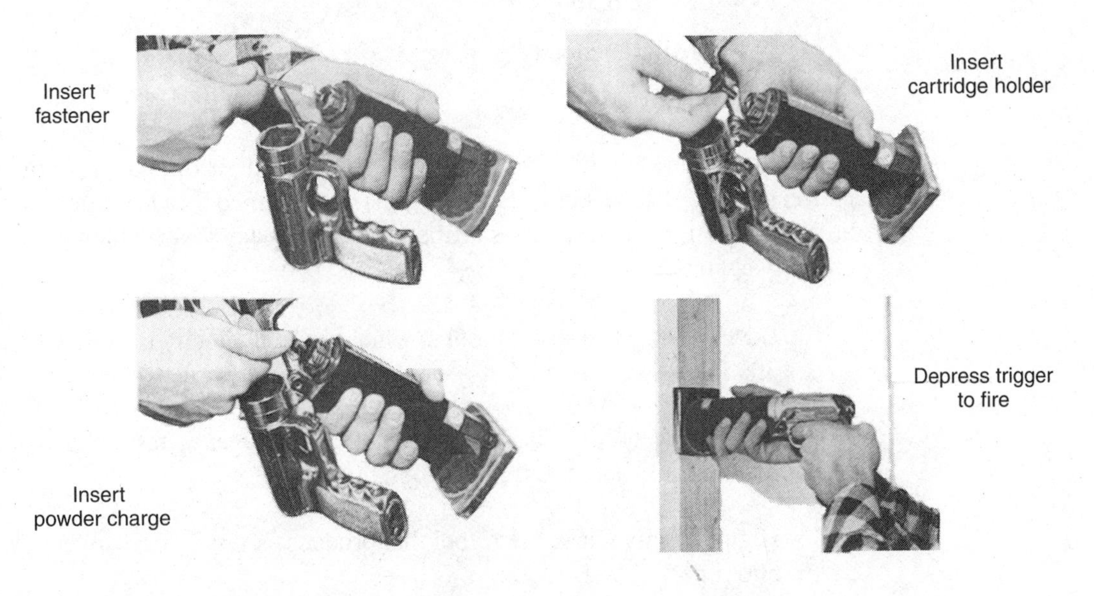

Figure 14-30: Basic operating steps for a powder-actuated fastening system.

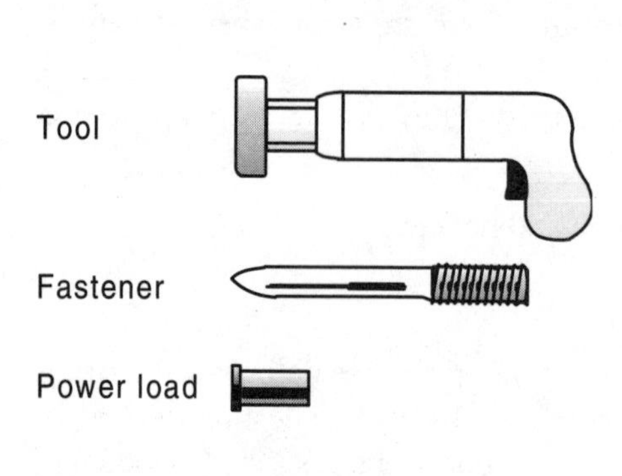

Figure 14-31: Basic parts of a powder-actuated fastening system.

its destination. Although this system is simple to use, there are precautions and safeguards that must be observed.

The system (*see* Figure 14-31) consists of:

- Tool
- Fastener
- Power Load

Tools

The tools are identified by types and classes. There are two types of tools:

Direct-Acting Type — A tool in which the expanding gas of a power load acts directly on the fastener to be driven into the work.

Indirect-Acting Type — A tool in which the expanding gas of a power load acts on a captive piston which in turn drives the fastener into the work.

There are three classes of tools:

- Low velocity
- Medium velocity
- High velocity

Low, medium, and high velocity tools may be either direct-acting or indirect-acting. The velocity class of a tool is determined by a ballistic test utilizing the lightest fastener and the strongest power load which will properly chamber in the tool.

1. Low Velocity Class — A tool in which the test velocity is limited to 300 ft per second.

2. Medium Velocity Class — A tool that produces a test velocity between 300 and 500 ft per second.

3. High Velocity Class — A tool that produces a test velocity over 500 ft per second.

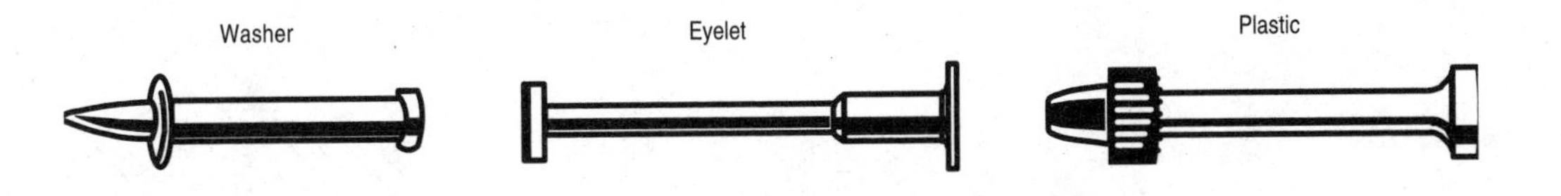

Figure 14-32: Typical alignment tips.

Fasteners

The fasteners used in powder actuated tools are not common nails. They are manufactured from special steel and heat-treated to produce a very hard, yet ductile fastener. These properties are necessary to permit the fastener to penetrate concrete or steel without breaking.

The fastener is equipped with some type of tip, washer, eyelet, or other guide member. This guide aligns the fastener in the tool as it is being driven and is usually used to retain the fastener in the tool. *See* Figure 14-32.

The two basic types of fasteners used are drive pins and threaded studs.

1. Drive Pin — A special nail-like fastener designed to permanently attach one material to another such as wood to concrete or steel. Head diameters are generally ¼ in, 5⁄16 in or ⅜ in. However, for additional head bearing in conjunction with soft materials, washers of various diameters are either fastened through or made a part of the drive-pin assembly. *See* Figure 14-33.

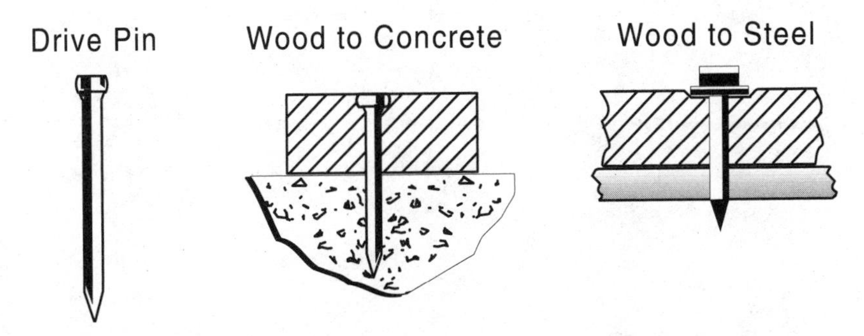

Figure 14-33: Application of powder-actuated drive pins.

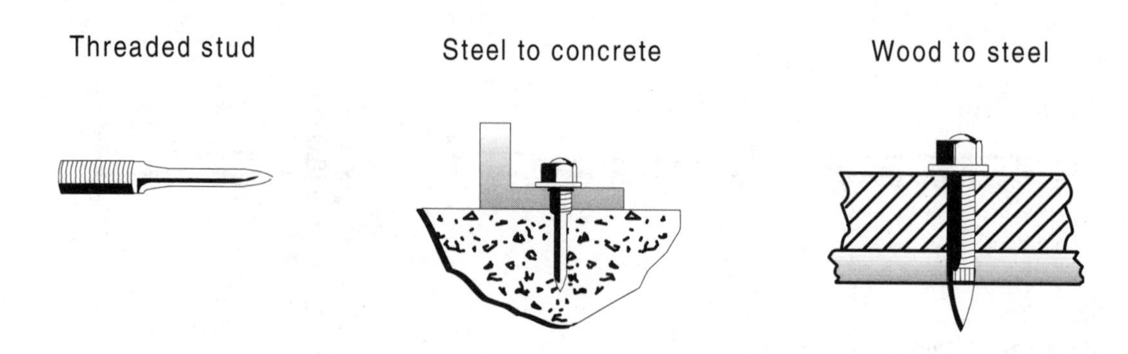

Figure 14-34: Application of threaded studs.

2. Threaded Stud — A fastener composed of a shank portion which is driven into the base material and a threaded portion to which an object can be attached with a nut. Usual thread sizes are 8-32, 10-24, ¼-20, 5⁄16-18, and ⅜-16. *See* Figure 14-34.

There are also other types of special fasteners designed for specific applications.

1. Eye Pin — A fastener with a hole through which wires, chains, etc., can be passed for hanging ceilings, light fixtures, etc. *See* Figure 14-35.

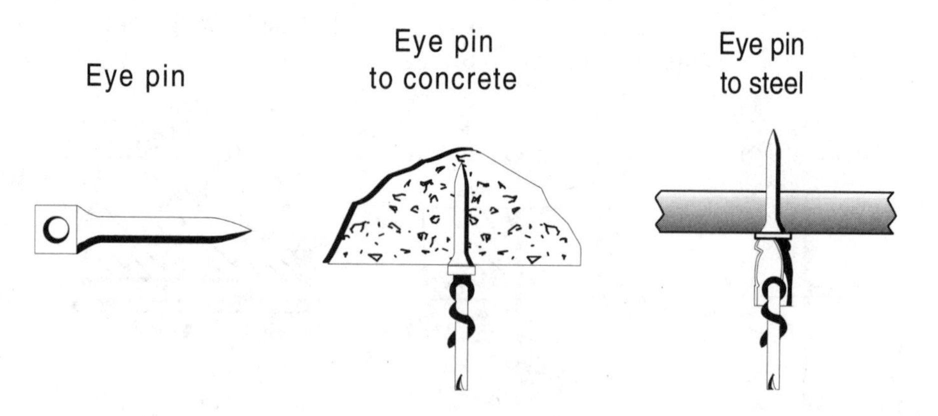

Figure 14-35: Application of eye pin.

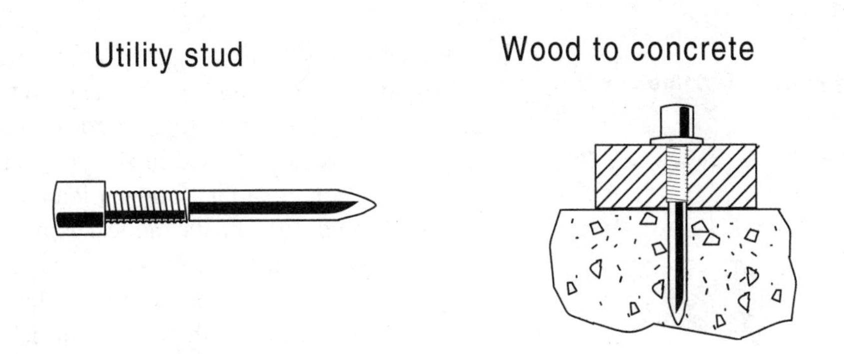

Figure 14-36: Application of utility stud.

2. Utility Stud— A threaded stud with a threaded collar which can be tightened or removed after the fastener has been driven into the work surface. *See* Figure 14-36.

Power Loads

The power load is a unique, portable, self-contained energy source used in powder actuated tools. These power loads are available in two forms: cased or caseless. The propellant in a cased power load is contained in a metallic case.

Cased power loads are available in various sizes ranging from .22 through .38 caliber, and caseless loads are available in various sizes and shapes.

The caseless power load does not have a case and the propellant is in a solid form. *See* Figure 14-37 for various types of power loads.

Types of Power Loads

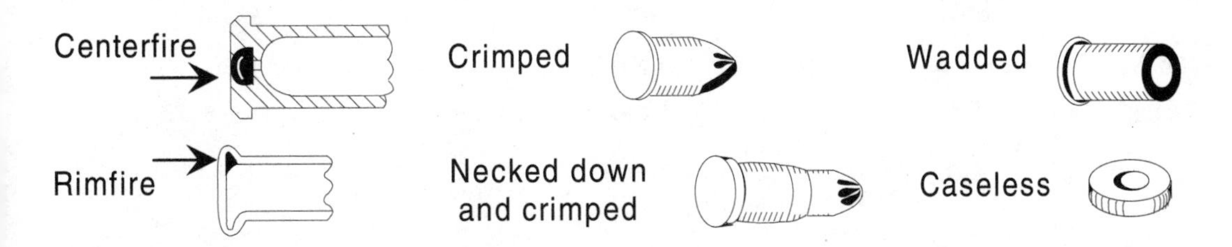

Figure 14-37: Various types of power loads.

Cased Power Load Terminology

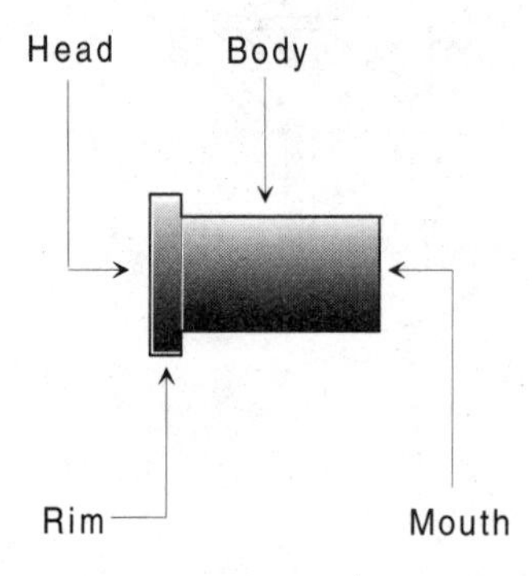

Figure 14-38: Cased powder load terminology.

Regardless of the type, caliber, size or shape, there is a standard number and color code used to identify the power level or strength of all power loads.

Cased power loads used in all types and classes of tools cover a range of 12 power load levels numbered 1 through 12, with the lightest being No. 1 load and the heaviest being No. 12 load. A basic six-color code of gray, brown, green, yellow, red, and purple is used twice because there are not 12 different readily distinguishable permanent colors. Power loads Nos. 1 through 6 are in brass colored cases and power loads Nos. 7 through 12 are in nickel colored cases. It is the combination of the case color and load color that defines the load level or strength. Each cased power load is clearly identified on one end by its power level color. *See* Figure 14-38.

The following chart shows this simple number and color identification code.

Power Color Identification

Power Level	Case Color	Load Color
1	Brass	Gray
2	Brass	Brown
3	Brass	Green
4	Brass	Yellow
5	Brass	Red
6	Brass	Purple
7	Nickel	Gray
8	Nickel	Brown
9	Nickel	Green
10	Nickel	Yellow
11	Nickel	Red
12	Nickel	Purple

Caseless loads are manufactured only in the Nos. 1 through 6 load levels and are color coded with the basic load colors gray, brown, green, yellow, red and purple.

In addition to the identification of the power load, each package is color coded and shows the load level number.

Power Load Selection

In selecting the proper power load to use for any application, it is important to start with the lightest power level recommended for the tool being used. Using the lightest load, if the first test fastener does not penetrate to the desired depth, the next higher power load should be tried. If necessary, continue increasing power levels by single steps until proper penetration is obtained.

Shields and Special Fixtures

Shields and special fixtures are important parts of the powder-actuated fastening system and are used for safety and tool adaption to the job.

Low velocity class tools are supplied with a shield to confine flying particles. This shield should be used whenever fastening directly into a base material such as driving threaded studs or eye pins into steel or concrete. In addition to confining flying particles, the shield also helps hold the tool perpendicular to the work. A shield or special fixture should be used when fastening one material to another whenever the material being fastened does not confine flying particles.

Medium and high velocity class tools are designed so that the tool cannot fire unless a shield or fixture is attached. Some standard shields are adjustable for fastening close to any obstruction. It is important that these adjustable shields be used only where the work provides safety equal to the full shield position. For other special applications, special fixtures should be used in place of the standard shield for the purpose of additional safety and positive fastener location. *See* Figure 14-39.

Summary

To choose the correct anchor or support for a job, avoid damage to parts and tools, and make long-lasting repairs or installations, you should be familiar with various types of anchors and supports. Furthermore, you should have a working knowledge on how they are used on the job site to accomplish your work. Sometimes a particular task could be accomplished

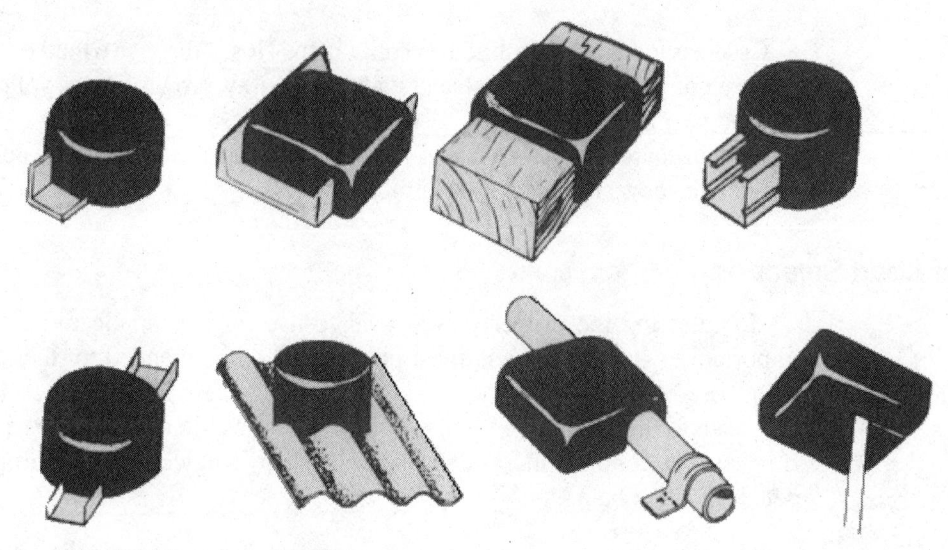

Figure 14-39: Shields and special fixtures.

by using several different types of anchors or supports, although selecting the correct anchor or support for the job maintains quality and consistency throughout the workers' trade. Electricians should ensure that they are familiar with the correct terms used for anchors and supports to avoid confusion and misselection of specific items due to slang terms that are sometimes applied to different types of anchors and supports.

Note: *One of the most important acts that a conscientious user of a powder-actuated fastener system can perform is to see that the tool being used is equipped with the proper safety shield or special fixture to assure both safety and good workmanship.*

Chapter 15

Electric Motors

The principal means of changing electrical energy into mechanical energy or power is the electric motor — ranging in size from small fractional-horsepower, low-voltage motors to the very large, high-voltage synchronous motors.

Electric motors are classified according to the following:

- Size (horsepower)
- Type of application
- Electrical characteristics
- Speed, starting, speed control, and torque characteristics
- Mechanical protection
- Method of cooling

In basic terms, electric motors convert electric energy into the productive power of rotary mechanical force. This capability finds many applications in unlimited ways in industrial establishments for powering machines, HVAC and refrigeration fans and compressors, gasoline and water pumps, power tools, and a host of other applications.

All of these and more represent the scope of electric motor participation in powering and controlling machines and equipment used in industrial establishments.

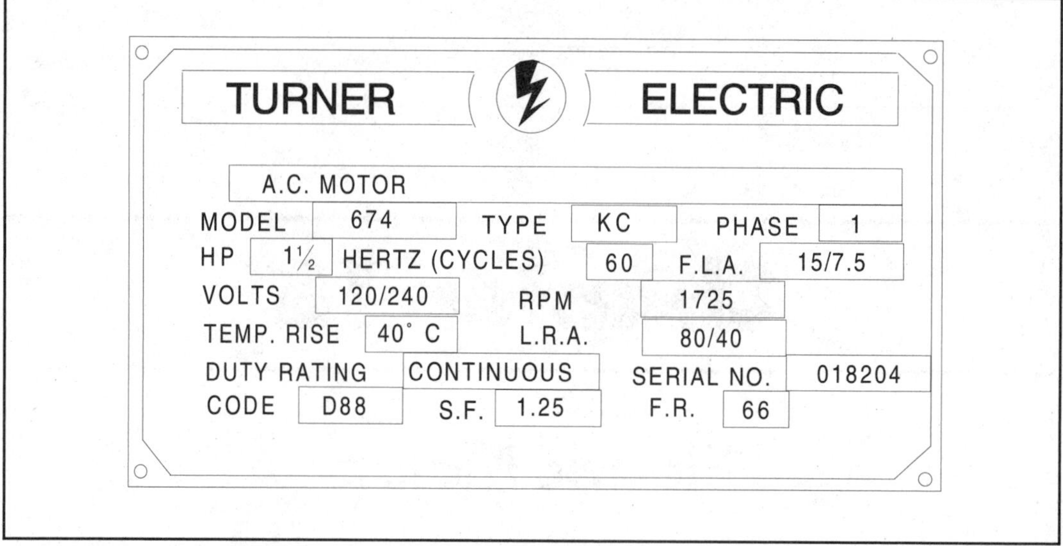

Figure 15-1: Typical motor nameplate.

Electric motors are machines that change electrical energy into mechanical energy. They are rated in horsepower. The attraction and repulsion of the magnetic poles produced by sending current through the armature and field windings cause the armature to rotate. The armature rotation produces a twisting power called torque.

Nameplate Information

A typical motor nameplate is shown in Figure 15-1. A nameplate is one of the most important parts of a motor since it gives the motor's electrical and mechanical characteristics; that is, the horsepower, voltage, rpms, etc. Always refer to the motor's nameplate before connecting it to an electric system. The same is true when performing preventative maintenance or troubleshooting motors.

Referring again to the motor nameplate in Figure 15-1, note that the manufacturer's name and logo is at the top of the plate; these items, of course, will change with each manufacturer. The line directly below the manufacture's name identifies the motor for use on ac systems as opposed to dc or ac-dc systems. The model number identifies that particular motor from any other. The type or class specifies the insulation used to ensure the motor will perform at the rated horsepower and service-factor load. The phase indicates whether the motor has been designed for single- or three-phase use.

Horsepower on the nameplate defines the rated output capacity of the motor; hertz (cycles) indicates the alternating current frequency at which the motor is designed to operate. The F.L.A. section gives the amperes of current the motor draws at full load. When two values are shown on the nameplate, the motor usually has a dual voltage rating. Volts and amps are inversely proportional; the higher the voltage, the lower the amperes, and vice versa. The higher amp value corresponds to the lower voltage rating on the nameplate. Two-speed motors will also show two ampere readings.

Voltage is the electrical potential "pressure" for which the motor is designed. Sometimes two voltages are listed on the nameplate, such as 120/240. In this case the motor is intended for use on either a 120- or 240-V circuit. Special instructions are furnished for connecting the motor for each of the different voltages. For example, Figure 15-2 shows connections for a three-phase motor for use on either 208/240- or 480-V systems.

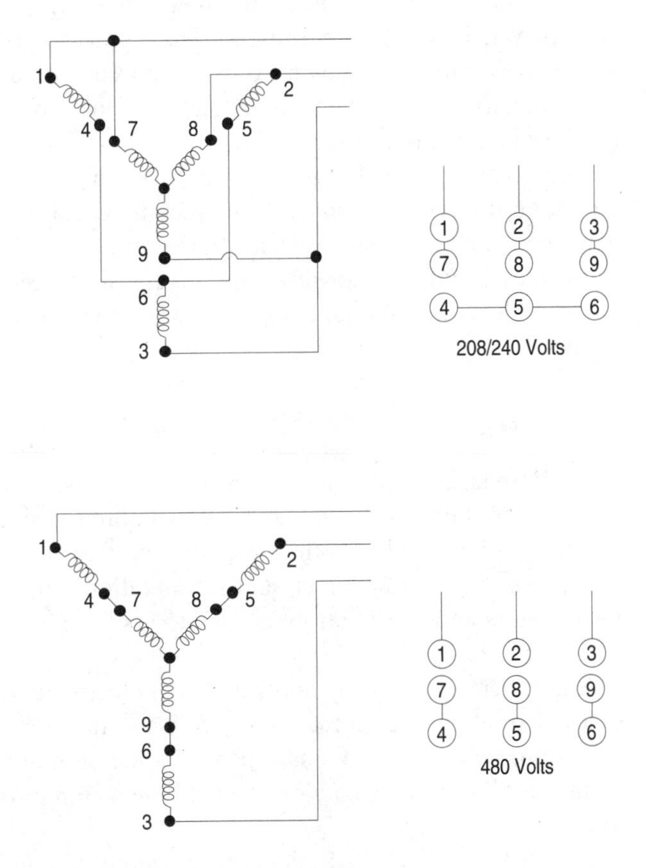

Figure 15-2: Three-phase, dual-voltage motor connections.

The rpm inscription represents revolutions per minute; that is, the motor speed. The rpm reading on motors is the approximate full-load speed. Temp. rise designates the maximum air temperature immediately surrounding the motor. Forty degrees centigrade is the NEMA maximum ambient temperature.

L.R.A. stands for locked rotor amps. It relates to starting current and selection of fuse or circuit breaker size. When two values are shown, the motor usually has a dual-voltage rating. The duty rating designates the duty cycle of a motor. "Continuous" means that the motor is designed for around-the-clock operation.

Each motor is usually given a different serial number for identification and tracking purposes. The code designation is a serial data code used by the manufacturer. In Figure 15-1, the first letter identifies the month and the last two numbers identify the year of manufacture (D88 is April 88).

A service factor (S.F.) is a multiplier which, when applied to the rated horsepower, indicates a permissible horsepower loading which may be carried continuously when the voltage and frequency are maintained at the value specified on the nameplate, although the motor will operate at an increased temperature rise.

The frame (F.R.) designation specifies the shaft height and motor mounting dimensions and provides recommendations for standard shaft diameters and usable shaft extension lengths.

Besides the information just mentioned, many larger motors also contain plates with wiring diagrams to facilitate connections, maintenance, and repairs.

SINGLE-PHASE MOTORS

Single-phase ac motors are usually limited in size to about two or three horsepower. For residential and small commercial applications, these motors will be found in both central and individual room air conditioning units, fans, ventilating units, and refrigeration units such as household refrigerators and the larger units used to cool produce and other foods in market places.

Since there are so many applications of electric motors, there are many types of single-phase motors in use. Some of the more common types are repulsion, universal, and single-phase induction motors. This latter type includes split-phase, capacitor, shaded-pole, and repulsion-induction motors.

Figure 15-3 shows the basic parts of a motor to familiarize you with its makeup.

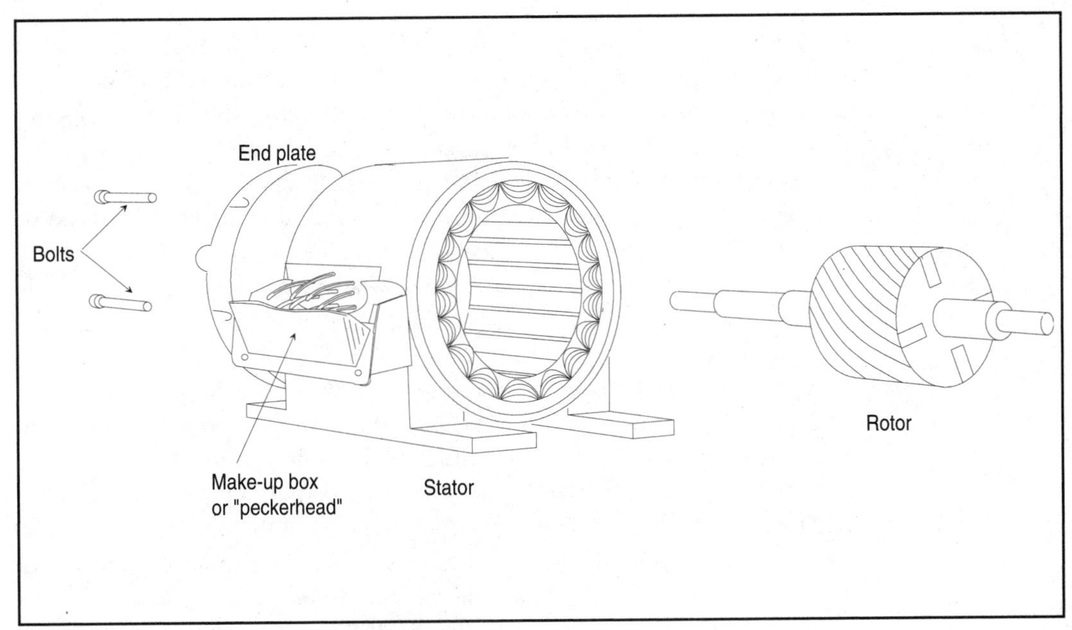

Figure 15-3: Basic parts of an induction motor.

Split-Phase Motors

Split-phase motors are fractional-horsepower units that use an auxiliary winding on the stator to aid in starting the motor until it reaches its proper rotation speed (*see* Figure 15-4). This type of motor finds use in small pumps, oil burners, and other applications.

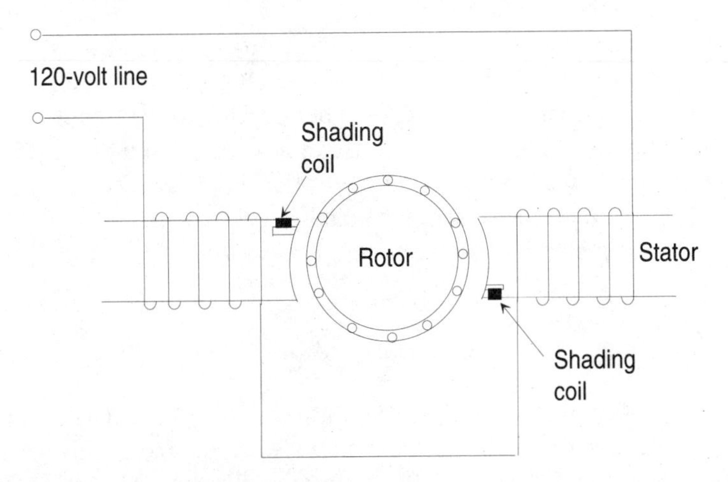

Figure 15-4: Wiring diagram of a split-phase motor.

In general, the split-phase motor consists of a housing, a laminated iron-core stator with embedded windings forming the inside of the cylindrical housing, a rotor made up of copper bars set in slots in an iron core and connected to each other by copper rings around both ends of the core, plates that are bolted to the housing and contain the bearings that support the rotor shaft, and a centrifugal switch inside the housing. This type of rotor is often called a squirrel cage rotor since the configuration of the copper bars resembles an actual cage. These motors have no windings as such, and a centrifugal switch is provided to open the circuit to the starting winding when the motor reaches running speed.

To understand the operation of a split-phase motor, look at the wiring diagram in Figure 15-4. Current is applied to the stator windings, both the main winding and the starting winding, which is in parallel with it through the centrifugal switch. The two windings set up a rotating magnetic field, and this field sets up a voltage in the copper bars of the squirrel-cage rotor. Because these bars are shortened at the ends of the rotor, current flows through the rotor bars. The current-carrying rotor bars then react with the magnetic field to produce motor action. When the rotor is turning at the proper speed, the centrifugal switch cuts out the starting winding since it is no longer needed.

Capacitor Motors

Capacitor motors are single-phase ac motors ranging in size from fractional horsepower (hp) to perhaps as high as 15 hp. This type of motor is widely used in all types of single-phase applications such as powering air compressors, refrigerator compressors, and the like. This type of motor is similar in construction to the split-phase motor, except a capacitor is wired in series with the starting winding, as shown in Figure 15-5.

The capacitor provides higher starting torque, with lower starting current, than does the split-phase motor, and although the capacitor is sometimes mounted inside the motor housing, it is more often mounted on top of the motor, encased in a metal compartment.

In general, two types of capacitor motors are in use:

- Capacitor-start motor
- Capacitor start-and-run motor

As the name implies, the former utilizes the capacitor only for starting; it is disconnected from the circuit once the motor reaches running speed, or at about 75 percent of the motor's full speed. Then the centrifugal switch opens to cut the capacitor out of the circuit.

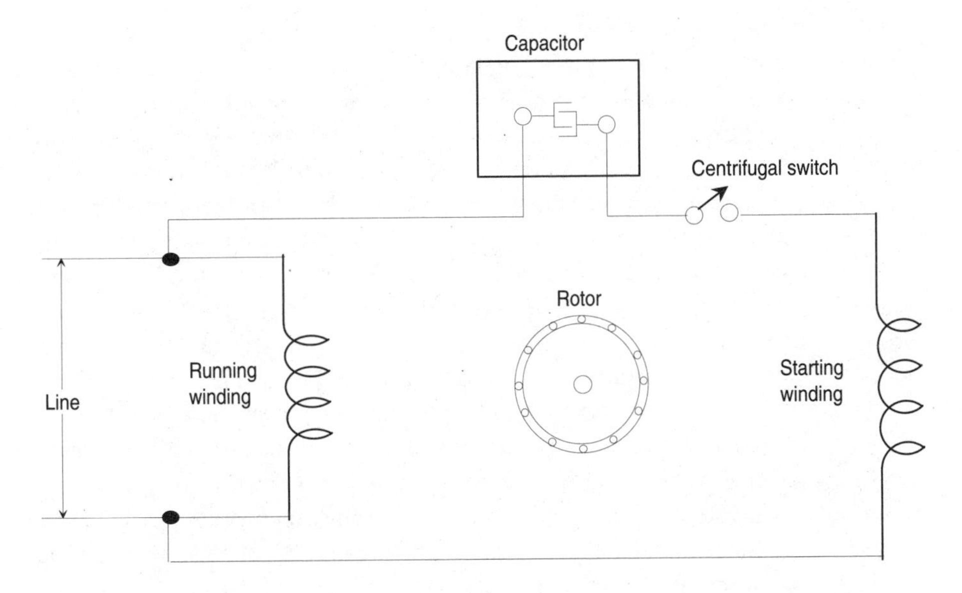

Figure 15-5: Diagram of a capacitor-start motor.

The capacitor start-and-run motor keeps the capacitor and starting winding in parallel with the running winding, providing a quiet and smooth operation at all times.

Capacitor split-phase motors require the least maintenance of all single-phase motors, but they have a very low starting torque, making them unsuitable for many applications. Their high maximum torque, however, makes them especially useful in HVAC systems to power slow-speed, direct-connected fans.

Repulsion-Type Motors

Repulsion-type motors are divided into several groups, including (1) repulsion-start, induction-run motors, (2) repulsion motors, and (3) repulsion-induction motors. The repulsion-start, induction-run motor is of the single-phase type, ranging in size from about $1/10$ hp to as high as 20 hp. It has high starting torque and a constant-speed characteristic, which makes it suitable for such applications as commercial refrigerators, compressors, pumps, and similar applications requiring high starting torque.

The repulsion motor is distinguished from the repulsion-start, induction-run motor by the fact that it is made exclusively as a brush-riding type and does not have any centrifugal mechanism. Therefore, this motor both starts and runs on the repulsion principle. This type of motor has high starting torque and a variable-speed characteristic. It is reversed by shifting the brush holder to either side of the neutral position. Its speed can be decreased by moving the brush holder farther away from the neutral position.

The repulsion-induction motor combines the high starting torque of the repulsion-type and the good speed regulation of the induction motor. The stator of this motor is provided with a regular single-phase winding, while the rotor winding is similar to that used on a dc motor. When starting, the changing single-phase stator flux cuts across the rotor windings and induces currents in them; thus, when flowing through the commentator, a continuous repulsive action on the stator poles is present.

This motor starts as a straight repulsion-type and accelerates to about 75 percent of normal full speed when a centrifugally operated device connects all the commutator bars together and converts the winding to an equivalent squirrel-cage type. The same mechanism usually raises the brushes to reduce noise and wear. Note that, when the machine is operating as a repulsion-type, the rotor and stator poles reverse at the same instant, and that the current in the commutator and brushes is ac.

This type of motor will develop four to five times normal full-load torque and will draw about three times normal full-load current when starting with full-line voltage applied. The speed variation from no load to full load will not exceed 5 percent of normal full-load speed.

The repulsion-induction motor is used to power air compressors, refrigeration (compressor and fans), pumps, stokers, and the like. In general, this type of motor is suitable for any load that requires a high starting torque and constant-speed operation. Most motors of this type are less than 5 hp.

Universal Motors

This type of motor is a special adaptation of the series-connected dc motor, and it gets its name "universal" from the fact that it can be connected on either ac or dc and operates the same. All are single-phase motors for use on 120 or 240 V.

In general, the universal motor contains field windings on the stator within the frame, an armature with the ends of its windings brought out to a commutator at one end, and carbon brushes that are held in place by the motor's end plate, allowing them to have a proper contact with the commutator.

When current is applied to a universal motor, either ac or dc, the current flows through the field coils and the armature windings in series. The magnetic field set up by the field coils in the stator react with the current-carrying wires on the armature to produce rotation. Universal motors are frequently used on small fans.

Shaded-Pole Motors

A shaded-pole motor is a single-phase induction motor provided with an uninsulated and permanently short-circuited auxiliary winding displaced in magnetic position from the main winding. The auxiliary winding is known as the shading coil and usually surrounds from one-third to one-half of the pole. The main winding surrounds the entire pole and may consist of one or more coils per pole.

Applications for this motor include small fans, timing devices, relays, instrument dials, or any constant-speed load not requiring high starting torque.

POLYPHASE MOTORS

Three-phase motors offer extremely efficient and economical application and are usually the preferred type for commercial and industrial applications when three-phase service is available. In fact, the great bulk of motors sold are standard ac three-phase motors. These motors are available in ratings from fractional hp up to thousands of hp in practically every standard voltage and frequency. In fact, there are few applications for which the three-phase motor cannot be put to use.

Three-phase motors are noted for their relatively constant speed characteristic and are available in designs giving a variety of torque characteristics; that is, some have a high starting torque and others a low starting torque. Some are designed to draw a normal starting current, others a high starting current.

The three main parts of a three-phase motor are the stator, rotor, and end plates. It is very similar in construction to conventional split-phase motors except that the three-phase motor has no centrifugal switch.

The stator consists of a steel frame and a laminated iron core with windings formed of individual coils placed in slots. The rotor may be a squirrel-cage or wound-rotor type. Both types contain a laminated core pressed onto a shaft. The squirrel-cage rotor is similar to a split-phase motor. The wound rotor has a winding on the core that is connected to three slip rings mounted on the shaft.

The end plates or brackets are bolted to each side of the stator frame and contain the bearings in which the shaft revolves. Either ball bearings or sleeve bearings are used.

Induction Motors

Induction motors, both single-phase and polyphase, get their name from the fact that they utilize the principle of electromagnetic induction. An induction motor has a stationary part, or stator, with windings connected to the ac supply, and a rotation part, or rotor, which contains coils or bars. There is no electrical connection between the stator and rotor. The magnetic field produced in the stator windings induces a voltage in the rotor coils or bars.

Since the stator windings act in the same way as the primary winding of a transformer, the stator of an induction motor is sometimes called the primary. Similarly, the rotor is called the secondary because it carries the induced voltage in the same way as the secondary of a transformer.

The magnetic field necessary for induction to take place is produced by the stator windings. Therefore, the induction-motor stator is often called the field and its windings are called field windings.

The terms primary and secondary relate to the electrical characteristics and the terms stator and rotor to the mechanical features of induction motors.

The rotor transfers the rotating motion to its shaft, and the revolving shaft drives a mechanical load or a machine, such as a pump, spindle, or clock.

Commutator segments, which are essential parts of dc motors, are not needed on induction motors. This simplifies greatly the design and the maintenance of induction motors as compared to dc motors.

The turning of the rotor in an induction motor is due to induction. The rotor, or secondary, is not connected to any source of voltage. If the magnetic field of the stator, or primary, revolves, it will induce a voltage in the rotor, or secondary. The magnetic field produced by the induced voltage acts in such a way that it makes the secondary follow the movement of the primary field.

The stator, or primary, of the induction motor does not move physically. The movement of the primary magnetic field must thus be achieved electrically. A rotating magnetic field is made possible by a combination of two or more ac voltages that are out of phase with each other and applied to the stator coils. Direct current will not produce a rotating magnetic field. In three-phase induction motors, the rotating magnetic field is obtained by applying a three-phase system to the stator windings.

The direction of rotation of the rotor in an ac motor is the same as that of its rotating magnetic field. In a three-phase motor the direction can be reversed by interchanging the connections of any two supply leads. This interchange will reverse the sequence of phases in the stator, the direction of the field rotation, and therefore the direction of rotor rotation.

Synchronous Motors

A synchronous polyphase motor has a stator constructed in the same way as the stator of a conventional induction motor. The iron core has slots into which coils are wound, which are also arranged and connected in the same way as the stator coils of the induction motor. These are, in turn, grouped to form a three-phase connection and the three free leads are connected to a three-phase source. Frames are equipped with air ducts, which aid in cooling of the windings, and coil guards protect the windings from damage.

The rotor of a synchronous motor carries poles that project toward the armature; they are called salient poles. The coils are wound on laminated pole bodies and connected to slip rings on the shaft. A squirrel-cage winding for starting the motor is embedded in the pole faces.

The pole coils are energized by direct current, which is usually supplied by a small dc generator called the exciter. This exciter may be mounted directly on the shaft to generate dc voltage, which is applied through brushes to slip rings. On low-speed synchronous motors, the exciter is normally belted or of a separate high-speed, motor-driven type.

The dimensions and construction of synchronous motors vary greatly, depending on the rating of the motors. However, synchronous motors for industrial power applications are rarely built for less than 25 hp or so. In fact, many are 100 hp or more. All are polyphase motors when built in this size. Vertical and horizontal shafts with various bearing arrangements and various enclosures cause wide variations in the appearance of the synchronous motor.

Synchronous motors are used in electrical systems where there is need for improvement in power factor or where low power factor is not desirable. This type of motor is especially adapted to heavy loads that operate for long periods of time without stopping, such as for air compressors, pumps, ship propulsion, and the like.

The construction of the synchronous motor is well adapted for high voltages, as it permits good insulation. Synchronous motors are frequently used at 2300 V or more. Its efficient slow-running speed is another advantage.

MOTOR ENCLOSURES

Electric motors differ in construction and appearance, depending on the type of service for which they are to be used. Open and closed frames are quite common. In the former enclosure, the motor's parts are covered for protection, but the air can freely enter the enclosure. Further designations for this type of enclosure include drip-proof, weather-protected, and splash-proof.

Totally enclosed motors have an air-tight enclosure. They may be fan cooled or self-ventilated. An enclosed motor equipped with a fan has the fan as an integral part of the machine, but external to the enclosed parts. In the self-ventilated enclosure, no external means of cooling is provided.

The type of enclosure to use will depend on the ambient and surrounding conditions. In a drip-proof machine, for example, all ventilating openings are so constructed that drops of liquid or solid particles falling on the machine at an angle of not greater than 15 degrees from the vertical cannot enter the machine, even directly or by striking and running along a horizontal or inclined surface of the machine. The application of this machine would lend itself to areas where liquids are processed.

An open motor having all air openings that give direct access to live or rotating parts, other than the shaft, limited in size by the design of the parts or by screen to prevent accidental contact with such parts is classified as a drip-proof, fully guarded machine. In such enclosures, openings shall not permit the passage of a cylindrical rod ½ in in diameter, except where the distance from the guard to the live rotating parts is more than 4 in, in which case the openings shall not permit the passage of a cylindrical rod ¾ in in diameter.

There are other types of drip-proof machines for special applications such as externally ventilated and pipe ventilated, which as the names imply are either ventilated by a separate motor-driven blower or cooled by ventilating air from inlet ducts or pipes.

An enclosed motor whose enclosure is designed and constructed to withstand an explosion of a specified gas or vapor that may occur within the motor and to prevent the ignition of this gas or vapor surrounding the machine is designated " explosionproof " (XP) motors.

Hazardous atmospheres (requiring XP enclosures) of both a gaseous and dusty nature are classified by the National Electrical Code (*NEC*) as follows:

- Class I, Group A: atmospheres containing acetylene.
- Class I, Group B: atmospheres containing hydrogen gases or vapors of equivalent hazards such as manufactured gas.

- Class I, Group C: atmospheres containing ethyl ether vapor.
- Class I, Group D: atmospheres containing gasoline, petroleum, naphtha, alcohol, acetone, lacquer-solvent vapors, and natural gas.
- Class II, Group E: atmospheres containing metal dust.
- Class II, Group F: atmospheres containing carbon-black, coal, or coke dust.
- Class II, Group G: atmospheres containing grain dust.

The proper motor enclosure must be selected to fit the particular atmosphere. However, explosionproof equipment is not generally available for Class I, Groups A and B, and it is therefore necessary to isolate motors from the hazardous area.

MOTOR TYPE

The type of motor will determine the electrical characteristics of the design. NEMA-designated designs for polyphase motors are given in the table in Figure 15-6.

An "A" motor is a three-phase, squirrel-cage motor designed to withstand full-voltage starting with locked rotor current higher than the values for a B motor and having a slip at rated load of less than 5 percent.

A "B" motor is a three-phase, squirrel-cage motor designed to withstand full-voltage starting and developing locked rotor and breakdown torques adequate for general application, and having a slip at rated load of less than 5 percent.

NEMA Design	Starting Torque	Starting Current	Breakdown Torque	Full-Load Slip
A	Normal	Normal	High	Low
B	Normal	Low	Medium	Low
C	High	Low	Normal	Low
D	Very High	Low	—	High

Figure 15-6: NEMA-designated motor designs.

A "C" motor is a three-phase, squirrel-cage motor designed to withstand full-voltage starting, developing locked rotor torque for special high-torque applications, and having a slip at rated load of less than 5 percent.

Design "D" is also a three-phase, squirrel-cage motor designed to withstand full-voltage starting, developing 275 percent locked rotor torque, and having a slip at rated load of 5 percent or more.

SELECTION OF ELECTRIC MOTORS

Each type of motor has its particular field of usefulness. Because of its simplicity, economy, and durability, the induction motor is more widely used for industrial purposes than any other type of ac motor, especially if a high-speed drive is desired.

If ac power is available, all drives requiring constant speed should use squirrel-cage induction or synchronous motors on account of their ruggedness and lower cost. Drives requiring varying speed, such as fans, blowers, or pumps may be driven by wound-rotor induction motors. However, if there are applications requiring adjustable speed or a wide range of speed control, it will probably be desirable to install dc motors on such equipment and supply them from the ac system by motor-generator sets of electronic rectifiers.

Practically all constant-speed machines may be driven by ac squirrel-cage motors because they are made with a variety of speed and torque characteristics. When large motors are required or when power supply is limited, the wound-rotor is used even for driving constant-speed machines. A wound-rotor motor, with its controller and resistance, can develop full-load torque at starting with not more than full-load amperes at starting, depending on the type of motor and the starter used.

For varying-speed service, wound-rotor motors with resistance control are used for fans, blowers, and other apparatus for continuous duty, and for other intermittent duty applications. The controller and resistors must be properly chosen for the particular application.

Cost is an important consideration where more than one type of ac motor is applicable. The squirrel-cage motor is the least expensive ac motor of the three types considered and requires very little control equipment. The wound-rotor is more expensive and requires additional secondary control.

NEC REQUIREMENTS

NEC Article 430 covers application and installation of motor circuits and motor-control connections, including conductors, short-circuit and ground-fault protection, starters, disconnects, and overload protection.

NEC Article 440 contains provisions for motor-driven equipment and for branch circuits and controllers for HVAC equipment.

All motors must be installed in a location that allows adequate ventilation to cool the motors. Furthermore, the motors should be located so that maintenance, troubleshooting, and repairs can be readily performed. Such work could consist of lubricating the motor's bearings, or perhaps replacing worn brushes. Testing the motor for open circuits and ground faults is also necessary from time-to-time.

When motors must be installed in locations where combustible material, dust, or similar material may be present, special precautions must be taken in selecting and installing motors.

Exposed live parts of motors operating at 50 V or more between terminals must be guarded; that is, they must be installed in a room, enclosure, or location so as to allow access by only qualified persons (electrical maintenance personnel). If such a room, enclosure, or location is not feasible, an alternative is to elevate the motors not less than 8 ft above the floor. In all cases, adequate space must be provided around motors with exposed live parts — even when properly grounded — to allow for maintenance, troubleshooting, and repairs.

The chart in Figure 15-7 on the next page summarizes installation rules from the 1996 *NEC*.

MOTOR INSTALLATION

The best motors on the market will not operate properly if they are installed incorrectly. Therefore, all personnel involved with the installation of electric motors should thoroughly understand the proper procedures for installing the various types of motors that will be used. Furthermore, proper maintenance of each motor is essential to keep it functioning properly once it is installed.

When an electric motor is received at the job site, always refer to the manufacturer's instructions and follow them to the letter. Failure to do so could result in serious injury or fatality. In general, disconnect all power before servicing. Install and ground according to *NEC* requirements and good practices. Consult qualified personnel with any questions or services required.

Application	*NEC* Regulation	*NEC* Section
Location	Motors must be installed in areas with adequate ventilation. They must also be arranged so that sufficient work space is provided for replacement and maintenance.	430-14
	Open motors must be located or protected so that sparks cannot reach combustible materials.	430-16
	In locations where dust or flying material will collect on or in motors in such quantities as to seriously interfere with the ventilation or cooling of motors and thereby cause dangerous temperatures, suitable types of enclosed motors that will not overheat under the prevailing conditions must be used.	
Disconnecting means	A motor disconnecting means must be within sight from the controller location (with exceptions) and disconnect both the motor and controller. The disconnect must be readily accessible and clearly indicate the ***Off/On*** positions (open/closed).	Article 430(H)
	Motor-control circuits require a disconnecting means to disconnect them from all supply sources.	430-74
	The service switch may serve to disconnect a single motor if it complies with other rules for a disconnecting means.	430-106
	The disconnecting means must be a motor-circuit safety switch rated in horsepower or a circuit breaker.	430-109
Wiring methods	Flexible connections such as Type AC cable, "Greenfield," flexible metallic tubing, etc., are standard for motor connections.	Articles 310 and 430
Motor-control circuits	All conductors or a remote motor control circuit outside of the control device must be installed in a raceway or otherwise protected. The circuit must be wired so that an accidental ground in the control device will not start the motor.	430-73

Figure 15-7: Summary of NEC requirements for motor installations.

Application	*NEC* Regulation	*NEC* Section
Guards	Exposed live parts of motors and controllers operating at 50 V or more must be guarded by installation in a room, enclosure, or location so as to allow access by only qualified persons, or elevated 8 ft or more above the floor.	Article 430
Motors operating over 600 V	Special installation rules apply to motors operating at over 600 V.	Article 430(J)
Controller grounding	Motor controllers must have their enclosures grounded.	430-144

Figure 15-7: Summary of NEC requirements for motor installations. (*Cont.*)

Uncrating: Once the motor has been carefully uncrated, check to see if any damage has occurred during handling. Be sure that the motor shaft and armature turn freely. This time is also a good time to check to determine if the motor has been exposed to dirt, grease, grit, or excessive moisture in either shipment or storage before installation. Motors in storage should have shafts turned over once each month to redistribute grease in the bearings.

The measure of insulation resistance is a good dampness test. Clean the motor of any dirt or grit.

Lifting: Eyebolts or lifting lugs on motors are intended only for lifting the motor and factory motor-mounted standard accessories. These lifting provisions should never be used when lifting or handling the motor when the motor is attached to other equipment as a single unit.

The eyebolt lifting-capacity rating is based on a lifting alignment coincident with the eyebolt centerline. The eyebolt capacity reduces as deviation from this alignment increases.

Guards: Rotating parts such as pulleys, couplings, external fans, and unusual shaft extensions should be permanently guarded against accidental contact with clothing or body extremities.

Requirements: All motors should be installed, protected, and fused in accordance with the latest *NEC*, NEMA Standard Publication No. MG-2, and any and all local requirements.

Frames and accessories of motors should be grounded in accordance with *NEC* Article 430. For general information on grounding, refer to *NEC* Article 250.

Thermal Protector Information: A space on the motor's nameplate may or may not be stamped to indicate the following:

- The motor is thermally protected
- The motor is not thermally protected
- The motor has an overheat-protective device

Troubleshooting Motors

Induction motors get their name from the fact that they utilize the principle of electromagnetic induction. An induction motor has a stationary part, or stator, with windings connected to the ac supply, and a rotating part, or rotor, which contains coils or bars. There is no electrical connection between the stator and rotor. The magnetic field produced in the stator windings induces a voltage in the rotor coils or bars.

Since the stator windings act in the same way as the primary winding of a transformer, the stator of an induction motor is sometimes called the *primary*. Similarly, the rotor is called the *secondary* because it carries the induced voltage in the same way as the secondary of a transformer.

The useful life of an induction motor depends largely upon the condition of its insulation, which should be suitable for the operating requirements.

When an induction motor malfunctions, the stator (stationary) windings will usually be defective, and these windings will then have to be repaired or replaced. Stator problems can usually be traced to one or more of the following causes:

- Worn bearings
- Moisture
- Overloading
- Operating single phase
- Poor insulation

Dust and dirt are usually contributing factors. Some forms of dust are highly conductive and contribute materially to insulation breakdown. The effect of dust on the motor temperature through restriction of ventilation is another reason for keeping the machine clean, either by periodically blowing it out with compressed air or by dismantling and cleaning. The compressed air must be dry and throttled down to a low pressure that will not danger the insulation.

One of the worst enemies of motor insulation is moisture. Therefore, motor insulation must be kept reasonably dry, although many applications make this practically impossible unless a totally enclosed motor is utilized. If a motor must be operated in a damp location, special moisture-resisting treatment should be given to the windings.

The life of a winding depends upon keeping it in its original condition as long as possible. In a new machine, the winding is snug in the slots and the insulation is fresh and flexible. This condition is best maintained by periodic cleaning, followed by varnish and oven treatments.

One condition that frequently hastens winding failure is movement of the coils due to vibration during operation. After insulation dries out, it loses its flexibility and the mechanical stresses caused by starting and plugging, as well as the natural stresses in operation under load, will tend to cause short circuits in the coils and possibly failures from the coil to ground — usually at the point where the coil leaves the slot. The effect of periodic varnish and oven treatments properly carried out so as to fill all air spaces caused by drying and shrinkage of the insulation will maintain a solid winding, and also provide an effective seal against moisture.

Troubleshooting Charts

The troubleshooting charts that follow give most motor problems found in the industry, along with the cause and remedy.

Symptoms	Probable Cause	Action or Items to Check
Hot Bearings — General	Bent or sprung shaft.	Straighten or replace shaft.
	Excessive belt pull.	Decrease belt tension.
	Pulley too far away.	Move pulley closer to bearing.
	Pulley diameter too small.	Use larger pulleys.
	Misalignment.	Correct by realignment of drive.

Figure 15-8: Basic motor troubleshooting chart.

Symptoms	Probable Cause	Action or Items to Check
Hot Bearings — Sleeve	Oil grooving in bearing obstructed by dirt.	Remove bracket or pedestal with bearing and clean oil grooves and bearing housing; renew oil.
	Bent or damaged oil rings.	Repair or replace oil rings.
	Oil too heavy.	Use a recommended lighter oil.
	Oil too light.	Use a recommended heavier oil.
	Insufficient oil.	Fill reservoir to proper level in overflow plug with motor at rest.
	Too much end thrust.	Reduce thrust induced by driven machine or supply external means to carry thrust.
	Badly worn bearing.	Replace bearing.
Hot Bearings — Ball	Insufficient grease.	Replace bearing.
	Deterioration of grease or lubricant contaminated.	Remove old grease, wash bearings thoroughly in kerosene and replace with new grease.
	Excess lubricant.	Reduce quantity of grease. Bearing should not be more than half filled.
	Heat from hot motor or external source.	Protect bearing by reducing motor temperature.
	Overloaded bearing.	Check alignment, side thrust, and end thrust.
	Broken ball or rough races.	Replace bearing; first clean housing thoroughly.

Figure 15-8: Basic motor troubleshooting chart. ***(Cont.)***

Symptoms	Probable Cause	Action or Items to Check
Oil leakage from overflow plugs	Stem of overflow plug not tight.	Remove, recement threads, replace and tighten.
	Cracked or broken overflow plug.	Replace the plug.
	Plug cover not tight.	Requires cork gasket, or if screw type, may be tightened.
Motor dirty	Ventilation blocked, end windings filled with fine dust or lint.	Clean motor will run 10 to 30° C cooler. Dust may be cement, sawdust, rock dust, grain dust, coal dust, and the like. Dismantle entire motor and clean all windings and parts.
	Rotor winding clogged.	Clean, grind, and undercut commutator. Clean and treat windings with good insulating varnish.
	Bearing and brackets coated inside.	Dust and wash with cleaning solvent.
Motor wet	Drenched condition.	Motor should be covered to retain heat and the rotor position shifted frequently.
	Subject to dripping.	Wipe motor and dry by circulating heated air through motor. Install drip or canopy type covers over motor for protection.
	Submerged in flood waters.	Dismantle and clean parts. Bake windings in oven at 105° C for 24 hours or until resistance to ground is sufficient. First make sure commutator bushing is drained of water.

Figure 15-8: Basic motor troubleshooting chart. ***(Cont.)***

Symptoms	Probable Cause	Action or Items to Check
Fails to start	Circuit not complete.	Switch open, leads broken.
	Brushes not down on commutator.	Held up by brush springs, need replacement. Brushes worn out.
	Brushes stuck in holders.	Remove and sand, clean up brush boxes.
	Armature locked by frozen bearings in motor or main drive.	Remove brackets and replace bearings or recondition old bearings if inspection makes possible.
	Power may be off.	Check line connections to starter with light. Check contacts in starter.
Motor starts, then stops and reverses direction of rotation	Reverse polarity of generator that supplies power.	Check generating unit for cause of changing polarity.
	Shunt and series fields are bucking each other.	Reconnect either the shunt or series field so as to correct the polarity. Then connect armature leads for desired direction of rotation. The fields can be tried separately to determine the direction of rotation individually.
Motor does not come up to rated speed	Overload.	Check bearing to see if in first class condition with correct lubrication. Check driven load for excessive load or friction.
	Starting resistance not all out.	Check starter to see if mechanically and electrically incorrect.
	Voltage low.	Measure voltage with meter and check with motor nameplate.
	Short circuit in armature windings or between bars.	For shorted armature inspect commutator for blackened bars and burned adjacent bars. Inspect windings for burned coils or wedges.

Figure 15-8: Basic motor troubleshooting chart. *(Cont.)*

Symptoms	Probable Cause	Action or Items to Check
Motor does not come up to rated speed	Starting heavy load with very weak field.	Check full field relay and possibilities of full field setting of the field rheostat.
	Motor off neutral.	Check for factory setting of brush rigging or test motor for true neutral setting.
	Motor cold.	Increase load on motor so as to increase its temperature, or add field rheostat to set speed.
Motor runs too fast	Voltage above rated.	Correct voltage or get recommended change in air gap from manufacturer.
	Load too light.	Increase load or install fixed resistance in armature circuit.
	Shunt field coil shorted.	Install new coil.
	Shunt field coil reversed.	Reconnect coil leads in reverse.
	Series coil reversed.	Reconnect coil leads in reverse.
	Series field coil shorted.	Install new or repaired coil.
	Neutral setting shifted off neutral.	Reset neutral by checking factory setting mark or testing for neutral.
	Part of shunt field rheostat or unnecessary resistance in field circuit.	Measure voltage across field and check with nameplate rating.
	Motor ventilation restricted causing hot shunt field.	Hot field is high in resistance; check causes for hot field in order to restore normal shunt field current.
Motor gaining speed steadily and increasing load does not slow it down	Unstable speed load regulation.	Inspect motor to see if off neutral. If series field has a shunt around the series circuit that can be removed, check series field to determine shorted turns.

Figure 15-8: Basic motor troubleshooting chart. *(Cont.)*

Summary

Always select the appropriate motor for the application. In selecting a location for the motor, the first consideration should be given to ventilation. It should be far enough from walls or other objects to permit a free passage of air.

The motor should never be placed in an area with a hazardous process or where flammable gasses or combustible material may be present unless it is specifically designed for this type of service.

- Drip-proof motors are intended for use where the atmosphere is relatively clean, dry, and noncorrosive. If the atmosphere is not like the preceding, then request approval of the motor for the use intended.
- Totally enclosed motors may be installed where dirt, moisture, and corrosion are present or in outdoor locations.
- Explosionproof motors are built for use in hazardous locations as indicated by the Underwriters' label on the motor.

The ambient temperature of the air surrounding the motor should not exceed 40°C or 104°F unless the motor has been especially designed for high-ambient-temperature applications. The free flow of air around the motor should not be obstructed.

After a location has been decided upon, the mounting proceeds as follows: For floor mounting, motors should be provided with a firm, rigid foundation, with the plane of four mounting stud pads flat within .010 in for a 56 to 210 frame and .015 in for a 250 to 680 frame. This may be accomplished by shims under the motor feet.

Chapter 16

Motor Controls

INTRODUCTION TO MOTOR CONTROLS

Electric motors provide one of the principal sources for driving all types of equipment and machinery. Every motor in use, however, must be controlled, if only to start and stop it, before it becomes of any value.

Motor controllers cover a wide range of types and sizes, from a simple toggle switch to a complex system with such components as relays, timers, and switches. The common function, however, is the same in any case, that is, to control some operation of an electric motor. A motor controller will include some or all of the following functions:

- Starting and stopping
- Overload protection
- Overcurrent protection
- Reversing
- Changing speed
- Jogging
- Plugging
- Sequence control
- Pilot light indication

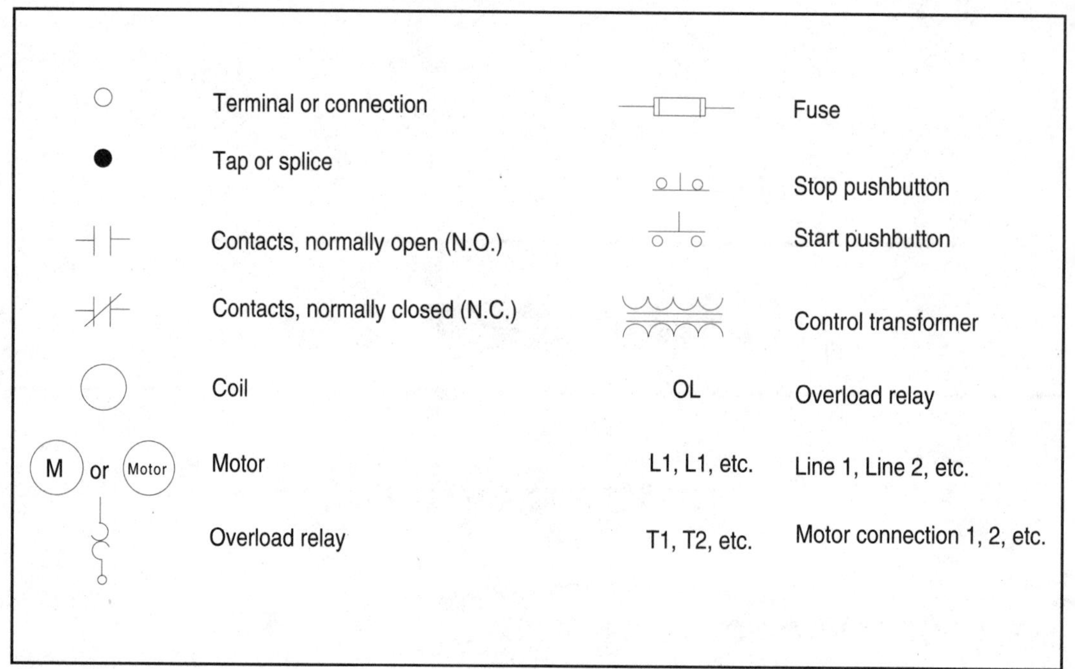

Figure 16-1: Symbols used for drawing wiring diagrams.

The controller can also provide the control for auxiliary equipment such as brakes, clutches, solenoids, heaters, and signals, and may be used to control a single motor or a group of motors.

The term *motor starter* is often used and means practically the same thing as a *motor controller*. Strictly, a motor starter is the simplest form of controller and is capable of starting and stopping the motor and providing it with overload protection.

See Figure 16-1 for a review of the electrical symbols used in the wiring diagrams in this chapter.

TYPES OF MOTOR CONTROLLERS

A large variety of motor controllers are available that will handle almost every conceivable application. However, all of them can be grouped in the following categories.

Plug-and-Receptacle

NEC Section 430-81 defines a controller as any switch or device normally used to start and stop a motor by making or breaking motor-circuit current.

The simplest form of controller allowed by the *NEC* is an attachment plug and receptacle. *See* Figure 16-2 on the next page. However, such an arrangement is limited to portable motors rated at ⅓ horsepower (hp) or less.

Referring again to Figure 16-2, note that drawing (A) is a pictorial view of a portable drill motor with a cord-and-plug assembly attached. If this motor is portable and less than ⅓ horsepower (hp), then the plug and receptacle may act as the motor's controller as permitted in *NEC* Section 430-81(c).

Drawing (B) in Figure 16-2 is the same circuit, but this time depicted in the form of a wiring diagram. Note that symbols have been used to represent the various circuit items rather than actually drawing the items in life form; yet they are arranged on the basis of their physical relationship to each other. This simplifies the drawing — both from a drafter's point of view and also those who must interpret the drawing.

Another form of drawing for this same circuit is shown in (C) of Figure 16-2. This type of drawing has become known as a *ladder diagram*; it's a schematic representation of the electrical circuit in question — the same as the drawing in (B). However, ladder diagrams are drawn in an "H" format, with the energized power conductors represented by vertical lines and the individual circuits represented by horizontal lines. Rather than physically representing the circuit items as in drawing (B), a ladder diagram arranges the conductors and electrical components according to their electrical function in the circuits; that is, schematically. Therefore, ladder diagrams merely represent the current paths (shown as the rungs of a ladder) to each of the controlled or energized output devices.

Where stationary motors rated at ⅛ hp or less and which are normally left running (clock motors, fly fans, and the like), and are so constructed that they cannot be damaged by overload or failure to start, the branch-circuit protective device may serve as the controller. Consequently, the branch-circuit breaker or fusible disconnect serves as both branch-circuit overcurrent protection and motor controller. Such a circuit appears in Figure 16-3.

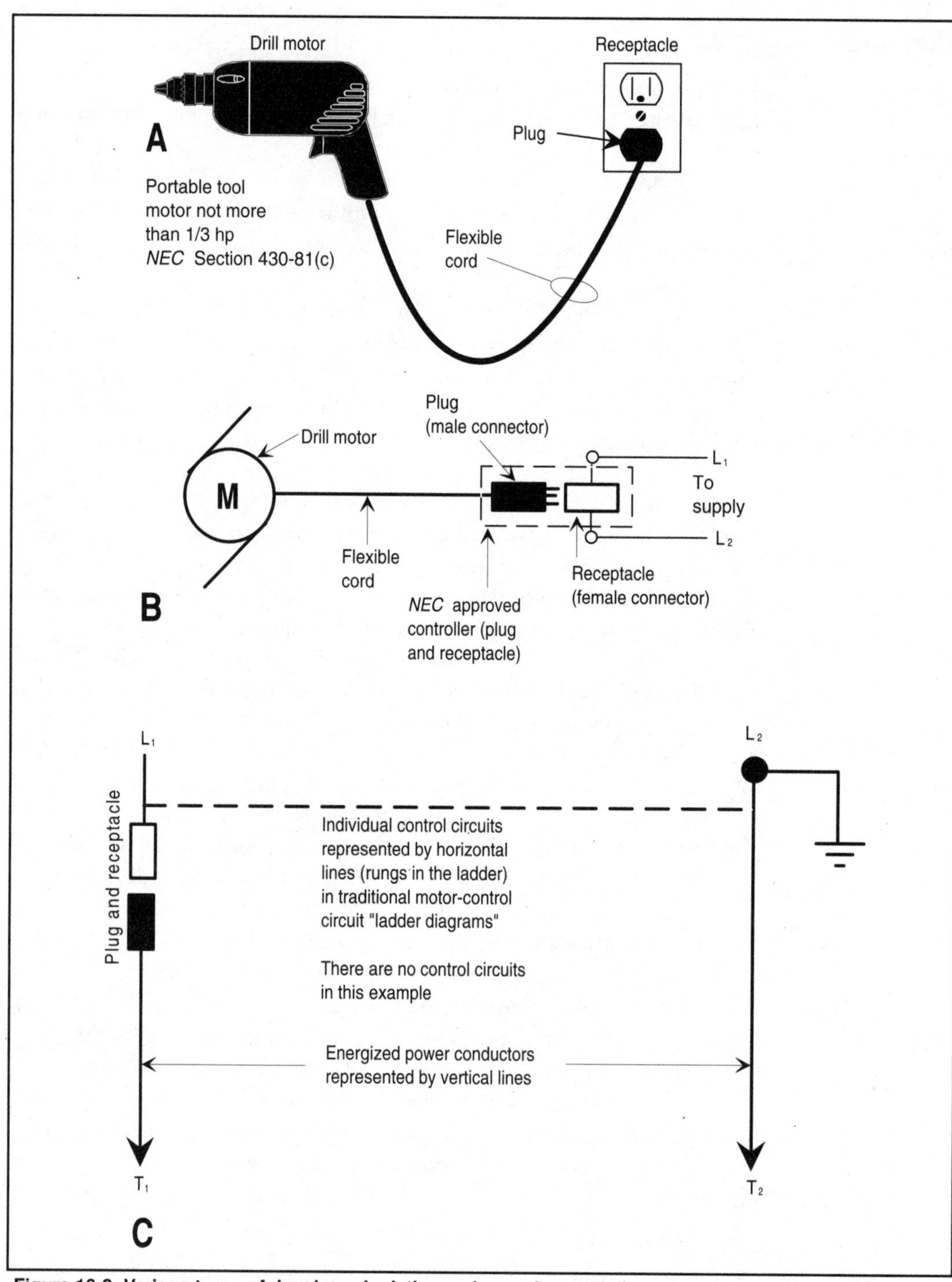

Figure 16-2: Various types of drawings depicting a plug-and-receptacle motor controller.

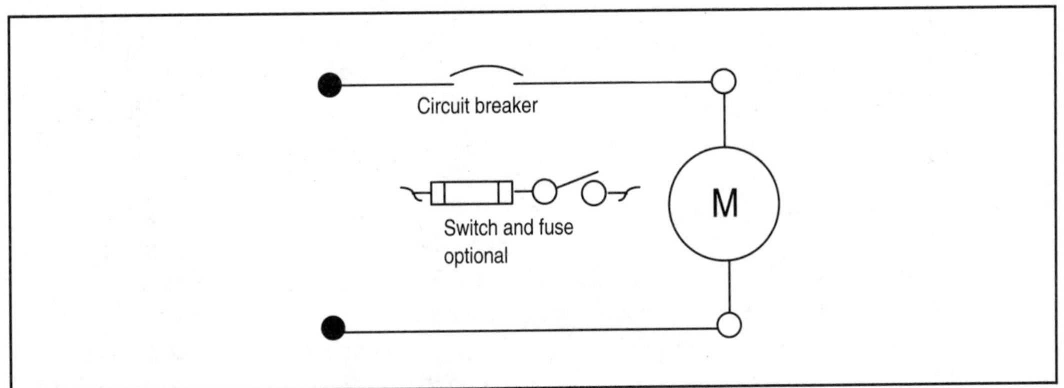

Figure 16-3: Branch-circuit protective device serving as the motor controller.

Manual Starters

A manual starter is a motor controller whose contact mechanism is operated by a mechanical linkage from a toggle handle or pushbutton, which is, in turn, operated by hand. A thermal unit and direct-acting overload mechanism provide motor running overload protection. Basically, a manual starter is an ON-OFF switch with overload relays.

Manual starters are used mostly on small machine tools, fans and blowers, pumps, compressors, and conveyors. They have the lowest cost of all motor starters, have a simple mechanism, and provide quiet operation with no ac magnet hum. The contacts, however, remain closed and the lever stays in the ON position in the event of a power failure, causing the motor to automatically restart when the power returns. Therefore, low-voltage protection and low-voltage release are not possible with these manually operated starters. However, this action is an advantage when the starter is applied to motors that run continuously.

Fractional-Horsepower Manual Starters

Fractional-horsepower manual starters are designed to control and provide overload protection for motors of 1 hp or less on 120- or 240-V single-phase circuits. They are available in single- and two-pole versions and are operated by a toggle handle on the front. When a serious overload occurs, the thermal unit trips to open the starter contacts, disconnecting the motor from the line. The contacts cannot be reclosed until the overload relay has been reset by moving the handle to the full OFF position, after allowing about 2 minutes for the thermal unit to cool. The open-type starter will fit into a standard outlet box and can be used with a standard flush plate. The compact construction of this type of device makes it possible to mount it directly on the driven machinery and in various other places where the available space is small. Figure 16-4 shows fractional-horsepower (FHP) manual motor-starter diagrams for both 120- and 240-V, single-phase motors.

Note that the single-pole FHP starter has only one contact to trip and disconnect the motor from the line; the grounded (neutral) conductor is not opened when the handle is in the OFF position. This single-pole starter also has one overload relay connected in series with the ungrounded conductor.

The two-pole FHP starter has two contacts to open both phases when connected to a 240-V circuit. When the toggle handle is in the off position, no current flows to the motor. However, only one overload relay is needed, since one will shut down the motor if the relay detects an overload and opens.

Manual Motor-Starting Switches

Manual motor-starting switches provide ON-OFF control of single- or three-phase ac motors where overload protection is not required or is separately provided. Two- or three-pole switches are available with ratings up to 10 hp, 600 V, three phase. The continuous current rating is 30 A at 250 V maximum and 20 A at 600 V maximum. The toggle operation of the manual switch is similar to the fractional-horsepower starter, and typical applications of the switch include pumps, fans, conveyors, and other electrical machinery that have separate motor protection. They are particularly suited to switch nonmotor loads, such as resistance heaters.

Integral Horsepower Manual Starters

The integral horsepower manual starter is available in two- and three-pole versions to control single-phase motors up to 5 hp and polyphase motors up to 10 hp, respectively.

Two-pole starters have one overload relay and three-pole starters usually have three overload relays. When an overload relay trips, the starter mechanism unlatches, opening the contacts to stop the motor. The contacts cannot be reclosed until the starter mechanism has been reset by pressing the STOP button or moving the handle to the RESET position, after allowing time for the thermal unit to cool.

Integral horsepower manual starters with low-voltage protection prevent automatic start-up of motors after a power loss. This is accomplished with a continuous-duty solenoid, which is energized whenever the line-side

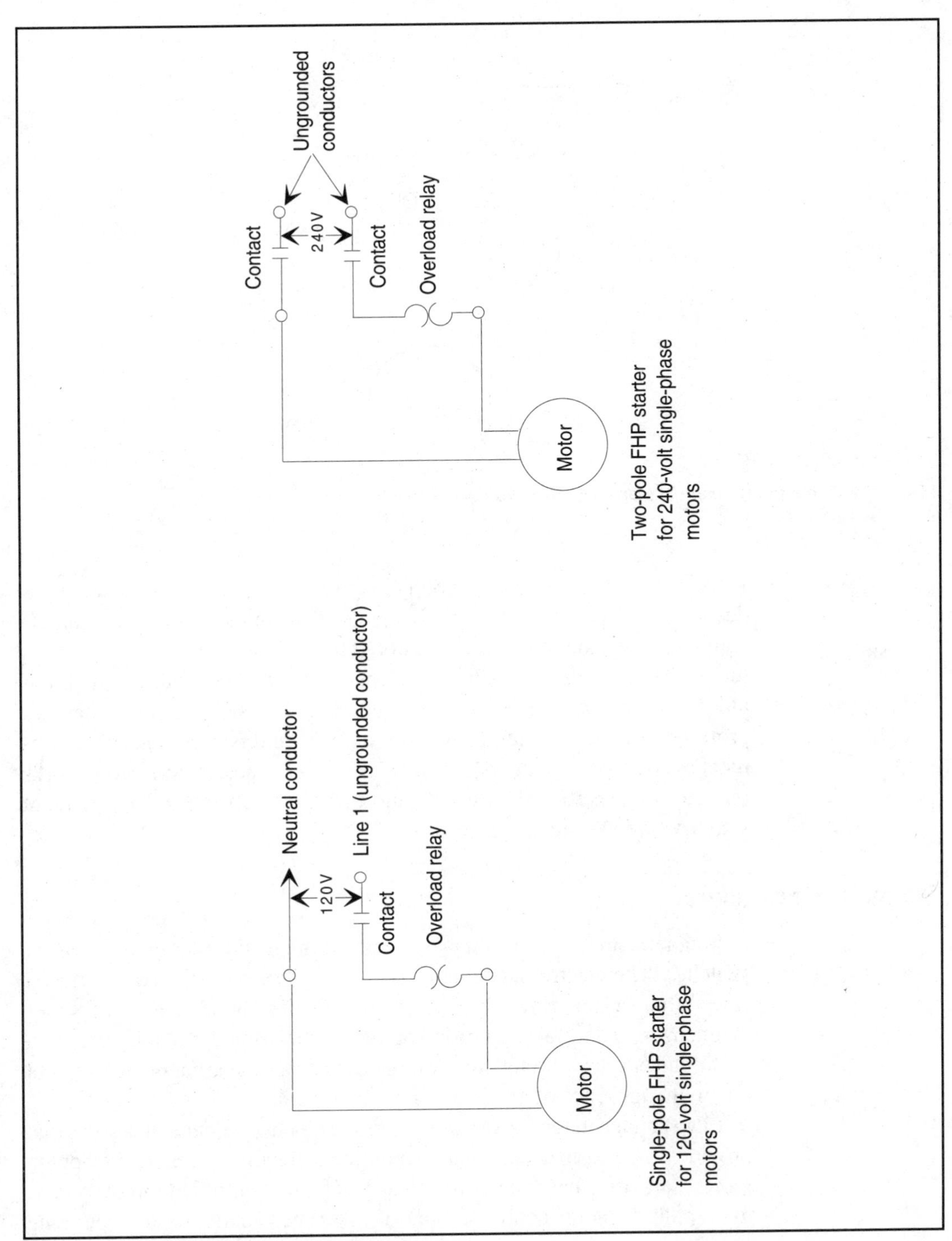

Figure 16-4: Wiring diagrams of FHP manual starters.

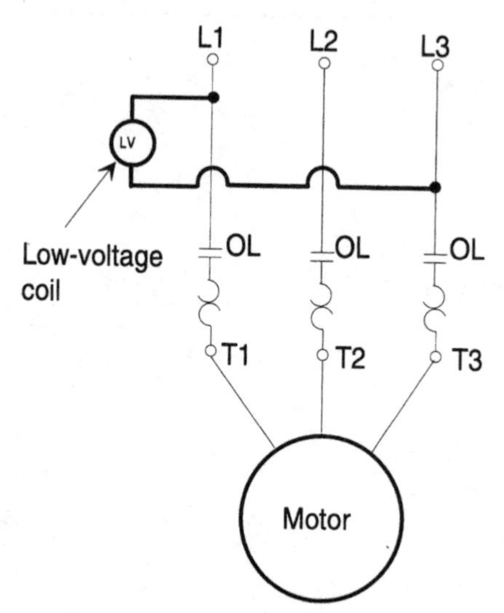

Figure 16-5: Integral HP manual starter with low-voltage protection.

voltage is present. If the line voltage is lost or disconnected, the solenoid deenergizes, opening the starter contacts. The contacts will not automatically close when the voltage is restored to the line. To close the contacts, the device must be manually reset. This manual starter will not function unless the line terminals are energized. This is a safety feature that can protect personnel or equipment from damage and is used on such equipment as conveyors, grinders, metal-working machines, mixers, woodworking, etc. Figure 16-5 shows a wiring diagram of an integral HP manual starter with low-voltage portection.

Magnetic Controllers

Magnetic motor controllers use electromagnetic energy for closing switches. The electromagnet consists of a coil of wire placed on an iron core. When current flows through the coil, the iron of the magnet becomes magnetized and attracts the iron bar, called the armature. An interruption of the current flow through the coil of wire causes the armature to drop out due to the presence of an air gap in the magnetic circuit.

Line-voltage magnetic motor starters are electromechanical devices that provide a safe, convenient, and economic means for starting and stopping motors, and they have the disadvantage of being controlled remotely. The great bulk of motor controllers are of this type. Therefore, the operating principles and applications of magnet motor controllers should be fully understood.

In the construction of a magnetic controller, the armature is mechanically connected to a set of contacts so that, when the armature moves to its closed position, the contacts also close. When the coil has been energized and the armature has moved to the closed position, the controller is said to be picked up and the armature is seated or sealed-in. Some of the magnet and armature assemblies in current use are as follows:

- *Clapper type:* In this type, the armature is hinged. As it pivots to seal in, the movable contacts close against the stationary contacts.
- *Vertical action:* The action is a straight line motion with the armature and contacts being guided so that they move in a vertical plane.
- *Horizontal action:* Both armature and contacts move in a straight line through a horizontal plane.
- *Bell crank:* A bell crank lever transforms the vertical action of the armature into a horizontal contact motion. The shock of armature pickup is not transmitted to the contacts, resulting in minimum contact bounce and longer contact life.

These four types of assemblies are shown in Figure 16-6 on the next page.

The magnetic circuit of a controller consists of the magnet assembly, the coil, and the armature. It is so named from a comparison with an electrical circuit. The coil and the current flowing in it causes magnetic flux to be set up through the iron in a similar manner to a voltage causing current to flow through a system of conductors. The changing magnetic flux produced by alternating currents results in a temperature rise in the magnetic circuit. The heating effect is reduced by laminating the magnet assembly and armature by placing a coil of many turns of wire around a soft iron core, the magnetic flux set up by the energized coil tends to be concentrated; therefore, the magnetic field effect is strengthened. Since the iron core is the path of least resistance to the flow of the magnetic lines of force, magnetic attraction will concentrate according to the shape of the magnet.

The magnetic assembly is the stationary part of the magnetic circuit. The coil is supported by and surrounds part of the magnet assembly in order to induce magnetic flux into the magnetic circuit.

The armature is the moving part of the magnetic circuit. When it has been attracted into its sealed-in position, it completes the magnetic circuit.

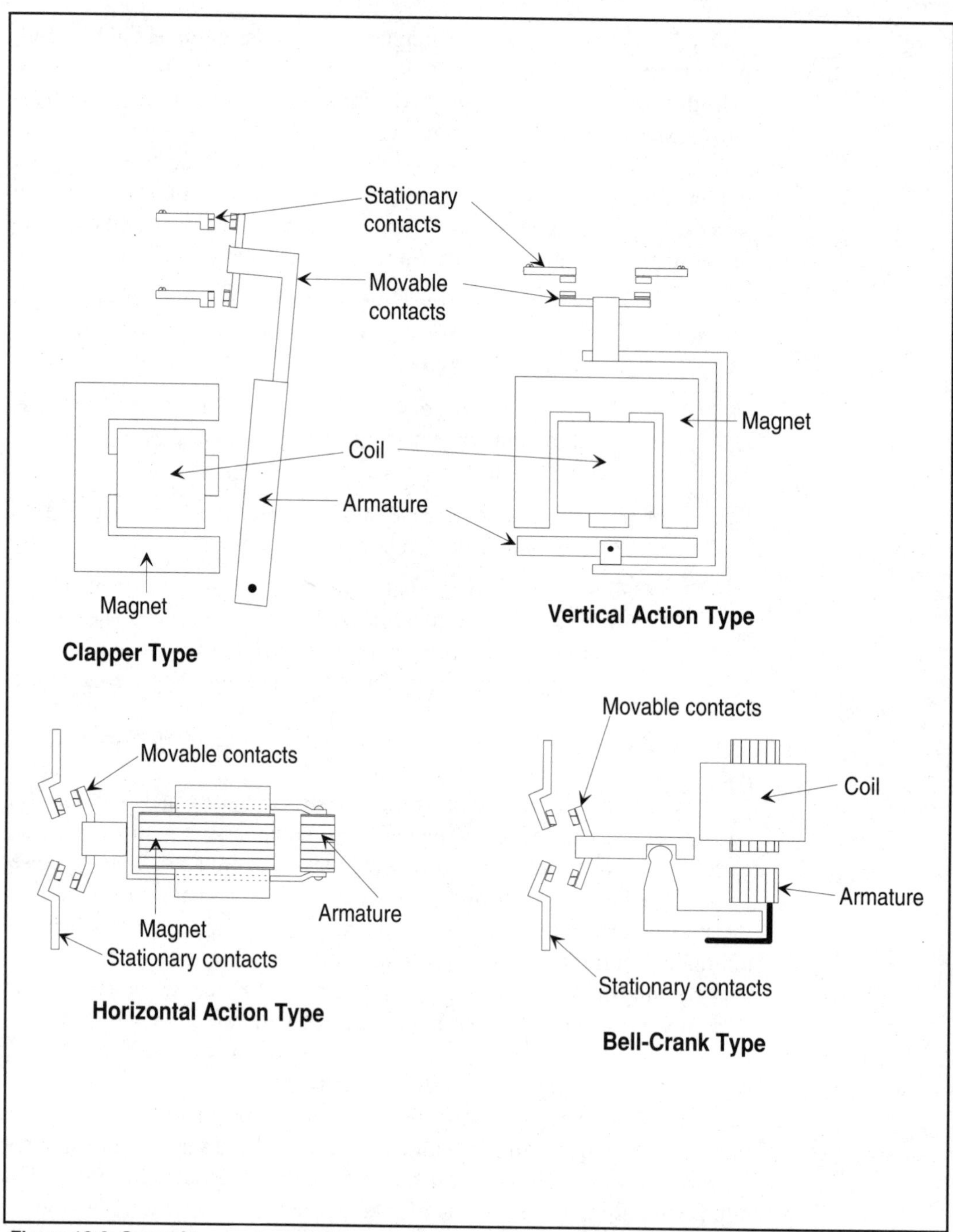

Figure 16-6: Several types of magnetic-armature assemblies.

To provide maximum pull and to help ensure quietness, the faces of the armature and the magnetic assembly are ground to a very close tolerance.

When a controller's armature has sealed-in, it is held closely against the magnet assembly. However, a small gap is always deliberately left in the iron circuit. When the coil becomes deenergized, some magnetic flux (residual magnetism) always remains, and if it were not for the gap in the iron circuit, the residual magnetism might be sufficient to hold the armature in the sealed-in position.

The shaded-pole principle is used to provide a time delay in the decay of flux in dc coils, but it is used more frequently to prevent a chatter and wear in the moving parts of ac magnets. A shading coil is a single turn of conducting material mounted in the face of the magnet assembly or armature. The alternating main magnetic flux induces currents in the shading coil, and these currents set up auxiliary magnetic flux that is out of phase from the pull due to the main flux, and this keeps the armature sealed-in when the main flux falls to zero (which occurs 120 times per second with 60-cycle ac). Without the shading coil, the armature would tend to open each time the main flux goes through zero. Excessive noise, wear on magnet faces, and heat would result. A magnet assembly and armature showing shading coils is shown in Figure 16-7.

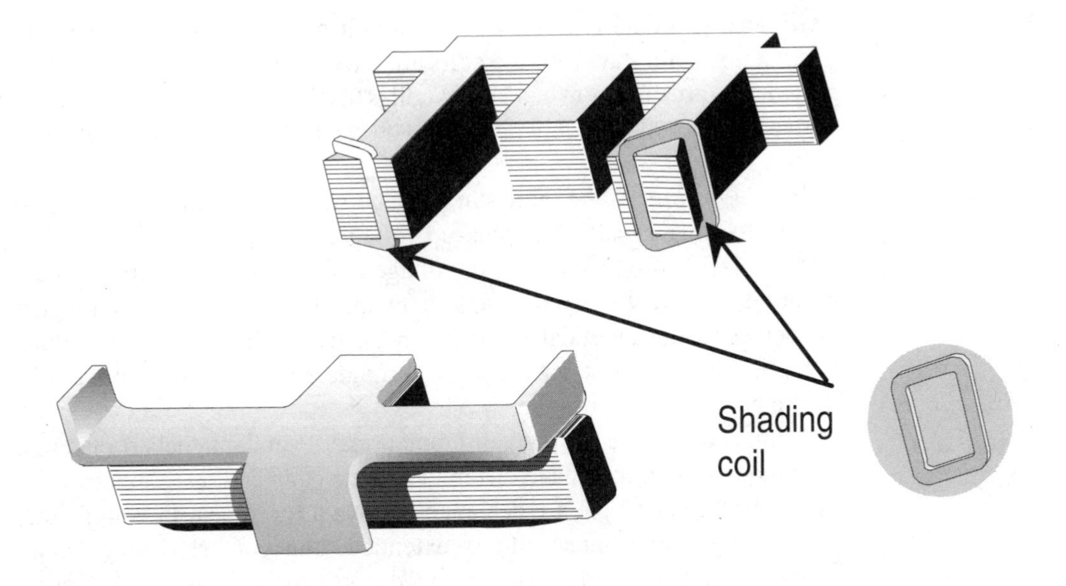

Figure 16-7: Magnet assembly and armature along with shading coils.

MAGNET COILS

The magnet coil used in motor controllers has many turns of insulated copper wire wound on a spool. Most coils are protected by an epoxy molding which makes them very resistant to mechanical damage.

When the controller is in the open position there is a large air gap (not to be confused with the built-in gap discussed previously) in the magnet circuit; this is when the armature is at its furthest distance from the magnet. The impedance of the coil is relatively low, due to the air gap, so that when the coil is energized, it draws a fairly high current. As the armature moves closer to the magnet assembly, the air gap is progressively reduced, and with it, the coil current, until the armature has sealed-in. The final current is referred to as the sealed current. The inrush current is approximately 6 to 10 times the sealed current. The ratio varies with individual designs. After the controller has been energized for some time, the coil will become hot. This will cause the coil current to fall to approximately 80 percent of its value when cold.

AC magnetic coils should never be connected in series. If one device were to seal-in ahead of the other, the increased circuit impedance will reduce the coil current so that the "slow" device will not pick up or, having picked up, will not seal. Consequently, ac coils are always connected in parallel.

Magnet coil data is usually given in volt-amperes (VA). For example, given a magnetic starter whose coils are rated at 600 VA inrush and 60 VA sealed, the inrush current of a 120-V coil is 600/120 or 5 A. The same starter with a 480-V coil will only draw 600/480 or 1.25 A inrush and 60/480 or .125 A sealed.

Pick-up voltage: The minimum voltage which will cause the armature to start to move is called the pick-up voltage.

Sealed-in voltage: The seal-in voltage is the minimum control voltage required to cause the armature to seat against the pole faces of the magnet. On devices using a vertical action magnet and armature, the seal-in voltage is higher than the pick-up voltage to provide additional magnetic pull to insure good contact pressure.

Control devices using the bell-crank armature and magnet arrangement are unique in that they have different force characteristics. Devices using this operating principle are designed to have a lower seal-in voltage than pick-up voltage. Contact life is extended, and contact damage under abnormal voltage conditions is reduced, for if the voltage is sufficient to pick-up, it is also high enough to seat the armature.

If the control voltage is reduced sufficiently, the controller will open. The voltage at which this happens is called the drop-out voltage. It is somewhat lower than the seal-in voltage.

Voltage Variation

NEMA standards require that the magnetic device operate properly at varying control voltages from a high of 110 percent to a low of 85 percent of rated coil voltage. This range, established by coil design, insures that the coil will withstand given temperature rises at voltages up to 10 percent over rated voltage, and that the armature will pick up and seal in, even though the voltage may drop to 15 percent under the nominal rating.

Effects of Voltage Variation

If the voltage applied to the coil is too high, the coil will draw more than its designed current. Excessive heat will be produced and will cause early failure of the coil insulation. The magnetic pull will be too high, which will cause the armature to slam home with excessive force. The magnet faces will wear rapidly, leading to a shortened life for the controller. In addition, contact bounce may be excessive, resulting in reduced contact life.

Low control voltage produces low coil currents and reduced magnetic pull. On devices with vertical action assemblies, if the voltage is greater than pick-up voltage, but less than seal-in voltage, the controller may pick up but will not seal. With this condition, the coil current will not fall to the sealed value. As the coil is not designed to carry continuously a current greater than its sealed current, it will quickly get very hot and burn out. The armature will also chatter. In addition to the noise, wear on the magnet faces result.

In both vertical action and bell-crank construction, if the armature does not seal, the contacts will not close with adequate pressure. Excessive heat, with arcing and possible welding of the contacts, will occur as the controller attempts to carry current with insufficient contact pressure.

AC Hum

All ac devices which incorporate a magnetic effect produce a characteristic hum. This hum or noise is due mainly to the changing magnetic pull (as the flux changes) inducing mechanical vibrations. Contactors, starters, and relays could become excessively noisy as a result of some of the following operating conditions:

- Broken shading coil.
- Operating voltage too low.
- Misalignment between the armature and magnet assembly—the armature is then unable to seat properly.
- Wrong coil.
- Dirt, rust, filings, etc. on the magnet faces—the armature is unable to seal in completely.
- Jamming or binding of moving parts so that full travel of the armature is prevented.
- Incorrect mounting of the controller, as on a thin piece of plywood fastened to a wall; such mounting may cause a "sounding board" effect.

POWER CIRCUITS IN MOTOR STARTERS

The power circuit of a starter includes the stationary and movable contacts, and the thermal unit or heater portion of the overload relay assembly. The number of contacts (or "poles") is determined by the electrical service. In a three-phase, 3-wire system, for example, a 3-pole starter is required. *See* Figure 16-8.

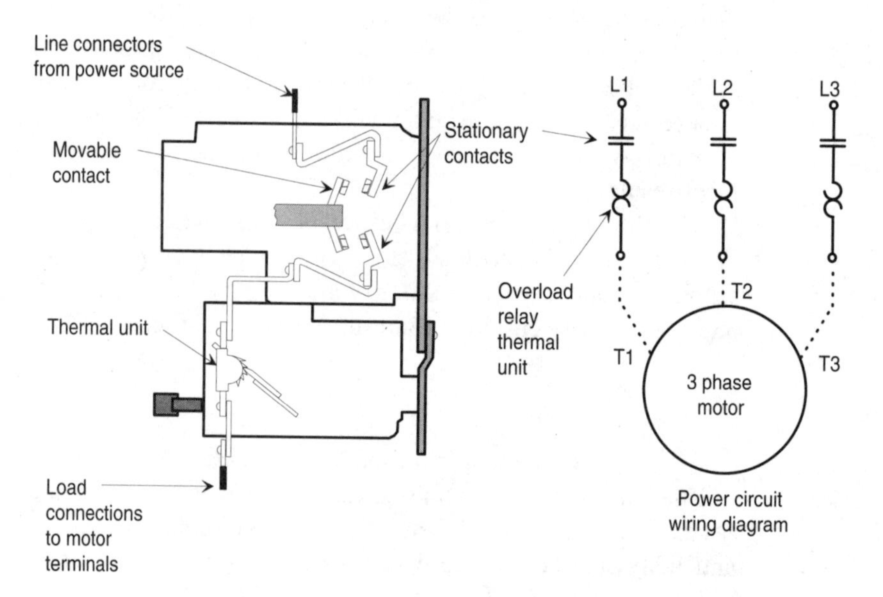

Figure 16-8: Power circuit in a typical 3-pole magnetic starter.

To be suitable for a given motor application, the magnetic starter selected should equal or exceed the motor horsepower and full-load current ratings. For example, let's assume that we want to select a motor starter for a 50-hp motor to be supplied by a 240-V, three-phase service, and the full-load current of the motor is 125 A. Referring to the table in Figure 16-9, it can be seen that a NEMA Size 4 starter would be required for normal motor duty. If the motor were to be used for jogging or plugging duty, a NEMA Size 5 starter should be chosen.

CAUTION!

For three-phase motors having locked-rotor kVA per horsepower in excess of that for the motor code letters in the right table (Figure 16-11), do not apply the controller at its maximum rating without consulting the manufacturer. In most cases, the next higher horsepower rated controller should be used.

Power circuit contacts handle the motor load. The ability of the contacts to carry the full-load current without exceeding a rated temperature rise, and their isolation from adjacent contacts, corresponds to NEMA Standards established to categorize the NEMA Size of the starter. The starter must also be capable of interrupting the motor circuit under locked rotor current conditions.

NEMA SIZE	Volts	Maximum HP Nonplugging and Nonjogging Duty		Maximum HP Rating Plugging and Jogging Duty		Continuous Current Rating in amperes 600 V Max.	Service-Limit Current Rating Amperes	Tungsten and Infrared Lamp Load, Amperes 250 V Max.	Resistance Heating Loads in KW, other than Infrared Lamps Loads		kVA Rating for Switching Transformer Primaries at 50 or 60 Cycles		3-Phase Rating for Switching Capacitors
		Single Phase	Poly-Phase	Single Phase	Poly-Phase				Single Phase	Poly-Phase	Single Phase	Poly-Phase	Kvar
00	115	⅓	—	—	—	9	11	5	—	—	—	—	—
	200	—	1½	—	—	9	11	5	—	—	—	—	—
	230	1	1½	—	—	9	11	5	—	—	—	—	—
	380	—	1½	—	—	9	11	—	—	—	—	—	—
	460	—	2	—	—	9	11	—	—	—	—	—	—
	575	—	2	—	—	9	11	—	—	—	—	—	—
0	115	1	—	½	—	18	21	10	—	—	0.9	1.2	—
	200	—	3	—	1½	18	21	10	—	—	—	1.4	—
	230	2	3	1	1½	18	21	10	—	—	1.4	1.7	—
	380	—	5	—	1½	18	21	—	—	—	—	2.0	—
	460	—	5	—	2	18	21	—	—	—	1.9	2.5	—

Figure 16-9: Electrical ratings for ac magnetic contactors and starters.

NEMA SIZE	Volts	Maximum HP Nonplugging and Nonjogging Duty		Maximum HP Rating Plugging and Jogging Duty		Cont. Current Rating in amperes 600 V Max.	Service-Limit Current Rating Amperes	Tungsten and Infrared Lamp Load, Amperes 250 V Max.	Resistance Heating Loads in KW, other than Infrared Lamps Loads		kVA Rating for Switching Transformer Primaries at 50 or 60 Cycles		3-Phase Rating for Switching Capacitors
1	115	2	—	1	—	27	32	15	3	5	1.4	1.7	—
	200	—	7½	—	3	27	32	15	—	9.1	—	3.5	—
	230	3	7½	2	3	27	32	15	6	10	1.9	4.1	—
	380	—	10	—	5	27	32	—	—	16.5	—	4.3	—
	460	—	10	—	5	27	32	—	12	20	3	5.3	—
	575	—	10	—	5	27	32	—	15	25	3	5.3	—
1P	115	3	—	1½	—	36	42	24	—	—	—	—	—
	230	5	—	3	—	36	42	24	—	—	—	—	—
2	115	3	—	2	—	45	52	30	5	8.5	1.0	4.1	—
	200	—	10	—	7½	45	52	30	—	15.4	—	6.6	11.3
	230	7½	15	5	10	45	52	30	10	17	4.6	7.6	13
	380	—	25	—	15	45	52	—	—	28	—	9.9	21
	460	—	25	—	15	45	52	—	20	34	5.7	12	26
	575	—	25	—	15	45	52	—	25	43	5.7	12	33
3	115	7½	—	—	—	90	104	60	10	17	4.6	7.6	—
	200	—	25	—	15	90	104	60	—	31	—	13	23.4
	230	15	30	—	20	90	104	60	20	34	8.6	15	27
	380	—	50	—	30	90	104	—	—	56	—	19	43.7
	460	—	50	—	30	90	104	—	40	68	14	23	53
	575	—	50	—	30	90	104	—	50	86	14	23	67

Figure 16-9: Electrical ratings for ac magnetic contactors and starters. *(Cont.)*

NEMA SIZE	Volts	Maximum HP Nonplugging and Nonjogging Duty		Maximum HP Rating Plugging and Jogging Duty		Continuous Current Rating in amperes 600 V Max.	Service-Lmit Current Rating Amperes	Tungsten and Infrared Lamp Load, Amperes 250 V Max.	Resistance Heating Loads in KW, other than Infrared Lamps Loads		kVA Rating for Switching Transformer Primaries at 50 or 60 Cycles		3-Phase Rating for Switching Capacitors
4	200	—	40	—	25	135	156	120	—	45	—	20	34
	230	—	50	—	30	135	156	120	30	52	11	23	40
	380	—	75	—	50	135	156	—	—	86.7	—	38	66
	460	—	100	—	60	135	156	—	60	105	22	46	80
	575	—	100	—	60	135	156	—	75	130	22	46	100
5	200	—	75	—	60	270	311	240	—	91	—	40	69
	230	—	100	—	75	270	311	240	60	105	28	46	80
	380	—	150	—	125	270	311	—	—	173	—	75	132
	460	—	200	—	150	270	311	—	120	210	40	91	160
	575	—	200	—	150	270	311	—	150	260	40	91	200
6	200	—	150	—	125	540	621	480	—	182	—	79	139
	230	—	200	—	150	540	621	480	120	210	57	91	160
	380	—	300	—	250	540	621	—	—	342	—	148	264
	460	—	400	—	300	540	621	—	240	415	86	180	320
	575	—	400	—	300	540	621	—	300	515	86	180	400
7	230	—	300	—	—	810	932	720	180	315	—	—	240
	460	—	600	—	—	810	932	—	360	625	—	—	480
	575	—	600	—	—	810	932	—	450	775	—	—	600
8	230	—	450	—	—	1215	1400	1080	—	—	—	—	360
	460	—	900	—	—	1215	1400	—	—	—	—	—	720
	575	—	900	—	—	1215	1400	—	—	—	—	—	900

Figure 16-9: Electrical ratings for ac magnetic contactors and starters. (*Cont.*)

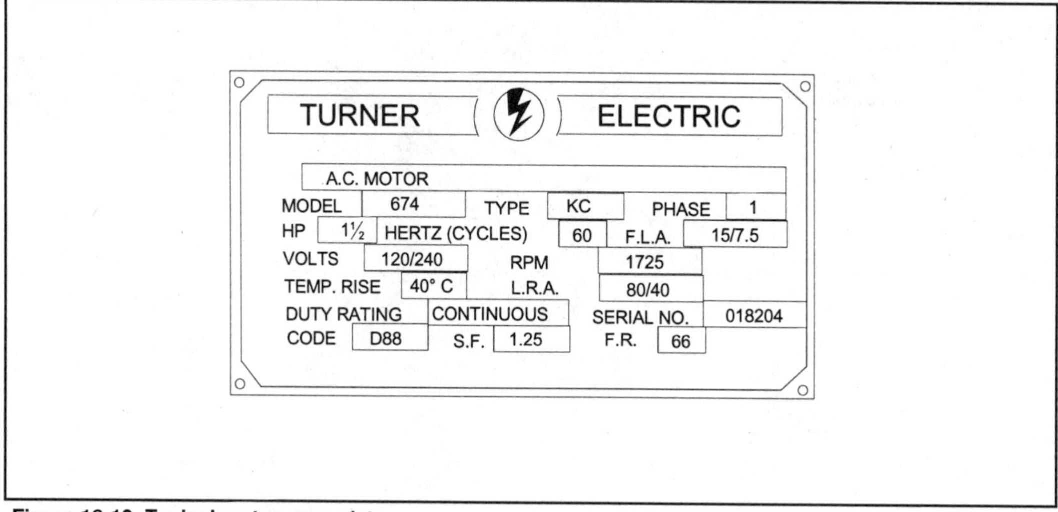

Figure 16-10: Typical motor nameplate.

Motor Nameplates

Much information about sizing motor starters can be found from the motor's nameplate. Consequently, a review of motor nameplates is in order.

A typical motor nameplate is shown in Figure 16-10. A nameplate is one of the most important parts of a motor since it gives the motor's electrical and mechanical characteristics; that is, the horsepower, voltage, rpms, etc. Always refer to the motor's nameplate before connecting it to an electric system or selecting the motor starter and related components. The same is true when performing preventative maintenance or troubleshooting motors.

Referring again to the motor nameplate in Figure 16-10, note that the manufacturer's name and logo is at the top of the plate; these items, of course, will change with each manufacturer. The line directly below the manufacturer's name identifies the motor for use on ac systems as opposed to dc or ac-dc systems. The model number identifies that particular motor from any other. The type or class specifies the insulation used to ensure the motor will perform at the rated horsepower and service-factor load. The phase indicates whether the motor has been designed for single- or three-phase use.

When selecting motor starters for a given motor, the motor code letter on the nameplate plays an important role. The chart in Figure 16-11 may be used as a guide when selecting motor controllers.

Controller HP Rating	Maximum Allowable Motor Code Letter
1½	L
3 – 5	K
7½ and above	H

Figure 16-11: Using motor code letters to size motor controllers.

OVERLOAD PROTECTION

Overload protection for an electric motor is necessary to prevent burnout and to ensure maximum operating life. Electric motors will, if permitted, operate at an output of more than rated capacity. Conditions of motor overload may be caused by an overload on driven machinery, by a low line voltage, or by an open line in a polyphase system, which results in single-phase operation. Under any condition of overload, a motor draws excessive current that causes overheating. Since motor winding insulation deteriorates when subjected to overheating, there are established limits on motor operating temperatures. To protect a motor from overheating, overload relays are employed on a motor control to limit the amount of current drawn. This is overload protection, or running protection.

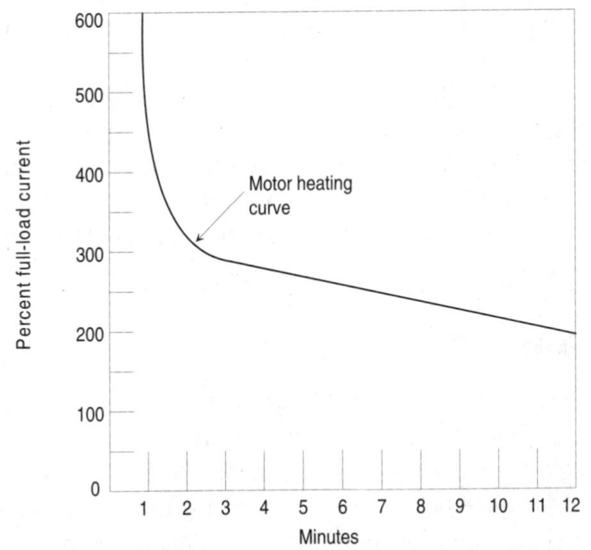

Figure 16-12: Sensing properties of overload protection should follow heating curve of the motor.

The ideal overload protection for a motor is an element with current-sensing properties very similar to the heating curve of the motor (*see* Figure 16-12), which would act to open the motor circuit when full-load current is exceeded. The operation of the protective device should be such that the motor is allowed to carry harmless overloads, but is quickly removed from the line when an overload has persisted too long.

Fuses are not designed to provide overload protection. Their basic function is to protect against short circuits (overcurrent protection). Motors draw a high inrush current when starting and conventional single-element fuses have no way of distinguishing between this temporary and harmless inrush current and a damaging overload. Such fuses, chosen on the basis of motor full-load current, would blow every time the motor is started. On the other hand, if a fuse were chosen large enough to pass the starting or inrush current, it would not protect the motor against small, harmful overloads that might occur later.

Dual-element or time-delay fuses can provide motor overload protection, but suffer the disadvantages of being nonrenewable and must be replaced.

The overload relay is the heart of motor protection. It has inverse trip-time characteristics, permitting it to hold in during the accelerating period (when inrush current is drawn), yet providing protection on small overloads above the full-load current when the motor is running. Unlike dual-element fuses, overload relays are renewable and can withstand repeated trip and reset cycles without need of replacement. They cannot, however, take the place of overcurrent protective equipment.

The overload relay consists of a current-sensing unit connected in the line to the motor, plus a mechanism, actuated by the sensing unit, that serves to directly or indirectly break the circuit. In a manual starter, an overload trips a mechanical latch and causes the starter contacts to open and disconnect the motor from the line. In magnetic starters, an overload opens a set of contacts within the overload relay itself. These contacts are wired in series with the starter coil in the control circuit of the magnetic starter. Breaking the coil circuit causes the starter contacts to open, disconnecting the motor from the line.

Overload relays can be classified as being either thermal or magnetic. Magnetic overload relays react only to current excesses and are not affected by temperature. As the name implies, thermal overload relays rely on the rising temperatures caused by the overload current to trip the overload mechanism. Thermal overload relays can be further subdivided into two types, melting alloy and bimetallic.

Melting Alloy Thermal Overload Relays

The melting alloy assembly of the heater element overload relay and solder pot is shown in Figure 16-13. Excessive overload motor current passes through the heater element, thereby melting a eutectic alloy solder pot. The ratchet wheel will then be allowed to turn in the molten pool, and a tripping action of the starter control circuit results, stopping the motor. A cooling off period is required to allow the solder pot to "freeze" before the overload relay assembly may be reset and motor service restored.

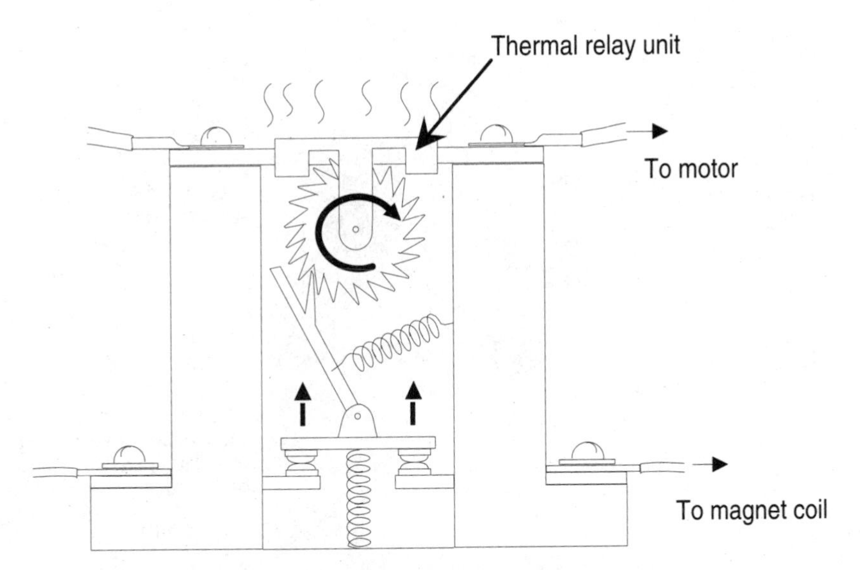

Figure 16-13: Operating characteristics of a melting-alloy overload relay.

Melting alloy thermal units are interchangeable and of a one-piece construction, which ensures a constant relationship between the heater element and solder pot and allows factory calibration, making them virtually tamper-proof in the field. These important features are not possible with any other type of overload relay construction. A wide selection of these interchangeable thermal units is available to give exact overload protection of any full-load current to a motor. An internal view of a melting-alloy relay is shown in Figure 16-14 on the next page. Again, as heat melts the alloy, the ratchet wheel is free to turn; the spring, in turn, pushes the contacts open and shuts down the motor.

Bimetallic Thermal Overload Relays

Bimetallic overload relays are designed specifically for two general types of application: the automatic reset feature is of decided advantage when devices are mounted in locations not easily accessible for manual operation and, second, these relays can easily be adjusted to trip within a range of 85 to 115 percent of the nominal trip rating of the heater unit. This feature is useful when the recommended heater size might result in unnecessary tripping, while the next larger size would not give adequate protection. Ambient temperatures affect overload relays operating on the principle of heat. Figure 16-15 shows a bimetallic overload relay with side cover removed.

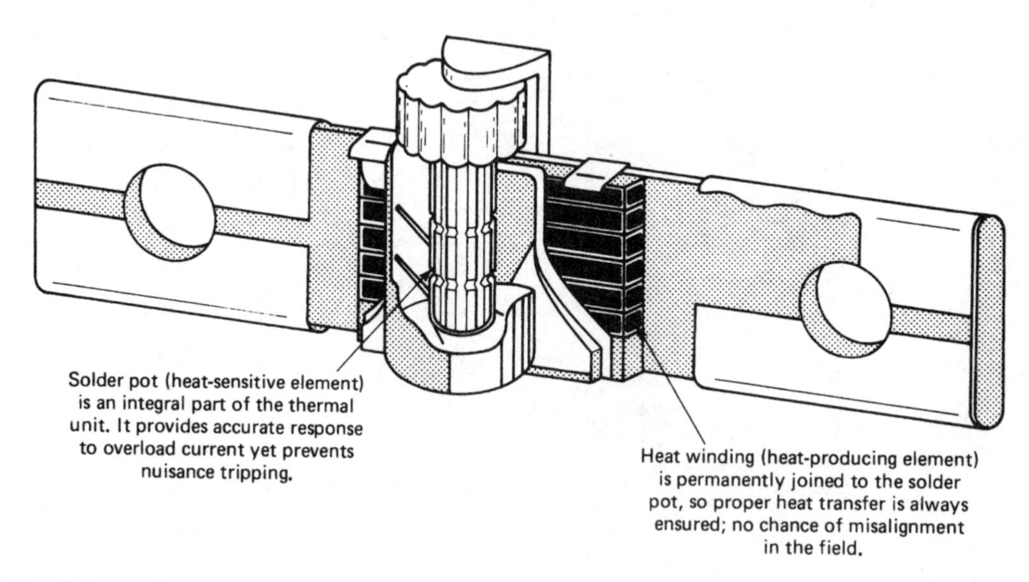

Figure 16-14: Melting alloy thermal overload relay.

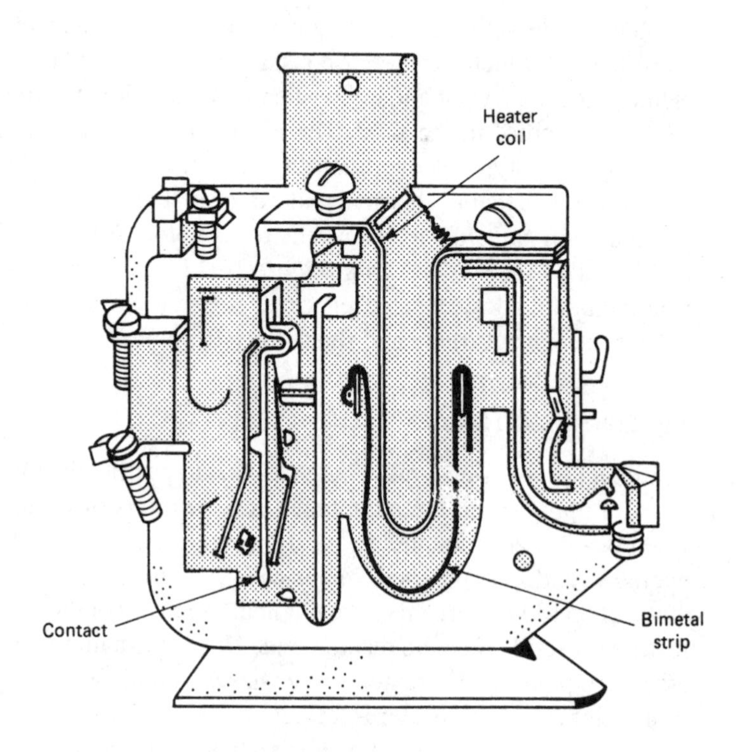

Figure 16-15: Bimetallic overload relay with side cover removed.

Ambient Compensation

Ambient-compensated bimetallic overload relays were designed for one particular situation, that is, when the motor is at a constant temperature and the controller is located separately in a varying temperature. In this case, if a standard thermal overload relay were used, it would not trip consistently at the same level of motor current if the controller temperature changed. This thermal overload relay is always affected by the surrounding temperature. To compensate for the temperature variations the controller may see, an ambient-compensated overload relay is applied. Its trip point is not affected by temperature and it performs consistently at the same value of current.

Melting alloy and bimetallic overload relays are designed to approximate the heat actually generated in the motor. As the motor temperature increases, so does the temperature of the thermal unit. The motor and relay heating curves (*see* Figure 16-16) show this relationship. From this graph, we can see that, no matter how high the current drawn, the overload relay will provide protection, yet the relay will not trip out unnecessarily.

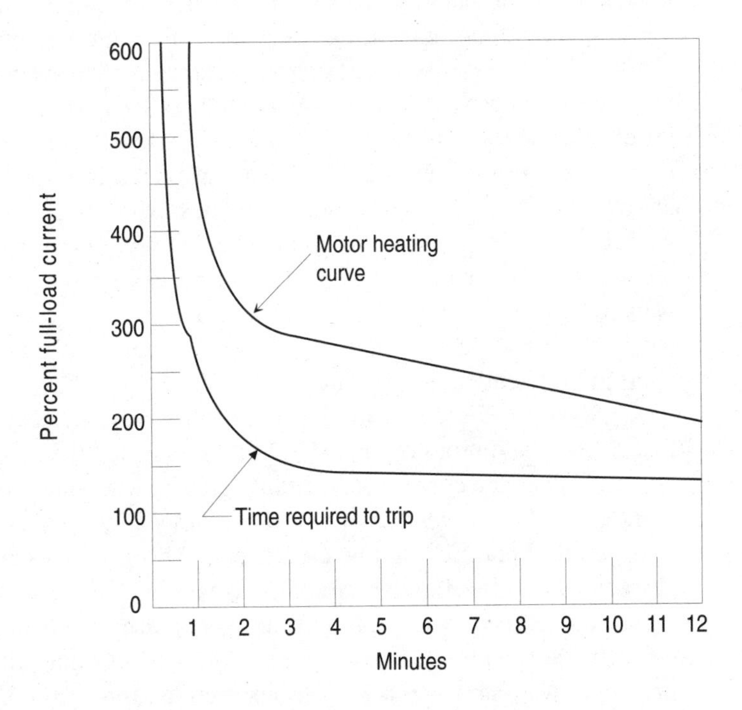

Figure 16-16: Comparison of motor heating curve and overload relay trip curve.

Selecting Overload Relays

When selecting thermal overload relays, the following must be considered:

- Motor full-load current
- Type of motor
- Difference in ambient temperature between motor and controller

Motors of the same horsepower and speed do not all have the same full-load current, and the motor nameplate must always be checked to obtain the full-load amperes for a particular motor. Do not use a published table. Thermal unit selection tables are published on the basis of continuous-duty motors, with 1.15 service factor, operating under normal conditions. The tables are shown in the catalog of manufacturers and also appear on the inside of the door or cover of the motor controller. These selections will properly protect the motor and allow the motor to develop its full horsepower, allowing for the service factor, if the ambient temperature is the same at the motor as at the controller. If the temperatures are not the same, or if the motor service factor is less than 1.15, a special procedure is required to select the proper thermal unit. Standard overload relay contacts are closed under normal conditions and open when the relay trips. An alarm signal is sometimes required to indicate when a motor has stopped due to an overload trip. Also, with some machines, particularly those associated with continuous processing, it may be required to signal an overload condition, rather than have the motor and process stop automatically. This is done by fitting the overload relay with a set of contacts that close when the relay trips, thus completing the alarm circuit. These contacts are appropriately called alarm contacts.

A magnetic overload relay has a movable magnetic core inside a coil that carries the motor current. The flux set up inside the coil pulls the core upward. When the core rises far enough, it trips a set of contacts on the top of the relay. The movement of the core is slowed by a piston working in an oil-filled dashpot mounted below the coil. This produces an inverse-time characteristic. The effective tripping current is adjusted by moving the core on a threaded rod. The tripping time is varied by uncovering oil bypass holes in the piston. Because of the time and current adjustments, the magnetic overload relay is sometimes used to protect motors having long accelerating times or unusual duty cycles.

PROTECTIVE ENCLOSURES

The correct selection and installation of an enclosure for a particular application can contribute considerably to the length of life and trouble-free operation. To shield electrically live parts from accidental contact, some form of enclosure is always necessary. This function is usually filled by a general-purpose, sheet-steel cabinet. Frequently, however, dust, moisture, or explosive gases make it necessary to employ a special enclosure to protect the motor controller from corrosion or the surrounding equipment from explosion. In selecting and installing control apparatus, it is always necessary to consider carefully the conditions under which the apparatus must operate; there are many applications where a general-purpose enclosure does not afford protection.

Underwriters' Laboratories has defined the requirements for protective enclosures according to the hazardous conditions, and the National Electrical Manufacturers Association (NEMA) has standardized enclosures from these requirements:

NEMA 1: general purpose. The general-purpose enclosure is intended primarily to prevent accidental contact with the enclosed apparatus. It is suitable for general-purpose applications indoors where it is not exposed to unusual service conditions. A NEMA 1 enclosure serves as protection against dust and light and indirect splashing, but is not dust-tight.

NEMA 3: dust-tight, raintight. This enclosure is intended to provide suitable protection against specified weather hazards. A NEMA 3 enclosure is suitable for application outdoors, such as construction work. It is also sleet-resistant.

NEMA 3R: rainproof, sleet resistant. This enclosure protects against interference in operation of the contained equipment due to rain, and resists damage from exposure to sleet. It is designed with conduit hubs and external mounting as well as drainage provisions.

NEMA 4: watertight. A watertight enclosure is designed to meet a hose test which consists of a stream of water from a hose with a 1-in nozzle, delivering at least 65 gals per minute. The water is directed on the enclosure from a distance of not less than 10 ft and for a period of 5 min. During this period, it may be directed in one or more directions, as desired. There should be no leakage of water into the enclosure under these conditions.

NEMA 4X: watertight, corrosion-resistant. These enclosures are generally constructed along the lines of NEMA 4 enclosures except that they are made of a material that is highly resistant to corrosion. For this reason, they are ideal in applications such as meatpacking and chemical plants, where contaminants would ordinarily destroy a steel enclosure over a period of time.

NEMA 7: hazardous locations, Class I. These enclosures are designed to meet the application requirements of the National Electrical Code for Class I hazardous locations: "Class I locations are those in which flammable gases or vapors are or may be present in the air in quantities sufficient to produce explosive or ignitible mixtures." In this type of equipment, the circuit interruption occurs in air.

NEMA 9: hazardous locations, Class II. These enclosures are designed to meet the application requirements of the *NEC* for Class II hazardous locations. "Class II locations are those which are hazardous because of the presence of combustible dust." The letter or letters following the type number indicates the particular group or groups of hazardous locations (as defined in the *NEC*) for which the enclosure is designed. The designation is incomplete without a suffix letter or letters.

NEMA 12: industrial use. This type of enclosure is designed for use in those industries where it is desired to exclude such materials as dust, lint, fibers and flyings, oil seepage, or coolant seepage. There are no conduit openings or knockouts in the enclosure, and mounting is by means of flanges or mounting feet.

NEMA 13: oil-tight, dust-tight. NEMA 13 enclosures are generally made of cast iron, gasketed, or permit use in the same environments as NEMA 12 devices. The essential difference is that due to its cast housing, a conduit entry is provided as an integral part of the NEMA 13 enclosure, and mounting is by means of blind holes rather than mounting brackets.

MOTOR-CONTROL CIRCUITS

A complete wiring diagram is best used when making the initial connections of a circuit or when tracing a fault in a circuit. It shows the device in symbol form and indicates the actual connections of all wires between the devices. Ladder diagrams use the same symbols to represent the individual devices, but indicate that these devices are in the same circuit by using only one line. Such schematic diagrams are simple and can be quickly prepared when a motor-control circuit must be prepared.

A motor-control circuit is represented by its complete wiring diagram in Figure 16-17, and by a ladder diagram in Figure 16-18. In the wiring diagram in Figure 16-17, the three supply conductors are indicated by L_1, L_2, and L_3, and the motor terminals by T_1, T_2, and T_3. Each line has a terminal overload protective device (O.L.) connected in series with the normally open line contactors M_1, M_2, and M_3, which are controlled by the magnetic coil (C). Each contactor has a pair of contacts that close or open during operation. The control station, consisting of start-stop push-

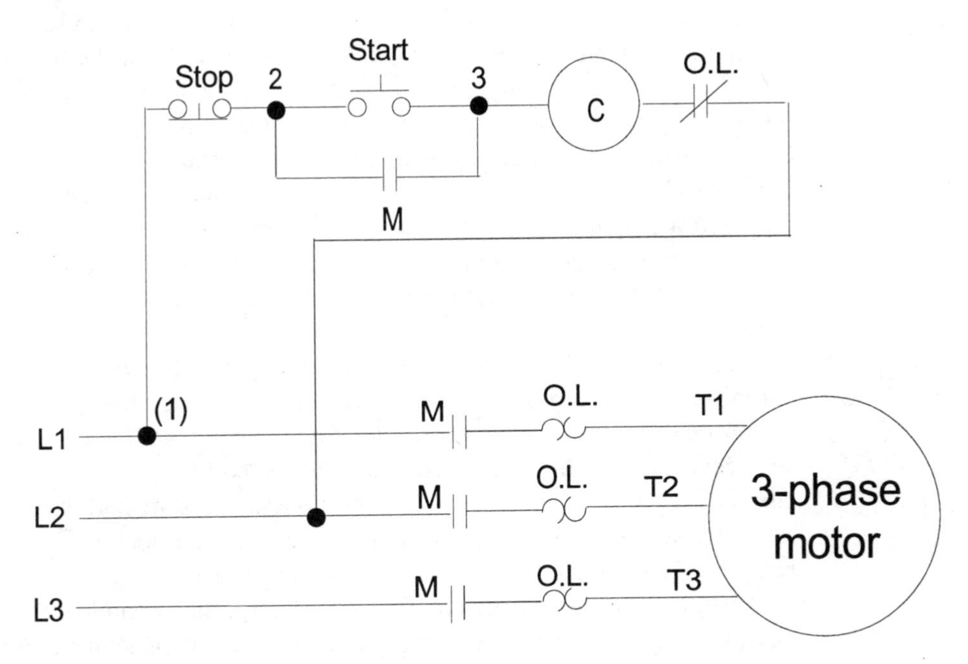

Figure 16-17: Complete wiring diagram of a simple motor-control circuit.

buttons is connected across line L_1 and L_2. An auxiliary contactor (M) is connected in series with the stop pushbutton and in parallel with the start pushbutton. The control circuit also has a normally closed overload contactor (O.C.) connected in series with the starter coil (C).

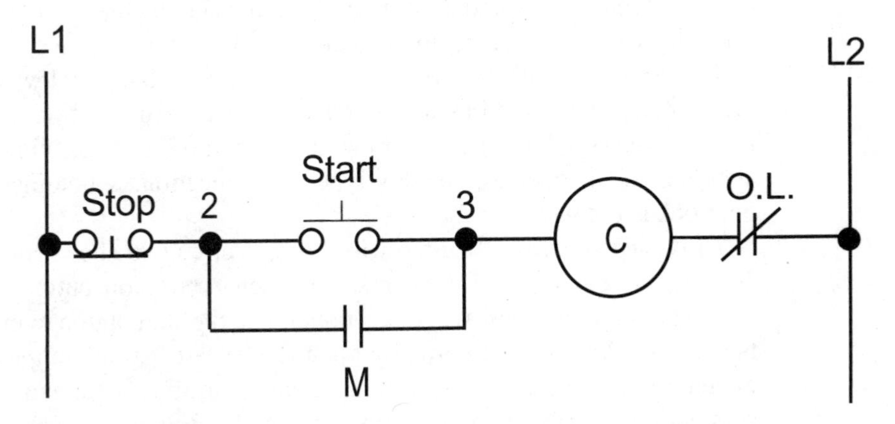

Figure 16-18: Ladder diagram of the control circuit in Figure 16-17.

The same connections are represented in Figure 16-18 by a ladder diagram. As mentioned previously, it is customary to draw the two main lines L_1 and L_2 vertically, and to connect them with a horizontal line representing the control circuit and a control station containing two pushbuttons, an auxiliary interlocking contactor, the normally closed overload contactor OC, and the starter coil C. The diagram depicts the situation when the control circuit is not energized and the normally open contactors are open. There is no complete path for the current unless the start button is pushed. Always read the ladder diagram from left to right; that is, from L_1 to L_2.

When the start pushbutton is momentarily pressed down, the path is complete from L_1 through the closed stop button, through the start button, the normally closed overload contactor, and the coil C to line L_2. Current will flow through this circuit and energize the coil C. Coil C closes the auxiliary, or sealing contactor. The spring-actuated start button may be released but the auxiliary contacts of contactor M interlock (or seal) the circuit and keep it closed as long as coil C is energized.

When the contacts of the control device close, they complete the coil circuit of the motor starter, causing it to pick up and connect the motor to the lines. When the control device contacts open, the starter is deenergized, stopping the motor.

The line contactors M_1, M_2, and M_3 in Figure 16-17 close when coil C is energized causing the circuit to the motor terminals to be completed. The wiring diagrams in Figures 16-17 and 16-18 do not show the motor starter, the speed controller, or similar control devices — only the pushbutton arrangement for starting and stopping the motor.

When the stop pushbutton is pressed momentarily, the circuit is opened, coil C is deenergized, and the auxiliary contact and the line contacts open. There is no path for current to the motor, so the motor stops.

Two-wire control provides low-voltage release but not low-voltage protection. When wired as illustrated, the starter will function automatically in response to the direction of the control device, without the attention of an operator. In this type of connection, a holding circuit interlock is not necessary.

Three-wire control: A three-wire control circuit is shown in Figure 16-19. This circuit uses momentary contacts, start-stop buttons, and a holding circuit interlock, wired in parallel with the start button to maintain the circuit. Pressing the normally open (NO) start button completes the circuit to the coil. The power circuit contacts in lines 1, 2, and 3 close, completing the circuit to the motor, and the holding circuit contact also closes. Once the starter has picked up, the start button can be released, as

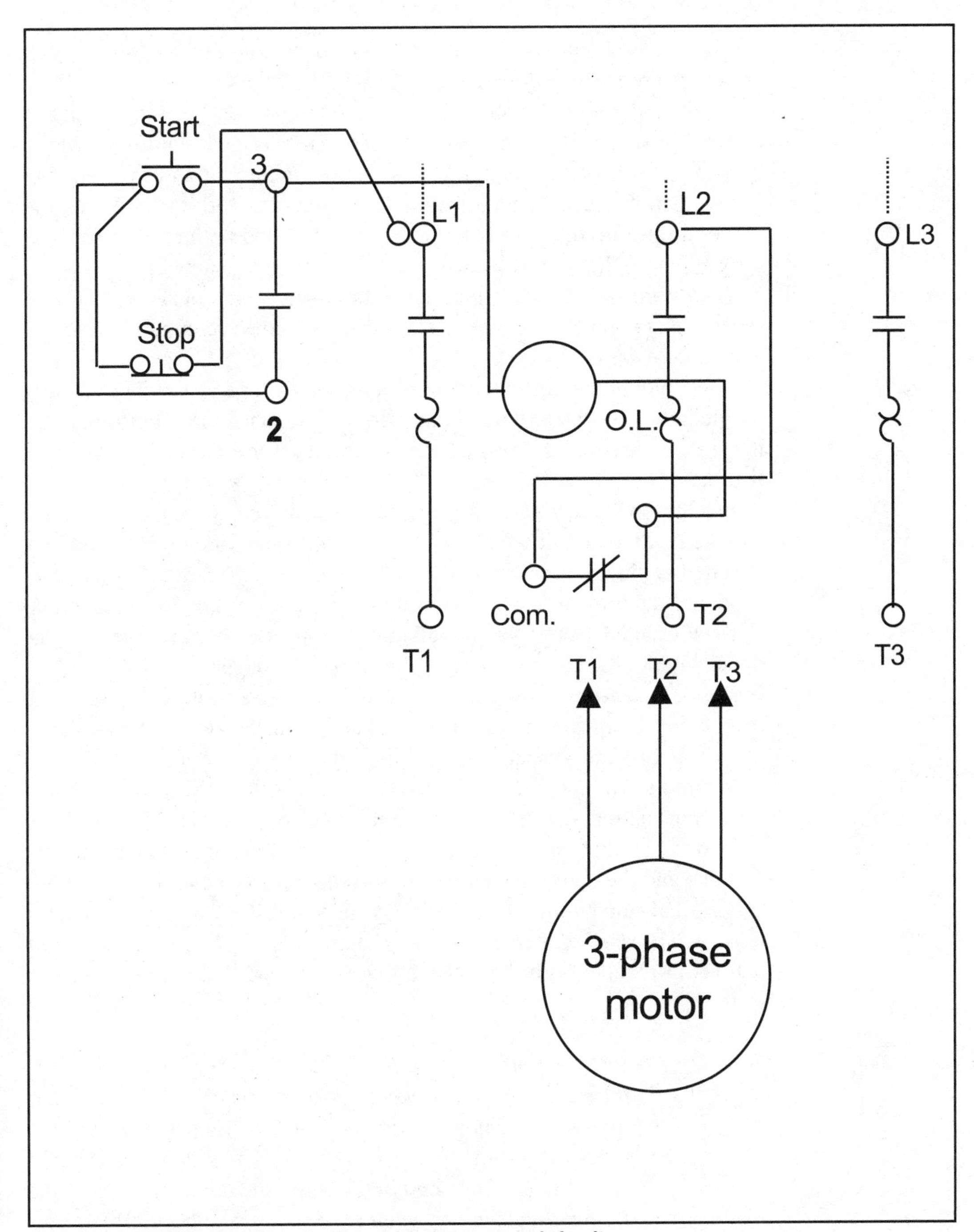

Figure 16-19: Wiring diagram of a three-wire motor-control circuit.

the now-closed interlock contact provided an alternative current path around the reopened start contact.

Pressing the normally closed (NC) stop button will open the circuit to the coil, causing the starter to drop out. An overload condition, which caused the overload contact to open, a power failure, or a drop in voltage to less than the seal-in value would also deenergize the starter. When the starter drops out, the interlock contact reopens, and both current paths to the coil, through the start button and the interlock, are now open.

Since three wires from the pushbutton station are connected into the starter — at points 1, 2, and 3 — this wiring scheme is commonly referred to as three-wire control.

The holding circuit interlock is a normally open auxiliary contact provided on the standard magnetic starters and contactors. It closes when the coil is energized to form a holding circuit for the starter after the start button has been released.

In addition to the main or power contacts which carry the motor current, and the holding circuit interlock, a starter can be provided with externally attached auxiliary contacts, commonly called electrical interlocks. Interlocks are rated to carry only control circuit currents, not motor currents. Both NO and NC versions are available. Among a wide variety of applications, interlocks can be used to control other magnetic devices where sequence operation is desired; to electrically prevent another controller from being energized at the same time; and to make and break circuits to indicating or alarm devices such as pilot lights, bells, or other signals.

If motors are required to be stated from more than one location, additional pushbutton stations may be connected to the circuit. In doing so, additional start buttons must be connected in parallel with the original start buttons, and the additional stop buttons must be connected in series with the original stop button as shown in Figure 16-20. The auxiliary contactor must also be starting. For three control stations, there should be three start buttons in parallel with the auxiliary contactor and three stop buttons in series.

Note: *Any control device connected in the control circuit to start the motor must be connected in parrallel with the start button and be of the normally open type. Every device that has the function of stopping the motor must be in series with the stop button and be normally closed. By the addition of elements to the circuit in this manner, a complex control circuit may be obtained.*

Reversing Motor Rotation

The circuit in Figure 16-20 shows a three-pole reversing starter used to control a three-phase motor. Three-phase, squirrel-cage motors can be reversed by reconnecting any two of the three line connections to the motor. By interwiring two contractors, an electromagnetic method of making the reconnection can be obtained.

As seen in the power circuit (Figure 16-20), the contacts (F) of the forward contactor — when closed — connect lines 1, 2, and 3 to the motor terminals T1, T2, and T3, respectively. As long as the forward contacts are closed, mechanical and electrical interlocks prevent the reverse contactor from being energized.

When the forward contactor is deenergized, the second contractor can be picked up, closing its contacts (R), which reconnect the lines to the motor. Note that by running through the reverse contacts, line 1 is connected to motor terminal T3, and line 3 is connected to motor terminal T1. The motor will now run in reverse.

A modified reversing-control circuit is shown in Figure 16-21, while a ladder diagram of the same circuit is shown in Figure 16-22. Instead of using two starting buttons for forward and reverse rotation, one selector switch is used with one start button. When the selector switch is first turned

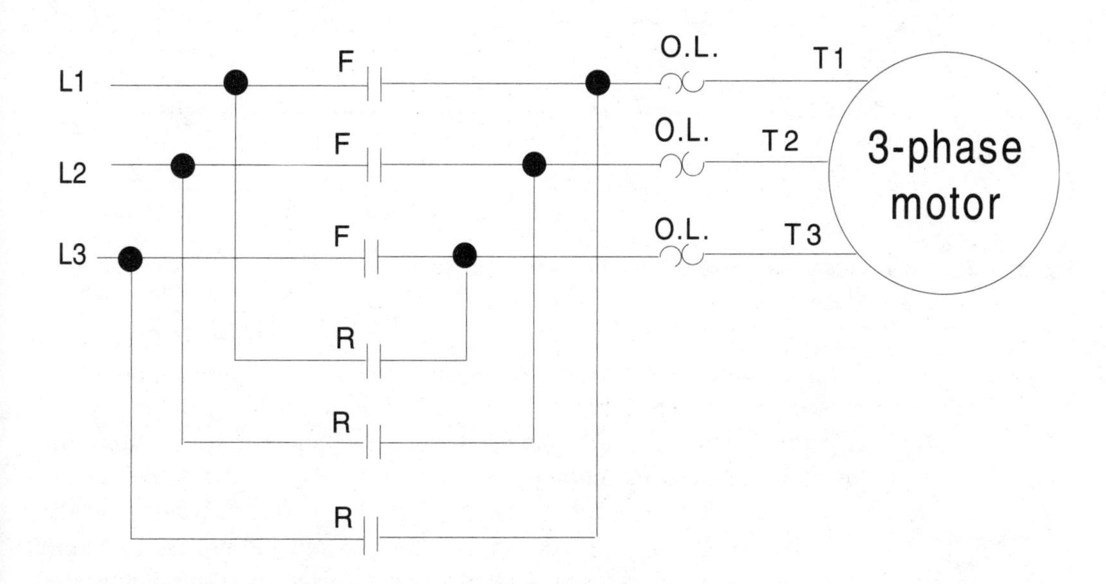

Figure 16-20: Three-pole reversing starter used to control a three-phase motor.

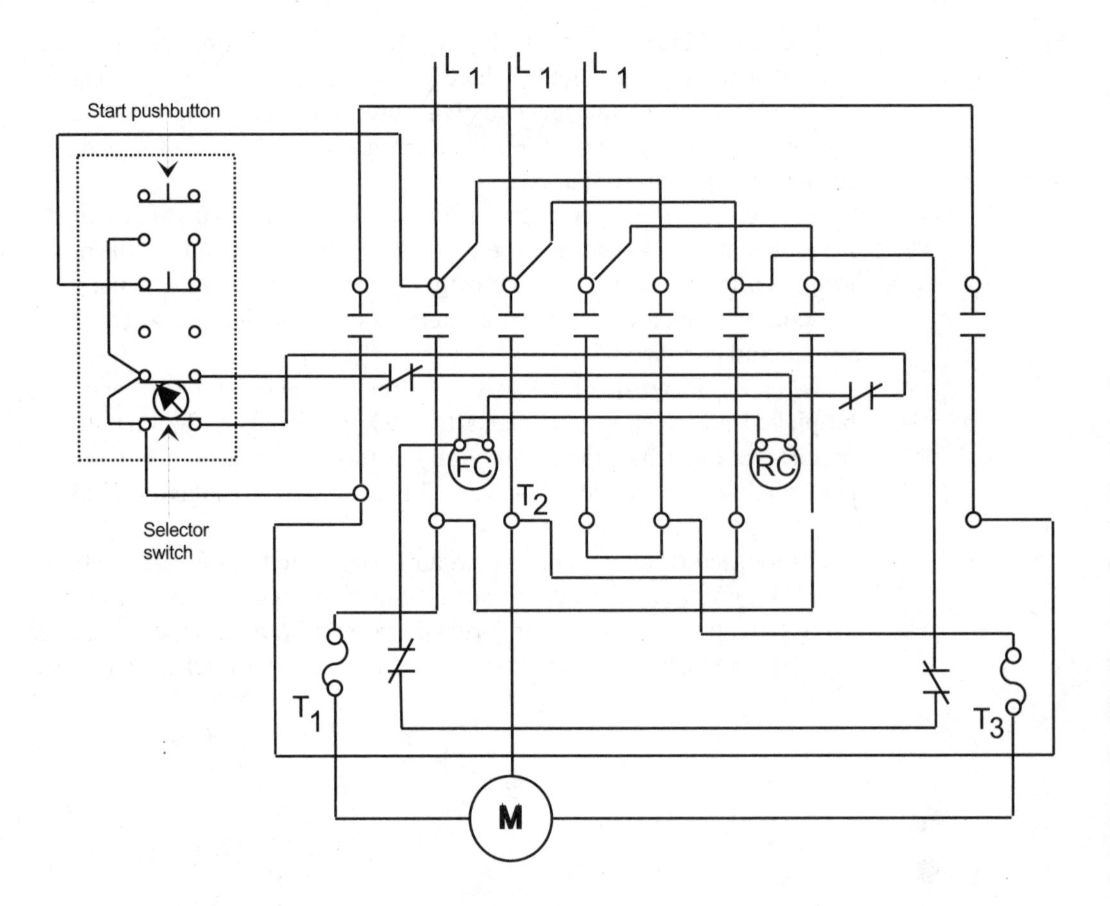

Figure 16-21: Wiring diagram of a reversing motor-control circuit.

to the forward (F) position and the start pushbutton is pressed, the motor runs in the forward direction. To reverse the direction of the motor, the selector switch is turned to the reverse (R) position. The stop button stops the motor in either direction.

Manual reversing starters (employing two manual starters) are also available. As in the magnetic version, the forward and reverse switching mechanisms are mechanically interlocked, but since coils are not used in the manually operated equipment, electrical interlocks are not furnished.

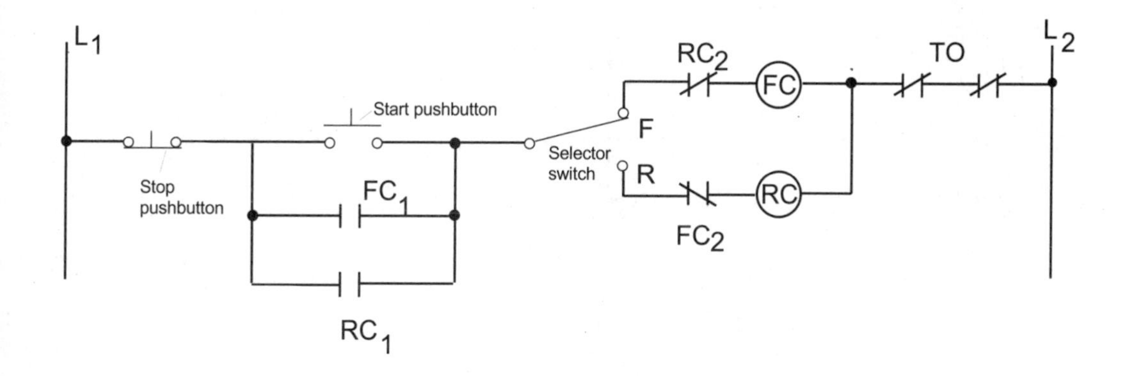

Figure 16-22: Ladder diagram of the control circuit in Figure 16-21.

MISCELLANEOUS STARTERS

There are several variations of the motor-control circuits previously described. There are also different types of motor starters. Some of these starters are becoming obsolete and are seldom used on new construction. However, electrical technicians often work on existing installations and a knowledge of the older starters is necessary.

Across-the-Line Motor Starters

Induction-type and synchronous ac motors have a high starting current; therefore, they are mostly started with a reduced voltage, although there are many quite large ac motors that are started across the line, that is, by applying the full rated voltage. A connection diagram for a full-voltage starter of a three-phase induction motor is shown in Figure 16-23 on the next page. Note that overload-protection relays (OL) are in two of the three lines and the motor is started by the basic control circuit, which uses a control station with start and stop buttons.

Reduced-Voltage Starting of AC Motors

The starter shown in Figure 16-24 uses autotransformers to provide reduced-voltage starting of a squirrel-cage induction motor. The motor (1) is connected to the three lines L_1, L_2, and L_3 by means of a movable lever (2). When the lever is moved from the middle, or OFF position, to the

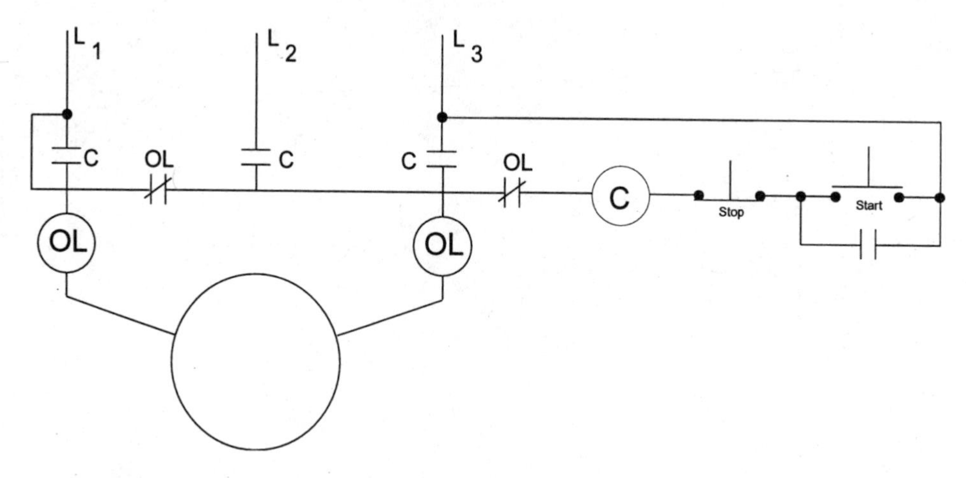

Figure 16-23: Full-voltage starter for a three-phase ac motor.

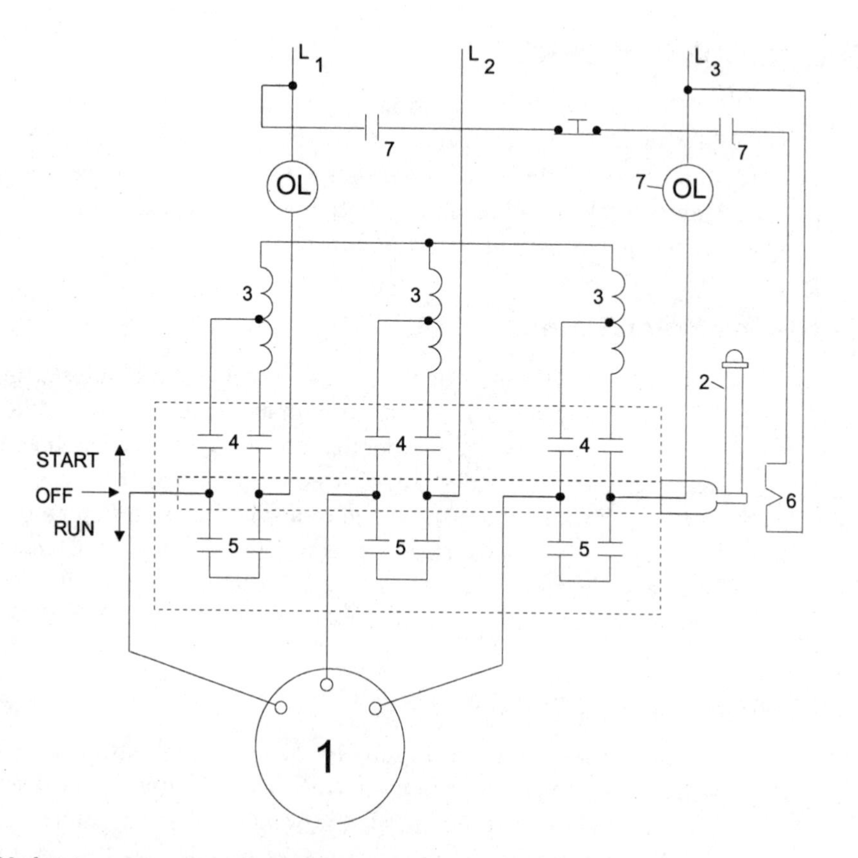

Figure 16-24: Autotransformers used to start a three-phase motor.

START position, to the RUN position, the autotransformers are disconnected and the supply is connected directly to the motor by the contacts (5).

The lever is held closed by the holding coil (6), which is connected across one phase of the supply through the relay contacts (7) between the relays and stop button. The relays are built with dashpots, which delay the operation of the relays during the starting period and for momentary overloads. Operation of the relays, or of the stop pushbutton, opens the holding coil circuit and allows the lever to come to the OFF position, breaking the circuit made by the relay contacts (7).

Autotransformers with several taps may be used for multispeed control of a squirrel-cage motor. All squirrel-cage motors use controllers in the primary or stator winding.

Resistance Magnetic Starters

A resistance-type starter is shown in Figure 16-25 as used with a three-phase, squirrel-cage induction motor. The motor is started by pressing the start button to energize coil C, which closes the three line contactors C_1, C_2, and C_3 and the interlock C_4 to maintain power on the coil after the start button is released.

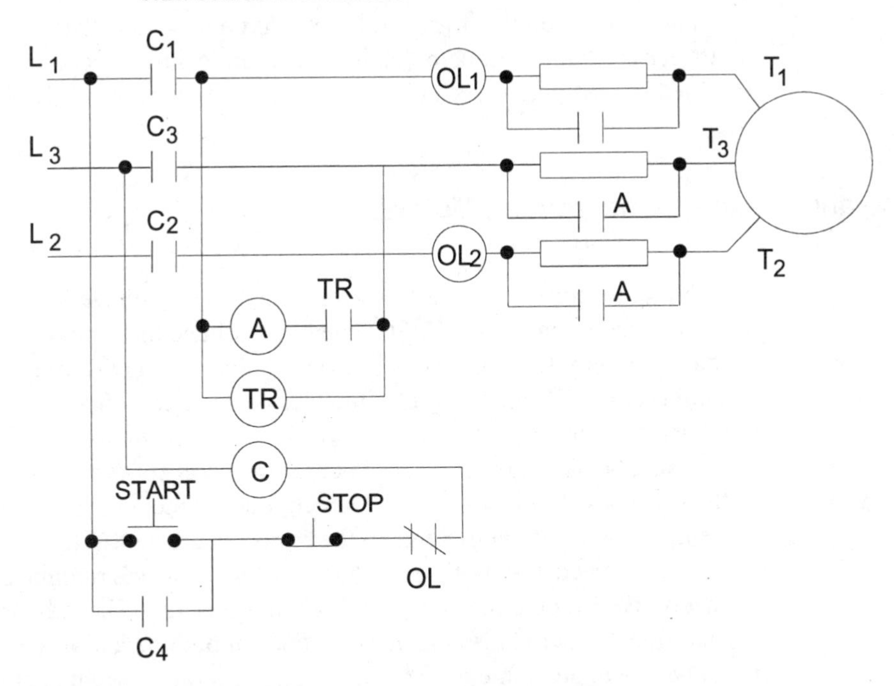

Figure 16-25: Resistance starter used to control a squirrel-cage induction motor.

When the contactor is closed, the timing relay (TR) becomes energized, and this relay immediately starts to measure the time for which it is adjusted. At the end of the timing period, the contacts (TR) close to energize coil A of the accelerating contactor (A). When the three-pole contactor (A) closes, the starting resistors will be short-circuited and the motor will be connected directly to the power lines. The heater elements of either of the thermal-type overload relays OL_1 and OL_2 open the normally closed contactor OL in case of the overload.

Starter for Wound Induction Motors

Wound three-phase induction motors differ from squirrel-cage induction motors in that the former use wound coils in the rotor instead of bars, and three sets of brushes to collect the current from the three collector rings; these are not used in squirrel-cage motors. To start a wound motor, it is necessary to connect the brushes to external resistors. These resistors increase the resistance of the rotor which, in turn, increases the torque of the rotor and decreases its speed. When the resistance is gradually cut out, the speed gradually increases and the torque decreases. With no secondary resistance, the wound motor operates at maxi multispeed. The starter and controller are built with combinations of basic control circuits. Manuals provided by manufacturers show control circuits for each specific type.

Starting and Braking Synchronous Motors

The main electrical connections for a large high-voltage synchronous motor are shown in Figure 16-26. The high-voltage three-phase lines L_1, L_2, and L_3 supply 13,800 volts to the synchronous motor (1) through a main (running) circuit breaker (2). This breaker is electrically interlocked with the starting breakers (3) and (4), and when the starting breakers are closed, the running breaker must be open. The starting breakers connect to the line of the starting autotransformer (5), which reduces the voltage to about 60 percent, or 6900 volts, and energizes the motor for the start. When the motor comes up to speed, the starting breaker (4) is open and the running breaker connects the motor directly to the high-voltage line. The primary of a current transformer (6) is connected to one supply line and its secondary is connected to an ammeter (7) to measure the motor current at any time. The field winding is energized by the exciter (9) through the field rheostat

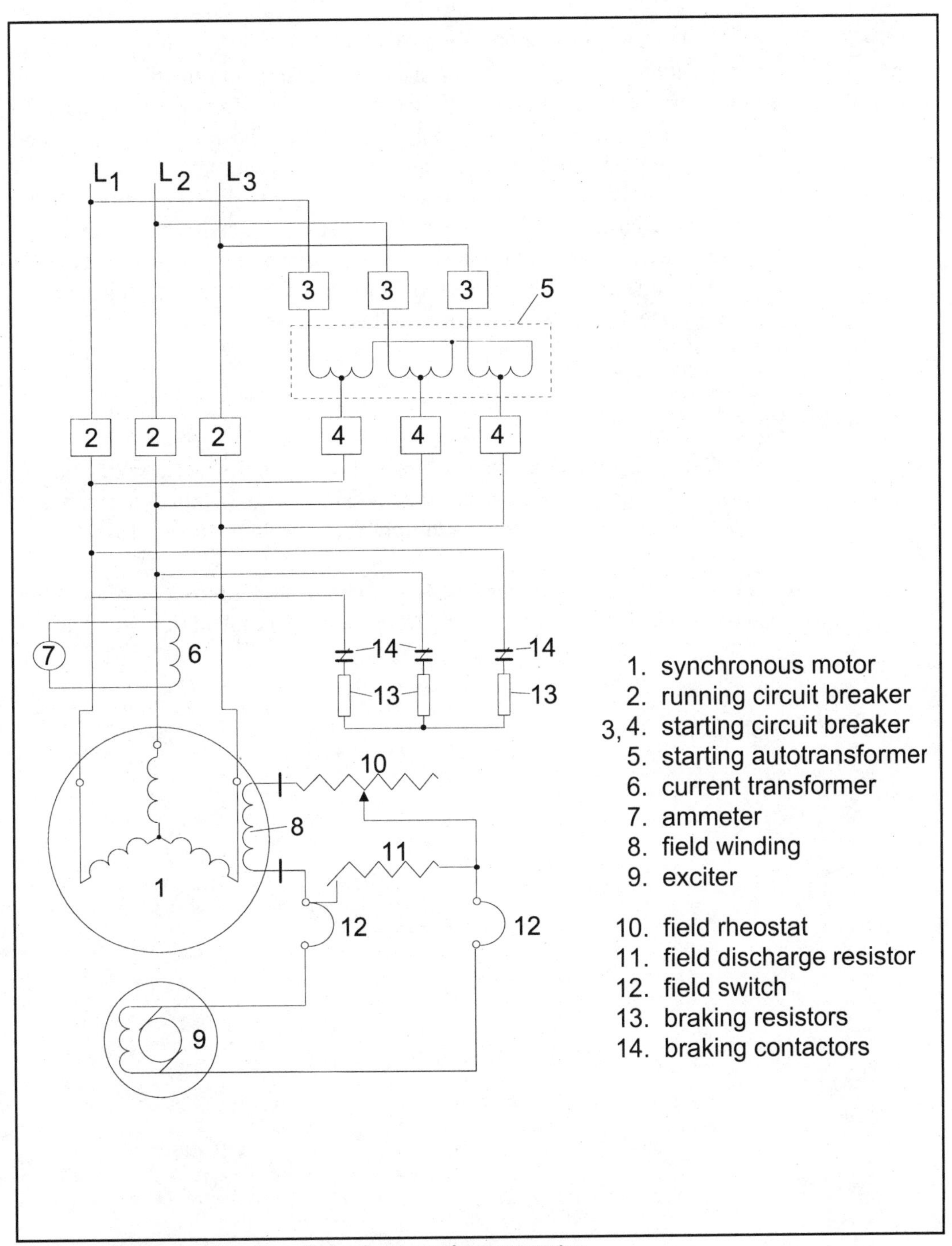

Figure 16-26: Connections for a three-phase synchronous motor.

(10). The exciter may be a shunt-wound dc generator with its own field rheostat (not shown). The field-discharge resistor (11) and the field switch (12) serve a double purpose in that they provide a low-resistance path for the induced currents in the motor field when the motor is started and discharge the induced currents when the field breaker is opened.

Dynamic braking is obtained by braking resistors (13) through the braking contactors (14). The field is left energized and the motor is disconnected from the line and connected to the braking resistors. By using quick-acting control sequences, the motor can be stopped within approximately one second without shock or high stresses.

Control Relays

A relay is either an electromagnetic or electronic device with contacts that are used to control various circuits. They are generally used to amplify the contact capability or to multiply the switching functions of a pilot device.

The wiring diagram in Figure 16-27 demonstrates how a relay amplifies contact capacity. Figure 16-28(A) represents a current amplification. Relay

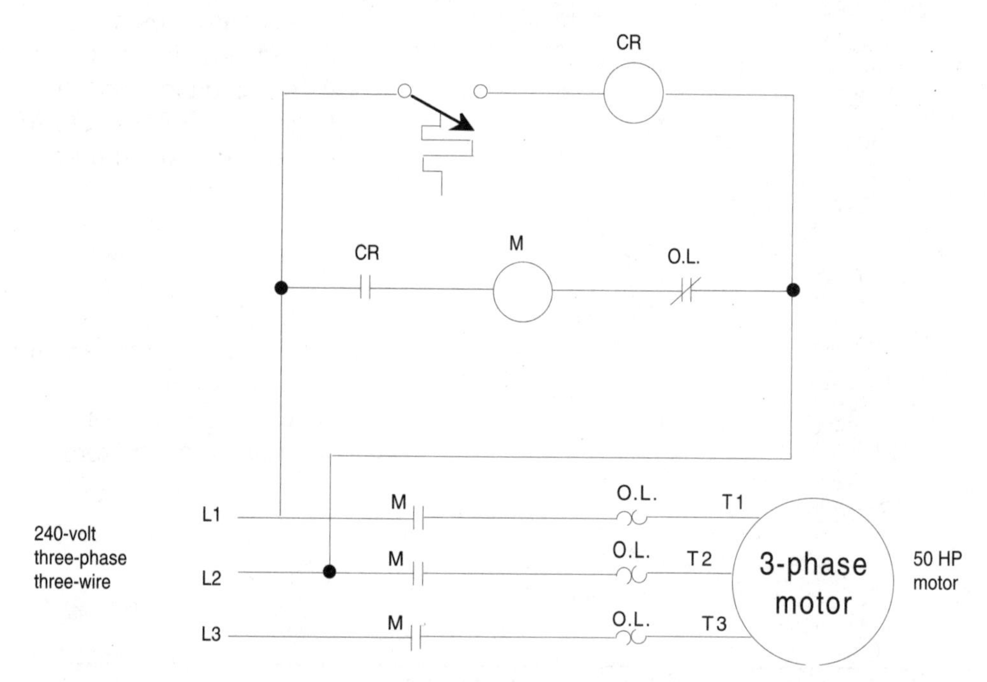

Figure 16-27: Relay amplifying contact capacity.

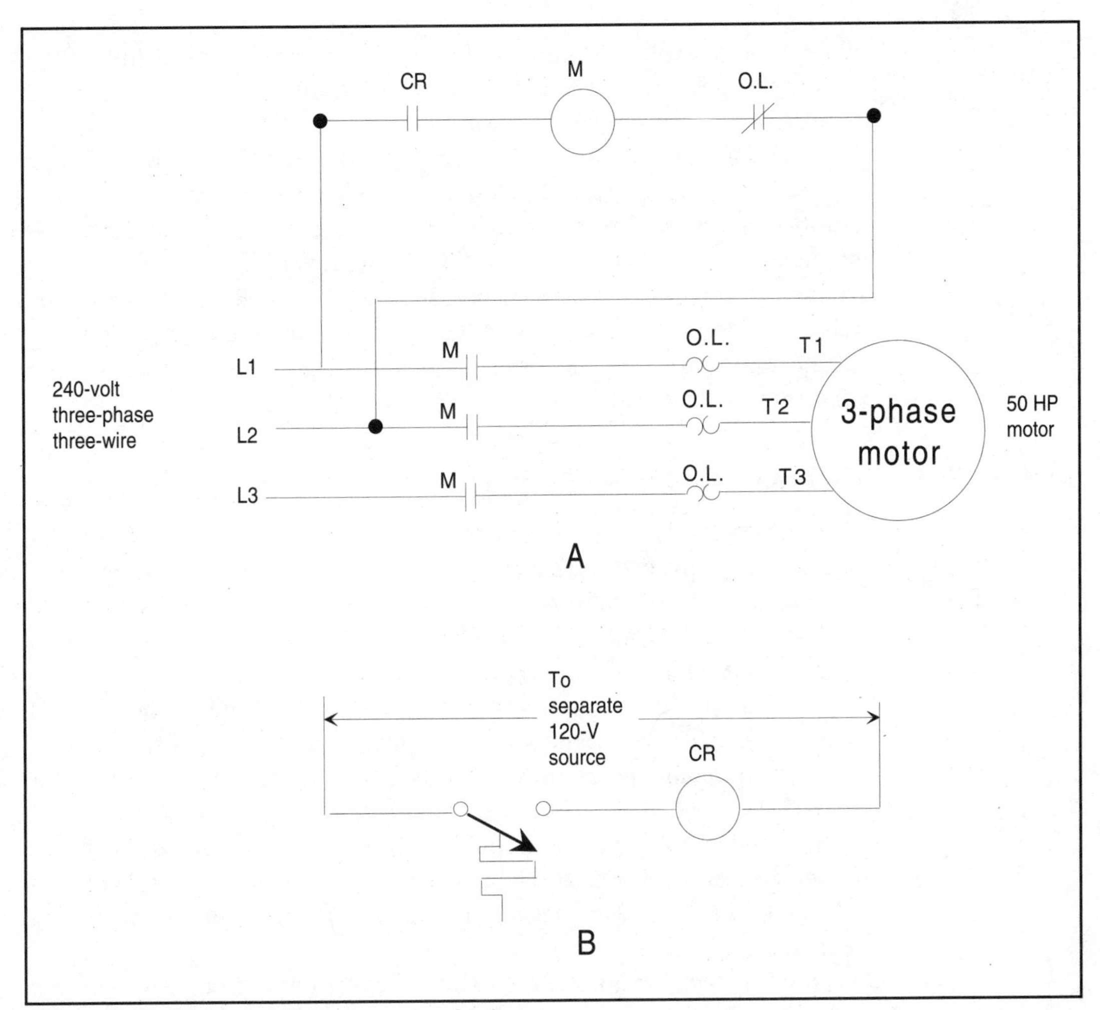

Figure 16-28: Circuit amplifying current and voltage.

and starter coil voltages are the same, but the ampere rating of the temperature switch is too low to handle the current drawn by the starter coil (M). A relay is interposed between the temperature switch and the starter coil. The current drawn by the relay coil (CR) is within the rating of the temperature switch, and the relay contact (CR) has a rating adequate for the current drawn by the starter coil.

Figure 16-28(B) represents a voltage amplification. A condition may exist in which the voltage rating of the temperature switch is too low to permit its direct use in a starter control circuit operating at a higher voltage. In this application, the coil of the interposing relay and the pilot device are wired to a low-voltage source of power compatible with the rating of the pilot device. The relay contact, with its higher voltage rating, is then used to control the operation of the starter.

Relays are commonly used in complex controllers to provide the logic or "brains" to set up and initiate the proper sequencing and control of a number of interrelated operations. In selecting a relay for a particular application, one of the first steps should be a determination of the control voltage at which the relay will operate. Once the voltage is known, the relays that have the necessary contact rating can be further reviewed, and a selection made, on the basis of the number of contacts and other characteristics needed.

ADDITIONAL CONTROLLING EQUIPMENT

Timers and timing relays: A pneumatic timer or timing relay is similar to a control relay, except that certain kinds of its contacts are designed to operate at a preset time interval after the coil is energized or deenergized. A delay on energization is also referred to as *on delay*. A time delay on de-energization is also called *off delay*.

A timed function is useful in such applications as the lubrication system of a large machine, in which a small oil pump must deliver lubricant to the bearings of the main motor for a set period of time before the main motor starts.

In pneumatic timers, the timing is accomplished by the transfer of air through a restricted orifice. The amount of restriction is controlled by an adjustable needle valve, permitting changes to be made in the timing period.

Drum switch: A drum switch is a manually operated three-position, three pole switch which carries a horsepower rating and is used for manual reversing of single- or three-phase motors. Drum switches are available in several sizes and can be spring-return-to-off (momentary contact) or maintained contact. Separate overload protection, by manual or magnetic starters, must usually be provided, as drum switches do not include this feature.

Pushbutton station: A control station may contain pushbuttons, selector switches, and pilot lights. Pushbuttons may be momentary- or maintained-contact. Selector switches are usually maintained-contact, or can be spring-return to give momentary-contact operation.

Stand-duty stations will handle the coil currents of contactors up to size 4. Heavy-duty stations have higher contact ratings and provide greater flexibility through a wider variety of operators and interchangeability of units.

Foot switch: A foot switch is a control device operated by a foot pedal used where the process or machine requires that the operator have both hands free. Foot switches usually have momentary contacts but are available with latches which enable them to be used as maintained-contact devices.

Limit switch: A limit switch is a control device that converts mechanical motion into an electrical control signal. Its main function is to limit movement, usually by opening a control circuit when the limit of travel is reached. Limit switches may be momentary-contact (spring-return) or maintained-contact types. Among other applications, limit switches can be used to start, stop, reverse, slow down, speed up, or recycle machine operation.

Snap Switch: Snap switches for motor control purposes are enclosed, precision switches which require low operating forces and have a high repeat accuracy. They are used as interlocks and as the switch mechanism for control devices such as precision limit switches and pressure switches. They are also available with integral operators for use as compact limit switches, door operated interlocks, and so on. Single-pole, double-throw and two-pole, double-throw versions are available.

Pressure Switch: The control of pumps, air compressors, and machine tools requires control devices that respond to the pressure of a medium such as wire, air, or oil. The control device that does this is a pressure switch. It has a set of contacts which are operated by the movement of a piston, bellows, or diaphragm against a set of springs. The spring pressure determines the pressures at which the switch closes and opens its contacts.

Chapter 17

Heat Tracing and Freeze Protection

Heat-tracing systems are a critical component of many plant systems. They can provide freeze protection for piping systems, or maintain critical temperatures in processing plants.

A variety of forms of heat tracing are available for use, each with its own benefits and disadvantages. Steam, oil, and electricity can all be used to facilitate heat tracing. Different methods of controlling the heat tracer temperature are also available with varying degrees of accuracy.

Proper installation and maintenance is important to ensure proper operation of the heat tracing equipment. Improper installation can lead to catastrophic failure of not only the heat tracers, but the process piping itself.

Besides heat-tracing systems, electrical engineers, designers, drafters, and workers will also encounter other types of heating systems used for freeze protection:

- Snow-melting cable
- Snow-melting mats
- Freeze protection electric heating cable used on pipe lines

All of these heating methods, along with their *NEC* installation requirements are covered in this chapter.

Reasons for Heat Tracing

Freeze Protection: A major consideration in heat tracing is the ambient temperature. Freezing of the liquid inside the piping must be prevented for several reasons. Primarily, liquids lowered below their freezing point will form a solid obstruction in the piping system thereby stopping fluid flow. In addition, as a liquid turns to a solid it also expands causing severe damage to the surrounding piping. This is the primary reason most automobile engines are equipped with freeze plugs, which will burst prior to engine block damage. An exception can be made for those fluid pipes installed below the ground frost level. The temperature in this area remains relatively constant regardless of the surroundings. Freeze protection is the reason many water systems have heat tracing installed.

Dew point: The dew point is that temperature at which a mixture of water vapor and a gas is saturated. If this mixture's temperature drops below its dew point, condensation of the water vapor will occur. Moisture can cause operating difficulty, especially in gas burners where the element may become fouled by the liquid. In addition natural gas containing moisture may cause freeze up of not only control valves but possibly the entire system. Moisture is especially harmful in compressors, and in the presence of hydrogen sulfide it can produce harmfully corrosive sulfuric acid. Hydrogen sulfide is produced in some oil refineries during the crude oil distillation process.

Viscosity Control: Many systems may require heat tracing in order to maintain the correct viscosity of the liquids involved. With extremely high viscosity liquids, large high-horsepower pumps are required to move the fluid. By heat tracing the system piping, the viscosity of the liquid inside may be lowered, thus a smaller pump may be used to move the fluid. This is a major concern in many lubricating oil, hydrocarbon, heavy alcohol, and caustic fluid systems.

Aside from easing pumping requirements, lubricating properties of oil can be affected by oil temperatures. A high speed motor or engine usually requires a lower viscosity lubricant. This lower viscosity can be achieved by heating the lubricating oil using tracers.

Many diesel engines have a form of heat tracing on their fuel lines to prevent the fuel from turning to a gel at lower temperatures. Increased fuel temperatures also add to increased efficiency during the combustion process.

Note: *In each of the above listed considerations, care must be taken to ensure the temperature range of the heat-tracing system is compatible with the system heat requirements.*

TYPES OF HEAT-TRACING SYSTEMS

Steam Heat Tracing

The first form of heat tracing to be utilized was steam tracing. The physical properties of steam make it ideal for this use. Only a small amount of steam is required to carry a high-heat load, it can transfer heat quickly, and does not need to be pumped to the process piping requiring heat tracing. However, electricians will seldom be involved with this method — other than perhaps connecting some steam-heating controls.

Hot Liquid Tracers

Hot oil or circulating media heat tracers are the most expensive type of heat-tracing systems. The benefit of using this type of system is that system temperature may be maintained at values above and below those achievable using steam tracers. In addition to hot oil, these types of systems may employ the use of glycols with antifreeze properties when used in especially cold climates. System design is similar to that used for steam tracing, except that a heat exchanger replaces the boiler and steam traps are not required. *See* Figure 17-1.

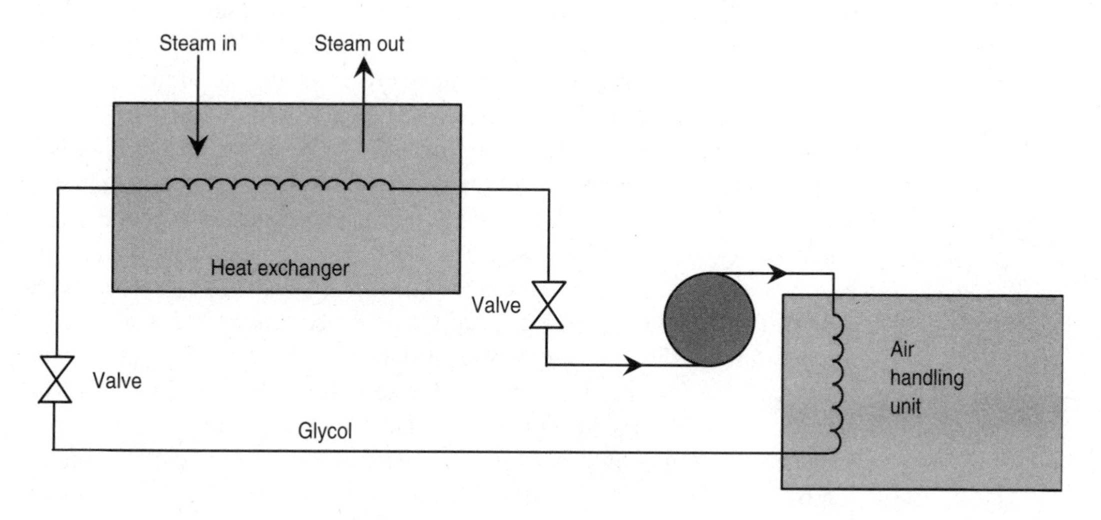

Figure 17-1: Hot liquid tracer system.

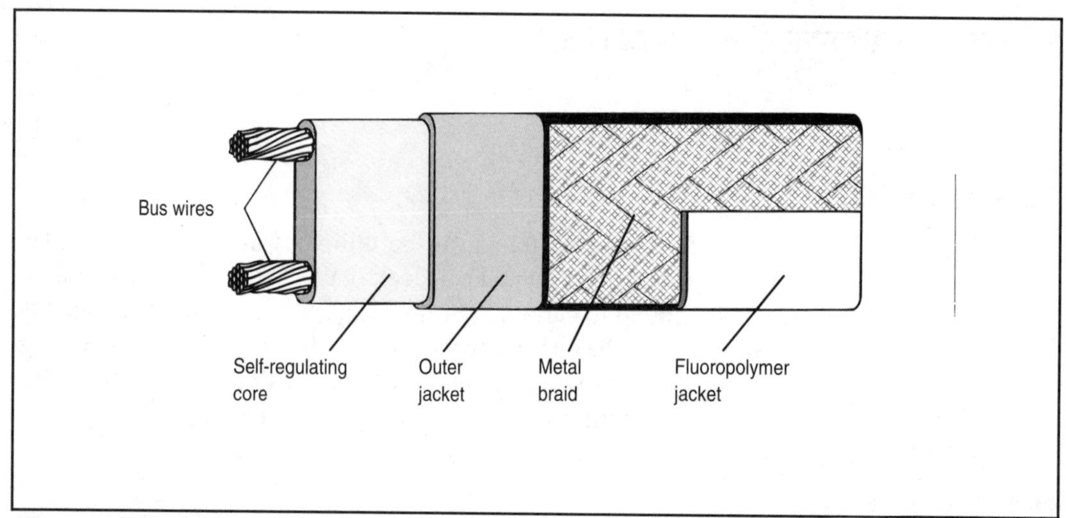

Figure 17-2: Electric resistance heat tracer.

Electric Resistance Heat Tracing

Electrical heat tracing provides an economic alternative to steam tracing in most applications. Early uses of electric tracers were in the 1950s on oil, asphalt, and wax systems. Electric tracers are especially useful on long piping systems as steam tracers are generally limited to between 100 and 200 ft in length.

The most popular form of electrical resistance heat tracers is the self-regulating heater. The use of this type of tracer eliminates the possibility of heater burnout inherent in other designs as the heat is dissipated internally. Heater burnout is the most common failure among electric tracers. Figure 17-2 shows a typical resistance heat tracer. Self-regulating tracers are usually produced in the form of a heater strip consisting of two parallel wires embedded in a polymer core which acts as a heating element. The entire element is then surrounded by a polymer insulator, metal braid, and possibly a fluoropolymer jacket. The heater core consists of carbon particles embedded in a polymer matrix. Heat is generated by resistance to current flowing through the conductive polymer heating element. As the temperature of the core increases, so does the electrical resistance. The result is a lower heat output for each rise in temperature. This acts as a self-regulating thermostat for the element protecting it from damage due to high or low temperatures. Figure 17-3 shows a graph of resistance versus temperature for a self-regulating heat tracer.

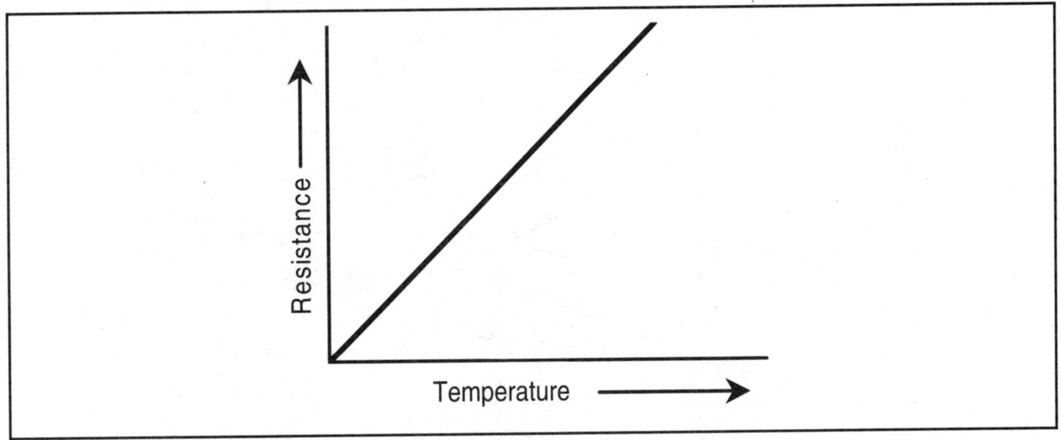

Figure 17-3: Resistance versus temperature for self-regulating heat tracer.

Induction Heat Tracing

Induction heat tracing uses the system piping itself as a heating element by placing it in a magnetic field of an alternating-current source. Low-resistance wire is wound around a conductive pipeline or vessel, and the alternating current flowing through the coils generates a rapidly changing magnetic field that induces eddy currents and hysteresis losses. This produces a large amount of heat, much in the same fashion that heat is produced in a transformer. Induction heating is used in high-temperature, high-power applications where very rapid heating is needed. Induction heat-tracing systems do not easily lend themselves to uniform heating and produce only moderate efficiency at best.

NEC Article 427 Part E covers the installation of line frequency induction heating equipment and accessories for pipelines and vessels. However, the *NEC* requirements are very brief and the *NEC* merely states the following:

- Induction coils that operate or may operate at a voltage greater than 30 V ac shall be enclosed in a nonmetallic or split-metallic enclosure, isolated or made inaccessible by location to protect personnel in the area. *NEC* Section 427-36.
- Induction coils shall be prevented from inducting circulating currents in surrounding metallic equipment, supports, or structures by shielding, isolation, or insulation of the current paths. Stray current paths shall be bonded to prevent arcing.

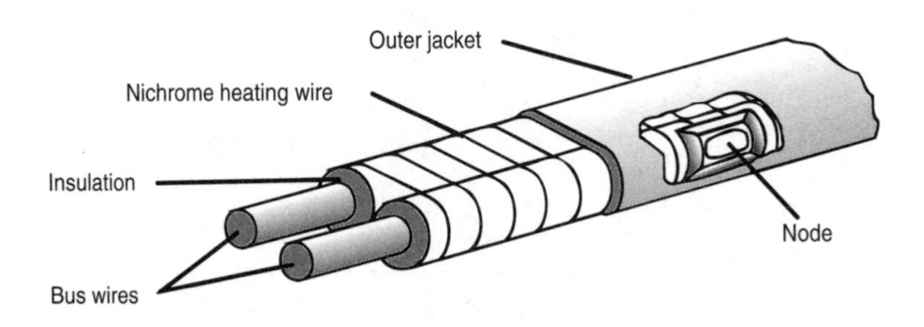

Figure 17-4: Zone heating cable.

Zone Heaters

Zone heaters have been extremely popular in the past, however, most current applications use self-regulating heaters. A zone heater is made up of two bus wires wrapped with a smaller gauge heating wire. The heating wire is connected to the bus wire every 2 to 3 ft. The distance between the bus wire/heating wire connection is called a *zone*. Heat is generated by current flowing between the bus wires through the heating wires. Zone heaters may be cut to any length as needed. It should be noted that the portion of cable between the cut and the next connection point between the heating and bus wires will not receive electricity, and therefore, will not provide any heating. Figure 17-4 shows a zone heater.

The biggest drawback to zone heaters is the high rate of heater burnout. Zone heaters commonly overheat and destroy themselves, thus the replacement by self-regulating heaters.

The table in Figure 17-5 lists various heat-tracing elements along with their advantages and disadvantages.

WARNING!

Do not exceed maximum temperature ratings of the process and process control equipment.

Heat-Tracing Method	Maximum Operational Temperature	Maximum Exposure Temperature	Advantages	Disadvantages
Self-regulating	150 – 300°F (65 – 149°C)	185 – 420°F (85 – 215°C)	Long life; most reliable electric heating cable; energy efficient	Limited temperature range
Zone	150 – 400°F (65 – 204°C)	250 – 1000°F (150 – 538°C)	Can be field cut; if a heating element fails, circuit is maintained	Relatively fragile; can self-destruct from its own heat; can burn out if crossed over itself
Inductance	Up to curie point	None	High-temperature capability; high heat-transfer rates	Very expensive; difficult custom design, not commercially exploited

Figure 17-5: Comparison of electric heat-tracing methods.

Control of Heat-Tracing Systems

Control of temperature inside the system requiring heat tracing can be accomplished by a variety of methods. The most widely used methods are mentioned below.

No Control or Self-Regulating Control

The heater is constantly supplied with full power and the self-regulating characteristics of the heater control the pipe temperature as in an electrical resistance heat tracer.

Ambient-Sensing Control

A thermostat measures ambient temperature and engages the heating system when the surrounding temperature drops below a predetermined value. The cost involved with this type of system is significantly less than with other methods as only one thermostat is required. Seventy to 80 percent of all freeze protection systems are of this type.

Line-Sensing Control

A thermostat is installed which measures the temperature of the pipe. Each heating circuit is turned on and off independently when signaled by its associated thermostat. Line sensing units offer the highest degree of temperature control and lowest energy use, but it has the highest installation cost due to the multiple thermostats.

Dead Leg Control

A thermostat measures a dead leg of pipe that does not have fluid flow. The entire heat-tracing system is designed to maintain the temperature in this dead leg. Dead leg control has some cost reduction as only one thermostat is required, but it is hard to control temperature in high flow conditions as dead leg temperature is the temperature determining factor.

INSTALLATION OF HEAT-TRACING SYSTEMS

Preinstallation Checks

Prior to insulating any heat-tracing system certain checks should be performed.

- Cables should be visually checked for signs of damage and electrical insulation integrity.
- A continuity check should be performed on all wiring by properly trained personnel.
- The resistance of the insulation should be measured between conductors and the metallic outer covering, or a specially applied conductive metal tape, or braid, by means of dc voltage of 600 V. The measured value should not be less than 60 megohms. This test should be performed by properly trained personnel.
- Individual controls should be bench tested to ensure proper calibration. This includes checking for the correct operating temperature range, proper span, and set points.

INSTALLATION

The following is a generalized procedure for installation of a heat-tracing system. Manufacturer's technical information should be used and strictly adhered to when the actual installation occurs.

Note: Use glass tape for anchoring heat trace to piping.

Electrical

- Cables should be installed on a clean, smooth portion of the pipeline, avoiding any sharp bends or jagged edges. Cables should be oriented to avoid damage due to impact, abrasion, or vibration, while still maintaining proper heat transfer.
- Cables should be applied in a manner to facilitate the removal of valves, small in-line devices, and instruments without the complete removal of cables, excessive thermal insulation, or cutting of the heating cable.
- Some valves and complicated in-line devices may need to have their surface irregularities covered with a metal foil or heat-transfer medium to prevent damage to the heat tracing.
- Overlapping heating cables can cause excessive temperatures at overlap points. This is the leading cause of damage to all but self-regulating resistance heat tracers.
- Heating cable cold leads should be positioned to facilitate penetrating the thermal insulation in the lower 180° segment to minimize water entrance.

The following conditions should be considered in relation to the location of the temperature sensors:

- Where two or more cables meet or join, the sensors should be mounted 3 ft to 5 ft from the junction.
- The sensor should be mounted to avoid the direct temperature effects of the heating cable.
- Where a pipeline runs through areas with different ambient conditions, such as inside and outside a heated building, two sensors and associated controls may be required to properly control pipeline temperature.
- Sensors should be located 3 ft to 5 ft from any heat source or heat sink in the system.

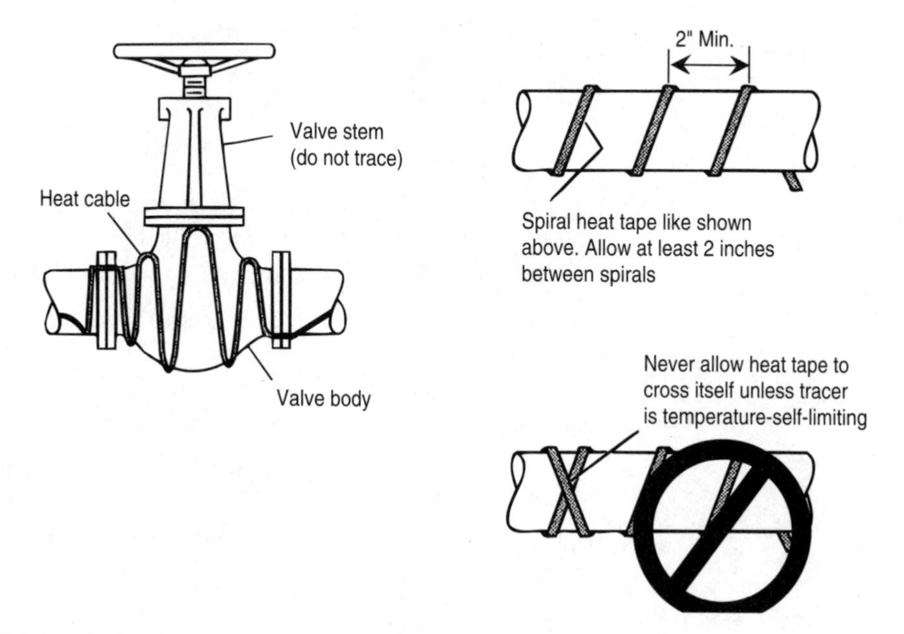

Figure 17-6: Installation tips for electric heat tracers.

Figure 17-6 shows several installation tips for electric heat tracers. In addition, the following general safety rules should be followed:

- Work closely with any other craft workers involved with the installation process.
- Follow the manufacturer's installation instructions carefully. Do not skip any steps.

Insulation of Heat-Tracing Systems

Insulation of heat tracing is important in order to reduce heat loss to the surrounding atmosphere and to increase efficiency. The following are important aspects to be considered when selecting an insulation material:

- Thermal characteristics
- Mechanical properties
- Chemical compatibility
- Moisture resistance
- Personal safety characteristics

- Fire resistance
- Cost

Insulating materials commonly used are:

- Expanded silica
- Mineral fiber
- Cellular glass
- Urethane
- Fiber glass
- Calcium silicate
- Perlite silicate

Piping systems using heat-tracing devices are insulated in the same manner as other systems without heat tracing. The only exception is that the heat tracers are included under the layer of insulation. In some cases, the traced line is wrapped with aluminum foil and then covered with shaped insulation. The foil increases the radiation heat transfer. It is essential that the space between the pipeline and the tracer be kept free of particles of insulating material.

Heat transfer capabilities of the tracer may be improved by putting a layer of heat-conducting cement (graphite mixed with sodium silicate or other binders) between the tracer and the pipeline. Figure 17-7 shows various tracer/insulation configurations.

Proper operation of heat-tracing systems depends upon the insulation being dry. Electric tracing normally does not have sufficient heat output to dry wet insulation. Some insulating materials, even though removed from the piping and force-dried, never regain their initial integrity after once being wet. Straight piping may be weather protected with metal jacketing, polymeric, or a mastic system. When metal jacketing is used, it should be sealed with closure bands and supplied with sealant on the outer edge where they overlap as seen in Figures 17- 8 and 17-9.

Post-Installation Inspection

Prior to installation of thermal insulation, the insulation resistance should be measured under normal dry conditions and before connection of the device to the associated wiring or control equipment. The resistance reading should not be less than 20 megohms at 500 Vdc.

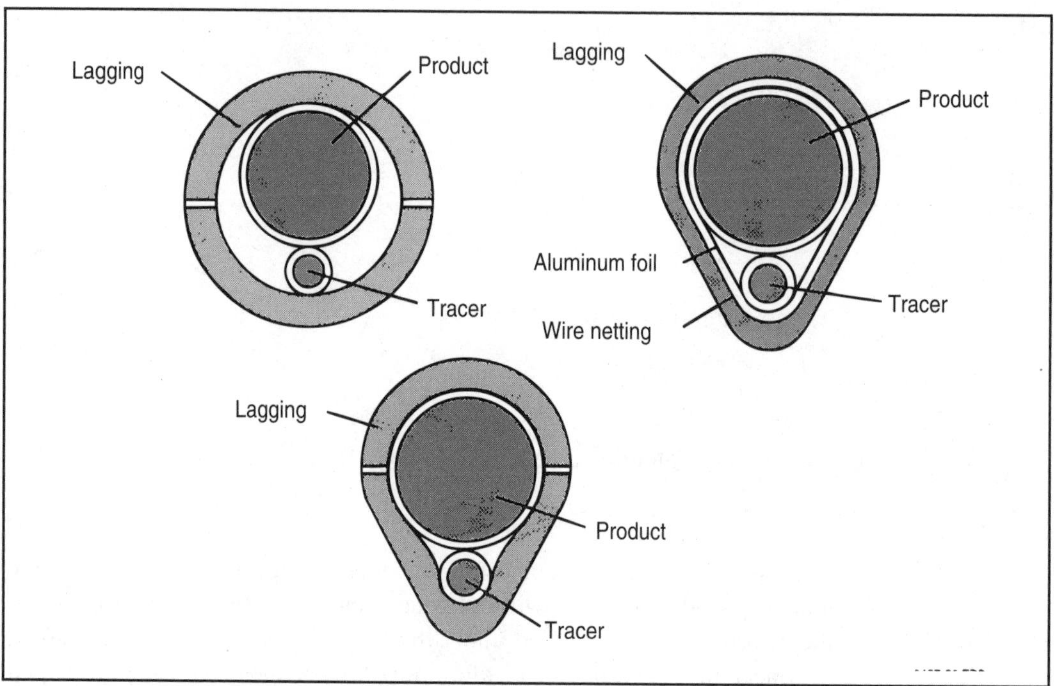

Figure 17-7: Tracer/insulation configurations.

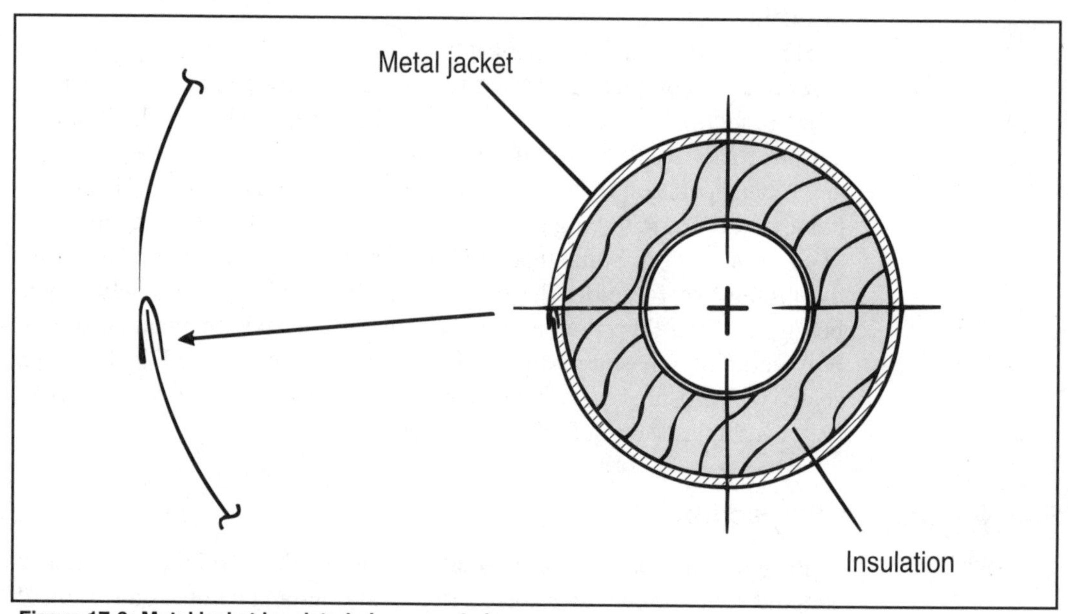

Figure 17-8: Metal jacket insulated pipe — end view.

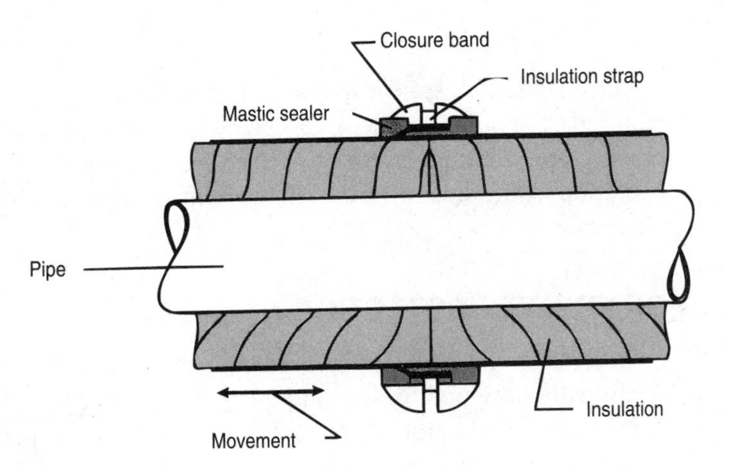

Figure 17-9: Sealing with metal jacketing — section.

Electrical wiring should be measured for continuity to ensure the wiring was not damaged during installation. Some control circuits have continuity monitors which send a low-voltage signal down the heat-tracing line when it is shut down. The alarm sounds if a break in system continuity is sensed, indicating damage to a portion of the heat-tracing system.

All forms of heat tracing systems require some form of predicted maintenance to ensure proper operation. In the case of heat tracers being used as freeze protection, these tests are usually carried out during the fall season. All circuits and controls should be checked for proper operation. A visual inspection should be performed with particular attention being paid to the following areas:

- Evidence of leaks, corrosion, or foreign matter
- Looseness of electrical connections
- Thermostats and control cabinets properly sealed and free of moisture
- Discoloration or damaged insulation due to overheating

When the heat tracing is used for normal temperature maintenance or critical process control, an operational check should be carried out on a more frequent basis. Resistance checks and continuity checks should be performed on electrical tracers following any mechanical maintenance on pipelines, vessels or equipment that has tracers installed. Improper reinstallation following maintenance is a major cause of heat-tracer failure.

Operation

Once started most heat-tracing systems are self-sustaining. All that is required is a periodic check to ensure that all thermostats and control circuits are working properly. In the case of steam tracers, the proper operation of the steam traps in the system is important to ensure system efficiency.

HEAT CABLE FOR FREEZE PROTECTION

Automatic and manually controlled heat tapes are available which are specifically designed for use on piping systems and related valves. Most are manufactured in ratings from 7 to 10 or more watts per foot for use in various applications.

The method used for a pipe-heating installation will be determined largely by the heat requirements and the length of pipe to be heated. Mineral-insulated cable, for example, can be run straight along the pipe as shown in Figure 17-10. Note that both cold leads terminate in the same junction box and that heat-transfer cement is used over the entire length of the cable and thermostat sensing bulb.

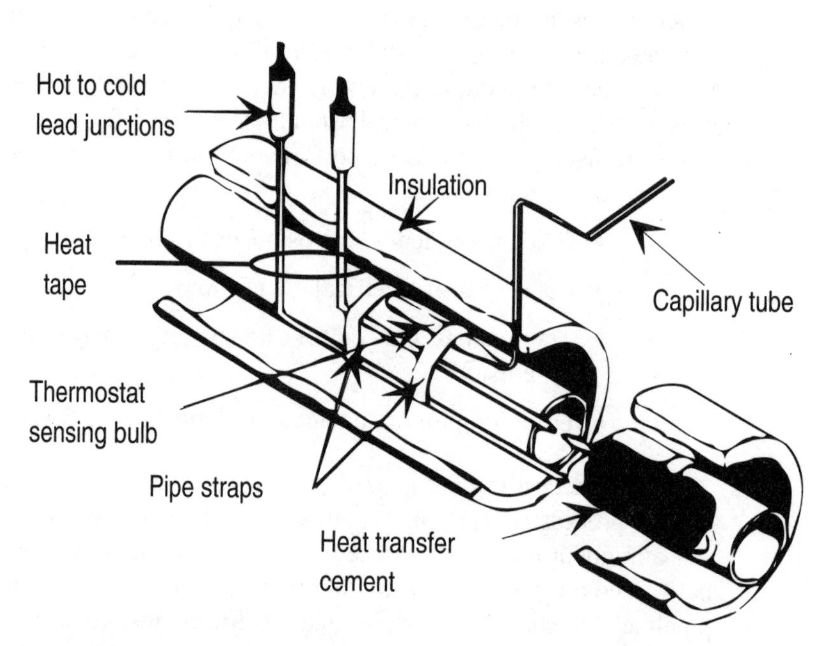

Figure 17-10: Heat cable run straight along the pipe.

Method of Wrapping	Nominal Pipe Size	3/8"	1/2"	3/4"	1"	1 1/4"	1 1/2"	2"	3"	4"	6"
Straight along pipe	Degrees of Protection (F)	-50°	-40°	-27°	-13°	-3°	-0°	+6°	+15°	+20°	+25°
	Length required per foot of pipe	1'	1'	1'	1'	1'	1'	1'	1'	1'	1'
Spiraled around pipe 3 turns per foot	Degrees of Protection (F)	—	—	-50°	-40°	-33°	-30°	-27°	-20°	-17°	-13°
	Length required per foot of pipe	1'3"	1'4"	1'6"	1'8"	1'10"	2'1"	2'4"	2'11"	3'11"	5'
Spiraled around pipe 6 turns per foot	Degrees of Protection (F)	—	—	—	-50°	-50°	-50°	-50°	-50°	-50°	-50°
	Length required per foot of pipe	1'9"	2'	2'5"	2'8"	3'1"	3'6"	4'2"	5'8"	7'7"	10'3"

Figure 17-11: Degree of protection and cable length required for common pipe sizes.

For short lengths of pipe, and where highly concentrated heat is required, it may be necessary to spiral the heating cable around the pipes; cable length required when spiraled around pipe is shown in the table in Figure 17-11.

Regardless of the type of system used, all branch circuits and feeders, including disconnects and overcurrent protection, must be in accordance with the *NEC*. Heat tapes designed especially for small projects are available with built-in thermostats.

First determine the size of the pipe to be protected and the lowest ambient temperature to which the pipe will be subjected. Then check the table in Figure 17-11 to determine the method of wrapping and the length of heat tape required for the particular application. Spirally wrap or apply straight as indicated in the tables for the specified condition. However, be sure not to let the heater wires touch, cross, or overlap. The unit may overheat and burn out if any of these conditions occur. Neither can the unit be made longer or shorter. If it is made shorter it will overheat; if made longer, it will not give off enough heat.

Continue to wrap the unit evenly, spacing each wrap until the frost line is reached; friction tape may be used to hold the end in place while you are

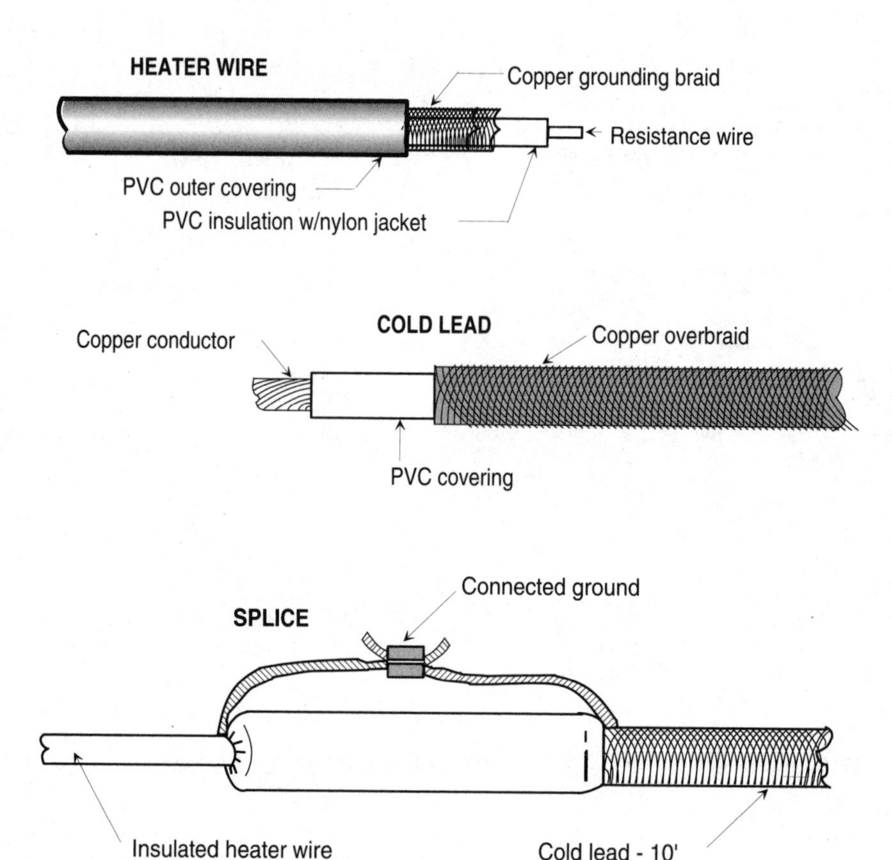

Figure 7-12: Splicing details for heat cable.

wrapping the heat cable. If there is a portion of the tape left unused, do not overlap the wires, but either rewrap the cable — spacing the wires closer together — or let the excess hang loose.

Most heat-cable assemblies come from the factory with the cold to hot lead junctions already made up. However, it may sometimes become necessary to make these splices or junctions in the field. If so, follow the manufacturers instructions exactly. To do otherwise is certain to cause problems. The details in Figure 17-12 show one manufacturer's recommendations for splicing cold lead to heat cable.

Many control devices for freeze-protection heat cable are furnished with detailed instructions to facilitate rough-in and connections. The wiring diagram for a power switch is shown in Figure 17-13. This control is designed to automatically operate an electric freeze-protecting system. The

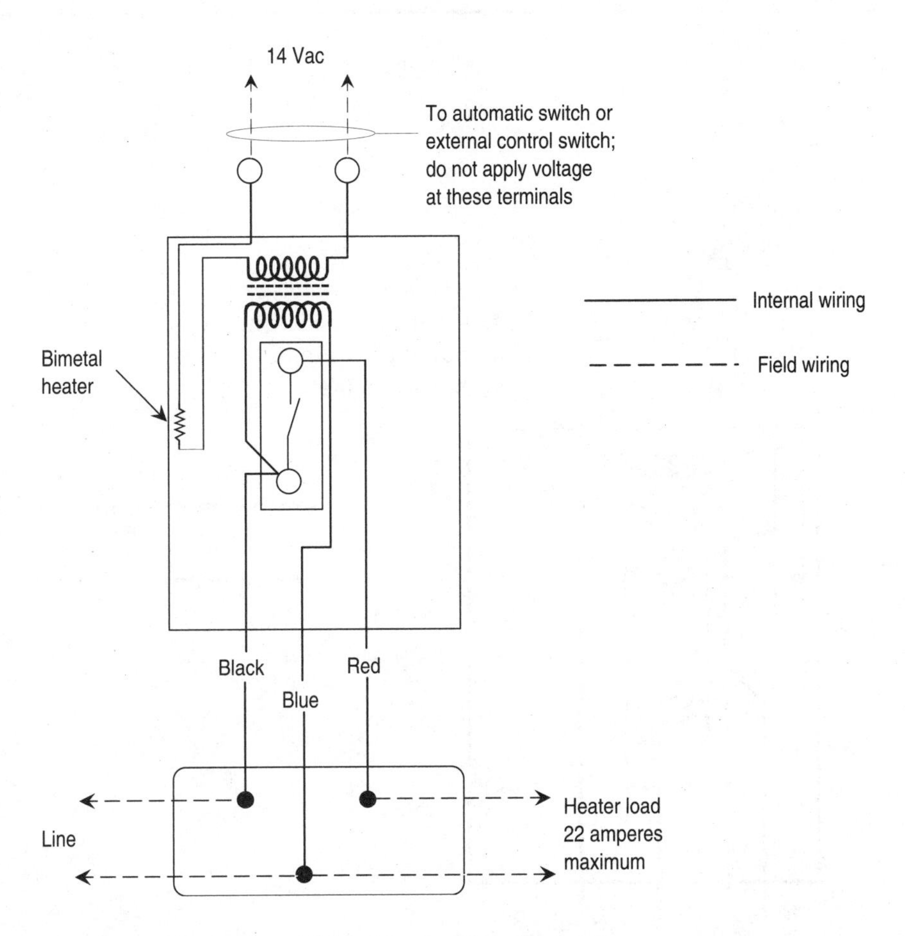

Figure 17-13: Internal schematic and field wiring for 208-volt models.

power switch consists of a line to low-voltage transformer; a low-voltage heater; a single-pole, single-throw, normally open line-voltage switch; and a compensating bimetal heater. On a call for heat, the control closes the circuit to the bimetal heater. In a short period of time, the warping action of the heater bimetal operates the switch to complete the circuit to the electric-heating load. When the controls open, the bimetal heater circuit is broken and the switch opens to break the circuit to the heating load. Typical field wiring diagrams are shown in Figures 17-14, 17-15, and 17-16.

Dimension drawings are also very useful to the workers for laying out and roughing-in circuits, junction boxes, and so on, for electric heating equipment. Figure 17-17 shows the dimensions and knockouts of an

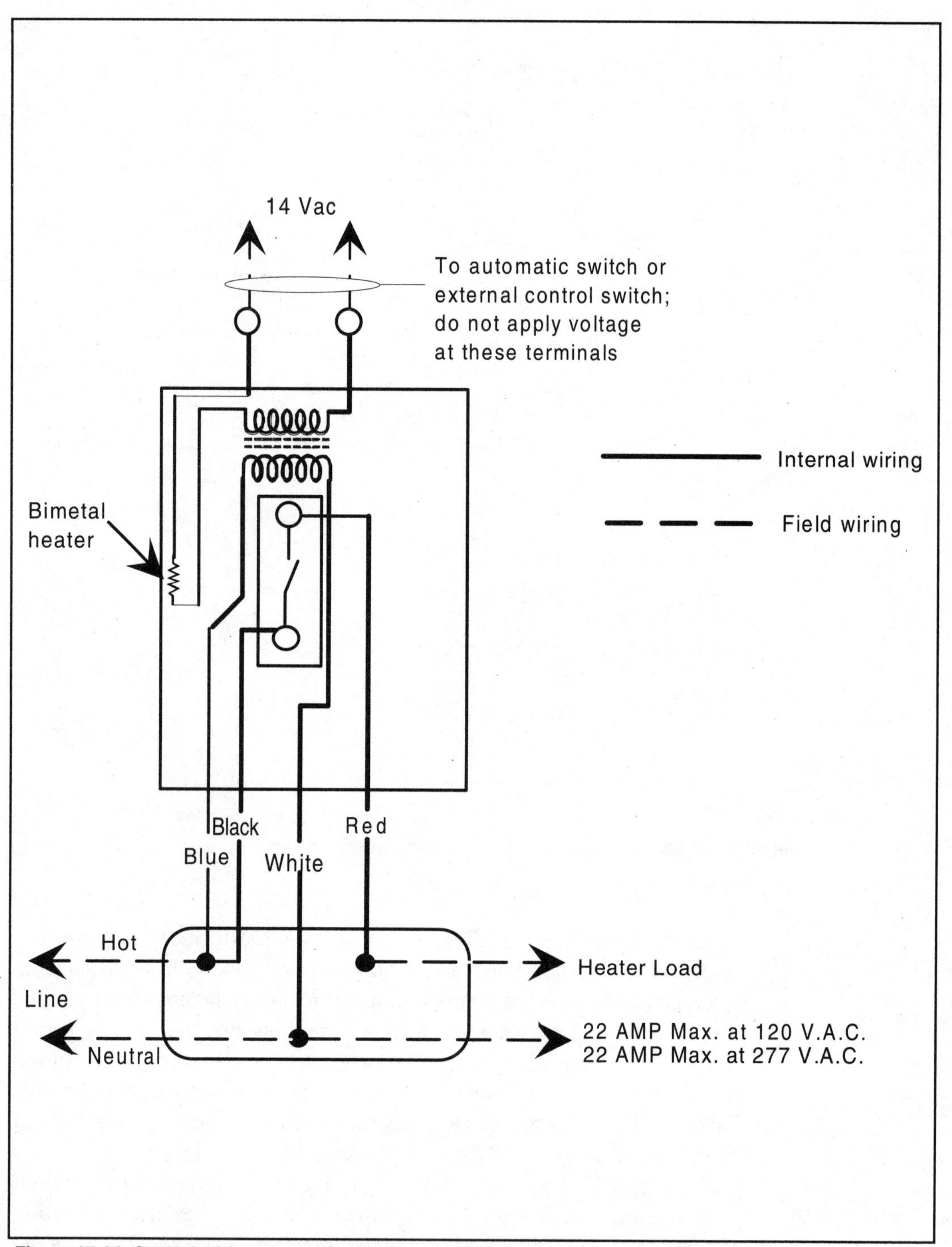

Figure 17-14: Control-wiring diagram for freeze-protection systems.

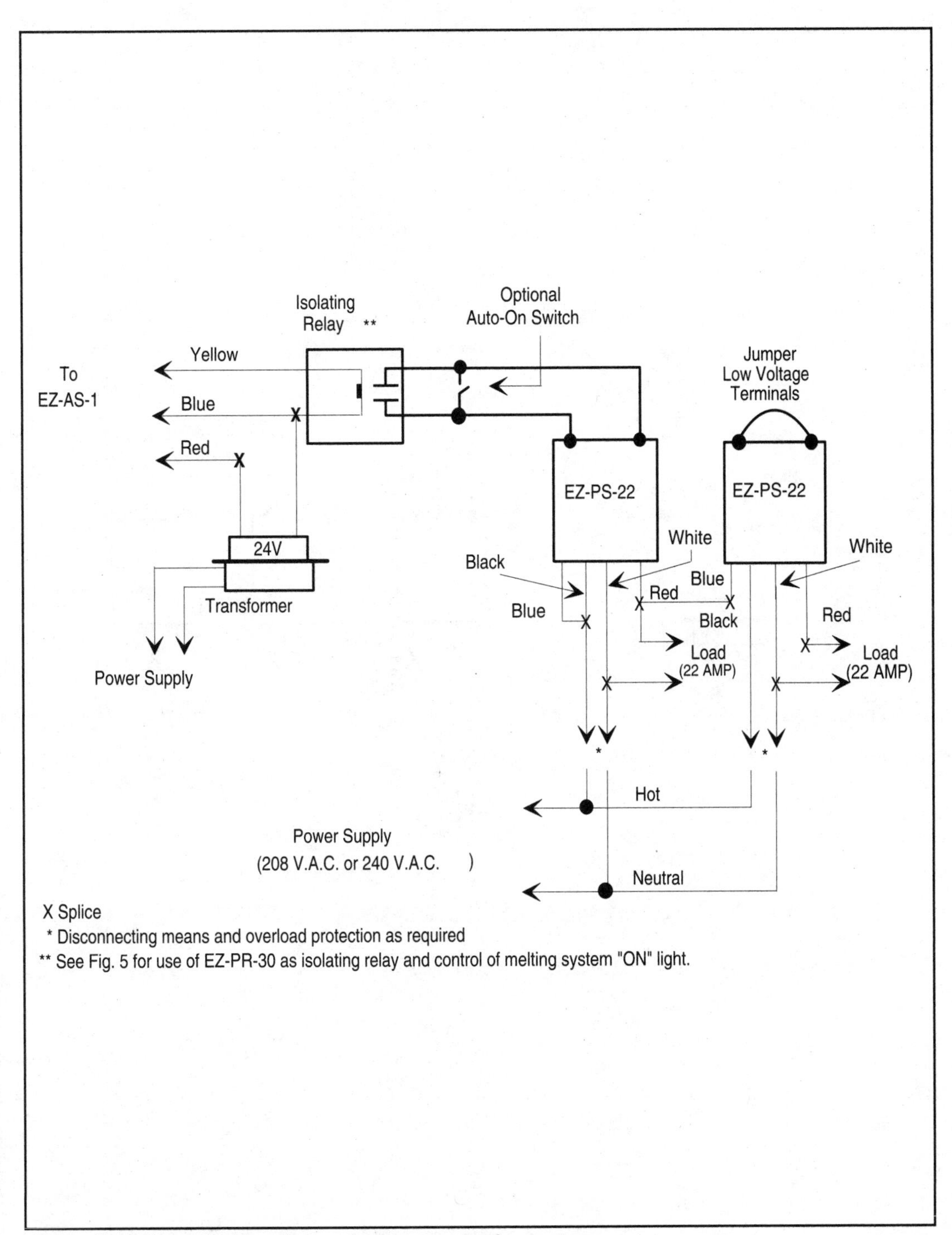

Figure 17-15: Control-wiring diagram for freeze-protection systems.

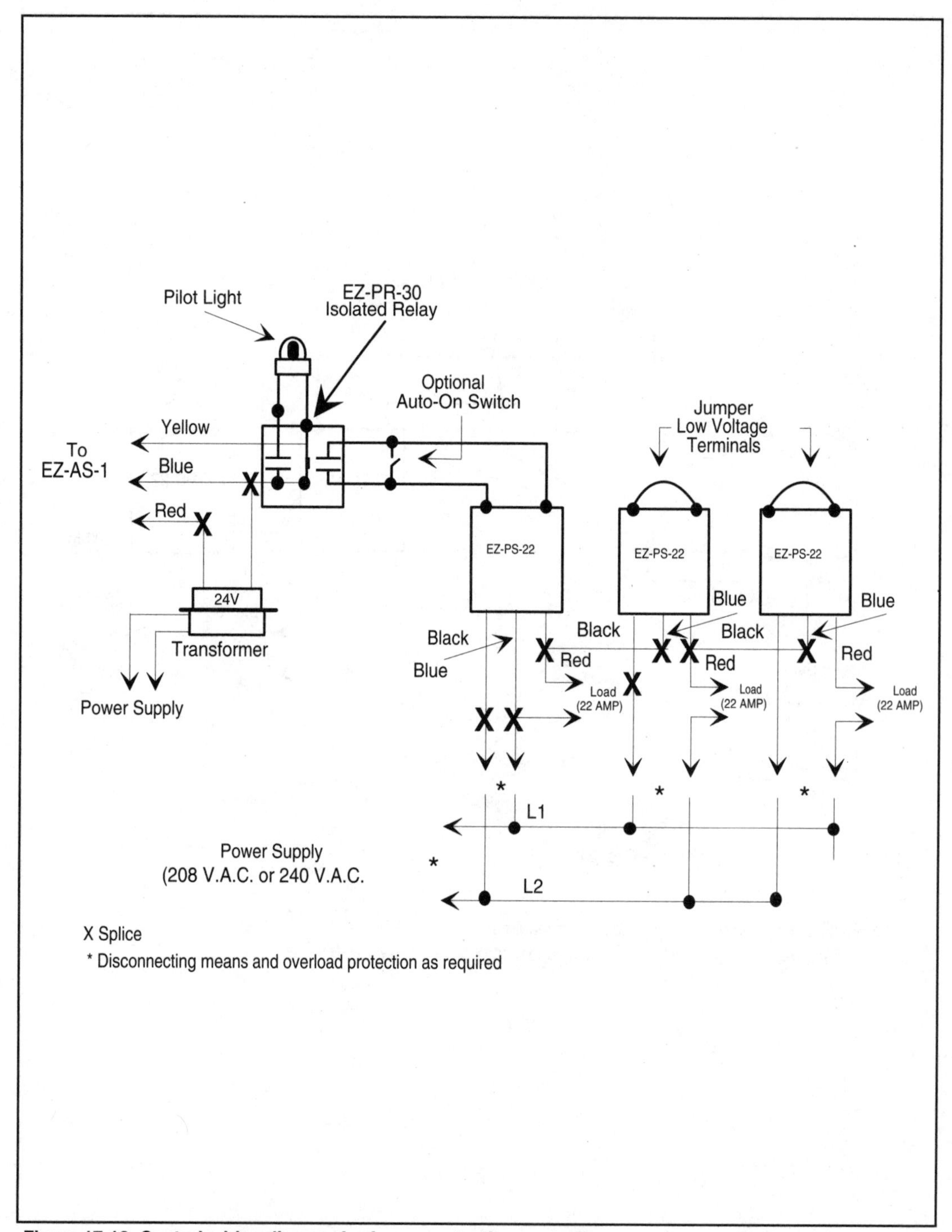

Figure 17-16: Control-wiring diagram for freeze-protection systems.

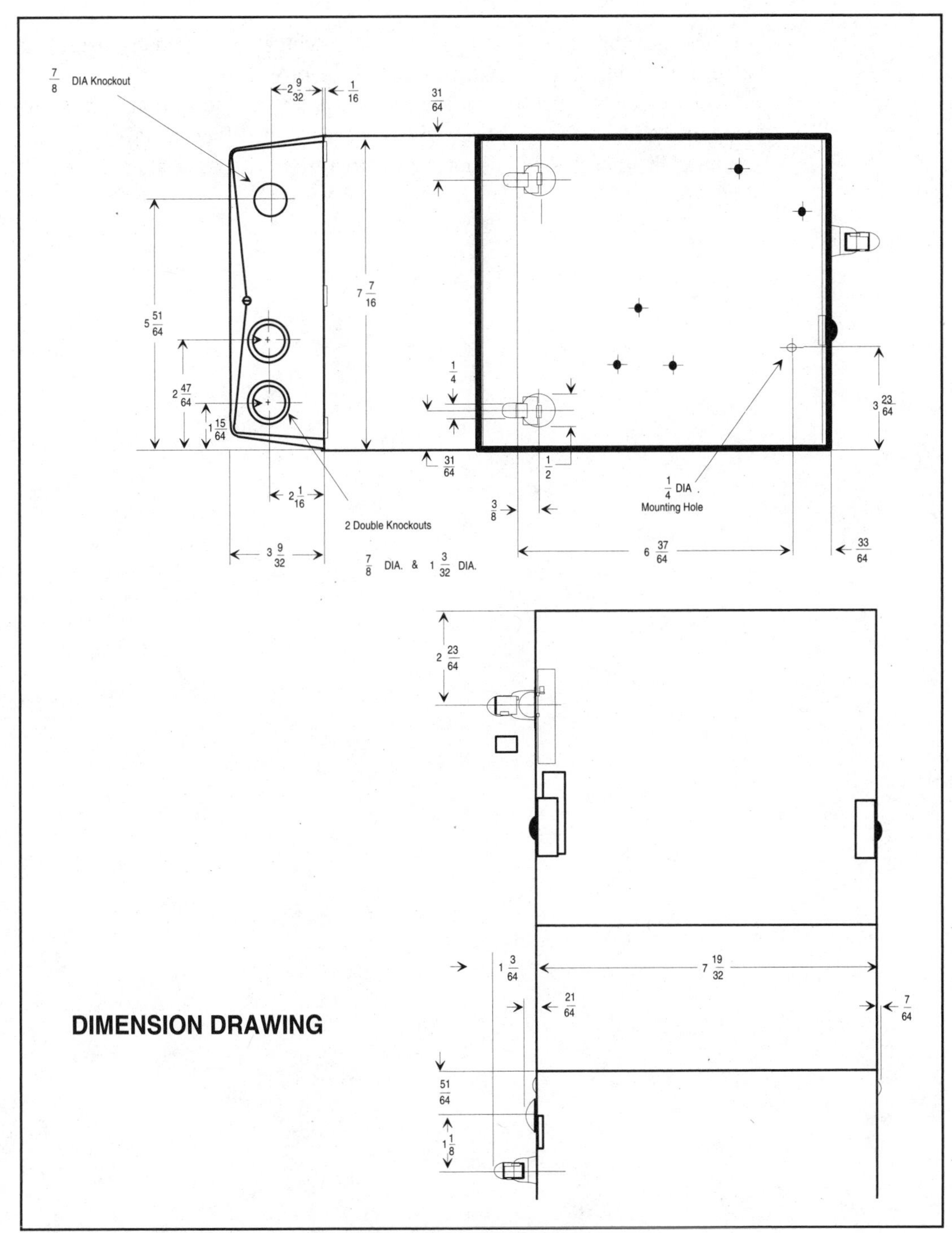

Figure 17-17: Dimensions of an electric heating control housing.

electric heating control housing. Such drawings are extremely useful to electricians during the rough-in stage of the project.

For models with a built-in thermostat, place the thermostat parallel to the pipe and hold it in place with friction tape. When installed in this manner, the thermostat will automatically turn the cable on at 35°F and off at 45°F.

Chapter 18

Wiring in Hazardous Locations

NEC Articles 500 through 503 cover the requirements of electrical equipment and wiring for all voltages in locations where fire or explosion hazards may exist. Locations are classified according to the properties of the flammable vapors, liquids, gases, or combustible dusts and fibers that may be present, as well as the likelihood that a flammable or combustible concentration or quality is present.

Any area in which the atmosphere or a material in the area is such that the arcing of operating electrical contacts, components, and equipment may cause an explosion or fire is considered a hazardous location. In all cases, explosionproof equipment, raceways, and fittings are used to provide an explosionproof wiring system.

Hazardous locations have been classified in the *NEC* into certain class locations. Furthermore, various atmospheric groups have been established on the basis of the explosive character of the atmosphere for the testing and approval of equipment for use in the various groups.

Class I Locations

Class I locations are those in which flammable gases or vapors are or may be present in the air in quantities sufficient to produce explosive or ignitible mixtures. Examples of this type of location would include interiors of paint spray booths where volatile flammable solvents are used;

inadequately ventilated pump rooms where flammable gas is pumped, and drying rooms for the evaporation of flammable solvents.

Class I atmospheric hazards are further divided into two Divisions (1 and 2) and also into four groups (A, B, C, and D). The classification involves the maximum explosion pressures, maximum safe clearance between parts of a clamped joint in an enclosure, and the minimum ignition temperature of the atmospheric mixture.

In general, the hazardous properties of the substances are greater for Group A. Group B is the next most hazardous, then Group C, with Group D the least hazardous. However, all four groups are extremely dangerous. Equipment to be used in these atmospheres must not only be approved for Class I, but also for the specific group of gases or vapors that will be present.

Class I, Division 2 covers locations where flammable gases, vapors, or volatile flammable gases, vapors, or volatile liquids are handled either in a closed system, or confined within suitable enclosures, or where hazardous concentrations are normally prevented by positive mechanical ventilation. Areas adjacent to Division 1 locations, into which gases might occasionally flow, would also belong in Division 2.

Class II Locations

Class II locations are those that are hazardous because of the presence of combustible dust. Class II, Division 1 locations are areas where combustible dust, under normal operating conditions, may be present in the air in quantities sufficient to produce explosive or ignitable mixtures; examples are working areas of grain-handling and storage plants and rooms containing grinders or pulverizers. Class II, Division 2 locations are areas where dangerous concentrations of suspended dust are not likely, but where dust accumulations might form.

Besides the two Divisions (1 and 2), Class II atmospheric hazards cover three groups of combustible dusts. The groupings are based on the resistivity of the dust. Group E is always Division 1. Group F, depending on the resistivity, and Group G may be either Division 1 or 2. Since the *NEC* is considered the definitive classification tool and contains explanatory data about hazardous atmospheres, refer to *NEC* Section 500-6 for exact definitions of Class II, Divisions 1 and 2.

Class III Locations

These locations are those areas that are hazardous because of the presence of easily ignitable fibers or flyings, but such fibers and flyings are not likely to be in suspension in the air in these locations in quantities sufficient to produce ignitable mixtures. Such locations usually include some parts of rayon, cotton, and textile mills, clothing manufacturing plants, and woodworking plants.

Once the class of an area is determined, the conditions under which the hazardous material may be present determines the division. In Class I and Class II, Division 1 locations, the hazardous gas or dust may be present in the air under normal operating conditions in dangerous concentrations. In Division 2 locations, the hazardous material is not normally in the air, but it might be released if there is an accident or if there is faulty operation of equipment.

PREVENTION OF EXTERNAL IGNITION/EXPLOSION

The main purpose of using explosionproof fittings and wiring methods in hazardous areas is to prevent ignition of flammable liquids or gases and to prevent an explosion.

Sources of Ignition

In certain atmospheric conditions when flammable gases or combustible dusts are mixed in the proper proportion with air, any source of energy is all that is needed to touch off an explosion.

One prime source of energy is electricity. Equipment such as switches, circuit breakers, motor starters, pushbutton stations, or plugs and receptacles, can produce arcs or sparks in normal operation when contacts are opened and closed. This could easily cause ignition.

Other hazards are devices that produce heat, such as lighting fixtures and motors. Here surface temperatures may exceed the safe limits of many flammable atmospheres.

Finally, many parts of the electrical system can become potential sources of ignition in the event of insulation failure. This group would include wiring (particularly splices in the wiring), transformers, impedance coils, solenoids, and other low-temperature devices without make-or-break contacts.

Nonelectrical hazards such as sparking metal can also easily cause ignition. A hammer, file, or other tool that is dropped on masonry or on a ferrous surface can cause a hazard unless the tool is made of nonsparking material. For this reason, portable electrical equipment is usually made from aluminum or other material that will not produce sparks if the equipment is dropped.

Electrical safety, therefore, is of crucial importance. The electrical installation must prevent accidental ignition of flammable liquids, vapors, and dusts released to the atmosphere. In addition, since much of this equipment is used outdoors or in corrosive atmospheres, the material and finish must be such that maintenance costs and shutdowns are minimized.

Combustion Principles

Three basic conditions must be satisfied for a fire or explosion to occur:

- A flammable liquid, vapor, or combustible dust must be present in sufficient quantity.
- The flammable liquid, vapor, or combustible dust must be mixed with air or oxygen in the proportions required to produce an explosive mixture.
- A source of energy must be applied to the explosive mixture.

In applying these principles, the quantity of the flammable liquid or vapor that may be liberated and its physical characteristics must be recognized.

Vapors from flammable liquids also have a natural tendency to disperse into the atmosphere, and rapidly become diluted to concentrations below the lower explosion limit particularly when there is natural or mechanical ventilation.

EXPLOSIONPROOF EQUIPMENT

Each area that contains gases or dusts that are considered hazardous must be carefully evaluated to make certain the correct electrical equipment is selected. Many hazardous atmospheres are Class I, Group D, or Class II, Group G. However, certain areas may involve other groups, particularly Class I, Groups B and C. Conformity with the *NEC* requires the use of fittings and enclosures approved for the specific hazardous gas or dust involved.

The wide assortment of explosionproof equipment now available makes it possible to provide adequate electrical installations under any of the various hazardous conditions. However, the electrician must be thoroughly familiar with all *NEC* requirements and know what fittings are available, how to install them properly, and where and when to use the various fittings.

For example, some workers are under the false belief that a fitting rated for Class I, Division 1 can be used under any hazardous conditions. However, remember the groups! A fitting rated for, say, Class I, Division 1, Group C cannot be used in areas classified as Groups A or B. On the other hand, fittings rated for use in Group A may be used for any group beneath A; fittings rated for use in Class I, Division 1, Group B can be used in areas rated as Group B areas or below, but not vice-versa.

Explosionproof fittings are rated for both classification and groups. All parts of these fittings, including covers, are rated accordingly. Therefore, if a Class I, Division 1, Group A fitting is required, a Group B (or below) fitting cover must not be used. The cover itself must be rated for Group A locations. Consequently, when working on electrical systems in hazardous locations, always make certain that fittings and their related components match the condition at hand.

Intrinsically Safe Equipment

Intrinsically safe equipment is equipment and wiring that are incapable of releasing sufficient electrical energy under normal or abnormal conditions to cause ignition of a specific hazardous atmospheric mixture in its most easily ignited concentration.

The use of intrinsically safe equipment is primarily limited to process control instrumentation, since these electrical systems lend themselves to the low energy requirements.

Installation rules for intrinsically safe equipment are covered in *NEC* Article 504. In general, intrinsically safe equipment and its associated wiring must be installed so they are positively separated from the nonintrinsically safe circuits because induced voltages could defeat the concept of intrinsically safe circuits. Underwriters' Laboratories Inc., and Factory Mutual list several devices in this category.

Explosionproof Conduit and Fittings

In hazardous locations where threaded metal conduit is required, the conduit must be threaded with a standard conduit cutting die that provides ¾-in taper per ft. The conduit should be made up wrench tight in order to minimize sparking in the event fault current flows through the raceway system (*NEC* Section 500-2). Where it is impractical to make a threaded joint tight, a bonding jumper shall be used. All boxes, fittings, and joints shall be threaded for connection to the conduit system and shall be an approved, explosionproof type (Figure 18-1). Threaded joints must be

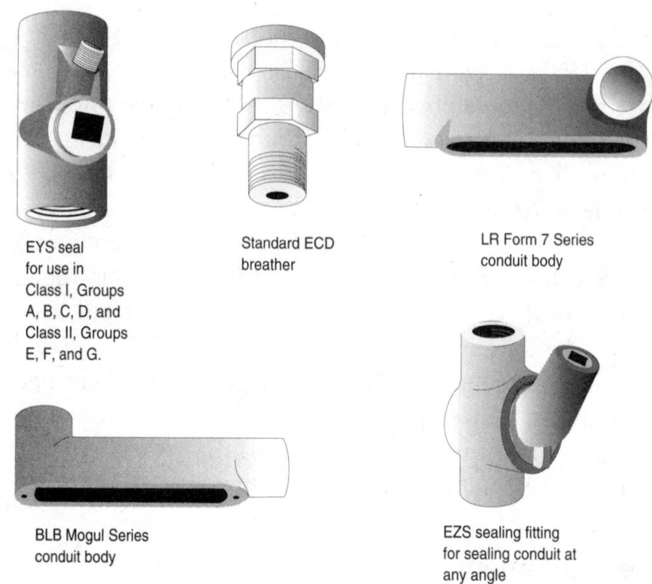

Figure 18-1: Typical fittings approved for hazardous areas.

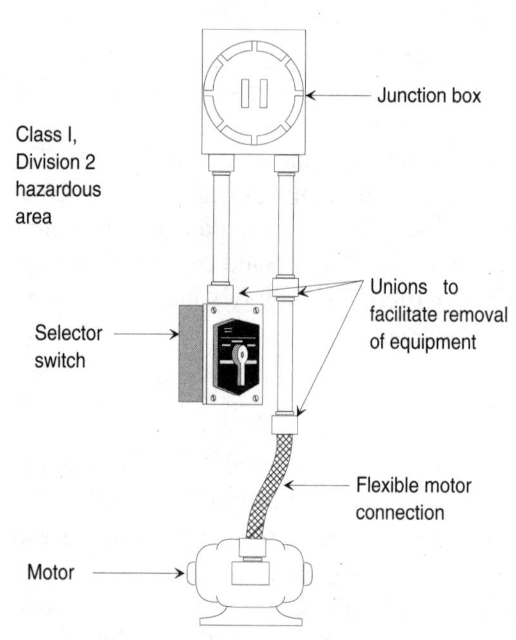

Figure 18-2: Explosionproof flexible connectors are frequently used for motor terminations.

made up with at least five threads fully engaged. Where it becomes necessary to employ flexible connectors at motor or fixture terminals (Figure 18-2), flexible fittings approved for the particular class location shall be used.

Seals And Drains

Seal-off fittings (Figure 18-3 on the next page) are required in conduit systems to prevent the passage of gases, vapors, or flames from one portion of the electrical installation to another at atmospheric pressure and normal ambient temperatures. Furthermore, seal-offs (seals) limit explosions to the sealed-off enclosure and prevent precompression of "pressure piling" in conduit systems. For Class I, Division 1 locations, the *NEC* [Section 501-5(1)] states:

In each conduit run entering an enclosure for switches, circuit breakers, fuses, relays, resistors, or other apparatus which may produce arcs, sparks, or high temperatures, seals shall

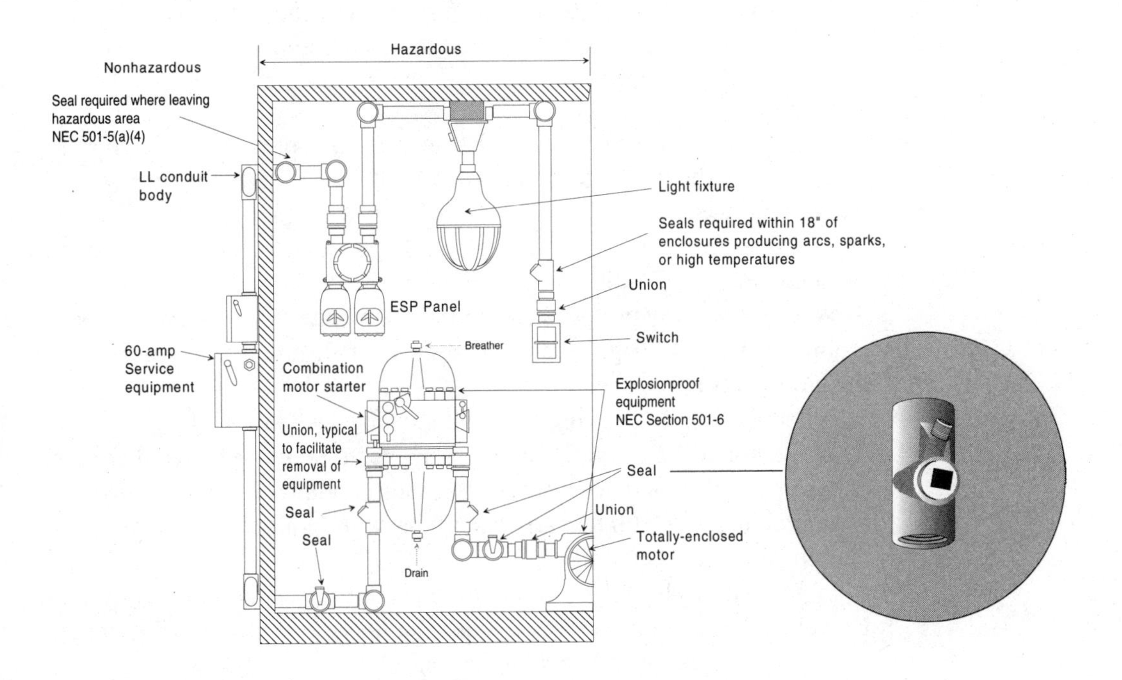

Figure 18-3: Seals must be installed at various locations in Class I, Division 1 locations.

> *be installed with 18 inches from such enclosures. Explosionproof unions, couplings, reducers, elbows, capped elbows and conduit bodies similar to "L," "T," and "cross" types shall be the only enclosures or fittings permitted between the sealing fitting and the enclosure. The conduit bodies shall not be larger than the largest trade size of the conduits.*

There is, however, one exception to this rule:

> Conduits $1\frac{1}{2}$ in and smaller are not required to be sealed if the current-interrupting contacts are either enclosed within a chamber hermetically sealed against the entrance of gases or vapors, or immersed in oil in accordance with Section 501-6 of the *NEC*.

Seals are also required in Class II locations under the following condition (*NEC* Section 502-5):

- Where a raceway provides communication between an enclosure that is required to be dust-ignitionproof and one that is not.

A permanent and effective seal is one method of preventing the entrance of dust into the dust-ignitionproof enclosure through the raceway. A horizontal raceway, not less than 10 ft long, is another approved method, as is a vertical raceway not less than 5 ft long and extending downward from the dust-ignitionproof enclosure.

Where a raceway provides communication between an enclosure that is required to be dust-ignitionproof and an enclosure in an unclassified location, seals are not required.

Where sealing fittings are used, all must be accessible.

While not an *NEC* requirement, many electrical designers and workers consider it good practice to sectionalize long conduit runs by inserting seals not more than 50 to 100 ft apart, depending on the conduit size, to minimize the effects of "pressure piling."

In general, seals are installed at the same time as the conduit system. However, the conductors are installed after the raceway system is complete and *prior* to packing and sealing the seal-offs.

Drains

In humid atmospheres or in wet locations, where it is likely that water can gain entrance to the interiors of enclosures or raceways, the raceways should be inclined so that water will not collect in enclosures or on seals but will be led to low points where it may pass out through integral drains.

Frequently, the arrangement of raceway runs makes this method impractical — if not impossible. In such instances, special drain/seal fittings should be used, such as Crouse-Hinds Type EZDs as shown in Figure 18-4. These fittings prevent harmful accumulations of water above the seal and meet the requirements of *NEC* Section 501-5(d).

Figure 18-4: Typical drain seal.

In locations which usually are considered dry, surprising amounts of water frequently collect in conduit systems. No conduit system is airtight; therefore, it may "breathe." Alternate increases and decreases in temperature and/or barometric pressure due to weather changes or due to the nature of the process carried on in the location where the conduit is installed will cause "breathing."

Outside air is drawn into the conduit system when it "breathes in." If this air carries sufficient moisture, it will be condensed within the system when the temperature decreases and chills this air. The internal conditions being unfavorable to evaporation, the resultant water accumulation will remain and be added to by repetitions of the breathing cycle.

In view of this likelihood, it is good practice to insure against such water accumulations and probable subsequent insulation failures by installing drain/seal fittings with drain covers or fittings with inspection covers even though conditions prevailing at the time of planning or installing do not indicate their need.

Selection of Seals and Drains

The primary considerations for selecting the proper sealing fittings are as follows:

- Select the proper sealing fitting for the hazardous vapor involved; that is, Class I, Groups A, B, C, or D.
- Select a sealing fitting for the proper use in respect to mounting position. This is particularly critical when the conduit runs between hazardous and nonhazardous areas. Improper positioning of a seal may permit hazardous gases or vapors to enter the system beyond the seal, and permit them to escape into another portion of the hazardous area, or to enter a nonhazardous area. Some seals are designed to be mounted in any position; others are restricted to horizontal or vertical mounting.
- Install the seals on the proper side of the partition or wall as recommended by the manufacturer.
- Installation of seals should be made *only* by trained personnel in strict compliance with the instruction sheets furnished with the seals and sealing compound.
- It should be noted that *NEC* Section 501-5(c)(4) prohibits splices or taps in sealing fittings.
- Sealing fittings are listed by U.L. for use in Class I hazardous locations with CHICO® A compound only. This compound, when properly mixed and poured, hardens into a dense, strong mass which is insoluble in water, is not attacked by chemicals, and is not softened by heat. It will withstand with ample safety factor, pressure of the exploding trapped gases or vapor.

- Conductors sealed in the compound may be approved thermoplastic or rubber insulated type. Both may or may not be lead covered.

Types of Seals and Fittings

Certain seals, such as Crouse-Hinds EYS seals, are designed for use in vertical or nearly vertical conduit in sizes for ½- through 1-in. Other styles are available in sizes ½-in through 6 in for use in vertical or horizontal conduit. In horizontal runs, these are limited to face-up openings.

Seals ranging in sizes from 1¼-in through 6-in have extra large work openings, and separate filling holes, so that fiber dams are easy to make. However, the overall diameter of these fittings are scarcely greater than that of unions of corresponding sizes, permitting close conduit spacing.

Crouse-Hinds EZS seals are for use with conduit running at any angle, from vertical through horizontal.

EYD drain seals provide continuous draining and thereby prevent water accumulation. EYD seals are for vertical conduit runs and range in size from ½-in to 4-in inclusive. They are provided with one opening for draining and filling, a rubber tube to form drain passage, and a drain fitting.

EZD drain seals provide continuous draining and thereby prevent water accumulation. The covers should be positioned so that the drain will be at the bottom. A set screw is provided for locking the cover in this position.

EZD fittings are suitable for sealing vertical conduit runs between hazardous and nonhazardous areas, but must be installed in the hazardous area when it is above the nonhazardous area. They must be installed in the nonhazardous area when it is above the hazardous area.

EZD drain seals are designed so that the covers can be removed readily, permitting inspection during installation or at any time thereafter. After the fittings have been installed in the conduit run and conductors are in place, the cover and barrier are removed. After the dam has been made in the lower hub opening with packing fiber, the barrier must be replaced so that the sealing compound can be poured into the sealing chamber.

EZD inspection seals are identical to EZD drain seals to provide all inspection, maintenance, and installation advantages except that the cover is not provided with an automatic drain. Water accumulations can be drained periodically by removing the cover (when no hazards exist). The cover must be replaced immediately.

Sealing Compounds and Dams

Poured seals should be made only by trained personnel in strict compliance with the specific instruction sheets provided with each sealing fitting. Improperly poured seals are worthless.

Sealing compound shall be approved for the purpose; it shall not be affected by the surrounding atmosphere or liquids; and it shall not have a melting point of less than 200° F. (93° C.). The sealing compound and dams must also be approved for the type and manufacturer of fitting. For example, Crouse-Hinds CHICO® sealing compound is the only sealing compound approved for use with Crouse-Hinds ECM sealing fittings.

To pack the seal-off, remove the threaded plug or plugs from the fitting and insert the fiber supplied with the packing kit. Tamp the fiber between the wires and the hub before pouring the sealing compound into the fitting. Then pour in the sealing cement and reset the threaded plug tightly. The fiber packing prevents the sealing compound (in the liquid state) from entering the conduit lines.

Most sealing-compound kits contain a powder in a polyethylene bag within an outer container. To mix, remove the bag of powder, fill the outside container, and pour in the powder and mix.

In practical applications, there may be dozens of seals required for a particular installation. Consequently, after the conductors are pulled, each seal in the system is first packed. To prevent the possibility of overlooking a seal, one color of paint is normally sprayed on the seal hub at this time. This indicates that the seal has been packed. When the sealing compound is poured, a different color paint is once again sprayed on the seal hub to indicate a finished job. This method permits the job supervisor to visually inspect the conduit run, and if a seal is not painted the appropriate color, he or she knows that proper installation on this seal was not done; therefore, action can be taken to correct the situation immediately.

The seal-off fitting in Figure 18-5 is typical of those used. The following procedures are to be observed when preparing sealing compound:

- Use a clean mix vessel for every batch. Particles of previous batches or dirt will spoil the seal.

CAUTION!

Always make certain that the sealing compound is compatible for use with the packing material, brand, and type of fitting, and also with the type of conductors used in the system.

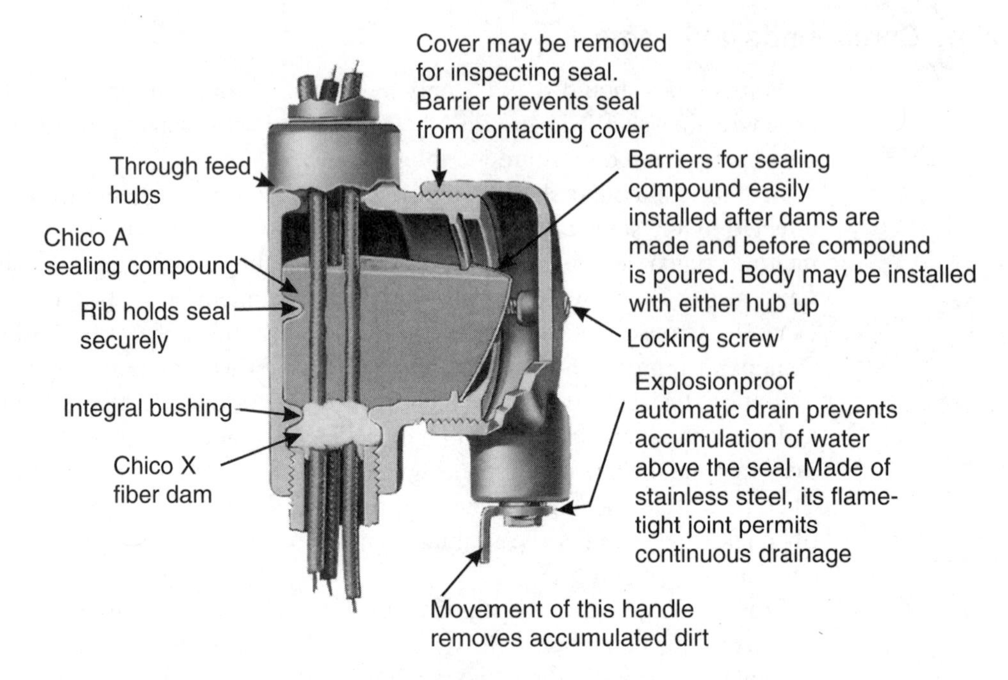

Figure 18-5: Cross-sectional view of a typical seal for Class I, Division 1 locations.

- Recommended proportions are by volume—usually two parts powder to one part clean water. Slight deviations in these proportions will not affect the result.
- Do not mix more than can be poured in 15 minutes after water is added. Use cold water. Warm water increases setting speed. Stir immediately and thoroughly.
- If batch starts to set do not attempt to thin it by adding water or by stirring. Such a procedure will spoil seal. Discard partially set material and make fresh batch. After pouring, close opening immediately.
- Do not pour compound in sub-freezing temperatures, or when these temperatures will occur during curing.
- See that compound level is in accordance with the instruction sheet for that specific fitting.

Most other explosionproof fittings are provided with threaded hubs for securing the conduit as described previously. Typical fittings include switch and junction boxes, conduit bodies, union and connectors, flexible couplings, explosionproof lighting fixtures, receptacles, and panelboard and motor starter enclosures. A practical representation of these and other fittings is shown in Figures 18-6 through 18-8.

PETRO/CHEMICAL HAZARDOUS LOCATIONS

Most manufacturing facilities involving flammable liquids, vapors, or fibers must have their wiring installations conform strictly to the *NEC* as well as governmental, state, and local ordinances. Therefore, the majority of electrical installations for these facilities are carefully designed by experts in the field — either the plant in-house engineering staff or else an independent consulting engineering firm.

Industrial installations dealing with petroleum or some types of chemicals are particularly susceptible to many restrictions involving many governmental agencies. Electrical installations for petro/chemical plants will therefore have many pages of electrical drawings and specifications which first go through the gambit for approval from all the agencies involved. Once approved, these drawings and specifications must be followed exactly because any change whatsoever must once again go through the various agencies for approval.

MANUFACTURERS' DATA

Manufacturers of explosionproof equipment and fittings expend a lot of time, energy, and expense in developing guidelines and brochures to ensure that their products are used correctly and in accordance with the latest *NEC* requirements. The many helpful charts, tables, and application guidelines are invaluable to anyone working on projects involving hazardous locations. Therefore, it is recommended that the trainee obtain as much of this data as possible. Once obtained, study this data thoroughly. Doing so will enhance your qualifications for working in hazardous locations of any type.

Manufacturers' data is usually available to qualified personnel (electrical workers) at little or no cost and can be obtained from local distributors of electrical supplies, or directly from the manufacturer.

Summary

Any area in which the atmosphere or a material in the area is such that the arcing of operating electrical contacts, components, and equipment may cause an explosion or fire is considered as a hazardous location. In all such

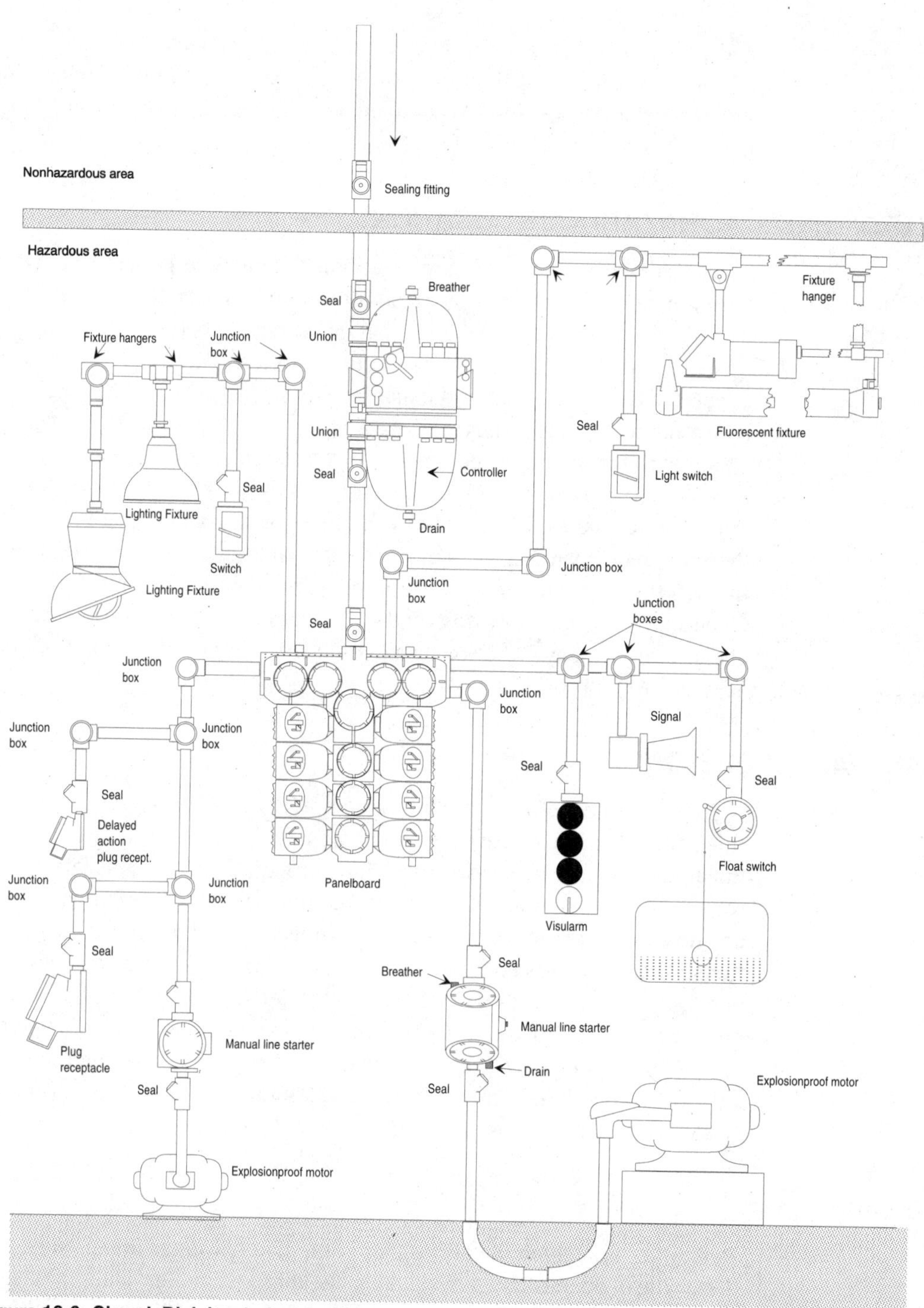

Figure 18-6: Class I, Division 1 electrical installation.

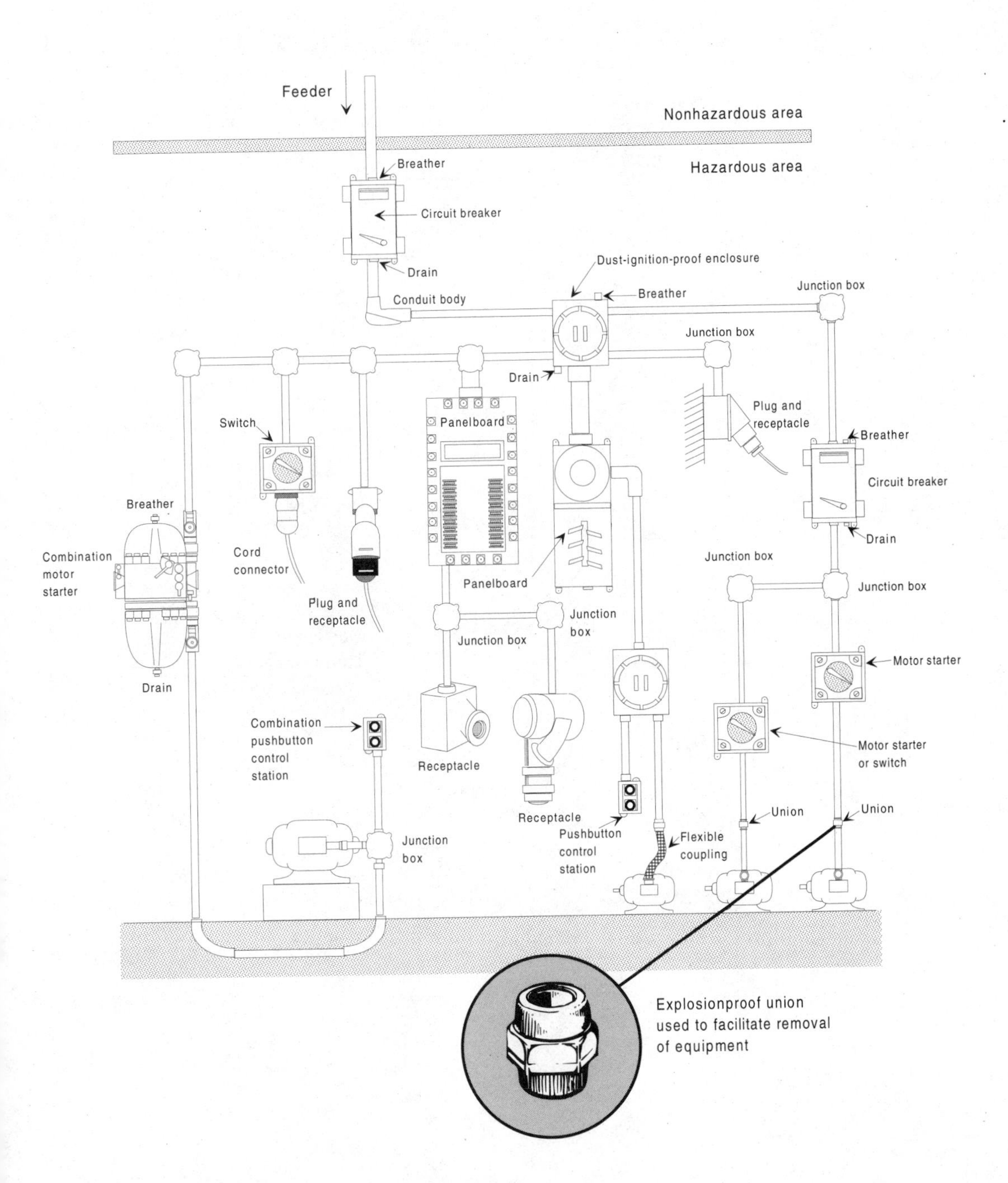

Figure 18-7: Class II, Division 1 electrical installation.

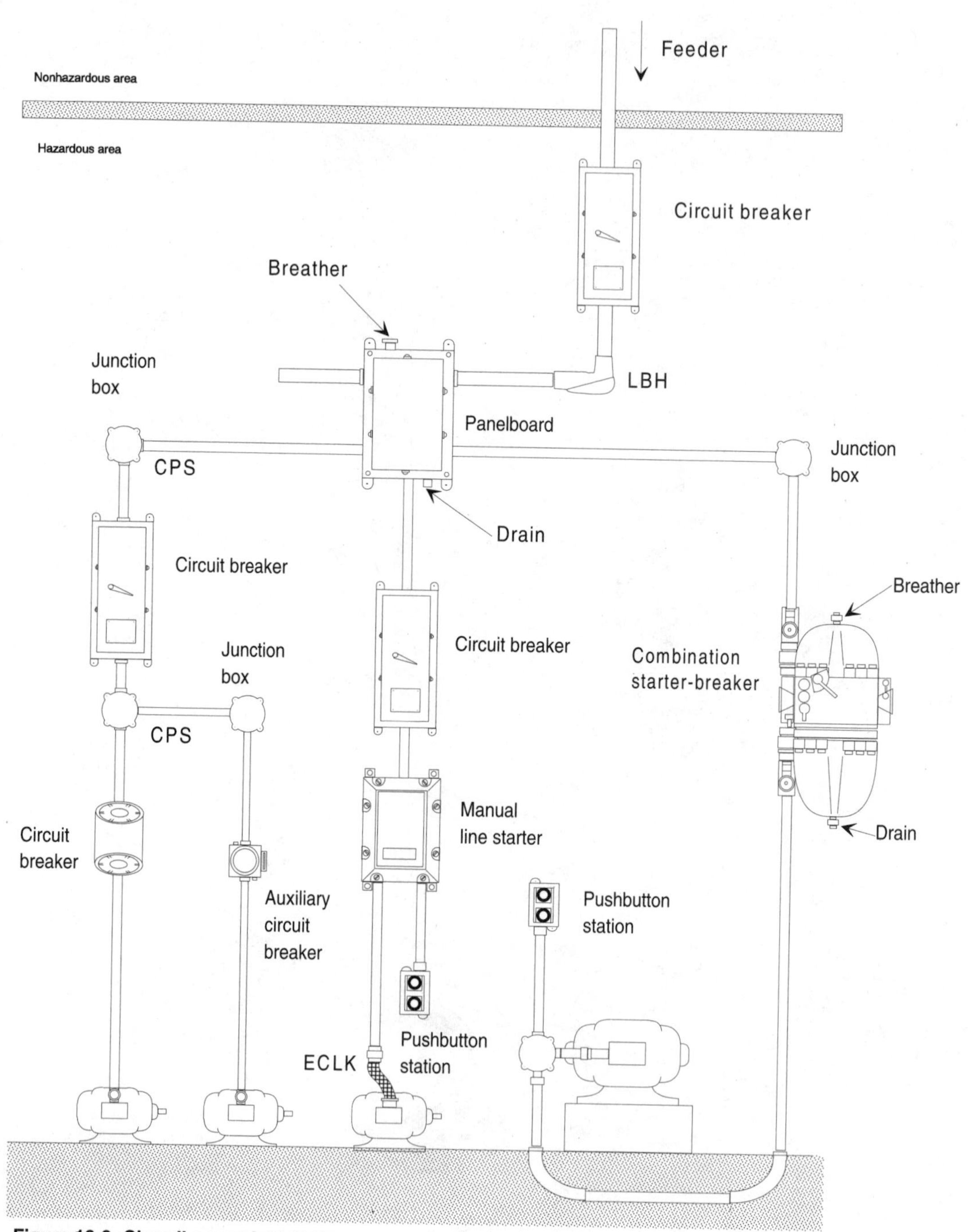

Figure 18-8: Class II power installation.

cases, explosionproof equipment, raceways, and fittings are used to provide an explosionproof wiring system.

The wide assortment of explosionproof equipment now available makes it possible to provide adequate electrical installations under any of the various hazardous conditions. However, the electrician must be thoroughly familiar with all *NEC* requirements and know what fittings are available, how to install them properly, and where and when to use the various fittings.

Many factors — such as temperature, barometric pressure, quantity of release, humidity, ventilation, distance from the vapor source, and the like — must be considered. When information on all factors concerned is properly evaluated, a consistent classification for the selection and location of electrical equipment can be developed.

Chapter 19

Industrial Lighting

Lighting layouts for industrial buildings should be designed to provide the highest visual comfort and performance that is consistent with the type of area to be lighted and the budget provided. However, since individual tastes and opinions vary, there can be many solutions to any given lighting application.

The data presented in this chapter are designed to give the reader a basic knowledge of lighting design and installation practices. The topics include:

- How to calculate lighting requirements
- How to lay out lighting schemes
- Selecting appropriate luminaires and lamp sources

LAMP SOURCES

Electric lamps are made in thousands of different types and colors, from a fraction of a watt to over 10 kW each, and for practically any conceivable lighting application.

Incandescent filament lamps, for example, consist of a sealed glass envelope containing a filament that produces light when heated to incandescence (white light) by its resistance to a flow of electric current. This type of light source is relatively inexpensive to install, is not greatly affected by ambient temperatures, is easily controlled as to direction and brightness, and gives a high color quality. Incandescent lamps, as compared to other lamps, are less efficient and result in a higher operating cost per lumen.

More heat is produced per lumen than electric-discharge lighting, causing the need for a larger air-conditioning system which in turn increases operating cost.

The quartz-iodine tungsten-filament lamp is similar to the basic incandescent lamp except that the glass envelope contains an iodine vapor, which prevents the evaporation of the tungsten filament. This increases the normal life to about twice that of a normal incandescent lamp.

Fluorescent lighting has a high efficiency as compared to incandescent lighting. To illustrate this fact, the average 40-W inside-frosted incandescent lamp delivers approximately 470 initial lumens, while a cool-white fluorescent of the same wattage delivers over 3200 initial lumens. This power efficiency not only saves on the cost of power consumed, but also lessens the heat produced by lamps, which in turn reduces air-conditioning loads. Further, fluorescent lighting provides a linear source of light, long lamp life, and a means of relatively low surface luminance. However, the initial installation cost is normally higher due to the required auxiliary equipment (ballast, and so on). Also, fluorescent lighting is temperature and humidity sensitive, produces radio interference, and does not lend itself to critical light control.

The most popular lamp types are cool white and deluxe warm white. The cool-white lamp is often selected for offices, factories, and commercial areas where a psychologically cool working atmosphere is desirable. It is also one of the most efficient fluorescent lamps manufactured today.

Deluxe warm-white lamps produce a more flattering color to the complexion; the color is very close to incandescent in that they impart a ruddy or tanned hue to the skin. They are generally recommended for application in homes and for commercial use where flattering effects on people and merchandise are considered important. Whenever a warm social atmosphere is desirable, this is the color of fluorescent lamp to use.

HIGH-INTENSITY DISCHARGE LAMPS

The term high-intensity discharge lamps describes a wide variety of lighting sources. Their common characteristic is that they consist of gaseous discharge arc tubes, which, in the versions designed for lighting, operate at pressures and current densities sufficient to generate desired quantities of radiation within their arcs alone.

Mercury vapor lamps contain arc tubes, which are formed of fused quartz. This has resulted in great improvements in lamp life and maintenance of output through life. These arcs radiate ultraviolet energy as well as light, but the glass used in the outer bulbs is generally of a heat-resisting type that absorbs most of the ultraviolet. Some mercury lamps have outer bulbs that are internally coated with fluorescent materials which, when activated by the ultraviolet, emit visible energy at wavelengths that modify the color of light from the arc. The General Electric deluxe-white mercury lamps, for example, have color characteristics well suited to many commercial and industrial lighting applications that could not have been considered for mercury until recently. General-lighting mercury lamps are available in wattages from 50 to 3000.

Multivapor lamps generate light with more than half again the efficiency of the mercury arc, and with better color.

Some years ago, General Electric introduced the Lucalox® lamp, which has the highest light-producing efficiency of any commercial source of white light. This lamp was made possible by the invention of a means of effectively sealing metal ends and electrodes to a tube of Lucalox ceramic in a combination that could withstand temperatures and corrosive effects produced by intensely hot vapors of the alkaline metals. The arc is principally made of metallic sodium, which yields much better color quality and compactness, with substantially higher luminous efficacy than has been available before for white light.

The outer bulbs of high-intensity discharge lamps are designed to provide, as nearly as possible, optimum internal environments for arc tube performance. For example, the rounded shapes labeled E and BT in the sketches in Figure 19-1 were devised to maintain uniform temperatures of the bulb walls for better performance of phosphor coatings. The E-bulb improves manufacturing efficiency and eliminates the clear bulb-end on phosphor-lined bulbs.

In some cases, special considerations dictate the bulb shape. The R and PAR contours have been selected to achieve desired directional distribution of light. Some of the smaller T-bulbs are made of highly specialized glasses, which are more economically formed in these simple contours.

Most of the general contours of high-intensity discharge lamps are shown in Figure 19-1, with verbal descriptions of the code used for the shapes. The complete description of a bulb also includes a number that represents the maximum diameter of the bulb in eighths of an inch. The E-37 bulb, therefore, is elliptical in shape and $4\frac{5}{8}$ in in diameter at its widest point; the R-80 is a reflector bulb with 10-in maximum diameter.

A typical mercury lamp consists of the parts schematically illustrated in Figure 19-2, enclosed in an outer bulb made of borosilicate glass, which can withstand high temperatures, and which is resistant to thermal shocks such as those created when cold raindrops strike a hot bulb. The outer bulb contains a small quantity of nitrogen, an inert gas; this atmosphere main-

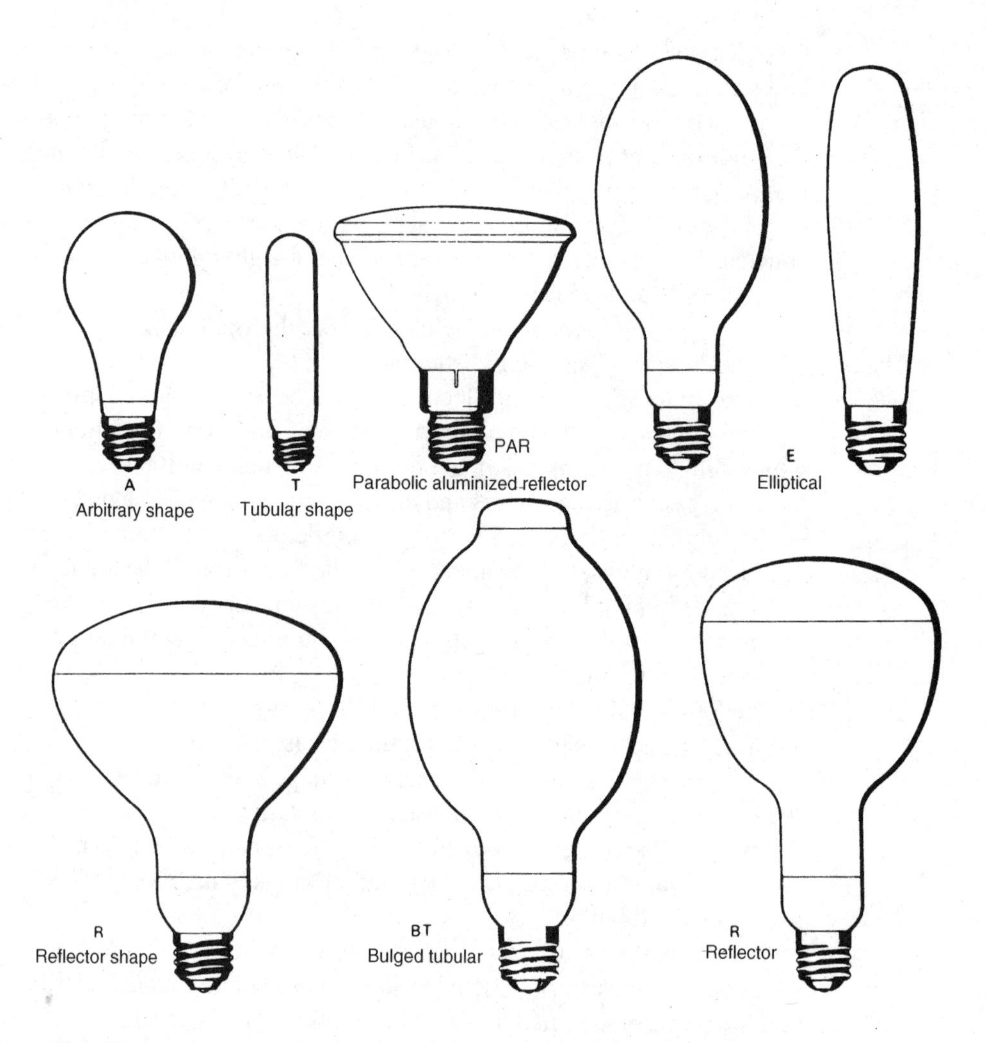

Figure 19-1: The general contour of high-intensity discharge lamps.

tains internal electrical stability, provides thermal insulation for the arc tube, and protects the metal parts from oxidation. The quartz arc tube contains a small quantity of high-purity mercury, and a starting gas, argon.

Most mercury lamps operate on ac circuits, and the ac-circuit ballast usually consists of a transformer to convert the distribution voltages of the lighting circuit to the required starting voltage for the lamp, and inductive or capacitive reactance components to control lamp current and, in some ballasts, to improve power factor.

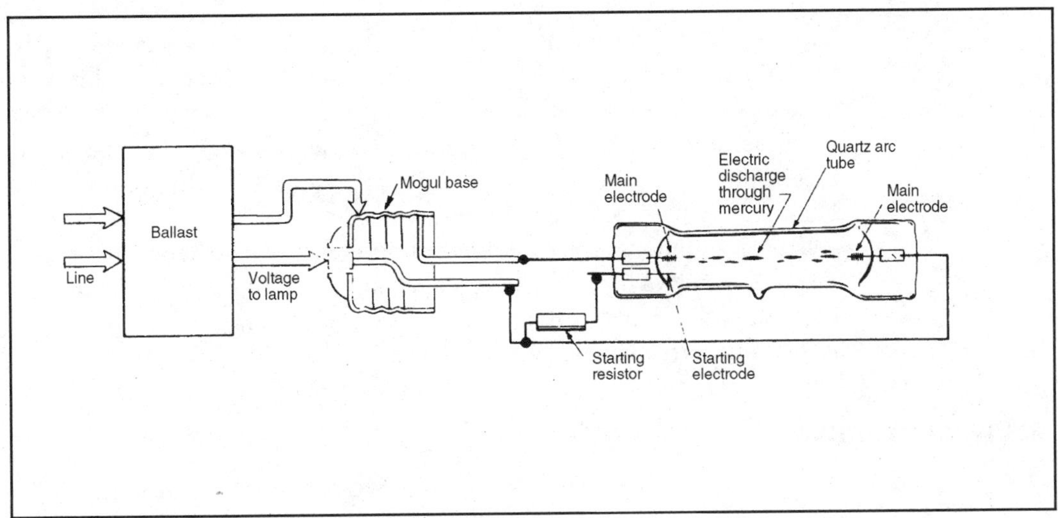

Figure 19-2: Basic parts of a typical mercury lamp.

Most mercury lamps start and operate equally well in any burning position. However, light output and maintenance of output through life generally are slightly higher with vertical than with horizontal operation.

The operating life of mercury lamps is very long, which accounts for much of their popularity in recent years. Most general lighting lamps of 100 to 1000 W have rated lives in excess of 24,000 hr, while the 50-, 75-, and 100-W lamps with medium screw bases are rated at 10,000 hr. Ratings are based on operation with properly designed ballasts, with five or more burning hours per start. More frequent starting may reduce life somewhat.

Stroboscopic Effect

The arc of a mercury lamp operating on a 60 Hz ac circuit is completely extinguished 120 times per second. Consequently, there is a tendency for the eye to see in flashes, with the result that a rapidly moving object may appear to move in a series of jerks (stroboscopic effect). This effect often goes unnoticed, and in most installations it is not a serious disadvantage. Where necessary, the stroboscopic effect may be greatly reduced by operating pairs of lamps on a lead-lag, two-lamp ballast, or three lamps on separate phases of a three-phase system. The use of some incandescent lamps in combination with mercury lamps also lessens the stroboscopic effect.

Where high output is more important than color, such as the lighting of steel mills, aircraft plants, and foundries, high-output white or standard white semireflector lamps are often the most economical choice — particularly those types that may be operated directly from 460-V circuits with inexpensive choke-type ballasts. Wholly coated lamps have lower surface brightness than semireflector lamps, and at relatively low mounting heights, the coated lamps may provide a more comfortable lighting installation, along with better color rendition.

For installations where color is important, deluxe white lamps are recommended. Where the unavoidable slight hum created by the ballasts might be objectionable, the ballasts are sometimes located remotely; that is, outside of the area where the lights are installed.

Multivapor Lamps

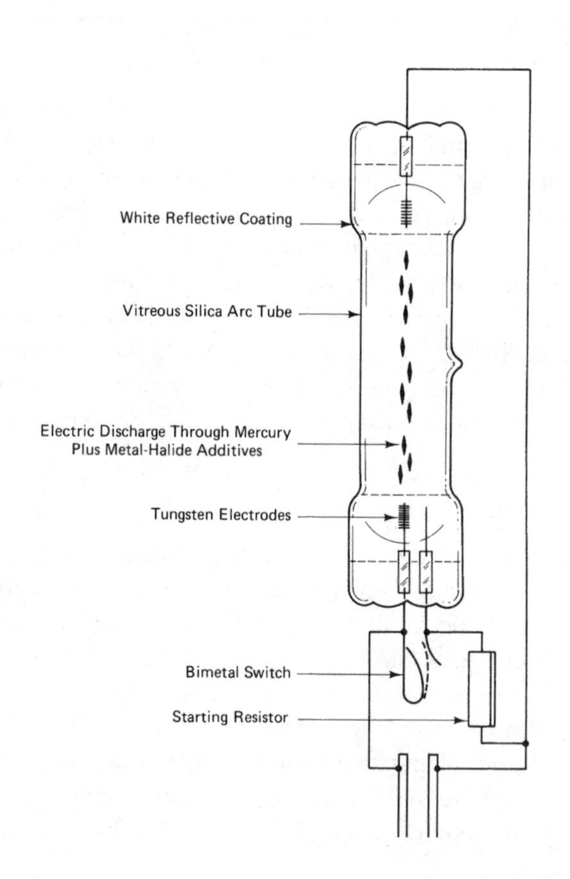

Figure 19-3: Internal components of a multivapor lamp.

Multivapor lamps are quite similar in physical appearance to conventional clear mercury lamps. The major differences in internal construction and appearance can be seen by comparing the sketch in Figure 19-3 with the one in Figure 19-2. At the present time, two sizes of multivapor lamps are available: 400 and 1000 W. The 1000-W size has been used most widely, largely because it is capable of delivering so much light from a single luminaire. Applications for multivapor to date include industrial lighting, street lighting, building floodlighting, and the floodlighting of several major stadiums to provide the amount and color quality of light needed for color telecasts, without resorting to huge increases in power requirements.

Lucalox Lamps

The construction, operation, and radiation characteristics of Lucalox lamps are quite unlike those of the other high-intensity discharge lamps.

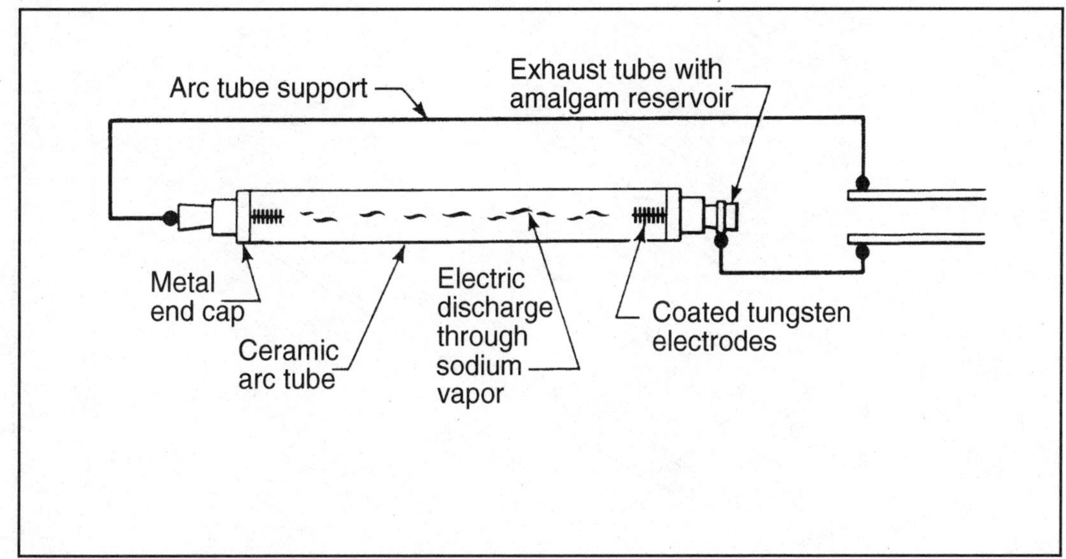

Figure 19-4: Basic components of a Lucalox lamp.

The illustration in Figure 19-4 shows that the essential internal components are fewer.

In addition to the more common mercury, multivapor, and Lucalox lamps, which are generally used for lighting purpsoes, there are a number of types of discharge lamps for special applications. Some of these include the following:

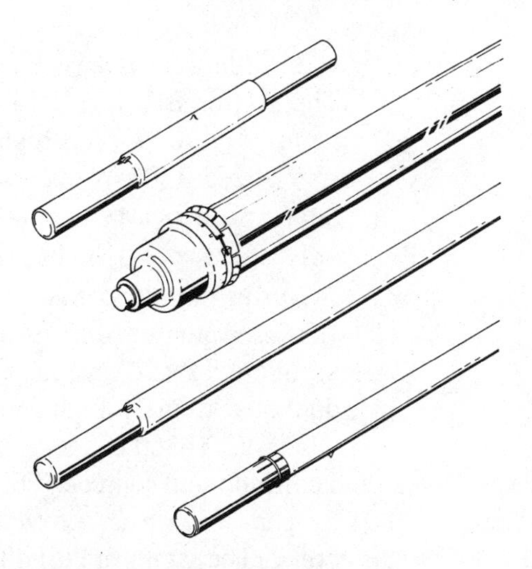

Figure 19-5: Typical uviarc lamps.

Sunlamps and black light lamps: Many conventional mercury lamps are good sources of "black light" or near ultraviolet.

Uviarc lamps: Uviarc lamps (Figure 19-5) are tubular mercury lamps designed and manufactured for use in diazo printing machines for processing engineering drawings. They are also used for copyboard lighting in some printing operations, for simulated weather tests, and for some special chemical processes. Wattages range from 250 to 7500. Some of the lamps have jacketed arc tubes for special performance or radiation characteristics.

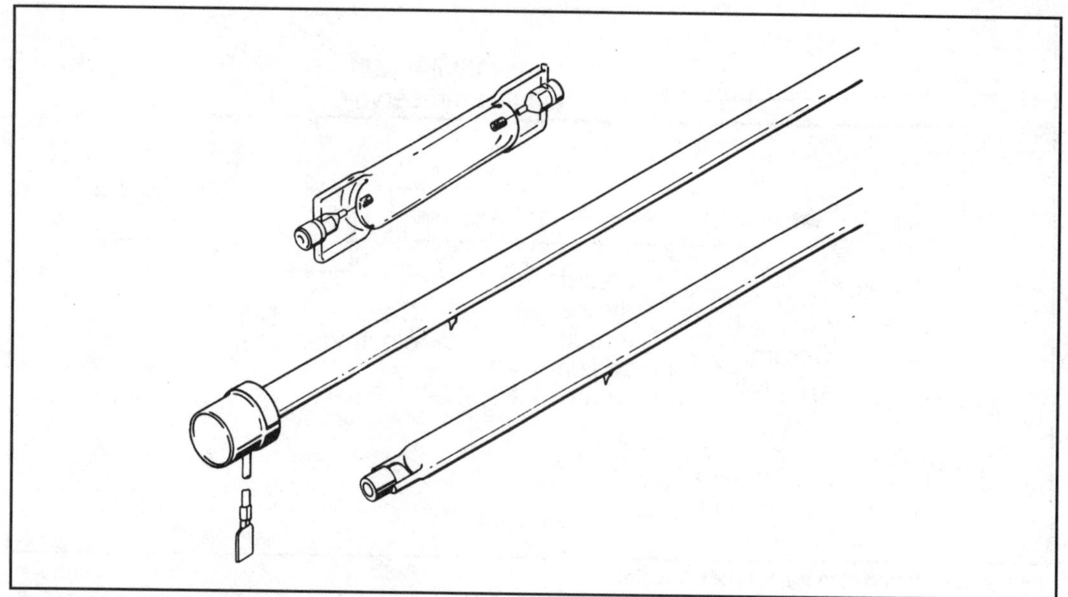

Figure 19-6: Tubular reprographic lamps.

Tubular reprographic lamps: A series of tubular mercury lamps (Figure 19-6) with quartz arc tubes is made in wattages from 250 to 1440; bulbs range from T-3 to T-7 in diameter, $4\frac{1}{2}$ to 26 in in length.

These lamps are primarily designed as ultraviolet radiation sources for common office copying and duplicating machines. They are also used as light sources in making lithographic plates.

Figure 19-7: Capillary mercury lamps.

Capillary mercury lamps: The AH-6, BH-6, and FH-6 lamps (Figure 19-7) are highly specialized 1000-W mercury lamps with quartz arc tubes only $\frac{1}{4}$ in × $3\frac{1}{4}$ in in size, including their brass base sleeves. Actual arc dimensions are about 1.5 × 20-25 mm, producing extremely high luminances (300-360 candelas per square millimeter). They are very potent and concentrated sources of both visible and ultraviolet radiation. The AH-6 operates with forced water cooling, while the BH-6 and FH-6 are air cooled. The lamps are used for a number of specialized applications, including the following: manufacture

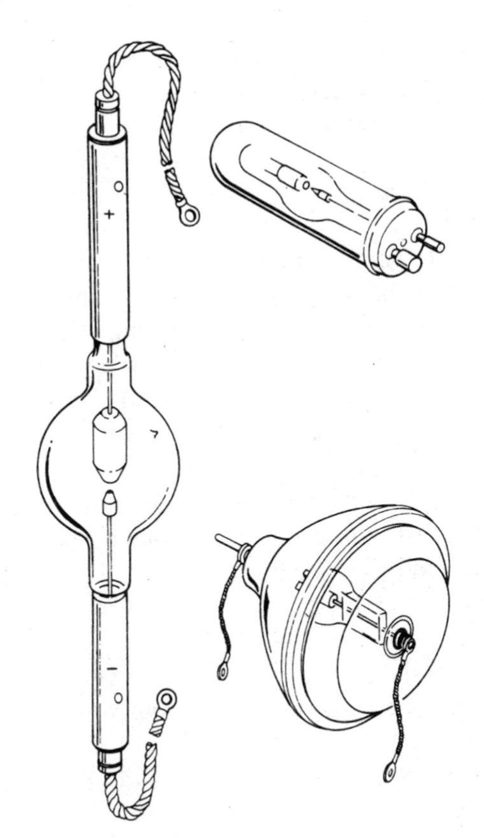

Figure 19-8: Xenon compact arc lamps.

of color television picture tubes, measuring "ceiling height" at airports, catalyzing chemical reactions by radiation, instrumentation in wind tunnels.

Xenon compact arcs: Three xenon arc lamps (Figure 19-8) are currently available. Two are 500-W lamps, featuring a compact arc only 2.5 mm long, and one 5000-W lamp has a 7.2-mm arc. They are designed for operation on dc circuits in enclosed equipment. One of the 500-W lamps has an ellipsoidal cool-beam reflector, designed to deliver maximum light with minimum infrared into special optical imaging systems. The 500-W lamp with tubular jacket is used in a number of optical devices, including one for the inspection of plate glass. The 500-W lamp is used in special optical devices, including solar radiation simulation for space research.

Sodium lab-arc: The NA-1 lamp is a small 500-lm source of low-pressure sodium radiation. Special laboratory equipment is available, including lamp and transformer. Principal application of the lamp is to generate modest concentrations of energy at the wavelength of the sodium resonance doublet, and 589 nm, for optical and other research purposes requiring essentially monochromatic light.

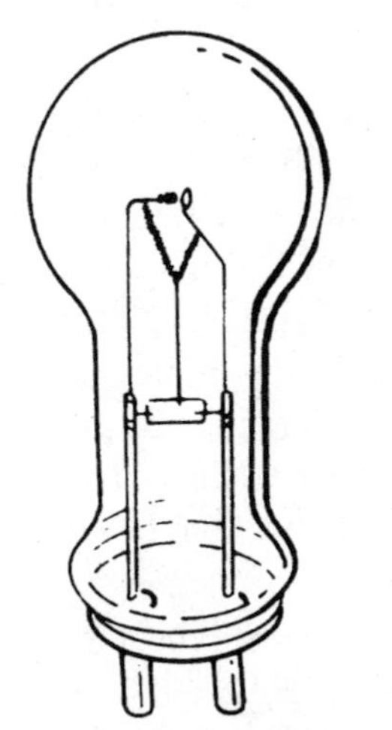

Figure 19-9: A typical tungsten-arc lamp.

Tungsten arc: The 30A/PS22 lamp (Figure 19-9) contains a tungsten arc with argon starting gas. The arc is used to heat a small, button-shaped tungsten electrode to incandescence, providing a sharply defined disc or relatively uniform luminance, and a color temperature of about 3000°K. It is used in a number of applications, which include photomicrography and optical comparators.

Germicidal lamps: Germicidal lamps are actually low-pressure mercury lamps similar to fluorescent lamps in electrical characteristics, but made with special glass designed to pass for ultraviolet energy.

LUMINAIRE DATA

A luminaire is a complete lighting unit: lamp, sockets, and equipment for controlling light distribution and comfort. For example, the lens controls high-angle luminance where a specific light distribution light pattern is desired; diffusers are used where general diffusion of light is desired; shielding — in the form of louvers, baffles, and reflectors — is used to reduce glare and excessive brightness; and reflectors can be used to direct light in useful directions.

A lighting fixture may be classified by its distribution of light, type of lamp used, or description. The distribution of light is based on the percentage of lumens emitted above and below the horizontal.

Direct (0% to 10%): Ninety to 100% down. This type is most efficient from the standpoint of getting the maximum amount of light from the source to the working plane. On the other hand, this type may produce the greatest luminance differences between ceiling and luminaire, and produce the most shadows and glare.

Semi-direct (1% to 40% up): Sixty to 90% down. Most of the light is still down, but some is directed up to the ceiling.

General diffuse (40% to 60% down): Forty to 60% up. This type makes light available about equally in all directions. A modification of the general diffuse is the direct-indirect fixture, which is shielded to emit little light in the zone near the horizontal.

Semi-indirect (60% to 90% up): Ten to 40% down. Here, the greater percentage of the light is directed toward the ceiling and upper walls. The ceiling should be of high reflectance in order to reflect the light.

Indirect (90% to 100% up): Zero to 10% down. Totally indirect reflectors direct all of the light up to the ceiling. Some types are slightly luminous to offset the luminance difference between the luminaire and the bright ceiling. Shadows are at a minimum, although glare may be present due to a bright ceiling. Inside-frosted lamps should be used instead of clear-glass incandescent lamps to prevent streaks and striations on the ceiling. Low-luminance fluorescent lamps are recommended.

The classification of fixture according to source will be any of the three described previously — incandescent, fluorescent, or high-intensity discharge lighting. Applications will include industrial, commercial, institutional, residential, and special-purpose applications such as for use in duct-tight, vapor-tight, and explosionproof areas.

LIGHTING DESIGN

The basic objectives of good industrial lighting design are the same as in other seeing areas: to provide adequate illumination and to make that illumination comfortable to the eyes. In order to obtain this condition, the following lighting demands must be met:

- The work plane must be adequately illuminated.
- Shadows falling on the work plane must be eliminated or at least minimized.
- Objectionable glare or reflections from machines, equipment, and other nearby objects or surfaces must be avoided.
- The contrast between the brightness of the work plane and immediate surroundings should not differ greatly. Research has demonstrated that, for best seeing, the brightness of the surrounding areas should approach that of the work.

Industrial interiors generally fall into two classifications:

- High-bay areas, where mounting height is 25 ft and higher.
- Low-bay areas, where mounting height is 25 ft and lower.

High-bay lighting systems usually consist of mercury vapor or high-output (800 to 1500 mA) fluorescent fixtures, although high-pressure sodium lamps are also popular. In selecting a lighting system, economy factors are usually weighed against all other factors.

If mercury vapor lamps are the chief source for high-bay areas, they are usually used in combination with some incandescent lamps. This combination has definite advantages over a system using mercury lamps alone.

First, mercury lamps go out if there is even a momentary power interruption, and it may be 10 to 15 minutes before they will restart and come up to full light output. This not only causes loss of time for the workers, but such an interruption could be highly dangerous in some industrial areas. The light from the incandescent system will provide light should such a power interruption occur.

If the system is designed for approximately equal initial lumens for each type of lamp, the incandescent lamp will have to be of higher wattage than the mercury lamps. Quartz lamps would be a good choice.

In low-bay areas, fluorescent lighting is almost the universally accepted system. High-output fluorescent lamps that are 8 ft long are the most economical and, therefore, are the ones normally used.

Industrial plants usually have office areas which should be treated in the same manner as the commercial lighting covered previously.

Supplementary Lighting

Supplementary lighting, as the name implies, supplements the general lighting by providing light for difficult seeing tasks or inspection processes. According to the specific situation, supplementary lighting may be installed for any of the following purposes:

- For a directional controlled beam of light.
- For additional properly shielded light with no special beam requirements.
- For a diffused light source, relatively large in area and uniform in brightness.
- For more specialized applications involving black light, polarized light, stroboscopic light, monochromatic light, lighted magnifier, and critical color matching.

Explosiveproof Equipment

Since explosive dusts and vapors are present in many industrial processes, application of proper lighting equipment for such areas must be considered. Dusttight lighting fixtures are necessary at various locations in grain elevators, flour and feed mills, and in plants in which cornstarch, sulfur, aluminum powder, and like substances are manufactured. The fine dust from these manufacturing processes, suspended in the atmosphere, is highly explosive.

In areas where there are explosive vapors, such as in paint spray booths and petroleum-processing areas, the lighting fixtures must be completely sealed with a gasket to prevent any explosive vapors from being sucked into the fixture.

NEC Sections 600 through 616 should be studied for a complete understanding of the various requirements that must be met in hazardous locations.

Light Distribution

In high-bay areas of industrial buildings, the work to be done commonly involves rather large three-dimensional objects that have diffuse reflecting characteristics. Under these circumstances, the seeing task is not severe and reflected glare presents no problem.

For these applications, a light source that has a high lumen output is desirable. Such sources in direct reflectors produce light with a directional component that causes slight shadows and mild highlights that aid in seeing. High-intensity discharge lamps are usually the most economical sources for high-bay lighting. In most cases, a few filament lamps are added to the HID installations to provide some light that will be available immediately following an electric service interruption. The nature of the work being performed and the reliability of the electric service govern the necessity of installing filament lamps for this purpose.

Intermediate Distribution Ventilated Reflector

This type of lighting fixture is best suited for use in high narrow rooms where it is necessary and more economical to produce illumination on the horizontal plane. Where the seeing task is inclined at an angle exceeding approximately 45°, lighting fixtures with a spread or widespread distribution should be used, even though somewhat less light reaches the horizontal plane. *See* Figure 19-10 for a comparison between narrow and wide beams.

Wide Distribution Ventilated Reflector

In wide high-bay areas, lighting fixtures with a wide distribution provide greater overlapping of light beams than is economical in narrow rooms, with resulting reduction in shadow intensity and higher vertical surface illumination. In rows of lighting fixtures near the walls, narrower distribution equipment may be used to minimize loss through wall and window absorption.

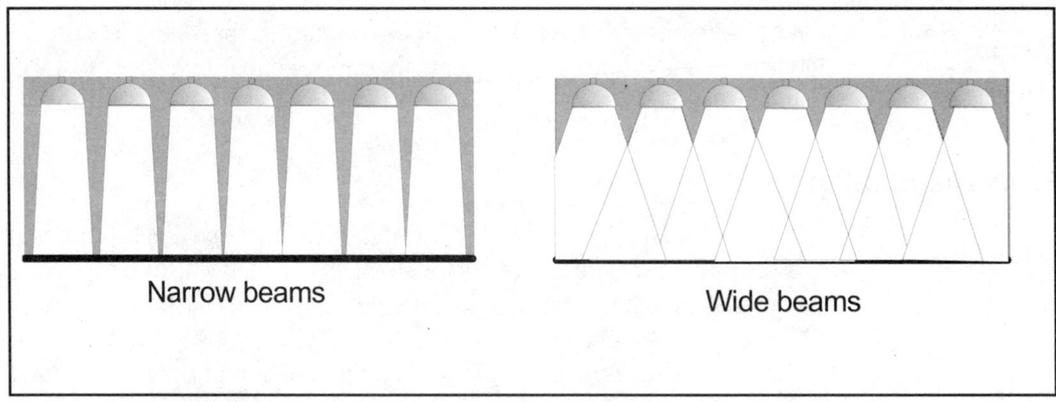

Figure 19-10: Wide distribution fixtures provide greater overlapping of light beams.

Channel Systems

A channel system offers a means of simultaneously providing for the electrical feed and the mechanical support of lighting and other equipment. It further assures true and rigid alignment and lends itself to systematic mass-assembly methods which economize on labor. Figure 19-11 shows a channel system with a fluorescent lighting fixture.

There are varying sizes of channel available. The hanger spacing is often determined by the type of building construction. The deflection then will determine the proper channel since this deflection should not exceed 1/240 of the span.

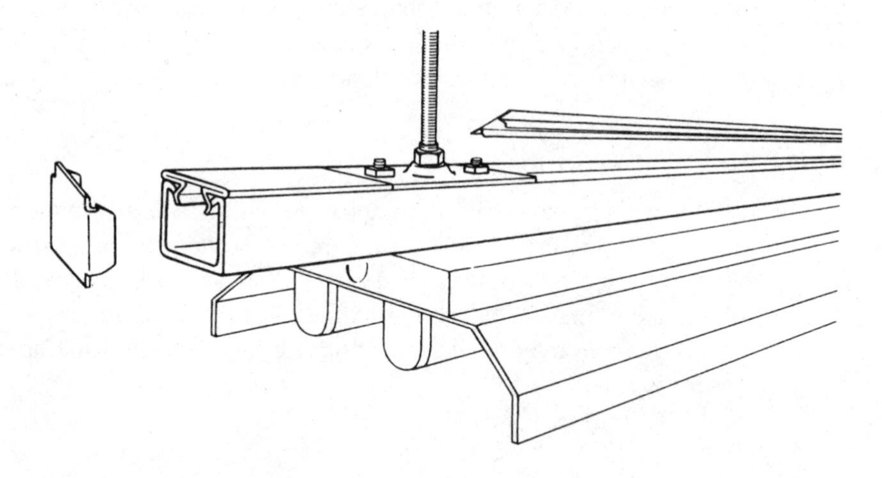

Figure 19-11: Application of channel systems.

DEFLECTION CONSTANTS FOR CONTINUOUS RUN, 4-FOOT FLUORESCENT FIXTURES							
Span Feet	B-906, B-956	B-900-M, G-975-M	G-953	B-900, G-975	B-901, G-950, G-955	B-900-2A	B-902, G-955
6	.004	.000	.000	.000	.000	.000	.000
8	.009	.002	.001	.001	.001	.001	.001
10		.005	.004	.003	.001	.000	.000
12		.010	.007	.006	.004	.001	.001
14				.012	.007	.002	.002
16				.020	.011	.004	.004
18					.018	.007	.006
20						.010	.009

Figure 19-12: Table of deflection constants for 4-ft fluorescent lighting fixtures.

To estimate the deflection at the center of an intermediate span in long continuous runs, multiply the weight of a single fixture times the applicable deflection constant (*see* table in Figure 19-12).

Multitap Ballasts

Current-limiting ballasts have been designed for each type of mercury lamp to furnish proper lamp voltage and current ballasting through the inductance of the windings. The electrical characteristics of ballasts, when used in conjunction with discharge lamps, are such as to produce a low power factor. This situation is commonly corrected by the addition of capacitance in the form of a condenser, generally built into the ballast. Uncorrected ballasts have power factors of 60 to 60%, whereas the corrected ones achieve 90% or better.

Two-lamp ballasts only slightly larger than the single-lamp type operate one lamp on a leading current and the other on a lagging current, producing an overall power factor of about 90% and reducing stroboscopic effect. All ballasts must be designed for the specific voltage and frequency of the supply with which they are to be used. For dependable starting and good lamp life, mercury lamps must be operated within rather narrow voltage limits, and the primary of each ballast is provided with taps for several voltages. Mercury lamps may also be operated from regulated-output (constant-wattage) ballasts or in straight series circuits without individual lamp ballasts, provided power safeguards are applied. Most equipment is designed for a frequency of 60 cycles. Operation on lower frequencies to 25 cycles is possible, although larger ballasts are required and stroboscopic effect is greater. Since the arc is actually extinguished each time the current reverses, at frequencies below 26 cycles, the mercury vapor may have time between cycles to deionize and the electrodes to cool sufficiently enough to prevent restriking of the arc. The main consideration on multitap ballasts is to read the packing instructions with each ballast or ballast/fixture and make absolutely certain that the tap used is compatible with the voltage to the circuit to which the fixture/ballast is connected. For example, to mistakenly connect the 208-V taps to a 277-V circuit could damage both the ballast and the lamp.

Zonal-Cavity Method

The zonal-cavity method is the latest recommended method for determining the average maintained illumination level on the work plane in a given lighting installation, and also for determining the number of lighting fixtures required in a given area to provide a desired or recommended illumination level.

The interior illumination form in Figure 19-13 is recommended when the zonal-cavity method is used. The illustration at the bottom of this form shows that a room or area is separated into three areas: (1) ceiling cavity, (2) room cavity, and (3) floor cavity. The cavity ratios are determined as follows:

$$\text{Room-Cavity Ratio (RCR): RCR} = \frac{5h_{rc}(L+W)}{(L+W)}$$

$$\text{Ceiling-Cavity Ratio (CCR):} \frac{5h_{cc}(L+W)}{(L+W)}$$

$$\text{Floor-Cavity Ratio (FCR):} \frac{5h_{fc}(L+W)}{(L+W)}$$

INTERIOR ILLUMINATION

FLOOR ______________ PROJECT ______________

ROOM ______________ FT. C. REQ'D ______________

WIDTH ______________ X LENGTH ______________ = AREA (A) ______________ SQ. FT.

CEILING HT. ______________ MOUNTING HT. ______________ RCR ______________

REFLECTION FACTORS: CEILING ______________ % WALLS ______________ % FLOOR ______________ %

COEF. OF UTILIZATION (CU) ______________ MAINTENANCE FACTOR (MF) ______________

FIXTURE TYPE ______________ LAMPS/FIX. ______________ WATTS/FIX. ______________

LAMP TYPE ______________ LUMENS/LAMP ______________

$$\text{TOTAL LUMENS} = \frac{\text{FC} \times \text{A}}{\text{CU} \times \text{MF}} = \text{\_\_\_\_\_\_\_\_} = \text{\_\_\_\_\_\_\_\_}$$

$$\text{NO. FIXTURES} = \frac{\text{TOTAL LUMENS}}{\text{LUMENS/LAMP} \times \text{LAMPS/FIX.}} = \text{\_\_\_\_\_\_\_\_} = \text{\_\_\_\_\_\_\_\_}$$

ACTUAL NO. FIXTURES DESIGNED ______________

$$\text{REVISED FC} = \frac{\text{TOTAL FIX.} \times \text{LAMPS/FIX.} \times \text{LUMENS/LAMP} \times \text{CU} \times \text{MF}}{\text{AREA}} = \text{\_\_\_\_\_\_\_\_}$$

$$\text{WATTS/SQ. FT.} = \frac{\text{NO. FIXTURES} \times \text{WATTS/FIX.}}{\text{AREA}} =$$

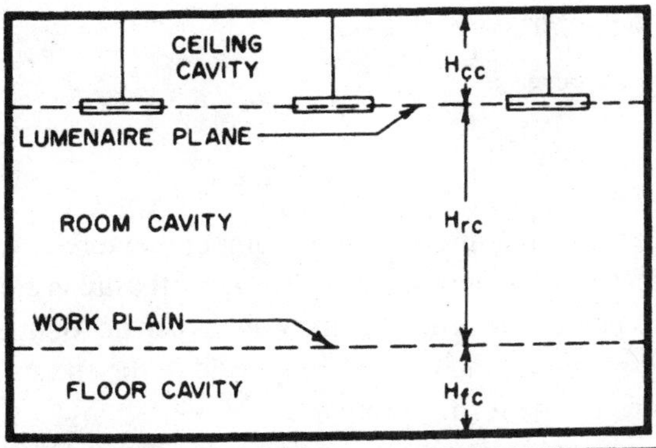

Figure 19-13: Interior illumination calculation form.

The cavity heights are represented by h_{rc}, h_{cc}, and h_{fc} and are shown in the illustration. *L* is the room length, and *W* is the room width.

In using the form, record the room width and length, the area of the room, the ceiling height, and the mounting height of the lighting fixtures above the floor as well as the data required in the form.

After this information is inserted, calculate the cavity ratios as described previously and insert the resulting data in the space designated RCR.

Now select the effective ceiling reflectance (P_{cc}) from Figure 19-14 for combination of ceiling and wall reflectances. Note that for a surface-mounted or recessed lighting fixture, CCR-O and the ceiling reflectance may be used as the effective cavity reflectance.

The next step is to select the effective floor-cavity reflectance (*PfC*) for the combination of floor and wall reflectances from Figure 19-14.

Next the coefficient of utilization is determined by referring to a *coefficient of utilization* table for the lighting fixture under consideration (supplied by most lighting fixture manufacturers). This figure is entered in the proper space on the form (CU). The maintenance factor (MF) is determined by the estimated amount of dirt accumulation on the fixtures prior to cleaning. This figure can vary and it takes some experience to select the right one. However, the following will act as a guide:

Very clean surroundings (industrial laboratories, etc.)	0.80
Clean surroundings (electronic assembly, lunch rooms, etc.)	0.75
Average surroundings (offices, etc.)	0.70
Below-average surroundings	0.65
Dirty surroundings	0.55

The manufacturer's catalog number is entered in the space marked Lamp Type, the number of lamps used in this fixture is entered next, and the total watts per fixture is next. Lamp lumens can be found in lamp manufacturers' catalogs under the type of lamp used in the fixture. The remaining calculations on the form are obvious.

Lighting fixture locations depend on the general architectural style, size of bays, type of lighting fixtures under consideration, and similar factors. However, in order to provide even distribution of illumination for an area, the permissible maximum spacing recommendations should not be ex-

Effective Ceiling Cavity Reflectance (%)	80			70			50			10		
Wall Reflectance (%)	50	30	10	50	30	10	50	30	10	50	30	10
Room Cavity Ratio												
1	1.08	1.08	1.07	1.07	1.06	1.06	1.05	1.04	1.04	1.01	1.01	1.01
2	1.07	1.06	1.05	1.06	1.05	1.04	1.04	1.03	1.03	1.01	1.01	1.01
3	1.05	1.04	1.03	1.05	1.04	1.03	1.03	1.03	1.02	1.01	1.01	1.01
4	1.05	1.03	1.02	1.04	1.03	1.02	1.03	1.02	1.02	1.01	1.01	1.00
5	1.04	1.03	1.02	1.03	1.02	1.02	1.02	1.02	1.01	1.01	1.01	1.00
6	1.03	1.02	1.01	1.03	1.02	1.01	1.02	1.02	1.01	1.01	1.01	1.00
7	1.03	1.02	1.01	1.03	1.02	1.01	1.02	1.01	1.01	1.01	1.01	1.00
8	1.03	1.02	1.01	1.02	1.02	1.01	1.02	1.01	1.01	1.01	1.01	1.00
9	1.02	1.01	1.01	1.02	1.01	1.01	1.02	1.01	1.01	1.01	1.01	1.00
10	1.02	1.01	1.01	1.02	1.01	1.01	1.02	1.10	1.01	1.01	1.01	1.00

Figure 19-14: Effective floor-cavity reflectance table.

ceeded. These recommendation ratios are usually supplied by the fixture manufacturers in terms of maximum spacing to mounting height. The fixtures, in some cases, will have to be located closer together than these maximums, in order to obtain the required illumination levels.

Point-by-Point Method of Calculating Illumination

The method for determining the average illumination of a given area was just described. However, it is sometimes desirable to know what the illumination level will be from one or more lighting fixtures upon a specified point within the area.

The point-by-point method accurately computes the level of illumination, in footcandles, at any given point in a lighting installation. This is accomplished by summing up all the illumination contributions, except surface reflection, to that point from every fixture individually. Since reflection from walls, ceilings, floors, and so on are not taken into consideration by this method, it is especially useful for calculations dealing with very large areas, outdoor lighting, and areas where the room surfaces are dark or dirty. With the aid of a candlepower distribution curve, we may calculate footcandle (ft c) values for specific points as follows:

$$\text{ft. c} = \text{candlepower} \quad H/d^2$$

or

$$\text{candlepower} \quad \cos^3 \theta/H^2$$

For vertical surface, use the following equations:

$$\text{ft. c} = \text{candlepower} \quad R/d^2$$

or

$$\text{candlepower} \cos {}^3\theta \quad \sin \theta/h^2$$

In using the point-by-point method, a specific point is selected at which it is desired to know the illumination level; for example, point *P* in Figure 19-15. Once the seeing task or point has been determined, the illumination level at the point can be calculated.

It is obvious that the illumination at point *P*, or at any point in the area, is due to light coming from all of the lighting fixtures. In this case, the calculations must be repeated to determine the amount of light that each fixture contributes to the point; the total amount is the sum of all the contributing values.

Before attempting any actual calculations using the point-by-point method, a knowledge of candlepower distribution curves and a review of trigonometric functions is necessary.

A candlepower distribution curve or graph consists of lines plotted on a polar

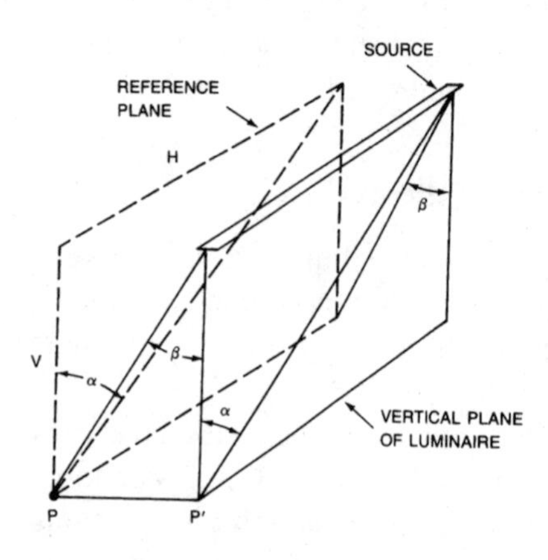

Figure 19-15: Principles of the point-by-point illumination method.

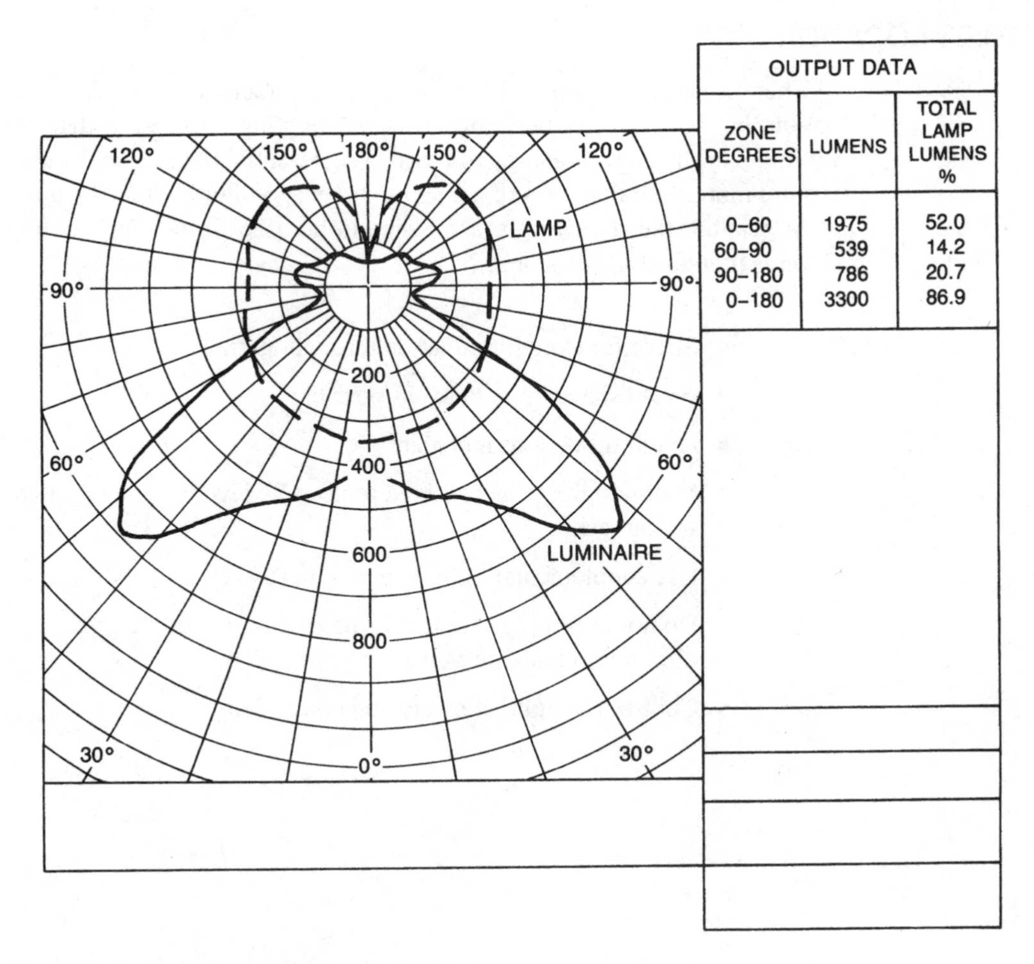

OUTPUT DATA

ZONE DEGREES	LUMENS	TOTAL LAMP LUMENS %
0–60	1975	52.0
60–90	539	14.2
90–180	786	20.7
0–180	3300	86.9

Figure 19-16: Typical candlepower distribution curve.

diagram, which show graphically the distribution of the light flux in some given plane around the actual light source. It also shows the apparent candlepower intensities in various directions about the light source. Figure 19-16 illustrates a typical candlepower distribution curve.

A table of trigonometric functions may be used in determining the degrees of the angle from the light source to the point in question. However, most inexpensive electronic pocket calculators have all trigonometric functions built in, and the sin, cosin, tangent, etc. may be found instantly merely by pressing a few keys. These figures are necessary to pick off the candlepower from the photometric distribution curve and also to use in the equations.

OUTLET LOCATION

There is no general rule for locating lighting outlets for general illumination, but it is usually desirable to locate lighting fixtures so that the illumination in a given area is uniform. Where certain seeing tasks may require more illumination, supplemental direct spot lighting may be used.

In general, the following scheme should be used in locating lighting fixtures in industrial work areas:

- Strive for even illumination throughout the area
- Avoid shadows as much as possible
- Avoid reflections and glare
- Maintain the recommended footcandle level for the required seeing task
- Use supplemental lighting where necessary
- Do not let supplemental lighting interfere with the area illumination or vice-versa
- Control the lighting safely and conveniently

Index

— D —

— E —

— F —

— G —

— H —

—O—

—P—

—Q—

—R—

—S—

— T —

— U —

— V —

— W —

— Y —

— Z —